EXPLORATORY EXPERIMENTS

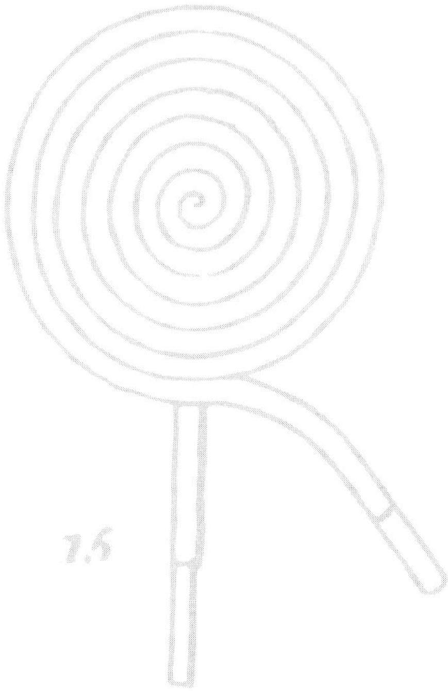

15

16

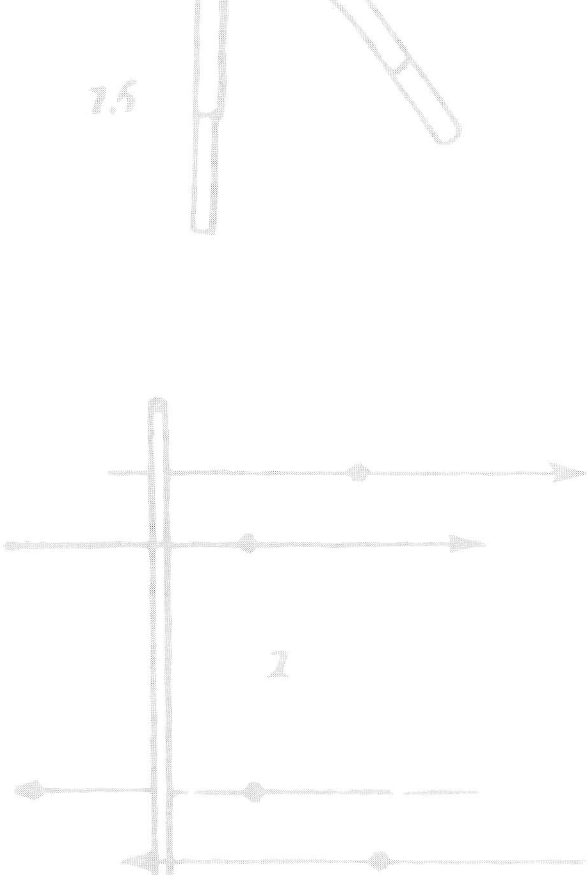

1

EXPLORATORY EXPERIMENTS

Ampère, Faraday, *and the* Origins *of* Electrodynamics

Friedrich Steinle

Translated by Alex Levine

UNIVERSITY OF PITTSBURGH PRESS

This English edition has been translated from the original German publication:

Friedrich Steinle, *Explorative Experimente* © 2005 by Franz Steiner Verlag, Stuttgart (Germany)

The translation of this work was funded by Geisteswissenschaften International—Translation Funding for Work in the Humanities and Social Sciences from Germany, a joint initiative of the Fritz Thyssen Foundation, the German Federal Foreign Office, the collecting society VG WORT, and the Börsenverein des Deutschen Buchhandels (German Publishers & Booksellers Association).

Published by the University of Pittsburgh Press, Pittsburgh, Pa., 15260

Manufactured in the United States of America
Printed on acid-free paper
10 9 8 7 6 5 4 3 2 1

Cataloging-in-Publication data is available from the Library of Congress

ISBN 13: 978-0-8229-4450-8
ISBN 10: 0-8229-4450-2

To Lorenz Krüger

CONTENTS

Acknowledgments ix

INTRODUCTION
Research Practice, Experiment, and Concept Formation
1

CHAPTER 1
Electricity and Galvanism in the Early Nineteenth Century
18

CHAPTER 2
Electromagnetism
52

CHAPTER 3
Ampère's First Studies of Electromagnetism
Entering a New Field
85

CHAPTER 4
Competing Pursuits in Paris, 1820–1821
143

CHAPTER 5
Electromagnetism in London
202

CHAPTER 6
Faraday's First Studies
Electromagnetic Rotation
228

CHAPTER 7
Experiment and Concept Formation
301

APPENDIX A
Source Materials for the Early Phases of Ampère's Research in Electrodynamics
339

APPENDIX B
Reconstruction of the Manuscript of Ampère's Lectures of September 18 and 25, 1820
384

Notes
423

References
461

Index of Names
491

ACKNOWLEDGMENTS

This book aims at putting detailed historical research to twofold use. It elucidates a particularly significant and momentous episode in the development of science—the origin of electrodynamics—and sharpens our analytic view of one of the most important epistemic tools of modern science: the experiment. The close relationship between historical and philosophical analyses that I have attempted here has itself an experimental character. Should my work awaken further interest in this combination of historical and epistemological approaches, it will have achieved one of its most important goals.

A great number of people have contributed to this effort with discussions both critical and supportive and with support and friendship. Since a list of names would be too long and necessarily incomplete, I shall mention just two: Hans-Jörg Rheinberger and—much earlier—Lorenz Krüger were extremely supportive in critical phases of my work. Besides giving institutional support, they provided examples—notwithstanding differences in approach—of how historical studies and epistemological analyses can fruitfully be connected. To them, and to all others who accompanied me in my work, I want to express my heartfelt thanks.

This study was originally published in German. Preparing the English version also served as a welcome opportunity to include recent research and discussion, in particular on the question of experiment. The intriguing observations of the reviewers of the German version and of the advisors of the University of Pittsburgh Press were very helpful in preparing the English version. My archival research was made possible by a research grant from the German Academic Exchange Service (DAAD) and the Maison des Sciences de l'Homme (MSH) in Paris, where I was able to thoroughly investigate Ampère's papers. Additional funding from the Deutsche Forschungsgesellschaft (DFG), along with time spent at the Max Planck Institute for the History of Science in Berlin and a one-year research fellowship at the Dibner Institute for the History of Science and Technology in Cambridge, Massachusetts, not only made the extended research that went into this project possible but also provided invaluable opportunities to make contacts and gain further support. I received extraordinarily helpful support at the archives of the Académie des Sciences (Paris), the École Polytechnique (Paris), the Royal Institution (London), and the Royal Society (London). I wish to express my thanks to all of these colleagues and institutions, both for their help and for permission to plumb and present some of the rich treasures of these collections.

The translation from German and cross-checking of my English were excellently performed by Alex Levine. Cynthia Miller and Abby Collier from the University of Pittsburgh Press provided constant help and encouragement. Teresa Hollerbach provided essential assistance in preparing the final texts. Many thanks to all of them; I can only hope that the result justifies all their combined efforts.

INTRODUCTION

Research Practice, Experiment, and Concept Formation

> And when you look more closely at "what scientists do," you might be surprised to find that research actually comprises both the so-called day science and night science. Day science calls into play arguments that mesh like gears, results that have the force of certainty. Its formal arrangement is as admirable as that of a painting by da Vinci or a Bach fugue. You can walk about in it as in a French garden. Conscious of its progress, proud of its past, sure of its future, day science advances in light and glory. By contrast, night science wanders blind. It hesitates, stumbles, recoils, sweats, wakes with a start. Doubting everything, it is forever trying to find itself, question itself, pull itself back together. Night science is a sort of workshop of the possible where what will become the building material of science is worked out. Where hypotheses remain in the form of vague presentiments and wooly impressions. Where phenomena are still no more than solitary events with no link between them. Where the design of experiments has barely taken shape.
>
> —François Jacob (1998, 126)

Research Practice

What molecular geneticist François Jacob has here so colorfully and accurately described is the contrast between the image of research science shows to the outside world and the actual research practice of the laboratory. It is no coincidence that Jacob, like Ludwik Fleck, whom he invokes, was active in the life sciences. What he describes as the night side of the sciences has traditionally been more clearly recognized within the biosciences and chemistry—we need only recall (physical) chemist Michael Polanyi—than in the exact sciences, thought to be more advanced, or to have been lent greater rigor in their research process by dint of mathematization.[1]

As Jacob's book makes clear, his depiction is not meant to suggest that the research practice of science is merely fortuitous, or even chaotic, and thus resists closer study. Jacob's distinction aims rather at drawing attention to the fact that representations of science are typically geared toward only one of the two sides: toward the side of systematic accumulation of knowledge. The actual practice of research, however, reveals the complexity of scientific investigation. In the experimental sciences, investigations demand both thought and the interaction with and experimental manipulation of instruments and materials. These activities depend on resources, technical assistance, cooperation with others, equipment and consumables, considerations of space and time, and of course also on money. All such things weigh heavily in the daily life of the laboratory, and sometimes experiments are performed in a particular way because only certain resources are available and not others, while other experiments, perhaps of equal or even greater theoretical interest, are never performed at all, and for similar reasons.

What's more, the laboratory is not an isolated place. Researchers must communicate, and in doing so they face a plethora of additional considerations: with whom to communicate, why, and by what means; what to share and what to hold back; at what stage to publish and in what format; and how to secure priority and ensure the best career outcome. Attempts at communication and the responses they provoke frequently factor into the planning of particular experimental trials and even into the very direction of research. Communication or, more generally, social activity, is an essential component of research practice.

Research is driven not only by the quest for knowledge but also by the biographical situation of its agents, their place within a community and the safeguarding thereof, social and cultural conventions and conditions, and not least by sheer coincidence—in short, by everything that drives human action in any other context. The human agents at work in this process also have lives and interests besides science. It would be naïve to simply assume that such matters have no effect on the course of research. To be sure, in typical first-person accounts, they recede into the background or disappear altogether, leaving only the systematic quest for knowledge and the necessities this quest strictly entails.

But even this remaining aspect, when it comes up in public presentations, finds itself dressed up and organized from a point of view that has been reached as the outcome of scientific research. Many of the criteria and concepts that now structure the presentation were unavailable during the research process itself and were instead developed within that process. Jacob's talk of tapping in the dark, stumbling, retreating, doubting, and resuming is particularly applicable to such developments. So long as we attend only to accounts published by the scientists themselves, we will have little prospect for understanding the generation of knowledge in all its complexity. To be sure, the point is not to accuse natural scientists of falsifying their reports. Their published accounts are always aimed at a particular audience and necessarily shaped by that goal. In most cases, and for good reason, their primary purpose is not that of historical narrative but rather the systematic presentation of results from a perspective whose attainment was made possible by the very research leading to those results. And even if a historical account would be attempted, it would be difficult to describe the previous situations of deep conceptual uncertainty from a later position in which everything has been sorted in its proper place, often by means of concepts that were not available at the time.

In focusing on research *practice,* I am not concerned with contrasting such practice with *theory* but rather with the picture that science presents to the public about its process. Nor am I exclusively concerned with *experimental* practice, for research practice can also be studied with regard to *theoretical* disciplines, such as theoretical physics and mathematics.[2] Research practice, thus understood, has only recently come within the scope of historical investigation.[3] In doing so, historians have become highly aware of the particular kinds of source materials required here. Research practice cannot be simply read off of documents; it can only be historically reconstructed. To be sure, published first-person accounts by relevant agents can serve as a useful starting point, though they often reveal more about the expectations and clichés of their authors than they do about experimental activity and the motives behind it.[4] By contrast, materials originally meant to assist the agents in their own work, with no thought of presenting them to others, provide more insight. Examples include laboratory records, sketches, and loose notes, some of which may display an idiosyncratic or even cryptic character. But there are no firm boundaries between the genres; in larger laboratories, records might also be intended as a medium of communication. Other sorts of documents, such as purchase orders, laboratory access logs, and so on, can enrich the picture.[5] To what extent and in what respects such sources enable the construction of a detailed picture of research practice always depends on the disposition of the sources in the particular instance. Because the resulting picture often diverges in characteristic ways from first-person accounts, such investigations can offer important insight into the conventions of these accounts.

In this book, studies of research practice form a core element. I reconstruct historical episodes and try to learn some more general lessons from their analysis. The expense and complexity of producing reconstructions of this kind typically constrain them to relatively narrow time periods and specific lines of inquiry. As in every historical narrative, it is especially important to account for the specific interests that guide the investigation and motivate the selection of time period and source materials. These issues are briefly addressed in the sections that follow.

Experiment

The past three decades have seen increased attention to the role of experiments in the sciences.[6] The so-called standard view, in which the role of experiment, as handmaiden to theory, is confined to the testing of hypotheses and theories, has been found wanting. Numerous studies have identified a plethora of other, equally important roles of experiments, leading Hacking (1983, 166) to speak even of a "Baconian fluster."[7] Most of these roles are important in the *formation* of theories, as well as in the conceptual, linguistic, and visual media of research. For a long time, the philosophy of science has ignored such "generative functions," as Heidelberger calls them, considering them irrelevant. The divide between the context of discovery and context of justification seemed to many to imply that the only epistemically important role of experiment was that of testing theories that had themselves been developed by other means.[8] But as insight into the inadequacy of this dichotomy (or at least of its specific interpretation) grew (Schickore and Steinle 2006), the prospect of taking other roles for experiment came prominently back into view. There is, after all, a broad field between theory testing and mindless fumbling about with an apparatus. Despite a few attempts at taxonomy, that field remains largely unexplored. This field is one of my primary research interests.

One characteristic of recent studies of experimentation, even those of primarily philosophical thrust, is that they are based on the analysis of historical or recent episodes in science. For philosophical reflection to proceed, a wide range of "empirical" materials seems to be required, without which we cannot hope to become aware of otherwise unanticipated issues.[9] In this regard, the study of research practice has much to offer. The fact that, within the philosophy of science, the received view of the role of experiment, or the context distinction, could become so entrenched and last so long has much to do with the fact that its originators, who were certainly well versed in contemporary science, were inclined to defer to the published self-representations of participating scientists. But these representations, in turn, were often shaped by philosophical prejudice or by general expectations concerning the proper conduct of science. The investigation of research practice allows us to see past these self-representations and enables us to have a more direct view of research activity.

Uncovering the manifold roles of experiment in the research process, and understanding their significance, is a chief aim of the present study. Even after numerous studies (e.g., Gooding 1990a; Hacking 1992; Hentschel 1997; Graßhoff, Casties, and Nickelsen 2000), the question of the various different epistemic goals pursued through experimentation and their consequences for experimental endeavors have still not been studied in appropriate breadth. This question constitutes the central focus of my work. The study of scientific practice will, for the first time, allow us to sketch a differentiated picture of those issues. Among other consequences, it will allow experimental endeavors of the kind I call "exploratory" to be characterized in detail and their far-reaching, if heretofore unrecognized epistemic significance, exposed.

The Language of Research and Formative Periods

Scientific thought takes place on many different levels, involving items such as categories, classifications, modes of presentation, concepts, conceptual schemes, empirical rules, theories, and theoretical entities. Wildly divergent "epistemic things" emerge, sometimes stabilizing, sometimes being discarded.[10] Sometimes the boundaries between them are not drawn clearly, and sometimes indeed they cannot be. But that does not mean such distinctions ought to be abandoned, subsuming all of this diversity under the heading "theory," as has sometimes been done. After all, there remain clear cases. For example, it is one thing to classify effects as electrical or magnetic on the basis of their observed characteristics and quite another to investigate the prospect of some hidden, unobservable emanation behind them all. The morphological classification of plants or animals is different in kind from ruminations on a possible life force, much as the establishment of regularities in spectral lines is different from their quantum-mechanical explanation at the atomic level. To ignore such differences would be to sacrifice the potential for significant differentiation and to undermine any attempt to come to grips with the range of possible epistemic goals that drive scientific research. In this work, as in other efforts in the recent literature, I shall be defending and exploiting a distinction between concepts, empirical laws, and theories.[11] This distinction is suggested by the practices and reflections of historical actors and becomes, once critically deployed, an important tool with which to grasp the differing epistemic goals in play within a given scientific practice.

I have conducted this study with a view toward a particular kind of epistemic situation, one in which there is insecurity at the basic conceptual level and in which, consequently, the reliability of not only special theories but also established conceptual schemes, forms of thought, and modes of representation has been profoundly shaken. Such situations of conceptual insecurity, even speechlessness, can be brought about by unexpected experimental results that, while clearly observable and reliably repeatable, remain resistant to treatment by means of customary concepts and are sometimes frankly ineffable. Such situations have arisen repeatedly in the history of the sciences; in Thomas Kuhn's conceptual framework they would count as serious anomalies. The developments traced within this book took one such situation or anomaly as their point of departure. Research conducted in such situations typically aims at articulating and developing a conceptual framework required to stabilize engagement with the new effects. I shall call such phases "formative periods." Often, though not always, the conceptual innovation of such periods goes hand in hand with intense experimentation. In such cases, experiments take on a special role that cannot be captured under received conceptions of experimentation, for the testing of hypotheses or expectations can take place only against the backdrop of a reliable language.[12] Within that language, such hypotheses can be formulated, a particular experimental question posed, and the experiment itself designed and assessed. But when such a language—such a conceptual framework—is lacking or when the reliability of the extant framework has been shaken, experimental activity must necessarily take on a very different character. Its study under such conditions is thus of particular interest.

The detailed case studies of experimental work presented in this book concern such formative periods. In the constant give-and-take between experimental activity and conceptualization, new concepts are formed and stabilized—or destabilized. With successful stabilization, a new perspective on the field is gained, along with new terminology, permitting reliable experimental engagement and thus fundamentally shaping all future work in the field. In subsequent efforts, the new concepts are no longer subject to revision; instead, they may simply be used in the formulation of further research questions. While concepts may be flexible and provisional in the formative period, they later take on a stable, fixed character. Formative periods are like branching nodes or, to adapt Reinhart Koselleck's phrase, saddle periods, in that in a relatively short span of time they determine the long-term direction of scientific research. The case studies in this book display this feature very clearly. The full generality of this phenomenon might be illustrated by further examples, such as the first clear classification of electrical and magnetic effects by William Gilbert, or Charles François de Cisternay du Fay's early eighteenth-century introduction of the distinction between two different electricities. In all such cases, intensive experimentation was at the core of the effort, not for theory testing but in an essentially constructive role.

Electromagnetism in 1820–1821

By the mid-nineteenth century, and in the wake of the rapid expansion of electrical telegraphy, electromagnetism and electrodynamics became prominent fields of academic research.[13] It became apparent at that point that there were already two competing theories. Based on very different concepts, they were thus not amenable to any effort at direct comparison. On the one hand stood electrodynamics, first developed by André-Marie Ampère in Paris and since widely disseminated. It dealt with the mutual attraction and repulsion of infinitesimal elements of electrical current in a fully mathematical way. The forces involved were conceived as central, acting at a distance along the lines connecting the pointlike centers of force. In 1846, Wilhelm Weber greatly expanded this theory by extending it to induction effects. On the other hand, Michael Faraday in London had qualitatively, while still very precisely, captured electrical and magnetic effects and the connection between them and developed completely different concepts. His central notions were those of electrical and magnetic lines of force whose interactions with each other and with matter gave rise to all electrical, magnetic, and electromagnetic phenomena. There was no action at a distance; it occurred only at points of immediate contact. Faraday's framework, unlike Ampère's, encompassed a huge range of extraordinarily diverse phenomena, from electrochemical decomposition to electrical discharges in rarefied gases, electromagnetic induction, the magnetic properties of various materials, and the effect of magnetism on light. Despite the universal acclaim with which Faraday's numerous experimental discoveries had been greeted, his concepts and explanations had found very little resonance. Among his contemporaries who knew of them at all, most found them nearly unintelligible. This began to change only in the late 1850s, when James Clerk Maxwell began to develop a mathematical formulation of Faraday's principles, in the end presenting field theory, as it was now called, as a viable alternative to Ampère-Weber action at a distance. With this development, the stark contrast between the two frameworks became especially striking. During the second half of the nineteenth century, the divide between the two theories and their respective frameworks, along with a third, Carl Neumann's potential theory, would be one of the most important debates in physics.

The roots of this controversy extended to a much earlier date. Ampère worked out his theory between 1820 and 1826, and he had already formulated its basic conceptual structure in the fall of 1820. Faraday, for his part, entered the field of electromagnetism in the fall of 1821 and quickly developed the essential structures, concepts, and procedures that would shape the course of all of his future work. The formative period of both of the major theories and conceptual schemes of electrodynamics was thus 1820–21, beginning with studies conducted in response to Hans Christian Ørsted's discovery in July 1820 of an electromagnetic effect.

Ørsted's discovery had been a sensation, sparking far-reaching research efforts across Europe. Almost no one had been prepared for an electromagnetic interaction, and, though none had any trouble replicating the effect, most researchers were left profoundly puzzled. It was not merely that they could not explain it; even capturing regularities or reporting individual experimental results proved enormously challenging. They simply lacked the concepts and language needed in order to coherently express their results and the complex relationships among them. In particular, the experimental results resisted formulation in terms of the concepts of attraction and repulsion, at least at first blush. But these were the very concepts that had shaped scientific thought at a fundamental level, and not only in physics. As a result, early reports of experiments took on a baroque character, as the mere statement of an experimental result demanded a meticulous accounting of the positions of battery and experimenter, couched in terms of such elementary spatial concepts as the cardinal points of the compass. The description of the experimental arrangement (which, after all, consisted only of battery, wire, and compass needle) often took half a page. Yet, such a description applied only to particular experiments and did not allow generalizations to be formulated. Early electromagnetism thus offers a distinctive example of the kind of situation described above, one in which an accepted conceptual scheme has been shattered.

It is therefore not surprising to find that a significant proportion of the many research activities undertaken within this situation were geared toward the construction of adequate concepts. Even a cursory sampling of reactions from across Europe reveals astounding similarities in the reaction of the researchers who found themselves in this epistemic situation. To be sure, there were also significant differences, which in turn reveal how diverse particular local contexts and traditions of scientific research remained throughout Europe in the early nineteenth century. A broad-based comparative study of reactions to Ørsted's discovery (of which I shall provide only a sketch) might yield an insightful panorama of early nineteenth-century scientific cultures and their diversity.[14]

Such a study, however, is not my primary objective. With the later dispute between field theory and action-at-a-distance theory in mind, I am concerned with the formative period in which the conceptual foundations for both theories were laid down. Because Ampère and Faraday are the two principals of this period, I concentrate on work undertaken in Paris and London, respectively. It is remarkable that such different, incompatible conceptual schemes were developed in response to one and the same discovery. The fact that this development took place in such a short time and, as I show, under conditions of at least partial mutual awareness demonstrates all the more clearly the significance of this episode for the study of formative phases.

Two Research Objectives

My goals are twofold. First and foremost, I attempt a reconstruction of working scientific practice in the two episodes under study. For Ampère, the relevant period runs from September 1820 through January 1821, whereas for Faraday it spans four months almost exactly a year later. There were other noteworthy congruencies: both projects dealt with the same problem domain, both researchers undertook their reflections amid intensive experimental labors, and both were newcomers to the fields of electricity and magnetism. At the same time, while Ampère began his efforts as a well-established professor of mathematics and academy member, Faraday was a mere laboratory assistant and scientific autodidact, albeit one who had already begun to gain a bit of prominence following some minor publications in chemistry. There are further points of contrast. Ampère's milieu was French physics, with its strong tradition of mathematical treatment, while Faraday's was the chemical laboratory of the Royal Institution, along with several self-organized groups for the dissemination of scientific knowledge to the broader public. Ampère conducted his research in great haste and under strong competitive pressure, when Ørsted's circular contained all that was known about electromagnetism. By the time Faraday's effort was under way, a scant year later, a great deal had already been published. Finally, Ampère had no easy access to experimental facilities, whereas Faraday had unlimited access to one of the best-equipped laboratories of his time. Weighed together, the similarities and differences strike me as balanced enough to sanction significant points of comparison. Such comparison turns out to be extraordinarily fruitful, making possible a clear portrait of the two research practices.

Both episodes have already been the subject of historical research.[15] Though authors such as Christine Blondel, L. Pearce Williams, James Hofmann, and David Gooding have contributed much, and my own study would have been impossible without them, there remain important gaps with respect to the two episodes, and they impede an adequate understanding of the formative phases. In Ampère's case, this has to do with the very challenging disposition of the necessary sources. My work was made possible only by recourse to previously unknown archival materials, laboriously uncovered. In Faraday's case, prior studies were confined to only a small portion of the whole episode. Partly as a result of this fragmentary sampling, they have led to misinterpretations. The comprehensive comparison of the two episodes thus demanded a new analysis of both, with broader temporal horizons and greater attention to the wider historical context.

The second goal of this book is an analysis both of the relationship between experimental activity and concept formation during a formative phase and of the open-ended process whereby new concepts are formulated and stabilized. Such analysis rests on the foundation of a detailed reconstruction of scientific practice. I should state from the outset that one of my central results is the conclusion that experimentation can proceed systematically, in accordance with clearly recognizable guidelines, even when it is not strictly driven by theories. The procedure of "exploratory experimentation," as I have called it, can be explicated and systematized in considerable detail. Attention to its features clearly reveals how concepts are questioned, transformed, or replaced, and eventually stabilized in the context of experimental activity—a remarkable insight for the philosophy of experimentation. For history of science, moreover, there is the surprising historical discovery that even Ampère had his exploratory phase. It was during this phase, brief though it may have been, that decisive conceptual innovations took place. Furthermore, Ampère's procedures throughout this phase of concept formation are similar, in the main, to those discernible to a much greater degree in Faraday's work.

In this analysis, as much as in the reconstruction of scientific practice, I find myself in uncharted territory. Nothing of the sort has previously been attempted for Ampère. Williams's and Hofmann's studies are devoted to epistemological questions, and their depiction of Ampère's work now appears inadequate in several important particulars, as, for example, with respect to Ampère's use of equilibrium experiments. These studies have also entirely missed Ampère's exploratory phase. In Faraday's case, however, Gooding has certainly pursued goals similar to mine, though he has passed over several important points with surprising consistency. Most significant of these is the explanatory goal of "reduction to the simple case," not only discernible but downright prominent in Faraday's work. Gooding thereby misses a central aspect of Faraday's method of devising new concepts. His general epistemological picture is thus remarkably vague when it comes to concept formation. My study seeks a much more nuanced representation.

Overview of the Book

All scientific research is conducted within a particular situation, defined both by cultural and biographical circumstances and by the theoretical, experimental, and instrumental state of the art. Studies of scientific research practice must thus be framed by due attention to this background. In chapter 1, I survey the state of knowledge and experimental culture(s) in whose context early nineteenth-century studies of electricity took place. This survey is a necessary condition for the remainder of this work.[16] Besides sketching the general state of the research field, I selectively focus on Paris and London. As a result, some other developments, including those in the German-speaking world, which were never fully appreciated in those venues, will get less weight in the present study than they would in any truly comprehensive survey. Instead of employing a bird's-eye view, I aim to capture the perspective of actors in London and Paris from street level.

The core of chapter 2 is the discovery in 1820 of electromagnetic action. Ørsted announced his findings quickly, self-consciously addressing a wide audience across Europe. So as to convey a sense of the excitement aroused by his announcement, I extend my gaze, for a few sections, beyond Paris and London. Following a sketch of the contents of Ørsted's report, I offer an overview of reactions throughout Europe that brings to the fore significant features common to most localities. I then turn back to the main stage with a detailed accounting of events in Paris, where the academicians were surprised by the intensive involvement of an unexpected researcher, André-Marie Ampère.

Chapter 3 is a study of Ampère's scientific practice during the first weeks of this involvement. These first weeks are of particularly far-reaching importance, for they saw the formation of the concepts that would shape Ampère's work thereafter. Ampère did not have a clearly defined research program until the end of this period, a point that has garnered almost no attention in previous historical studies. This seeming neglect can be traced both to the challenging state of the sources and to the way in which Ampère would later describe his own work. As a result of my investigation, the received view, which has more or less explicitly shaped prior historical efforts, must be substantially revised. There is now a new and for the first time satisfactory answer to the old question as to how Ampère could have achieved such far-reaching innovations in a mere three weeks. The revised view was made possible by the discovery and reconstruction of source materials unknown to earlier scholars. For the first time, we now have rich materials at hand that Ampère produced not retrospectively but at the very moment of his research. In an effort to convey a sense of the nature and scope of the archival work necessary to carry out this study, I present some of these materials in the two appendices to this book.

Chapter 4 is devoted to Ampère's work from October 1820 through January 1821. Historians have agreed that Ampère's lecture to the Académie des Sciences in Paris in January 1821 represented a turning point in the development of his theory, which was then followed by a longer pause in his research efforts. I have dedicated a whole chapter to the period from October through January, despite the fact that it has been dealt with in prior historical studies, in order to show how Ampère's various activities hang together, with each other and with aspects of his particular situation in Paris. This is the phase in which his competition with Jean-Baptiste Biot is most keenly evident, and some of Ampère's moves can be understood only against this previously neglected backdrop. By taking account of Ampère's now transformed experimental practice, his intensive efforts toward precise measurement and the conclusions he drew from their failure, and his numerous outreach activities and consciously crafted public image, it is possible to shed new light on his research practice. In particular, important departures from those first three weeks, as discussed in chapter 3, become recognizable, exposing the peculiarities of the initial period with even greater clarity.

In chapter 5 I turn to the scientific milieu of London around 1820. It differed markedly from that of Paris, and not only in its institutional structure. Even in this context, however, Michael Faraday's career was rather unusual. My survey of the London responses to Ørsted's discovery reveals characteristic differences relative to those of their Parisian counterparts. These may even be observed in the meticulous efforts undertaken by Humphry Davy, the most important of the London researchers, efforts that were of great importance for Faraday. Only in the summer of 1821, when an extensive literature on the topic had already been published, did Faraday commence his own research activities. His motives, and the conditions under which he entered the field, were completely different from those of Ampère. I explore them in some detail toward the end of this chapter.

Faraday's first work on electromagnetism is the focus of chapter 6, which thus becomes the companion piece to chapters 3 and 4. Faraday quickly arrived at his spectacular discovery of electromagnetic rotation and, thus armed, immediately joined the first rank of researchers in the field, despite being a newcomer to physics. Previous historical studies of this finding have left a number of important questions open. Some of these concern the continuity, only now apparent, between his apparently sudden discovery and his earlier attempt to write a survey of the field. To a much greater extent than the authors of prior studies, I have drawn on Faraday's efforts *after* his great discovery. It is here that the exploratory character of Faraday's work becomes apparent—all the more so when we consider his later integration of the rotation effect into a classificatory system he had yet to devise. This picture is further fleshed out by attention to the studies he carried out that December, which have not received the scrutiny of historians. The net result is a clear and nuanced portrait of Faraday's research practice during the initial phase. With this picture in hand, I am in a position, in my concluding comparison between Ampère and Faraday, to render the distinguishing characteristics of both figures crisply.

As this overview already indicates, throughout this book, historical narratives are often interrupted by reflections on more general themes. The seventh and final chapter of this book brings these reflections together, systematizing and extending them. The subject of experimentation itself forms its core. I begin by reviewing the recent discussions of this issue, then turning to the implications of my own studies of scientific practice. They make possible a detailed account of the roles of experimentation outside the realm of theory testing. Of particular significance is the kind of experimentation I call exploratory, which I characterize in detail. I also provide at least a partial answer to the question of why such exploratory experimentation typically recedes into the background, or disappears entirely, in self-representations of science. In addition, based on my historical case studies, I argue for the fundamental epistemological significance of exploratory experimentation, which the philosophy of science, captivated by the distinction between context of discovery and context of justification, has thus far overlooked. Of crucial importance are the processes of concept formation, the delicate relationship between concept formation and experimentation, and the mutual stabilization or destabilization of concepts and experimental activity. A glance at other, similar processes in the development of the sciences demonstrates the generality of this perspective, while simultaneously pointing toward the irreducible complexity of scientific research, as an activity undertaken by human actors.

Micro- and Macrohistory

Large portions of this book are presented as historical narratives. In the various chapters, however, they unfold on very different levels and timescales. They run the gamut, from a continent-wide overview of two decades in the history of a scientific field to detailed case studies of the research practice of two individual agents over periods of one to three weeks. At medium resolution, they canvass the whole research activity of a particular place over several months. In short, their spectrum ranges from microhistory to the borders of macrohistory.

The relationship of micro- and macrohistorical treatments is not one of mere juxtaposition; to the contrary, chapters with different spatial and temporal resolutions complement each other in mutual dependence. The microhistories, studies of daily events in the laboratory, would be isolated, nonspecific, and sometimes even incomprehensible without due attention to the researchers' broader spatiotemporal frame, within which they self-consciously position themselves. Among other elements, this frame encompasses theoretical traditions, scholarly debates, material cultures, experimental resources, academic customs, training and career trajectories, and publication structures, as well as even larger national and international academic constellations. Without such macrohistorical elements, microhistory would have little to teach us. On the other hand, a pure macrohistory could hardly deliver what I am most interested in: insight into research practice. What's more, it is in the context of microhistorical studies that new and fruitful questions for macrohistorical analysis itself come into view. This book attempts to do justice to this mutual dependence.[17]

The narrative approach, which characterizes long stretches of the book, is the only way I could envision presenting an account of the relevant historical material that is both sufficiently detailed and readable. In my writing style I have tried to keep close to the historical actors' perspectives, especially in my detailed case studies, so as to capture and render plausible the many contingencies of the daily research process. Reflective distance is offered through many interpolated passages of varying lengths, as well as in the concluding summary in my final chapter. Taken together, my presentations will, I hope, illustrate my earlier observations regarding the complexity of scientific activity and the consequent need to include, beyond programmatic purity, a rich diversity of aspects in studies of research practice.

History and Philosophy of Science

Scientific development is driven by a heterogeneous variety of different factors. While they can often be isolated for the sake of specific historical analysis, an adequate account of particular historical episodes may resist such isolation, instead demanding an inclusive treatment. Such an account must rely as much on an understanding of the social, cultural, and biographical situation as on the particular scientific domain with which the researchers were concerned, as well as a nuanced appreciation for the epistemic process as such in all its ramifications. After all, to study the development of a science is to study the course of a social enterprise whose goal is the search for and generation of knowledge. Like other social enterprises—politics, industry, and the arts, for example—this one has its distinctive characteristics. For the behavior of an individual agent and for the scientific development of a given period and a given field, particular conceptual, theoretical, and experimental constellations, along with questions of methodology and epistemology, are every bit as important as cultural, social, and biographical situations. Any attempt at a comprehensive understanding of the scientific enterprise must do justice to this multiplicity.

In recent decades, the exploration of social, political, cultural, and economic aspects has gained increasing attention in the historiography of science. This turn constituted a reaction to the traditional, narrower focus on theoretical developments that predominated into the 1970s. Long-neglected questions of rhetoric, modes of presentation, and material culture are now being investigated, establishing the status of science as a part of, and significant factor within, broader cultural developments. The turn to general history and cultural history has led, and continues to lead, to important insights, profoundly transforming our image of scientific process. Nonetheless, it strikes me as running the risk of neglecting the complexities of science "from the other side," as it were, by losing sight of the epistemic dimension of science as such, and thus of the avowed target of the whole undertaking. On the other hand, the philosophy of science, to which this epistemic dimension is central, can still assert that, from the epistemological perspective, all the newer insights, interesting though they may be, remain a sideshow of no real relevance to the understanding of knowledge generation itself. A glance at the philosophical journals is enough to show that, even in recent years, many of the debates in the philosophy of science continue to be conducted in formalistic sterility, without making contact with new insights from the history of science.

Serious attempts to reintegrate epistemological and historical research and insights have only recently received wider visibility.[18] The promise of such integration for the understanding of science—the potential benefits of integrating historical and epistemological reflections, questions, and methods in order to study how scientific knowledge, in all its complexity and irreducible historical contingency, is formed and developed—thus is far from being fully realized. The many-layered character of the epistemic process, and especially its ineliminable historicity, becomes truly visible only by means of thorough and comprehensive historical investigation. Should the analytic scrutiny of philosophy of science begin to take it seriously, there may yet be the chance to achieve a new and integrated picture of the generation of scientific knowledge. It may well be that the conceptual tools required to formulate such a picture have yet to be developed. Such integrated studies are an admittedly laborious undertaking, for urgent methodological issues arise (for example, the question of how to deal with historical case studies).[19] In addition, the age-old problem of "truth and history" lurks always in the background. But such problems are typical, and unavoidable, if we seek to understand science, and increasing numbers of reflective approaches seek to address them.[20]

One point, however, seems to be obvious: such a picture can be achieved only with proper attention to scientific practice. Indeed, such a turn toward practice can be observed in many recent attempts at integration. Such efforts allow us to discern the scope and breadth required for a given historical case and may also offer a glimpse of characteristic structures. In the process, we may go well beyond the particular case, gaining intimations of broader historical processes and epistemological generalities. The central issue of this book, the problem of experimentation, is just an example, albeit one well suited to demonstrating how both historical insight and philosophical analyses may be enriched by an integrated study. Experimentation illustrates with particular poignancy the aptness of a slogan, often highlighted by my esteemed academic mentor, the late Lorenz Krüger, in homage to a famous Kantian dictum: "History of science without philosophy of science is blind; philosophy of science without history of science is empty."[21]

CHAPTER 1

Electricity and Galvanism in the Early Nineteenth Century

"On all of the customary routes, the doctrine of electricity is so well-trodden and well explored, that there is nothing left to be found on the highway; one must march cross-country and plumb the ditches" (Lichtenberg 1972, 2:472 [no. 384]).[1] Thus did Georg Christoph Lichtenberg, in the early 1790s, describe the state of a field to which he himself had made substantial contributions. He was not alone in this assessment.[2] By this period, the traditional topics of electrical research were receding into the background. In their place, at the beginning of the nineteenth century, two new research programs had begun to rise to prominence, differing greatly in their outlook and style, as well as in the number of individual efforts devoted to them. Each could be recognized by its own emblematic instrument. On the one hand, there was the voltaic pile, whose invention had caused a pan-European sensation and a wave of new research efforts proliferating in all directions. In the already established field of galvanism and electrochemistry, the pile opened up entirely new vistas of both theory and experiment. On the other hand, and somewhat independently, the quantitative treatment of electricity had begun to garner attention. At the center of such efforts stood Charles-Augustin de Coulomb's torsion balance, the state of the art in the precision measurement of tiny forces. Geographically, the two research programs were unevenly distributed. Whereas work surrounding the voltaic pile was conducted throughout Europe, quantitative efforts were almost entirely confined to Paris. The two research fields were also fairly independent and treated by nearly disjointed groups of researchers. Even in Paris there were only a very few researchers involved in both. This distinctive arrangement would have lasting consequences for the scientific cultures that are my focus in this book—the cultures of Paris and London—and most especially for Ampère and Faraday, respectively.

In this chapter I outline activities in both research fields, thereby framing the epistemic, technical, and instrumental state of the art in 1820, when the discovery of the new electromagnetic phenomenon provided a spectacular turning point. In the first section I consider the development of mathematical physics in France. After sketching the circumstances that allowed this tendency to thrive, I turn to Siméon-Denis Poisson's work on electricity, situating it in the broader European context. Finally, I review the various breakdowns that had begun to plague Laplacean physics even before electromagnetism posed further challenges. The second section gives pride of place to the voltaic pile. After a glance at the instrument itself, I canvass the various ways in which it was harnessed in London, in the German-speaking world, and in Paris. My attention to the German-speaking world is warranted partly by the intimate relationship of this milieu with that of London, but mostly by the fact that it was the German milieu that molded Hans Christian Ørsted, thus setting the stage for the discovery of electromagnetism in the summer of 1820.

Mathematical Physics and Centralized Science in Paris

Encouraged by the successes of Newtonian gravitation and Eulerian mechanics, researchers had made repeated attempts since the middle of the eighteenth century to turn electricity into a mathematically tractable field. The Saint Petersburg academician Franz Aepinus's efforts went the furthest. In his *Tentamen theoriae electricitatis et magnetismi* (1759) he sought to apply the mathematical tools of Newtonian mechanics to the interaction among the particles of distinct electrical and magnetic fluids.[3] Each particle was envisioned as a center of force, with particles exerting centrifugal forces on each other. Initially, Aepinus's work met with little interest. The "electricians" were interested in other problems, most especially in the immense diversity and variability of electrical effects.

Mathematical exertions were not renewed until the 1780s, when the idea

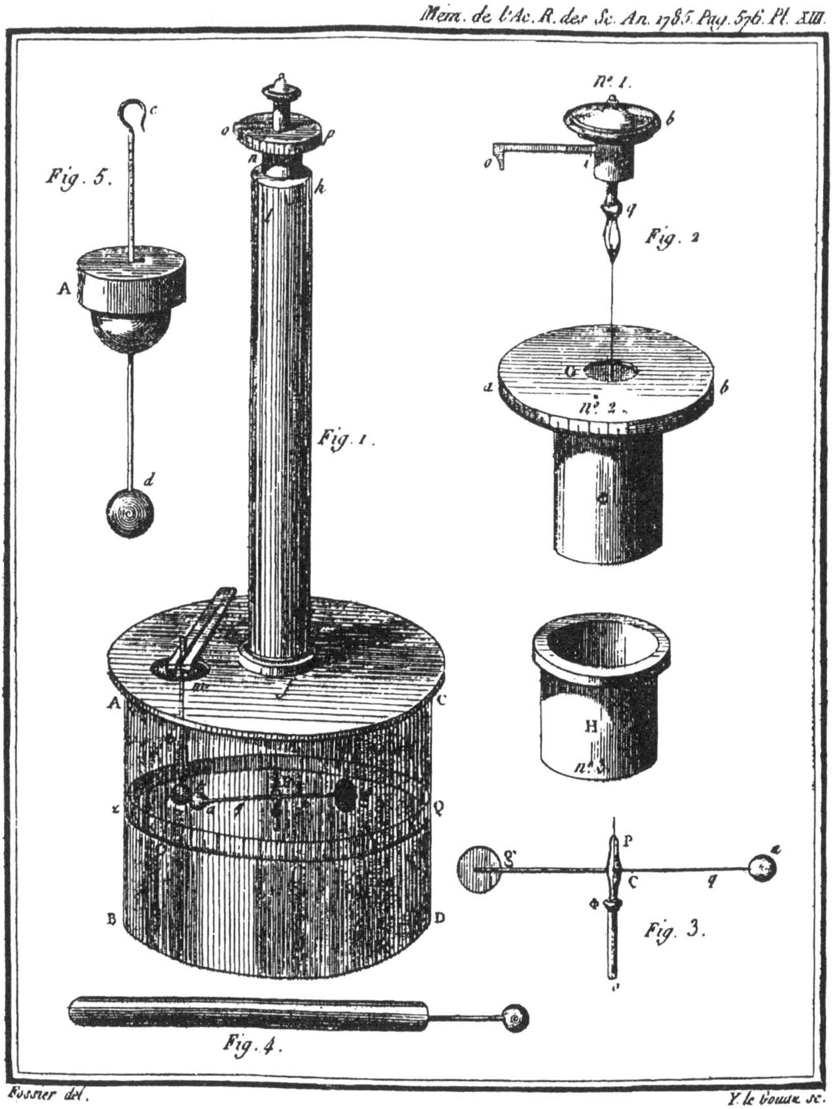

FIG. 1.1. Coulomb's torsion balance (Coulomb 1788b; Joubert 1884).

was taken up in Paris by the engineer and academician Charles Augustin Coulomb (Blondel 1994). On the basis of his experience with torsion instruments, such as the compass, toward the middle of the 1780s Coulomb began construction of an instrument for the measurement of extremely small forces. He applied it to the study of electricity and in particular to

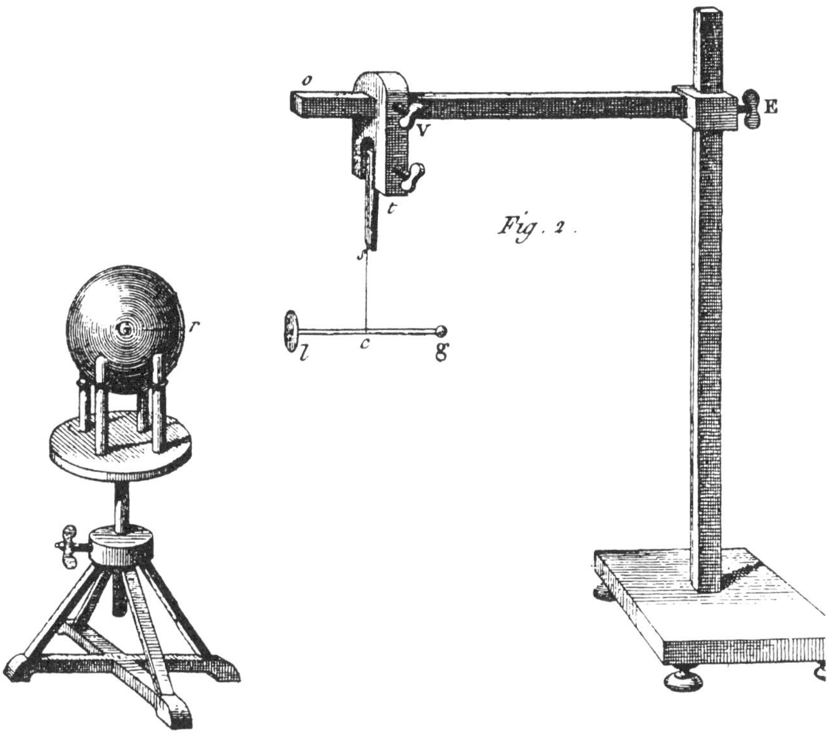

FIG. 1.2. Coulomb's apparatus for measuring the force of electrical attraction (Coulomb 1788a; Joubert 1884).

the old problem of the distance function associated with forces of attraction and repulsion.[4] By means of his now famous torsion balance (fig. 1.1), Coulomb established the inverse-square distance law for repulsion by analogy to the Newtonian law of gravitation.

Because the torsion balance could not be used to measure electrical attraction, as two oppositely charged bodies would too easily come into contact during oscillation, Coulomb developed a second instrument (fig. 1.2). This one allowed for the attractive force to be measured indirectly, calculated from the duration of the oscillations exhibited by a charged body (*l* in fig. 1.2) placed at various distances from an oppositely charged body (*r* in fig. 1.2) subjected to a slight blow.

Here, too, Coulomb deduced an inverse-square relation. These measurements were extraordinarily difficult to perform. Coulomb devoted much effort to achieving stable results, in 1785–91 repeatedly describing particular problems of measurement in a series of *mémoires*.[5] In addition, Coulomb investigated the dissipation all charged bodies eventually exhibited, as well as the transmission of electricity to conductive bodies.

Coulomb can well be regarded as the champion of exact measurement and the resulting laws, but any such characterization must be carefully hedged. For one thing, he envisioned a much more general law than he was ever able to justify by measurement. He simply assumed the proportionality of electrical forces to the "quantity" or "density" of electricity, this being presumably too self-evident to require any proof.[6] Here the analogy to Newtonian gravitation was decisive. In addition, Coulomb justified his inverse-square law by only *three* measurements, a fact that has frequently given rise to irritation.[7] Coulomb's hasty conclusion also made plain how dramatically his concepts of exact measurement and empirical justification fell short of any later standard.

The swift positive response with which Coulomb's results were met may be traced to French science's general inclination toward quantitative, mathematical approaches. In 1783, the mathematician Pierre-Simon de Laplace and the chemist Antoine Lavoisier had published their "Mémoire sur la chaleur," in which they had sought to make an exact science of the study of heat (Laplace and Lavoisier 1783).[8] At the center of their work stood another celebrated instrument, the ice calorimeter. Dovetailing with this development, Aepinus's work in the 1780s gained some belated recognition in Paris. In 1787, the mathematician René-Just Haüy published a French précis of the *Tentamen,* and the Académie Royale des Sciences empaneled a commission for its assessment. The resulting report offered high praise for Aepinus, describing his achievement as a true "epoch in the history of the sciences" (Haüy 1787, xxix).[9] Such praise came despite the fact that his treatise had appeared thirty years prior and had received scant attention since then. This programmatic zealotry was paired with pointed disparagement of earlier qualitative studies of electricity, as performed in the salons and cabinets of physical instruments and exemplified by the work of the Abbé (Jean Antoine) Nollet (Blondel 1994). Significantly, the aforementioned commission on Aepinus's work consisted entirely of mathematically inclined researchers, Coulomb and Laplace among them. By contrast, the traditional "electricians" had no representation, hinting at the shifting balance of power in the Paris academy.

Coulomb's series of mémoires was in perfect concert with this trend.[10] Familiar with Aepinus's treatise, he set out to back these mathematical reflections with exact measurements.[11] The torsion balance soon became emblematic of the entire program of mathematical natural science grounded in precise measurement, and Coulomb's indirect measurements by means of oscillation were to become, for many decades, simply standard procedure. In the wake of the triumph of measurement, laboratory practice changed dramatically. To cite one example, the torsion balance could be noticeably perturbed by the mere approach of an observer. The changed role of the experimenter came to light even more dramatically with Giovanni Domenico Cassini's comprehensive measurements of terrestrial magnetism at the Observatoire de Paris. Following Coulomb's advice, in the late 1780s Cassini switched from the pivot-point compass to the torsion compass. To attain stable measurements from these extraordinarily sensitive instruments, procedures for the proper handling of which were themselves a work in progress, the scientists found it increasingly necessary to shield them from any possible interference. In the end the instruments were mounted in the observatory cellar, shielded even from the interference of the observer. By the standard of all earlier experimental approaches to electricity, as they continued to be practiced elsewhere, the ideal of precision measurement entailed a radical departure in experimental procedures, as well as a transformation of the role of the experimenter. Up to this point, the field had never known a true culture of experts of the kind now emerging.[12] At the same time, more than ever before, the laboratory as a closed space, inaccessible to the public, was emerging within that process.

The program of mathematical formalization and exact measurement took its lead from Paris. This direction had to do with peculiarities of the environment and historical situation. France was especially steeped in the "quantifying spirit" of the late eighteenth and early nineteenth centuries, not least in consequence of political events. The revolution sought to imbue every aspect of life with rationality, in the process transforming quantification into a broader cultural movement, albeit one short-lived in many domains.[13] This factor combined with the particular structure of the school and university systems. During the revolution and under Napoleon, the entire French educational structure underwent thorough reform. The result was the very first education and research system organized entirely by the state. This system encompassed everything from primary and secondary schools to the newly founded *lycées* and the *grandes écoles* (universities and academies).[14] The organization of the sciences became centralized, highly regimented, and hierarchical. As successor to the Académie Royale des Sciences, the Institut de France took on the overarching supervisory role for research. The training of the scientific elite lay in the hands of the grandes écoles, including not only the leading École Polytechnique but also the École Normale, the École des Ponts et Chaussées, and the École des Mines. Napoleon, a proponent of scientific innovation and himself a member of the Institut de France, assigned himself a key role in this system, which allowed him to shape the general direction of research. One significant consequence of centralization and regimentation was that guidelines decreed from above could quickly penetrate even into elementary education. Preference was given to the natural sciences, whereas such fields as literature, considered potentially subversive, were given little support (Crosland 1992, 23). And in the natural sciences, the quantitative trend came to dominate education.

This thoroughly regimented scientific infrastructure was unique in all Europe. English natural science, receiving little organization from the state, was still mostly the province of well-to-do amateurs, as illustrated by the membership of the Royal Society (Hall 1984). The Royal Institution, a private and rather unusual initiative, became an important center for research only through the personal efforts of Humphry Davy. Scientific research had little presence at the universities. In the German-speaking world, while the universities and academies had many professors assigned to various scientific fields, these academics enjoyed little prestige and commanded at best modest resources for mounting experiments.[15] There were no research laboratories properly so called, and university collections more closely resembled cabinets of physical instruments. Whether and to what extent research took place alongside instruction depended entirely on the interest of the individual professor. The kind of coordination and bundling of research efforts imposed in France would have been inconceivable in the German-speaking world, a point that would eventually come to be explicitly acknowledged as a weakness.[16] It was in postrevolutionary France that the organization of scientific activity combined with political power in unprecedented ways.

Laplacean Physics

The French drive toward mathematical formalization of the natural sciences received an additional impetus from a very specific source. A special closeness with Napoleon allowed the mathematician Pierre-Simon Laplace and the chemist Claude Louis Berthollet to draw on generous financial support such that from 1806 on, despite the official policy of openness and transparency, they were able to raise their own elite research institution before the very gates of Paris: the Société d'Arcueil.[17] There they gathered the most promising young researchers they could find at the grandes écoles, fostering a level of exchange and collaboration whose intensity has led Robert Fox (1974, 133) to describe the Société d'Arcueil as "one of the most closely knit schools in the whole history of science." Eugene Frankel has expressed its well-defined research agenda as geared toward "four complementary activities":

- The quantification of new effects by precise measurement, including the development of new instruments, for which Lavoisier's ice calorimeter and Coulomb's torsion balance served as paradigm cases;
- The optimization of equipment with a view toward the minimization or elimination of error; unavoidable errors were "factored out" by subsequent data reduction;
- The expression of experimental data through analytical equations on physical magnitudes;
- The accommodation of these formulas to generalized action-at-a-distance theory. (Frankel 1977, 46–47)

Recent work on precision measurement, and on Coulomb in particular (e.g., Blondel and Dörries 1994; Wise 1995), yields a subtler view of the third and fourth activities. Equations relating measurable variables never follow necessarily or directly from empirical data, then any more than now. It would be more accurate to speak of the accommodation of data to specific, seemingly plausible formulas. The reasons why certain formulas and other mathematical formalisms were closer at hand than others can be traced to fundamental, dare I say metaphysical presuppositions regarding the underlying structure of matter. Chief among these was the conviction that all natural phenomena could be reduced to the action at a distance of forces of attraction and repulsion, operating along the lines connecting (microscopic) centers of force. These centers of force were conceived as the smallest particles of matter or of imponderable fluids—traditional notions that were now made rigorously mathematical. The famous trope of "Laplace's Demon," for whom knowledge of the state (location and motion) of every material particle at a given time, together with unbounded computational ability, was enough to lay bare all the past and future, serves as a compelling popular icon for the whole program.[18] It also makes clear to what extent the program was stimulated and shaped by the remarkable success of celestial mechanics, embodying what John T. Merz (1904, 1:347–48) so aptly called the "astronomical view of nature." On this approach, quantitative methods would not only be applied to previously formalized fields such as optics but also extended to the study of such domains as heat, electricity, and magnetism. The exemplary status of celestial mechanics became apparent even in the construction of specific mathematical tools and constituted the very ideal of precision measurement, for which the phrase "astronomical accuracy" serves as watchword.

The golden age of this Laplacean physics began before the turn of the nineteenth century and lasted through the middle of its second decade. In geographical terms it was mostly confined to Paris.[19] The catalog of researchers who passed through the Société d'Arcueil reads like a Who's Who of early nineteenth-century French physics and chemistry. Besides Laplace and Berthollet we find the names of Biot, Arago, Poisson, Dulong, Gay-Lussac, Malus, and Thenard. There were spectacular successes in optics (refraction indices, double refraction, corpuscular theory of light, and polarization), in fluid dynamics (capillary action, sound propagation in gases and solids), and in the caloric theory of heat, in electricity, and in magnetism. Biot's later physics textbook serves as a powerful illustration of the sheer breadth of the enterprise.[20] Outside the strict research context, the terminology and procedures of Laplacean physics became entrenched in the curriculum of the lycées and grandes écoles and especially at the École Polytechnique, the favored venue for training the next generation of scientists (e.g., Fox 1974, 107–8). The centralized institutional structure encouraged the swift occupation of many of Paris's key academic positions by Laplaceans. No other approach could make any headway against the rapid ascent of this research program. For ten to fifteen years, French physical science and Laplacean science were nearly the same thing.

In Paris, mathematical formalization and scientific specialization imposed rigorous disciplinary training for the first time. Physics was soon understood as something very different from what it had been as recently as the turn of the century. It was no longer a hodgepodge (Silliman 1974, 139–44), the methodologically heterogeneous study of all those aspects of nature that belonged neither to chemistry nor to natural history but the methodologically strict engagement with very specific natural domains, directed toward their quantification and mathematical formalization. It would be no exaggeration to call this a decisive formative phase in the emergence of physics as an independent discipline.

Poisson's Studies of Electricity

The field of electricity promised great contributions to the Laplacean program. By the turn of the century, it already had both precision instruments and a tradition of mathematical theory. In 1801, Laplace encouraged the then unknown Jean-Baptiste Biot to tackle Coulomb's theory with a view toward lending it a generalizable form. The inverse-square law itself was no longer in question. What were needed were mathematical techniques for applying that law to complex configurations. Biot attempted this task using a device Laplace had developed for and successfully applied in celestial mechanics: a mathematical function V, later known as the potential function.[21] It assigned every point in space a numerical value. When the function was expressed as an integral, the inverse-square law remained clearly recognizable:

$$V = \int dm\,(x', y', z') / r'$$

In this equation, dm stands for the mass of an infinitesimal element at point (x', y', z') and r' for its distance from a fixed origin point. The bounds of integration are defined by the entire body, of which dm is an element. This function provides an elegant representation of gravitational force or, in general, of any central force obeying an inverse-square law. Put in modern terms, the force exerted on a body is then proportional to the gradient of V at the body's location. In order to apply this device to electrical forces, Biot had made additional assumptions on the nature and behavior of the electrical fluid. Like Coulomb before him, he considered the distribution of charge across differently shaped conductive surfaces, with particular attention to ellipsoids.[22] But after his first publication (Biot 1801b), Biot pursued this strategy no further, his work having been interrupted and redirected by events surrounding the voltaic pile. We turn to these later.

It took ten years for the problem to be taken up again by another Laplacean. Siméon-Denis Poisson had joined the Société d'Arcueil at a young age and would prove one of the very few figures who remained devoted to the program his entire life. Hardly anyone else, besides Laplace himself, would pursue mathematical formalization in the Laplacean sense so consistently and in so many different fields (mechanics, heat theory, elasticity theory, optics, electricity, and magnetism among others).[23] After a series of studies on heat and elasticity, Poisson turned to the theory of electricity in 1810. Aware of his new direction, in January 1812 the Académie des Sciences announced a prize question clearly geared toward Poisson's research. Some of the academy's leading members were determined to have Poisson as the successor to the grievously ill Étienne-Louis Malus. It is thus hardly surprising to find Poisson submitting his first mémoire a scant two months later. It had attained the desired result, though by a somewhat different path than initially expected, and on March 23 he was elected to the academy. In August he presented a second mémoire on the same subject. Both were published quickly in the academy's *Mémoires* (Poisson 1812a, 1812b).[24]

Like Biot before him, Poisson studied the distribution of charge by means of the potential function, though he formulated his specific presuppositions more explicitly and precisely. He assumed that the electrical fluid, unlike gravitational masses, occurred not in stable aggregates but could freely distribute itself throughout charged bodies.[25] His task was to determine the distribution of the fluid in a given charged body by means of a potential function that depended on this distribution. Of particular significance were the boundary conditions: the fluid was constrained to a very fine layer on a body's surface, and there were no tangential forces operating along the surface. For the V-function, this meant that it remained constant throughout a given body, and all of its partial derivations along directions parallel to the surface could be assumed to vanish.

To simplify his calculations, Poisson additionally assumed the uniform "density" of the electrical fluid. Greater or lesser charges on particular areas of a surface could be expressed by assigning the layer greater or lesser thickness. Coulomb, by contrast, had assumed a fluid of variable density. Poisson took pains to assert the equivalence of the two different ways of expressing the "quantity of electricity gathered at a given point" (Poisson 1812a). Within the boundaries thus defined, Poisson's task was to determine the thickness of the fluid layer at every point on the body's surface. This problem could be solved only for particular, precisely stipulated arrangements of conductive bodies.

Poisson's first arrangement was a spheroid deviating only slightly from a true sphere, or in other words a generalization of the ellipsoid, which Biot had already investigated. Following Laplace's example, Poisson arranged the integrals appearing during the calculation of the V-function into a series, discounting higher-order terms. He thereby obtained the general result—that the thickness of the fluid layer y was always proportional to the force exerted in a direction normal to the surface:

$$-dV/dx = 4\pi\, y$$

For the special case of an ellipsoid, this term became proportional to the radius of the ellipsoid at the given point. This result contradicted Biot's, and, without mentioning Biot by name, Poisson provided a detailed diagnosis of the error in his calculations.[26] For the general case, the result is expressed by the integral equation

$$4\pi\, y = -dV/dx = -d/dn \int y\, \rho\, /\, r'\, dS'$$

where ρ is the thickness of the fluid layer, r' the distance of the surface element, dS' to the origin, and d/dn the derivative in the direction normal to the surface.

In this first example Poisson was simply describing mathematical methods, without any recourse to experimental data. To bring empirical data into his investigation, Poisson turned to a second arrangement: two spheres of differing radii. Coulomb had performed precise measurements for this configuration, for which he also offered a theoretical account. Poisson replaced Coulomb's cumbersome mathematical tools with his own more general treatment by means of the V-function. He succeeded, albeit with much greater effort than his first example had required, in calculating the distribution of charge on each sphere, even for the case in which the two spheres were in contact, as for the case in which they had been separated after coming into contact. A comparison with Coulomb's measurements, provided in the form of numerous tables, revealed no large deviations (Poisson 1812a).[27] Poisson took this as confirmation of the soundness of his mathematical approach.

Poisson's procedures were by no means taken off the shelf; every step had to be developed. This was true to an even greater degree in his second mémoire on electricity. There he greatly extended his mathematical techniques, which included approximation procedures, the evaluation of integrals, and the construction of series. In the eyes of many of his contemporaries, he had thereby convincingly demonstrated that Laplacean tools developed for celestial mechanics could be used in opening up new fields. To be sure, the circle of colleagues capable of following his arguments in any detail was rather small. To anyone who had not yet completely mastered such analytic techniques—and this included even most of the academy members in Paris, not to mention scientists at Europe's other centers of research—his work remained completely cryptic. Consequently, Poisson's studies met with scant reaction beyond his successful election to the Académie des Sciences. He did not pursue them further, turning for a long time to other topics (Grattan-Guinness 1990, §7.6.6).

Just as the mathematical methods were accessible to only a small group of specialists, so were their instrumental-experimental counterparts. Coulomb's torsion balance was an extremely sensitive instrument whose proper handling required great care and practice. His measurements of the distribution of charge across charged bodies were neither easily replicated nor easily extended to a broader domain of geometric shapes. It is thus hardly surprising that no further efforts were made in this direction, even less so as the problem was perceived by many as both specialized and of no special urgency. Indeed, Poisson had chosen this problem only because it seemed most congenial to mathematical treatment and because the experimental data already existed. He never conducted any electrical experiments himself. There were numerous other unsolved problems in the domains of electricity and magnetism, most of which seemed more important at the time and for which Poisson's mathematical approach offered no immediate solution. Though Poisson granted free movement to the electrical fluid, he had nothing to say about the hotly debated question of how such fluids interacted with ponderable matter and about how they came to remain within charged bodies and thus exert forces on them.[28] He also eschewed the entire problematic domain of galvanism, which was then one of the highest-priority topics in the study of electricity.

Poisson illustrates both the power of the Laplacean program and its limitations, for within that program, research outlook was shaped not only by the general goals of quantification and mathematical formalization but also by a very specific set of substantive assumptions and its precisely tuned mathematical toolbox. Such tools were suited only to very specific questions, while others were lost to view, and not necessarily because they were deemed uninteresting but simply because there was no clear way to deal with them using existing procedures. Furthermore, the emphasis on mathematical formalization and precise measurement brought with it the disparagement or exclusion of broad-based qualitative experimental research. It is thus no coincidence that the most significant studies of the voltaic pile, which led to equally original and innovative results, took place outside France. The focused quest for mathematical formalization had its price.

Alternative Pathways to Mathematization

The second decade of the nineteenth century saw the emergence of a Parisian countermovement. As early as 1807, the theory of heat transfer promulgated by Jean-Baptiste Joseph Fourier made no use of either caloric theory or any Laplacean terminology. Instead, it took the heat exchange between two adjacent surfaces as the basis for a mathematical analysis, without concerning itself with the microscopic nature of heat. Unsurprisingly, Fourier had no opportunity to publish this theory quickly in the proceedings of the academy—indeed, it was only seventeen years later (in 1824) that it was taken up there. But François Arago had published a detailed summary of it in 1815, in the widely read *Annales de chimie et de physique,* which he had just taken over as publisher and coeditor, along with Joseph Louis Gay-Lussac (Fourier 1824, 1826).[29] Though himself a beneficiary of the Arcueil school, Arago had become increasingly dissatisfied with the restrictions it imposed. His attitude had to do both with substantive issues and with Laplace's unsubtle patronage politics.

In the inaugural issue of the *Annales,* Arago also published the first version of the young Augustin-Jean Fresnel's (1816) un-Laplacean, or even anti-Laplacean, wave theory of light. Arago further supported this theory in another article, which he coauthored with Alexis Thérèse Petit, professor of physics (Arago and Petit 1816). The enormous challenge posed by Fresnel's theory was immediately clear. In response, the academy devoted its annual physics prize for 1817 to the question of diffraction phenomena, which Fresnel had cited as a significant point in his favor. Against expectations, it was Fresnel who won the prize, rather than any defender of the corpuscular theory. Now there was a real, lasting alternative to the Laplacean terminology.[30] Finally, in 1819, Pierre-Louis Dulong and Petit (1819) presented the law of specific heat that would be named after them, thereby issuing a sharp rebuke to the caloric theory of heat. By the end of the decade the earlier euphoria over the prospects of Laplacean physics had given way to the critical awareness that other approaches were not only possible but could also prove very fruitful.

The new mood in Paris also had much to do with the fact that the strong political and financial support that the Arcueil school had enjoyed under Napoleon had disappeared with the empire's fall. It was Arago, and not Laplace, Biot, or Poisson, who quickly secured political support in the post-1815 regime. The key professorships in physics and chemistry at the École Polytechnique were soon in the hands of the younger generation—those of Petit and Dulong. By the beginning of the 1820s, the curricular dominance of Laplacean physics had been systematically pared back (Fox 1974, 125–26). Significantly, in 1822 it was Fourier and not his competitor Biot who was elected as Jean-Baptiste Delambre's successor in the key post of *secrétaire perpétuel* of the academy.[31]

Remarkably, all those who were now questioning the old system—Arago, Fresnel, Dulong, and Petit—had once been closely associated with the Arcueil school and were thoroughly versed in Laplacean terminology. They maintained significant aspects of the Laplacean program entirely unmodified. By no means did they seek to question the ideals of quantification, precision measurement, or mathematical formalization; these remained lasting essential elements of the new discipline such ideals had come to constitute: scientific physics. What the new generation criticized were the narrow conceptual boundaries within which the Laplacean program was pursued. These boundaries had been chosen in part because they encompassed a whole suite of mathematical tools and concepts, tools whose efficacy had been demonstrated in celestial mechanics. Rejecting the boundaries while maintaining the ideal of mathematical formalization demanded the invention and refinement of new mathematical methods. Fourier and Fresnel offer well-known examples. It was in the context of the diminished authority of Laplacean physics that the discovery of electromagnetic interaction in 1820 arrived to pose a further challenge. Many of the Parisian reactions to this discovery can only be properly understood against this background. But before we turn to these reactions, we must first consider the other research field that so significantly shaped the early nineteenth-century study of electricity.

The Voltaic Pile

On March 20, 1800, in a detailed letter to Joseph Banks, president of the Royal Society of London, Alessandro Volta, a physics professor in Florence, reported on a new invention, soon to be commonly known as the "voltaic pile."[32] It consisted of a stack of individual cells (fig. 1.3), each of which was composed of a silver plate (labeled *A* for *argent,* or silver) and a zinc plate (labeled *Z*), with a layer of moistened cardboard or leather in between. Copper or brass was frequently substituted for silver and tin for zinc. Supported by rods, *m,* any number of such elements could be built into a column or into several shorter columns, connected by copper strips, *c.* Volta presented a number of different variations on this device. His "crown of cups" (*couronne des tasses,* fig. 1.4), in which metal wires were submerged in basins of saline solution, exhibited essentially the same effects as the pile.

The voltaic pile could operate in two fundamentally different ways. If the two ends of the pile were not connected, they could be made

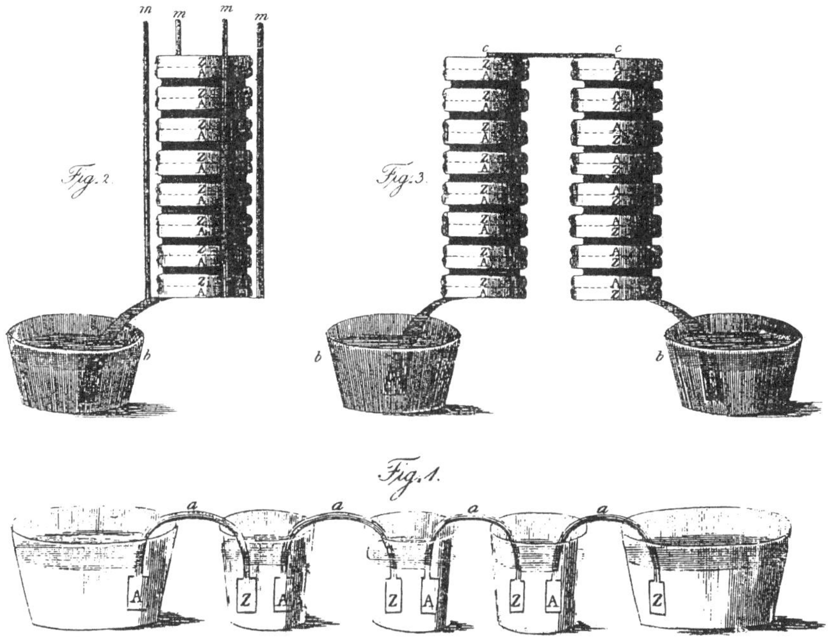

FIGS. 1.3 AND 1.4. The voltaic pile and the crown of cups (Volta 1800c).

to exhibit effects familiar from ordinary electricity: bipolarity, spark discharge, electrification of small bodies, and the activation of electrometers. In his letter, Volta argued that these were electrical phenomena in the ordinary sense, an interpretation that quickly gained general acceptance. The pile behaved quite differently when the two poles were connected with a wire, "closing" the pile (or, as we would now say, short-circuiting it). All of the aforementioned electrical effects vanished. Even the most sensitive electrometers, and Volta was famous for such, no longer registered anything. Instead, it exhibited what were known as galvanic phenomena. Animal tissue placed between the wires from either pole would be observed to twitch. In addition—and this effect was new—the connecting wire became noticeably heated and, with larger piles, might even begin to glow. Volta also ascribed such effects to electricity. He understood them as caused by a current of electrical fluid (*courant électrique*), describing his apparatus accordingly as an "electro-motive apparatus" (*appareil électro-moteur*) (e.g., Volta 1800a, 1800c). On this point, however, he met with some opposition, and the question of whether the effects of the closed column ought really to be attributed to electricity, or to some other agency, initially remained controversial.

For scientists of the day, the invention or discovery of this device was sensational. An arrangement of inorganic materials, mostly metals and liquids, could produce galvanic effects! In the minds of many (first and foremost Volta himself), this brought to a decisive conclusion the earlier, years-long debate between Luigi Galvani and Volta on the nature of galvanic electricity, in the context of which Volta had developed his apparatus in the first place. Because the voltaic pile required no animal tissue, many saw the problem of animal electricity as having been solved, and so for several decades it receded into the background (Trumpler 1992). Volta himself believed not only that he had refuted Galvani's account of animal electricity but also that he had confirmed his own theory, in which galvanic electricity arose at the points of contact between variously distributed metals. This point was often received skeptically. The numerous chemical changes that accompanied the operation of the voltaic pile were impossible to miss, and Volta, whom they had not escaped either, was too quick to dismiss them as inconsequential side effects.

Another characteristic of the voltaic pile that deeply impressed Volta's contemporaries was that it could evoke galvanism for a very long time. Volta even saw in his device the realization of the old dream of a *perpetuum mobile.* In light of Volta's report, even the Paris academy, which had long since decided not to pursue any further proposals for *perpetua mobilia,* saw itself as obliged to once again take the question seriously. And so the voltaic pile opened up an entirely new field of research questions and opportunities. Intensive research efforts began throughout Europe. Humphry Davy, one of the activists, would later retrospectively describe the discovery as an "alarm bell" for Europe's experimentalists (Knight 1992, 39).[33] In what follows, I sketch subsequent developments in England, the German states, and France. Although there were significant exchanges between these three regions, there were also strong local and national peculiarities.

Galvanism and Electrochemistry in London

Reports of Volta's discovery were circulating in London even before the reading of his letter to the Royal Society. Joseph Banks had shown it to his friend, the chemically inclined surgeon Anthony Carlisle, who, together with his own friend William Nicholson, immediately set about experimenting. Nicholson, who was active in the natural sciences and especially in chemistry, published the scientific *Journal of Natural Philosophy, Chemistry, and the Arts*.[34] Given their professional background, it was not surprising to find both of them drawn, more than Volta, to the pile's chemical effects. They were struck by the appearance of gas at the connecting wire when it came in contact with water. When the wire was cut and its two ends submerged, a whole series of effects were observed, especially the formation of hydrogen and oxygen on the ends. Nicholson and Carlisle interpreted this phenomenon as the decomposition of water into its elemental constituents—an interpretation that reflected their adherence to Lavoisier's new chemistry. The idea that galvanism might have chemical effects was not entirely new, but never had such effects included the decomposition of a substance. A report on the surprising discovery quickly appeared in Nicholson's journal (Nicholson and Carlisle 1800).

Besides Nicholson's own interest, the fact that the intensive debate that ensued took place mostly in the pages of his *Journal of Natural Philosophy, Chemistry, and the Arts* had to do with the fact that submissions could be published very quickly. Besides such well-established figures as William Hyde Wollaston, Henry Haldane, and William Cruikshank, participants in this debate included newcomers such as Humphry Davy, not to mention Nicholson and Carlisle themselves. Many contributions dealt with the pile and with the many variations on this apparatus that were rapidly devised. It was Cruikshank who proposed the much handier trough battery. Nicholson and Carlisle's results were a recurring theme. Their interpretation of the effect as decomposition was problematic because it failed to explain other associated phenomena. For example, while the environment surrounding the end of one wire became alkaline, that surrounding the other became acidic. What's more, the volume ratio of hydrogen to oxygen was unexpected. A further serious objection arose from the fact that the two gases appeared *separately*, each at one of the two submerged wire ends. But if the water molecule was being decomposed, the two gases ought to form together at the site of decomposition. Johann Wilhelm Ritter was particularly clear and articulate in voicing this objection, supporting his position through experiment. Ritter proposed a completely different interpretation of the phenomenon, treating it as the formation of a compound of water and electricity, or the synthesis of "galvanates." This explanation had its adherents, despite (or perhaps because of) the manifest challenge it posed to fundamental tenets of Lavoisier's chemistry.[35]

One of the most active participants in this debate was the young, still unknown Humphry Davy. Of humble background, lacking any formal training, Davy had taught himself chemistry, and in 1798 he became an assistant at a private chemical laboratory, the Pneumatic Institution of Clifton, outside Bristol. This venture belonged to the chemist Thomas Beddoes, known for his radical political views. Among other things, Davy had investigated the physiological effects of various gases (nitrogen oxides), performing a series of risky self-experiments.[36] Volta's discovery lent great support to Davy's conviction that there was an intimate connection between electricity and chemical processes. Davy's interest in the voltaic pile was presumably also fostered by the fact that the pile was among the most spectacular items of scientific interest in his day. Whoever succeeded in achieving novel results might gain greater recognition, and Davy was certainly very keen on becoming widely known, even famous. His interest in chemistry was one of many, and perhaps not even the most important among them. Well outside the bounds of natural science, he was devoted to literature, to the German romantics in particular. He was also a lifelong poet.[37] Forever brimming with ideas, Davy had an effusive temperament that profoundly shaped his scientific work.

Davy sought to understand the cause of the various voltaic effects. He was skeptical of Volta's contact theory, instead preferring to assume that effects produced by the pile were partly or entirely chemical in origin. Over the course of his investigations and many findings, he was able to develop a novel form of pile, containing only one kind of metal (zinc). Volta had always required two. London was quick to acknowledge the originality of Davy's work. After his first publications appeared in Nicholson's journal (Davy 1800a, 1800b, 1800c, 1800d), they were read to the Royal Society and reprinted in the *Philosophical Transactions* (Davy 1801). This was a most extraordinary commendation for a twenty-two-year-old.

Davy was a pivotal figure in English electrochemistry. His career was closely tied to a particular institution, beginning with his appointment there in 1801. The Royal Institution of London provided excellent working conditions. A private foundation spearheaded by Benjamin Thompson, Count Rumford, in 1799, the institution had as its goal the broad dissemination of scientific knowledge. Thanks to generous donations by its wealthy patrons, it boasted its own marvelously equipped building. Most visitors to the facilities came not from the ranks of the poor, whose conditions the founders had sought to improve, but from London's middle and upper classes. Significantly, a partially built second staircase, designed to allow "artisans and mechanics" to attend lectures without rubbing elbows with academics and nobility, was demolished before its completion.[38] Lecture admission fees were also set high enough to make participation by the "ordinary man" challenging. The Royal Institution quickly became a gathering place for interested members of loftier circles. Accordingly, the style of lectures was geared toward a reasonably well educated audience.[39]

In his quest for crowd-pleasing lecturers, Count Rumford's attention had been drawn to Davy, whom he engaged as assistant lecturer at the Royal Institution in the spring of 1801. Davy applied himself diligently, and his lecturing talents soon launched him on a remarkable career. A scant year later he was named professor of chemistry. The fate of the institution hinged on admission fees, and Davy promised success on this score. Consequently, the Royal Institution was perfectly prepared not only to pay him for industrial consultations but to invest in chemical research. Its public cachet and scientific reputation benefited greatly thereby. By 1803, on the strength of his work on tanning techniques and the voltaic pile, Davy was elected Fellow of the Royal Society, thus gaining the lasting international renown he had sought for himself, as well as firmly establishing the Royal Institution as a center of scientific research.

In 1806, Davy returned to the question of the proper interpretation of Nicholson and Carlisle's results, then still in dispute. He was convinced he was observing the decomposition of water, and he assumed that the unequal gas ratio could ultimately be attributed to the properties of the materials employed—vessels, electrodes, and the like. In pursuit of his quest, he performed very detailed analyses. He used especially pure (and very expensive) materials, as well as the Royal Institution's exceptionally powerful battery; resources were not a concern. After a laborious and inventive series of experiments he was not only able to explain the aforementioned unequal gas ratio in detail but to envision an experimental setup that would yield a precisely predicted ratio of hydrogen and oxygen. To explain the fact that the two gases formed separately, one on each submerged wire, he proposed a theory of processes taking place in water (or saline solution). Attempts in this direction had previously been made by the Rome-based amateur Theodor von Grotthus and the Swedish chemist Jöns Jakob Berzelius. Without naming them, Davy followed the trail they had blazed.[40] Like Grotthus, he began with the assumption that the actual decomposition was brought about not by contact with the electrical surface of the submerged wire but by a kind of polarization taking place within the liquid. He also had a somewhat vague, if plausible and experimentally sustainable notion of how the decomposition products migrated to the wires, only there becoming visible as gases.

Davy succeeded brilliantly, not only in securing broad acceptance for the decomposition interpretation, which would henceforth be taken as given, but also in demonstrating his own skill and inventiveness as a chemist. He was selected for the Royal Society's Bakerian Lecture of 1806, one of the society's highest honors, accompanied as it was by a substantial prize purse. His report was published in the *Philosophical Transactions* (Davy 1807), and his results also gained international acceptance and acclaim. Even the academy in Paris awarded him its annual prize in 1807 for best contribution to the study of electricity, a very clear signal of his prestige given the state of war between France and England.[41] Davy's success could hardly have been more unambiguous; it lent great impetus to his general speculations on the relationship between chemical and electrical forces, or even their identity. Departing from his usual, self-conscious practice of keeping his publications empirical, he went so far as to give explicit voice to these intimations, which could not but have aroused doubts in the minds of many of his readers (Davy 1807, 38ff).

Spurred on by his achievements, Davy went still further, turning to the study of alkaline minerals such as potash and soda, with whose chemical properties he had busied himself earlier in his career. These substances posed particular challenges to electrochemical investigation. In the solid state they were not conductive, whereas, in aqueous solution, what would decompose was the water rather than the mineral. But by once again developing wholly new experimental techniques, Davy was able to separate substances at the submerged ends of his battery leads: on one, oxygen, and on the other, a metal. This result, too, he interpreted as decomposition, in this case of a substance (soda or potash, or some other alkaline) composed of an alkalogenic metal and oxygen. He named the new metals "sodium" and "potassium." His interpretation was unexpected in the sense that it ascribed to oxygen, previously understood as an acidifying agent, an essential role in the formation of alkaline substances. Furthermore, the elemental status of the alkalogenes was by no means certain. Metals were elements, but were these substances really metals? They had very unusual properties, including low specific weight and aggressive reactivity. In light of such problems, it was hardly surprising for Davy's interpretation to be met with skepticism.

Over the course of these experiments, the role of the battery had changed. In addition to being a subject of research in its own right, it had now become an instrument for studying the composition of substances. Both Davy and his critics were well aware of this new role. Once again there were doubts as to the suitability of the voltaic pile as an analytic instrument; the new results might well turn out to be artifacts of the pile. Part of the problem was that, in these experiments, the pile was the *sole* analytic instrument, one still not completely understood. Knowledge of the instrument and of the effects it produced were equally insecure—a situation met with not infrequently in the development of a science, begging the all-important question of how stability can eventually be achieved.[42] Davy understood this situation, and he launched himself into the debate with fervor and rhetorical skill. His discovery finally gained broad-based international recognition, and he was once again invited to present his results in another Royal Society Bakerian Lecture.[43] Continuing this line of research, he was later able to isolate additional metals—the alkaline earth metals calcium, barium, strontium, and magnesium—though Berzelius in Stockholm beat him to the punch.

In England, his leading position in the natural sciences was now secure, and it is no coincidence that shortly following his latest success he was offered one of the Royal Society's two secretarial posts. As before, his personal triumph also accrued to the credit of the Royal Institution, which now joined the ranks of leading centers of scientific research. This, in turn, affected the resources devoted to such research. On the strength of Davy's successes, the Royal Institution launched a fund-raising drive, which soon bought Davy the most powerful battery in Europe at the time (Chilton and Coley 1980, 179). He thus transformed the laboratory of the Royal Institution, originally intended only to support experimental lectures, into a genuine research laboratory, one whose equipment was almost without equal in all Europe. To be sure, the state-financed École Polytechnique in Paris had comparable tools, but, unlike the Parisian chemists, Davy enjoyed almost exclusive use of and unrestricted access to his laboratory. The range of research resources at his disposal was unique.

Research laboratories on this scale represented a new kind of installation, one that remained quite rare into the middle of the century.[44] The new rules of scientific competition were illustrated by Napoleon's response to the news from London: he made a special allocation to the École Polytechnique for the construction of a still larger battery (Fischer 1988, 205–6). This sort of investment of national prestige in laboratory equipment was very unusual in its day (and would remain so for decades to come) and had to do not least with Napoleon's personal interest in the natural sciences and his aversion to speculative tendencies. The singular emergence of Davy's laboratory would have far-reaching consequences for the study of electricity, to which I return shortly.

Naturphilosophie and Galvanism in the German-Speaking World

Intensive studies of galvanism had been under way in the German-speaking world even before Volta's discovery, with principals including Alexander von Humboldt, Albrecht von Haller, Christoph Heinrich Pfaff, Friedrich Albrecht Carl Gren, and Johann Wilhelm Ritter (e.g., Trumpler 1992; Caneva 1978). Volta's announcement spurred them to further efforts, as testified so eloquently by the early volumes of the *Annalen der Physik*, published by Ludwig Wilhelm Gilbert in Leipzig.[45] The researchers already involved were soon joined by Professors Paul Ludwig Simon (Berlin), Karl Wilhelm Böckmann (Karlsruhe), and the Berlin academician Paul Erman, among others. Gilbert also contributed research of his own. These researchers could not count on universities to support their experiments with resources even vaguely comparable to those available in Paris or London. Beyond the standard equipment of a cabinet of physical instruments or small chemical laboratory, only very rarely did they have access to such expensive instruments as highly sensitive electrometers, large batteries, or the like.

Johann Wilhelm Ritter distinguished himself with especially innovative work. Like Davy, he had not been raised in an academic setting, but (also like Davy) he began his experimental research while still young, gaining the attention of Alexander von Humboldt, among others. He interacted with important romantic movement intellectuals and poets, including Novalis (pseudonym of Georg Philipp Friedrich Freiherr von Hardenberg) and the Schlegel brothers. As was common in these circles, he understood the study of nature not as something isolated but as always rooted in the quest for an all-embracing understanding of the world, one that encompassed the physical together with the spiritual and insensible. Strongly inspired by Friedrich Wilhelm Joseph Schelling, his study of nature nonetheless followed its own path, placing much greater emphasis on empirical work.[46] Though Schelling held Ritter in no particular esteem, he was perfectly prepared to make opportune use of his discoveries. Ritter's researches were very broad. For example, as early as 1801, inspired by John Frederick William Herschel's discovery of radiant heat and guided by his own reflections on symmetry, he discovered chemical effects caused by the other end of the spectrum: ultraviolet radiation. While his experimental results were always welcomed, Ritter's speculative tendencies often exposed him to criticism. In the field of galvanism, he was the unquestioned leader. But despite his scientific successes, his academic career was unstable. Only in 1804 was he able to settle in Munich, where he had been granted membership in the Bavarian Academy of Sciences. But even then he continued to experience constant financial distress, having as few inhibitions with money as with the use of his own body as experimental subject. He either funded his researches himself or sought sponsors for particular projects.[47]

In his work on galvanism, Ritter proceeded much more systematically than most, developing a generalized, abstract representational scheme (e.g., Trumpler 1999). By discovering the electrochemical series of the metals and providing stringent arguments in support of the widely suspected kinship of galvanic and chemical processes, he laid the groundwork for electrochemistry (Ritter 1798).[48] In the process, as he accumulated other achievements, he became convinced that galvanism without living tissue was possible. The invention of the pile thus came as no surprise to him; instead, he found it "unforgivable" to have been "so close to" this discovery, while still missing it (Ritter 1800b). He set to work, finding Nicholson and Carlisle's interpretation of their result as the decomposition of water entirely implausible and conducting numerous experiments in support of his own alternative proposal (e.g., Ritter 1800a).[49] Investigating the behavior of the pile itself, he reached the conclusion that the number and size of the plates were independent parameters that affected the pile's operation in different ways (Ritter 1805). He thus came closer than others to a proper distinction between tension and current, though without following through. Ritter also developed two novel forms of pile. In the "dry pile," the medium between the metal plates was mostly dry, making the instrument much easier to handle. Even more important was the "charging pile" variant, which could be recharged by connection with another pile after it had discharged, without replacing its metal and chemical constituents. When his student and friend Hans-Christian Ørsted brought news of this discovery to Paris, a copy was quickly built and presented to the academy, a clear sign of its phenomenal significance in the eyes of his contemporaries (e.g., Herrmann 1968, 84–93).

In the essay that Ritter dispatched to the Paris academy on this occasion, in September 1803, he also reported a discovery of still greater significance. An open, uncharged pile, he claimed, has an intrinsic charge of its own if it is aligned with specific compass points. Ritter regarded this charge as an indication of the existence of terrestrial *electrical* poles, analogous to magnetic poles, capable of affecting the electrification of correctly aligned bodies. This abrupt leap, from the properties of the pile to the large-scale condition of the earth itself, was typical of Ritter. However, despite the apparently concerted attempts of renowned Paris researchers (Biot, Coulomb), it proved impossible to replicate the effect. The definitive failure of these efforts would do lasting damage to Ritter's scientific reputation, even staining Ørsted. Ritter had, quite reasonably, entertained the hope of receiving the annual Volta Prize. It was now out of reach. For some time thereafter, the French academicians would remain suspicious of reports of new discoveries emanating from "romantic" circles.[50] To be sure, their suspicions cannot be attributed entirely to this one episode. In Paris, the blend of strictly empirical results with far-flung speculation typically found in such reports tended to be viewed as unacceptable.

One reason why research on galvanism was so common in the German-speaking world was that it could be undertaken with relatively modest equipment. The equally interesting field of (ordinary) electricity required electrification machines, large Leyden jars, sensitive electroscopes, and the like. By contrast, when it came to the effects of the closed pile, there were no detection methods available that permitted more than the coarsest quantitative comparison, even with complicated apparatus. Neither thermal effects, nor the sparks that appeared with the closing or opening of a circuit, nor the time-consuming and difficult-to-interpret observations of chemical changes were susceptible to greater quantitative precision. The most sensitive instrument for the detection of galvanic effects was the human body, whose use as such was widespread. As early as 1794, Volta had established his "electrochemical" series of metals on the basis of the intensity of the taste sensations they evoked in various combinations. Later in his dispute with Galvani, when he had succeeded in divorcing galvanism from physiology, he continued to rely on sensory stimuli in describing and comparing the pile's effects. Even Biot fell back on such methods in his early work on the pile, for lack of any more precise means of measurement.[51]

Though Ritter has become the history of science's poster boy for self-experimentation, it should be noted that the practice itself was by no means unusual or even new. What was unusual was the extent to which Ritter perfected it, even to the point of allowing him to make specific quantitative claims. More than a mere instrument of detection, his body became a genuine instrument of measurement.[52] In some instances, Ritter achieved real quantitative reliability. Thanks to his extraordinary willingness to exploit (or even torture) his own body, he had a unique tool at his disposal, allowing him to analyze the effects of the closed pile much more precisely than anywhere else. In his investigation of the dependence of voltaic effects on the number and size of plates, he relied on the perceived strength of electric shock as a third criterion, along with spark size and chemical phenomena (Ritter 1805).

The Voltaic Pile in Paris

News of Volta's discovery did not reach Paris until September 1800, when it arrived together with the report of Nicholson and Carlisle's results. The first reactions and experiments came from the leading chemists of two of the most important Parisian research institutions, Antoine François de Fourcroy (Muséum d'Histoire Naturelle), and Guyton de Morveau (École Polytechnique), both of whom had already concerned themselves with galvanism. When Ritter's and Pfaff's reports arrived shortly thereafter, the search for an adequate interpretation of the gas production quickly became central. Participants in the ensuing debate included the chemist Nicolas Gautherot, Claude Louis Berthollet, the mineralogist and physicist Jean Henri Hassenfratz, and the Haarlem experimentalist Martinus van Marum (Sutton 1981, 331–45).[53]

The operation of the pile itself was a further focus of interest. It was to this question that Jean-Baptiste Biot applied himself, conducting his very first work in experimental science well before the contributions to mathematical electrostatics discussed above. Having been encouraged in such investigations by Laplace, Biot naturally attempted a quantitative approach. Tellingly, he made no reference to the numerous qualitative studies released by chemists, instead citing Laplace—who to date had published nothing in the field.[54] Together with Frédéric Cuvier (brother of the famous zoologist Georges Cuvier), Biot first undertook to measure the oxygen consumption of the closed pile. The chief question here was whether the operation of the pile was necessarily associated with observed oxidation effects or might instead proceed without oxidation. After attempting to operate a pile in a vacuum, Biot cautiously endorsed the second option, but his results remained inconclusive, and not especially persuasive (Biot and Cuvier 1801). In parallel with his experimental work, Biot undertook far-reaching theoretical reflections. He attempted to apply Coulomb's account of the distribution of electrical charge across conductive bodies, to which he had just presented his own contribution, in the new field as well. On this basis, he predicted that galvanic fluid would travel more easily on the surface of a liquid than in its interior, that piles with smaller plates would yield stronger effects, and that electrical fluid would more easily escape from points or deeply creased surfaces than from flat surfaces. Even a charitably inclined contemporary like Gilbert could only ascribe such unsustainable assertions to the overly hasty application of a theory properly belonging to another domain (Biot 1801a).[55]

Parisian work on galvanism paused in October 1801 when Volta came to Paris and delivered a lecture at the Institut de France. It remains unclear whether he traveled on his own initiative or at Napoleon's behest.[56] That considerations outside science played a role is, however, beyond question. Volta acknowledged that one goal of his visit was to "cement an alliance of talent and science for the immortality of the two republics" (quoted in Heilbron 1976, 79), the French Republic and the Cisalpine Republic, the latter being the French client state in occupied Northern Italy. Napoleon, himself a member of the Institut de France, attended Volta's presentation, heaped praise on him, and proposed, to everyone's surprise, that he be awarded a medal. This procedure was most irregular, and thus it doubtless provoked some resentment on the part of the other academicians; here was a foreign researcher, visiting the global center of scientific research, being showered with honors not even the most prominent members of the academy had ever received. It was, however, effectively impossible to oppose any suggestion made by *this* member of the academy.[57] In yet another extraordinary measure, Napoleon announced two prizes in support of further discoveries: an annual prize of three thousand francs for the best work on galvanism and an unprecedented, jaw-dropping, onetime award of sixty thousand francs for a discovery whose contribution to the field rivaled those of Benjamin Franklin and Volta. Comparing this figure with the six-thousand-franc annual salary of a professor at the École Polytechnique should serve abundantly to convey the magnitude of Napoleon's intervention (Crosland 1967, 20–25; Fischer 1988, 142–51).[58]

With such incentives, the field rose suddenly to great prominence. Volta's text was finally published in a French journal (Volta 1800a). Laplace's protégés, Biot chief among them, now had essentially no choice but to apply themselves diligently to this domain. As a member of the commission charged with evaluating Volta's results, and the one tasked with drafting the report, Biot found himself in a privileged position. It presented an excellent opportunity to shape the course of research, and Biot knew how to exploit it. He included in the commission report a mathematical schema that he had developed, what would come to be known as the "electrostatic theory" of the voltaic pile.[59] Using his most sensitive condenser electrometers, Volta had established the appearance of contact tension between different metals. This result served as the point of departure for Biot's theory. He began by asking how electricity would distribute itself across individual copper and zinc plates as a pile was progressively assembled on an insulated base. His investigation (Biot 1804) assumed the following three conditions:

- When a zinc plate is in contact with a copper plate, its electrical state will always be higher, by a fixed amount, than that of the copper plate. For ease of exposition, this magnitude of this fixed amount is assigned to the unity.
- Because the moist intermediate layer conducts electricity, the two plates on either side of it must have the same electrical state.
- Because the pile rests on an insulating base, the net sum of all electrical states must always be equal to zero.[60]

For piles with differing numbers of cells, Biot's account can be presented by the following schema (table 1.1), where the lowermost plate is always copper, the insulating base is designated by "▪▪▪▪▪▪," and the moist intermediate layer by "———."

The schema could be extended indefinitely; for n cells, the topmost plate has electrical state +n/2, and the bottommost, −n/2. Should the pile be assembled on a grounded base, rather than an insulated base, the state of the lowermost plate would be 0 and that of the topmost plate, +n. Biot thus had a procedure for calculating the internal tensions of piles on the basis of measurements across the two outermost plates. It should be noted that Biot's notion of tension (and the notion prevalent in several decades to fol-

TABLE 1.1. Schematic representation of Biot's "electrostatic theory" of the voltaic pile

1 cell	2 cells	3 cells
		Zn: +3/2
		Cu: +1/2
		———
	Zn: +1	Zn: +1/2
	Cu: 0	Cu: −1/2
	———	———
Zn: +1/2	Zn: 0	Zn: −1/2
Cu: −1/2	Cu: −1	Cu: −3/2
▪▪▪▪▪▪	▪▪▪▪▪▪	▪▪▪▪▪▪

low) was neither particularly sharp nor congruent with our contemporary concept, as is indicated already by the fact that Biot speaks interchangeably of "quantity of electricity" (*quantité d'électricité*) and "electrical state" (*état électrique*). Like Volta, Biot assumed that all tension, and thus all activity in the pile, originated at the place where the metal plates came into contact. Biot had nothing further to say about the causes of this tension, except to dismiss the many chemical effects as mere by-products, albeit by-products that modified the conditions produced by metal-to-metal contact. This verdict presumably had less to do with direct experimental results than with the fact that such chemical effects were not susceptible to precise measurement.

In 1803, he made further strides toward more precise measurement techniques. In a mémoire on the role of oxidation he offered precise and definitive arguments for the irrelevance of chemical processes to the operation of the pile (Biot 1803). Toward that end, he availed himself of the most precise instrument available: Coulomb's torsion balance. Whereas in his earlier work he had characterized a given pile's activity by means of three criteria—strength of the imparted shock, thermal effect, and electrical attraction in the open state—he now had only one: the size of the charge imparted to a condenser. This alone could be measured with a torsion balance. To be sure, such experimental measurements were far from easy to obtain, and Biot had to exercise his mastery of the sensitive instrument to the fullest in order to attain stable data.[61] Confining the study to a single measurable parameter imposed severe limitations, for only an *open* pile could charge a condenser. Biot did not draw attention to this fact, however, but formulated his results in very general terms. Consequently, beyond its further refinement of the electrostatic theory, which would gain acceptance, the report of his research also contained gross errors. For example, he asserted that the fluid level in a trough device would have no effect on its operation. While this might have been true enough for an "open" trough battery, investigating the device in its closed condition would refute the claim immediately. In the absence of any prospect for further precision measurement, Biot ceased work on the pile.

The unsatisfactory state of Biot's account was apparent to many. The mathematician Haüy, for example, while a firm supporter of Biot's theory in general, pointed to the experimental finding that piles with the same number of plates, but with plates of different size, were in some respects similar (e.g., in the electric shocks they produced) but in others quite different (e.g., in the heat imparted to a wire when short-circuited). To accommodate this observation to theory, he spun far-reaching speculations on the details of the discharging process (Haüy [1803] 1806, 1:29ff).[62] A decade later, Biot incorporated a refined version of Haüy's account into the latest iteration of his electrostatic theory, which he included in his monumental physics textbook (Biot 1816, vol. 2). The great esteem in which this theory was long held in France had much to do with the fact that the effects of the closed pile were no longer intensely studied there. In Paris, research into the pile remained a marginal undertaking, yielding no contributions of significance comparable to those of England or the German states. We have here a remarkable instance in which early fixation on mathematical formalization and precise measurement severely constricted the scientific gaze, allowing it to linger only on a small subfield. Typical of this episode is the fact that Haüy considered the aforementioned commission report to be the beginning of "modern" research into the voltaic pile (Haüy [1803] 1806, 15). Given the restrictions under which French scientists labored, the full range of phenomena remained out of reach.

This is not to say that broader inquiries and approaches were wholly absent in Paris. As early as the spring of 1801, the chemist Nicolas Gautherot, mentioned earlier, cautioned that in such a complex new field of phenomena it was best to begin by gaining more experience, and observing "with care the modification of effects," before proceeding to theory. In light of the patchy state of empirical evidence, any theoretical effort would be premature (Gautherot 1801, 203–5). The kind of broad-based modification and variation of phenomena Gautherot (1801, chap. 7) was asking for is characteristic of the sort of experimentation I call "exploratory." Gautherot did not further elaborate his warning, but he may well have had in mind the sort of loss in experimental breadth that would later, in retrospect, become apparent. The fact that, in Paris, such voices of caution so quickly fell mute, along with every alternative to the single dominant research program, has much to do with the aforementioned historical and political circumstances. Geoffrey V. Sutton (1981) has shown in some detail how many chemists were drawn away from the field by other activities, without ever having planned such a move. As it was then unfolding, the academic milieu left little room for the emergence of qualitative, exploratory experimentation of the kind that might have led to the formation of broader concepts. A similar sequence would play itself out two decades later, in reaction to Ørsted's discovery of electromagnetism.

Voltaic Studies in the Second Decade

Toward the end of the first decade of the nineteenth century, the intensity of research on the voltaic pile began to trail off significantly, quickly in France and less hurriedly in England and the German states. For the moment, the time of broad leaps was over, though many uncertainties and open questions remained. One of the most important of these concerned the cause of the pile's activity. France was dominated by the electrostatic theory and thus by Volta's contact theory. Where electrochemical research was pursued more intensively, especially in London by Davy and in Stockholm by Berzelius, chemical theories of the pile were preferred. This problem would remain unsolved until the middle of the century (Ostwald 1896, chap. 12). The basic concepts with which the behavior of the pile was captured also remained uncertain or vague, especially the concepts of tension and current. And if there was uncertainty about what was happening inside the pile itself, there were even more questions about the processes at work in the "closing-wire." Volta had spoken of an "electric current" (courant électrique), and Biot had adopted the expression, though giving it a different meaning. In both cases it was bound up with speculative theories on the microscopic workings of the electrical fluid. Others thus avoided the expression entirely and continued into the 1820s to speak simply of the wire attached to the battery, arranged in this or that position.[63] Consequently, there were significant limits to scientific communication beyond the direct reporting of experiments. The mutual dependencies between various parameters (e.g., between the number of plates in a pile and its thermal effects) could be neither rigorously formulated nor expressed quantitatively. This situation was especially pronounced when it came to any observation regarding the closed pile, for which the classical notion of tension was manifestly useless.

Such fundamental conceptual indeterminacy had much to do with the lack of any instruments that would have allowed quantitative comparisons. Nicholson and Carlisle (1800, 187) allude to this problem at the end of their very first publication on the voltaic pile: "We are in want of a measure of the intensity of the action of these machines. Will this be derived from the quantities of water decomposed, or of gas extricated under like circumstances in given times? Or from any change of temperature? Or what other commensurate incident?" There were early, concrete suggestions on ways of exploiting the chemical effects of the pile, and gas formation in particular, for quantitative ends.[64] Thermal effects were also taken as an important indicator. So, for example, piles were calibrated by their ability to cause a platinum wire of a given diameter to glow. But such measures were extremely rough and had the decided disadvantage that the employment of those methods changed the very parameter they sought to measure: they required that a pile be placed in service, by which it quickly lost strength. Furthermore, weaker piles produced effects too small to sustain any quantitative description.

Nonetheless, undaunted by such conceptual uncertainties, investigators continued to make significant progress in developing equipment. The pile itself was further developed, and soon there appeared a whole range of different forms, all differing with respect to strength, endurance, manageability, and cost. Though the English word *pile* (or *Säule* in German, *pile* in French) was often retained—indeed, a flashlight battery is still called a *pile* in French today—the columnar pile was soon largely replaced by the more manageable trough apparatus. Sealing the cells, however, remained a problem for some time. In chemical laboratories and cabinets of physical instruments, small- or medium-sized voltaic devices soon became standard equipment and their use part of the daily routine. The construction of galvanic machines and the invention of new varieties were important activities throughout Europe, and the resulting experiences were a frequent subject of scientific communications. Professional instrument builders gained a great deal of expertise and began to keep inventories of completed devices. In 1820, this broad dispersal of equipment and expertise would be a significant factor in the universal, rapid replication of Ørsted's spectacular electromagnetic experiments.

Beyond its deployment in research and instruction, the pile found many widespread applications. The most important of these was in medicine. There had been medical applications of electricity and galvanism even before 1800, and the new equipment fostered a great expansion of such efforts. Attempts to cure deaf-mutes by means of the pile's effects gained attention throughout Europe (Ostwald 1896, 217–20).[65] The number of batteries built for such purposes may well have exceeded the total destined for academia. This demand also contributed significantly to the availability of the requisite materials and artisanal expertise. Here we have one of those historical cases in which technical improvements and applications outstrip academic research—cases that are hardly uncommon. Something similar would occur in early twentieth-century physics, for example, following Wilhelm Röntgen's discovery of X-rays. The emergence of such practical developments may eventually lead back into theoretical and conceptual work, with far-reaching consequences.

Another application of the voltaic pile received less attention. In 1809, the physician and anatomist Samuel Thomas Soemmering, a member of the Bayrische Akademie der Wissenschaften in Munich, proposed an electric telegraph. In his device, current from the voltaic pile was transmitted from a sender, through a many-wired cable, to a receiver, where it was detected by electrolytic gas production. Every letter of the alphabet had its own wire and electrolysis cell. Soemmering was able to transmit messages over several hundred meters and envisioned doing so over much longer distances. His description of the invention presented not only such important technological innovations as the continuous insulation of the individual wires by means of wound silk but also an estimate of costs for long-distance installation. However, his idea was neither applied nor further developed. Supposedly, Napoleon dismissed it, despite his openness to new technologies, as a "Germanic idea" (*idée gérmanique*) (Soemmering 1810).[66]

The broad dissemination of the voltaic apparatus faced one significant obstacle, for large devices were both very expensive and very difficult to handle. The aforementioned standard procedures were simply not equal to the task. This limitation applies both to the very strong currents such equipment could produce and to the operation of the battery itself. The latter produced large quantities of gas, for which appropriate precautions must be taken: "Devices with large-plated cells . . . are very costly, procurable only in a few places, and require knowledge, practice, and skill in their experimental employment" (Gilbert 1820b, 331–32, my translation). Gilbert's assessment aptly describes the European situation. His remarks were echoed in 1826 by London experimentalist William Sturgeon (1826, 357). Indeed, such large devices were rarely built and difficult to obtain. In Kiel, for example, Christoph Pfaff would repeatedly solicit Ørsted's support in acquiring a large battery; such equipment could apparently not be had in Kiel.[67] Even in Paris in 1820, Ampère would find it hard to obtain a very strong battery. Where very powerful batteries *were* to be found, it was always a point of pride. Facilities that boasted a large battery could produce truly spectacular effects, such as those demonstrated by the Geneva professor Gaspard de la Rive at a July 1820 meeting of the General Swiss Society for the Natural Sciences: "With it [his large galvanic device], he caused platinum wire to glow and melt, burnt iron, steel, gold, and silver, simultaneously decomposed both water and sulfuric sodium etc., and displayed the emission of an incredible amount of light between two coal stubs by which the voltaic circuit was closed. The eye can barely tolerate such bright light, which was not caused by burning—for it shone at least as brightly in an airless space as in the air" (Gilbert 1820a, 101–2; my translation).

The invention of the voltaic pile also revived interest in an issue that had arisen repeatedly in the history of electricity: the question of possible interaction between electricity and magnetism. In this regard, the "inexhaustible" character of the pile opened new experimental horizons. Ritter undertook a whole series of (unsuccessful) experiments as early as 1801, in an effort to detect possible galvanic properties of magnetic poles. In August 1802, the Parma amateur Gian Domenico Romagnosi reported in a local newspaper on the deviation of a magnetic needle by galvanism. But his report remained vague, as its author clearly failed to recognize the significance of the effect. In any case, he pursued neither the discovery itself nor its wider dissemination any further. Neither did he later claim priority over Ørsted's 1820 discovery. In 1804, the Genoa professor Benedetto Mojon reported on the magnetization of a steel needle he had incorporated into a voltaic circuit for twenty-two days. He also appears to have overlooked the significance of the finding, and he took it no further.[68] Also in 1804, Galvani's nephew, Giovanni Aldini, cited Romagnosi's and Mojon's results but without providing details. European science at large never became aware of them, and in any case the experiments they presented were neither particularly clear nor especially compelling.[69]

An entirely different level of publicity accompanied an experiment conducted in 1805 by the Parisian physicist Jean Nicolas Pierre Hachette and the chemist Charles Bernard Desormes. They positioned a very large pile containing a staggering 1,480 zinc-coated copper plates on a floating platform, to see whether it would be aligned by terrestrial magnetism. But as was typical in Paris, they left the pile open and obtained a negative result (Mottelay 1922, 376); had they also tried a closed pile, they would most probably have produced and observed an effect. The failure of such experiments left the prospect of electromagnetic interaction looking unlikely. Such skepticism was further reinforced by theoretical considerations, for imponderable fluid theory typically assumed the absence of any interaction between different kinds of fluid, and Poisson's theory of electricity did not envisage room for electromagnetic interaction. Under such conditions, Paris would witness no further efforts of this kind. Elsewhere, however, and especially in circles molded by German *Naturphilosophie,* the conviction that electricity and magnetism must somehow interact remained strong. The fact that some researchers were continuing to work on the problem became apparent only in the summer of 1820, when news of a positive result raced through Europe. This discovery would prove a watershed in the history of electricity and magnetism.

CHAPTER 2

Electromagnetism

Hans Christian Ørsted and the Discovery of Electromagnetism

The study of electricity and magnetism took a dramatic turn with the discovery of electromagnetic interaction in 1820. It triggered an avalanche of new work, such that for a few years the topic dominated the journals even more thoroughly than Volta's discovery twenty years prior. New questions were posed, new phenomena discovered, new instruments and concepts developed, and previously unknown actors stepped onto the stage—in short, a whole new field of research had begun. Such frantic episodes are not all that common in the history of science, and this one proved extraordinarily prolific in its consequences. Investigating those consequences lies at the center of this book. My goal in this chapter is to cast the initial conditions in sharp relief. I begin by sketching the background of the discoverer and the discovery, before canvassing the range of research projects launched so immediately in its wake throughout Europe. Finally, I consider what it was that made this particular finding so spectacular. In the last three sections I take a particularly close look at the situation in Paris, thus laying the groundwork for the inquiries in chapter 3.

Born in 1777 in rural Denmark, Hans Christian Ørsted showed an early interest in chemistry, nurtured for some time by an apprenticeship in his father's pharmacy.[1] While studying pharmacy in Copenhagen, he occupied himself with the philosophy of German idealism—with Kant, Schleiermacher, Schelling, and Fichte. He received his first academic prize in 1797, for a piece on the aesthetics of language, and was awarded his doctorate in 1799, with a dissertation on Kant's philosophy of nature. A sojourn of several years (1801–4) in the German states and France would shape the course of his future scientific work. During that time, Ørsted strengthened his ties to romantic *Naturphilosophie* and became close friends with Johann Wilhelm Ritter. In Paris, he became fully conversant with "modern" French physics, though without gaining much from the tendency toward mathematical formalization. As discussed in the previous chapter, it was he who brought Ritter's report of the discovery of the earth's electrical poles to Paris and who saw Ritter's reputation dwindle as attempts to replicate his results failed and his own name become associated with unreliable experiments. With such experiences behind him, his research would hew to a unique style for the rest of his life, one in which careful experimentation was always in the foreground. Though shaped by German idealism, Ørsted does not fit the stereotype of a "romantic physicist" sometimes applied to him.[2]

Upon his return to Denmark, Ørsted initially ran a Copenhagen pharmacy, but in 1806 he was awarded an extraordinary professorship in physics. In his position as secretary to the Royal Danish Academy of Sciences (Kongelige Danske Videnskabernes Selskab) from 1815 on, he became one of the most important and influential personalities in Danish natural science, especially with respect to its organization and public promotion, then undergoing transformation. His own research targeted issues in chemistry (among other things, he discovered aluminum), electricity, galvanism, magnetism, and electromagnetism, but he also dealt with acoustic figures, the compressibility of bodies, capillary action, heat theory, and the theory of light.[3] He was instrumental in introducing the new chemical terminology into Danish. For Ørsted, as for many others of this period, natural science was never a self-contained element of human spiritual life. His late, comprehensive magnum opus, *Aanden i Naturen* (published in Danish, and in German as *Der Geist in der Natur* in 1850, and in English translation, from the German, as *The Soul in Nature* [Ørsted 1852]) dealt with such topics as the relationship between natural science, poetry, and religion.

Ørsted had early become convinced that there must be some sort of interaction between electricity and magnetism, and he stated this explicitly in his *Ansicht der chemischen Naturgesetze* (View of the chemical laws of nature) (Ørsted 1812). But it appears that not until the spring of 1820 did he begin any detailed experimental investigation. He quickly observed the first hint of magnetic activity on the part of his wire but did not have the time to pursue it properly until he devoted several weeks to the problem that July. He was well aware of the failure of experiments performed on an open pile and thus used a closed pile from the outset. He was able to establish that a straight wire would exert an effect on a nearby compass needle when the ends of the wire were connected to a galvanic device. The needle assumed a different position from that imposed by terrestrial magnetism alone. A few of his laboratory sketches make it clear that this effect was difficult to achieve reliably, let alone make sense of (Meyer 1920, lxxiii–lxxxviii). Nonetheless, it was enough to show that galvanism could affect magnetism—that there was thus another deep connection between different forces of nature. Had Ørsted's friend Ritter still been living, he would have felt not only thrilled but also vindicated, for, having correlated the dates of important electrical discoveries and inventions (the Leyden jar, the electrophorus, the condenser, the voltaic pile) with oscillations in the inclination of the ecliptic, he had prophesied an important discovery on electricity for the year 1820.

An Experimental Report?

Ørsted's own later accounts of the investigations that led to his discovery varied widely, and not always consistently (e.g., Ørsted 1821b, 1830).[4] What is clear is that his eventual success was the result of a systematic search, a point of which most of his contemporaries were well aware (e.g., Faraday 1821–22, 107).[5] A full reconstruction of the course of his research would go beyond the scope of this book.[6] Instead, I shall content myself with extracting several substantive points from the text in which Ørsted (1820o) eventually presented his results to the public.[7] Even there it is possible to discern several characteristic problems that would later give rise to far-reaching innovations.

The experimental setup (fig. 2.1) called for the wire connecting the poles of a battery and fixed at points A and B to be positioned over a magnetic needle, mounted as in a compass. The position and orientation of the wire could be adjusted easily. When the connection with the galvanic device was closed, the needle was deflected. In Ørsted's first experiments, the wire ran horizontally, above and parallel to the needle, along the north-south axis. Ørsted (1820g, 274) described the results as follows: "Things

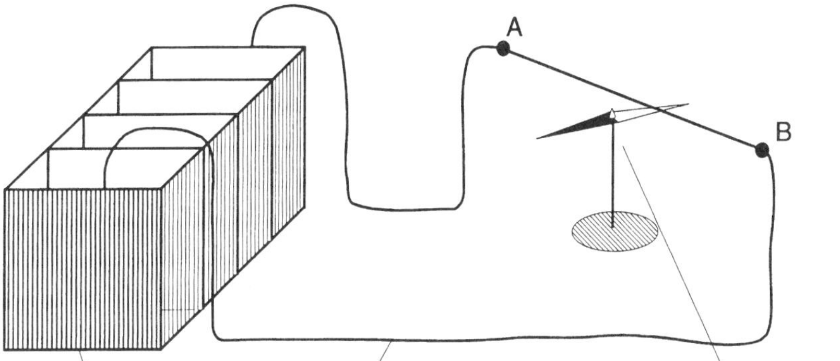

FIG. 2.1. Schematic representation of Ørsted's experimental setup.

being in this state, the magnetic needle will be moved; and the end of it next to the negative side of the battery will go westward."[8]

When the distance between wire and needle was about three-quarters of an inch, the deflection was 45°; it increased with greater proximity and decreased with greater distance. The angle also depended directly on the strength of the battery. The effect remained unchanged when the wire was positioned not directly above the needle but translated in parallel either to the east or to the west.

Ørsted's complicated and seemingly cumbersome language is already evident here. This can be traced, first, to his decision to avoid any use of the concept of current, forcing him to formulate his description of the direction of galvanic action by reference to the battery's polarity. Second, the deflection of the needle had to be expressed by reference to the compass points. Both peculiarities are also typical of many of the experimental reports that followed, pointing to a very specific difficulty associated with this effect: the lack of concepts with which the relevant complex spatial relations might be grasped more concisely. To put it in modern terms, the problem was to express the relation between three spatial directions: the direction of the wire, of the needle, and of the needle's deflection. There were no geometrical concepts available at that time for dealing with such constellations. This issue will arise frequently, and I return to it below.

Ørsted performed numerous variations on this experiment. The single wire could be replaced by several parallel wires or with one made of some other metal (platinum, gold, silver, brass, zinc, lead, or even mercury), without altering the effect. Even when the wire was interrupted by a stretch of water, some deflection was observed. Ørsted also investigated whether interposing various materials between needle and wire might block the effect. From this broad and rather challenging variation, involving metals, wood, wax, earthenware, stone, glass, and water, he obtained only negative results. From this finding he concluded that the effect he was observing was fundamentally different from other electrical and galvanic effects. Finally, he reported on experiments in which the magnetic needle was replaced by needles made of other materials: brass, glass, and shellac.[9] The connecting wire had no effect on any of these.

The question of how the effect depended on particular spatial configurations remained a lasting challenge. Ørsted observed that the direction of the needle's deflection reversed when the wire was placed beneath the needle rather than above it. Faced with this additional complication, he formulated his first general rule: "The pole *above* which the *negative* electricity enters is turned to the *west, under* it, to the *east*" (Ørsted 1820g, 275, emphasis added).[10] Rotating the wire out of alignment with the meridian (but keeping it horizontal) would strengthen or weaken the effect, depending on whether the wire was rotated in the direction of the deflection or against it. When the wire was in the same horizontal plane as the needle, then, rather than deflecting in the same plane, the needle would deflect from the plane itself, as in an inclination instrument. The pole closest to the "negative" end of the wire would point upward or downward, depending on whether the wire was west or east of the needle. When the wire was aligned with the east-west axis, either above the needle or below it, it had no effect. The vertical deflection of the needle pole occurred only when the distance between wire and needle was very small.

Ørsted also investigated conditions in which the wire was mounted vertically, though the needle remained horizontal (fig. 2.2). When one of the poles was close to the wire (the dotted line), it deflected toward the east. But when the wire was positioned between one pole and the pivot point (solid line), the pole deflected to the west. These directions of deflection were observed when the upper end of the wire was connected to the negative pole of the battery and were reversed when the polarity of the connection was reversed. Finally, Ørsted moved from experiments with a single wire to experiments with a double wire shaped into a very narrow U. The results were rather complicated.

Ørsted added to the report of his experiments proper a few general paragraphs of attempted explanation (Ørsted 1820o, 3–4; 1820g, 276) in

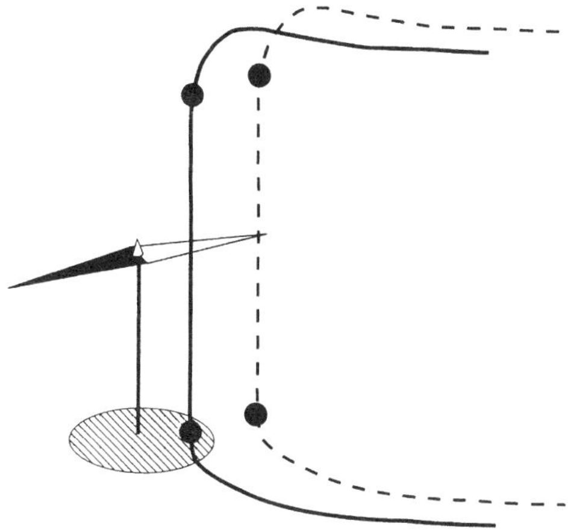

FIG. 2.2. Schematic representation of Ørsted's experiments with a vertical wire.

which his concept of "electrical conflict" played a central role. Initially, he used this expression to refer simply to the unknown process that took place in and around a wire when it was connected to a battery (Ørsted 1820o, 1; 1820g, 274). The term "current," which others were using, struck him as much too loaded with presuppositions for this purpose. His experiments seemed to him to sanction only a few inferences to the peculiarities of the "conflict." First, it struck him as obvious that the conflict was not confined to the wire alone but must also involve the surrounding space, in order to affect magnetic materials—and only those. Furthermore, the conflict must extend itself outside the wire in circles, for otherwise the fact that the deflection of the needle reversed itself when it was positioned above rather than below the wire would make no sense. More precisely, and in order to take into account the directedness of the effect, he supposed, it would have to extend itself not in circles, but in spirals. All of the needle's deflections could then be understood under the assumption that the "negative electrical force or matter" spread out from the wire in rightward-winding spirals (*spiralem dextrorsum flexam*), affecting only the north pole of the needle, not its south pole. Positive electricity would spread out in oppositely oriented spirals and affect only the south pole. To be sure, the claim that this account could really explain *all* of the experimental findings remained largely a promissory note. In its support, Ørsted could offer only instructions for a case-by-case replication of his experiments.

Ørsted's empirical generalizations were tightly linked with a well-developed theoretical background. As the strange expression itself implies, the concept of an "electrical conflict" was not just an instrumental, theory-neutral designation for whatever was going on in and around the wire. Rather, it was tied to a very specific conception of electricity and galvanism that Ørsted had developed early in his career and announced in his *Ansicht der chemischen Naturgesetze* (1812), as well as in other publications. In the case of the open voltaic pile, Ørsted subscribed to the standard view, in which the two electricities gathered at their respective poles. When the pile was closed, he imagined the two electricities flowing through the wire in opposite directions, each alternately revealing its own singular nature and immediately being neutralized and thus in a constant state of "conflict." The destruction and restoration of equilibrium occurred in such rapid succession that, from the outside, there appeared to be only a single, steady current. This view was embedded into Ørsted's general understanding of processes well beyond electricity and magnetism, a general vision of the fundamental forces operating in nature, their relationships, and their various expressions.[11] In his *Ansicht*, for example, he had surmised that light and heat also represented "electrical conflicts." Now that electromagnetic effects had brought such a conflict within reach, he returned to this supposition at the end of his report, suggesting that the spiral geometry of the conflict that had now become manifest might well explain the hotly debated phenomena of optical polarization.

Ørsted's conception of "electrical conflict" also explains his decision from the outset to describe the newly discovered effects as "*electro*-magnetic," rather than "galvano-magnetic." The choice was hardly self-evident, for no electrical phenomena in the traditional sense were involved, and in 1820 there was still no consensus on whether the effects of the closed pile should also be called "electrical." Other researchers would be much more cautious, and indeed Ørsted was later criticized for his choice of terminology (see below). A year later, he more or less conceded the point by acknowledging that his use of the term "electromagnetic" had really been justified only by Arago's later experiments, in which magnetic effects were evoked by conventional electricity (Ørsted 1821b, 323).

To any reader unfamiliar with Ørsted's prior work, the theoretical passages in his report must have seemed vague or even cryptic, for Ørsted had presented no example in support of the general power of his theoretical presuppositions. The text of the experimental part of his report, by contrast, was relatively clear. Uncertainties and ambiguities appeared at only a very few junctions, as, for example, in Ørsted's description of experiments using a wire aligned with the east-west axis and those using a vertically mounted wire. Here, the text left open whether the needle deflection had been observed for only one pole or for both equally. This particular uncertainty quickly attracted attention, and it was these experiments that would prompt far-reaching follow-up studies. Ørsted's report also drew attention to the problem that there were no suitable concepts for describing the complex spatial situation in which the needle deflections took place. Ørsted's use of the compass points had the consequence that the experimental apparatus must always have a particular alignment relative to them, making generalizations difficult. His concept of the circular or spiral geometry of the effects might have been helpful in this respect, but Ørsted did not pursue it further.[12] His early work in the field also did not contain any visual illustrations that might have facilitated general descriptions.

Announcing a Spectacular Finding

Ørsted was greatly interested in making his results public as quickly as possible, and he chose to announce them in a rather particular way. Most of his experiments appear to have taken place in July 1820, with some of them dated July 15 and the final series, July 19 (Meyer 1920, lxxxii–lxxxvii). It is thus striking that his summary report bears the date July 21, by which point it had already assumed its final, printed form. Ørsted obviously felt that he was under pressure, else he might have concluded his experiments and written his text in less haste. He had his report printed onto four-sided folds, which he began dispatching on July 22.[13] The fact that he then immediately returned to his research (see below) shows that he remained aware of further questions. However, cognizant of the sensation his discovery would cause, he was determined to announce it without delay.

He directed his circular to an audience of researchers across Europe, which is doubtless why he composed it in Latin. He did not address the academies as such—neither the Paris Académie des Sciences nor the Royal Society of London received official copies of the letter—but instead sent them to numerous renowned individual researchers. With some of these individuals he already had a history of personal correspondence, even friendship. Another target group consisted of journal editors, who might function as "multipliers." Well aware of the political situation, he omitted some, including Ludwig Wilhelm Gilbert in Leipzig, whose important *Annalen der Physik* circulated throughout the German-speaking world. Gilbert was well known as an opponent of any research bearing the stamp of *Naturphilosophie,* and Ørsted doubtless did not believe Gilbert would publish his missive.

He might easily have published his report as an ordinary magazine article; there was no shortage of suitable magazines. But that would not have yielded either the breadth or speed of distribution he achieved with his circular. And this was precisely his objective. He had no qualms about troubling scientific authorities and research centers across Europe with his letter. This strategy was certainly more laborious and undoubtedly more expensive (he may well have funded the print run out of his own pocket); it also demanded a much greater level of commitment, as well as greater risk, for if his discovery should turn out to be spurious, marginal, or old hat, the proactive strategy of his announcement would doubtless have haunted his international reputation for a long while to come. But while a magazine article would have been less risky, Ørsted was clearly very sure of the result and of its significance. He made this clear enough in the cover letter to his friend Christopher Hansteen, albeit with restrained understatement: "It seems to me that the consequences of my discovery may prove to be rather far-reaching."[14] Small wonder he moved swiftly to secure an unassailable priority claim.

Before I turn to reactions to Ørsted's circular, I shall briefly sketch the follow-up studies he conducted immediately after mailing his announcement. First, he investigated the role of the battery in greater detail than before, noting that it was not the "intensity" of its galvanic activity but the "quality" of it that affected electromagnetic phenomena. "Intensity" was evidently a function of the number of plates rigged in series, whereas "quality" depended on the size of the plates. Careful reflection on this point eventually led him to an arrangement with only a single pair of plates, which was much easier to handle, not to mention cheaper. But more importantly, given the lower weight of such devices, he could investigate the reverse effect; instead of the wire moving a magnetic needle, a magnet could move the wire. For that purpose he suspended an entire galvanic cell by a metal thread, leaving it free to rotate around its vertical axis, and rigged the connecting wire horizontally, between two booms. By bringing a strong magnet near these booms, he could indeed cause the device to rotate. Ørsted also attempted to allow his apparatus to align itself by terrestrial magnetism, ascribing his failure to its insufficient mobility.

Ørsted announced the results of these experiments a few weeks letter, in a second circular, this one composed in French.[15] (He had received unfavorable reactions to the Latin of his first letter.)[16] It's not clear whether he had the second letter printed or circulated it in manuscript. In any case, it reached European researchers toward the end of September and was immediately sent to press, generally by the same periodicals that had published his first letter.[17] Though it contained equally original and important findings, it did not attract nearly as much attention as the first essay. Its two chief novelties—the simplification of the battery and the reversal of the effect—had by this time been independently developed and studied by other researchers, who had set to work in response to the first announcement. These included Johann Schweigger in Halle, André-Marie Ampère in Paris, and Thomas Johann Seebeck in Berlin. After only a few weeks, Ørsted was no longer alone in the field. Some of the results presented in his second circular were instead associated with Schweigger's name. While the second circular was en route, Ørsted was asked by Tobias Mayer if he had received news of Schweigger's findings—a clear indication of the feverish level of activity sparked by the first letter.[18]

Activity across Europe after Ørsted's Report

Reactions to Ørsted's report were extraordinarily strong throughout Europe, though differing from place to place in characteristic ways.[19] A comparative history of the diverse reactions and their contexts has yet to be written.[20] In what follows, I simply recount the places and dates of various activities without detailed discussion. This narrative will at least convey a vivid impression of the severity of reactions.

The reaction of the French-speaking world was far and away the swiftest and most agitated. On August 19, Professors Marc-Auguste Pictet and Charles-Gaspard de la Rive replicated Ørsted's experiments in Geneva. Not only did François Arago, secretary of the Paris academy, happen to be in Geneva to witness them but he was joined by several of the most prominent researchers in the city and by a fellow guest, the physician and chemist Alexandre Marcet. The great significance that Pictet and de la Rive ascribed to Ørsted's report is made plain by the fact that they quickly had it translated into French for publication in the very next issue of their *Bibliothèque universelle* (Ørsted 1820e), quickly followed by an account of their own experiments (Pictet 1821).[21] Four weeks later, on September 18, the issue had been received in Paris and presumably elsewhere in Europe. But by this time there was already another French translation of Ørsted's report in print, in the *Annales générales des sciences physiques* out of Brussels. Its editor, Jean-Baptiste van Mons, had received a copy of the letter and translated it immediately (Ørsted 1820l).[22]

Even before Pictet's translation had been printed, Arago personally brought it back with him to Paris. At the first possible opportunity (September 4) he reported on the finding to the Paris academy, and, along with Laplace and Gay-Lussac, he was tasked with replicating the experiments. The following week he read out Ørsted's entire circular and demonstrated the experiment. He also arranged for the translation to be printed in the next issue (October 9) of the *Annales de chimie et de physique,* which he published together with Gay-Lussac. In his preface, he made it clear that, in this instance, he was ready to abandon his customary caution in the face of "extraordinary discoveries" (*découvertes extraordinaires*) (Ørsted 1820f).[23] Yet another translation of the circular appeared a few weeks later in the *Journal de physique, de chimie, et d'histoire naturelle et des arts,* whose publisher, the Paris professor of anatomy and zoology Henri Marie Ducrotay de Blainville, had received a copy from Ørsted. By this time Blainville had already received the second letter, which he translated and printed along with the first (Ørsted 1820b, 1820k).[24] He also forwarded the first letter to David Brewster in Scotland, editor of the *Edinburgh Philosophical Journal.*[25] The extraordinary race to publish Ørsted's circular was, if anything, surpassed by the rush of frenetic research set off by Arago's report of it in Paris, to which I return shortly.

If the London reaction was slower to build, it had to do in part with the fact that in September the Royal Society was still in its summer recess, which had begun in July. The influential researcher William Hyde Wollaston had received a copy of the letter directly from Ørsted, and, though he was certainly intrigued, he took no great pains to announce it further.[26] Thomas Young, secretary to the Royal Society, had also received the letter but apparently did not consider the topic of it to be his field. At any rate, he waited for others to replicate the experiments and confirm their results. By contrast, another recipient, Thomas Thomson, professor of chemistry in Glasgow and editor of the broadly circulated *Annals of Philosophy,* quickly commissioned a translation for publication in his October issue (Ørsted 1820g). Four weeks later, in the November issue, the first English translation of Ørsted's second letter appeared (Ørsted 1820i). When the Royal Society finally convened again on November 9, it had received not only both Ørsted letters in printed form but also a letter from Ampère describing his own preliminary results. Humphry Davy rushed to work and presented his results a week later, on November 16. The fact that it was decided, the very next day, to award Ørsted the Copley Medal underscores the importance now ascribed to his discovery in London.

Many of Ørsted's German-speaking addressees reacted very quickly. After only a few days, Christoph Heinrich Pfaff, a Kiel professor of natural science, wrote Ørsted seeking clarification of several passages whose meaning he had found obscure.[27] Pfaff lacked the large battery required to reproduce the experiments, and he entered the general discussion only later. Ørsted's friend in Berlin, Christian Samuel Weiss, was thoroughly enthused and immediately delivered the report to Paul Erman, secretary to the Royal Prussian Academy of Sciences, and to academy member Thomas Seebeck.[28] Seebeck began his own investigation immediately, observing by early September that the effect could be produced with simpler galvanic devices—the same result Ørsted would announce a few weeks later in his second circular.[29] Seebeck devoted several months to electromagnetism, lecturing at the Berlin academy on December 14, 1820, and again on February 8, 1821. The summary of his work, however, did not appear in print until May 1821, and his detailed report was published in 1822.[30] Erman, too, began work after only a brief delay, and in early 1821 he published what appears to have been the first German-language monograph in the new field.[31] At Erman's instigation, one of his students, Johann Christian Poggendorff, also began work on the new effects, soon proposing a detection instrument, which he called a "condenser" (*Condensator*), which would give rise to one of the several lineages of galvanometers (Poggendorff 1821; Erman 1821b, postscript).[32] In recognition of his discovery, the Berlin academy named Ørsted a corresponding member in December 1820.[33]

The Latin original of Ørsted's first circular appeared, with a delay much lamented by the publishers, in the *Journal für Chemie und Physik*, run by the Halle professors Schweigger and Meinecke. The same issue contained a German translation of the second circular.[34] Schweigger also set immediately to work on his own experiments, presenting to the Scientific Society (Naturforschende Gesellschaft) of Halle on September 16. Independently of Ørsted and Seebeck, he also observed that the effect could be achieved with a simpler galvanic device, and he reported on his (initially unsuccessful) experimental attempts to produce the hypothesized reverse effect.[35] Schweigger's lecture also envisioned a "doubling apparatus," later known as the "Schweigger Multiplier" (*Schweiggerischer Multiplikator*). As expected, Ludwig Wilhelm Gilbert in Leipzig, who published the widely read *Annalen der Physik,* was initially skeptical (e.g., Gilbert 1820c, 294). Only after reports from Geneva and Paris had persuaded him that something important was going on did he publish a (rather free) German translation of Ørsted's first circular in the November issue of his journal (Ørsted 1820n).[36]

He printed it along with translations of other articles on the subject, including those by Pictet, de la Rive, and Arago. But though acknowledging its existence and significance, Gilbert (1820c, 292) described Ørsted's finding as essentially the result of "happy chance," being unwilling to admit that the systematic pursuit of a research program for which he had such disdain might have led to such an important discovery.[37] It was perhaps precisely for this reason that he felt himself obligated—against his usual custom, as he stressed—to conduct his own experimental research. Alongside works by Biot and Félix Savart, as well as J. K. Yelin, the next issue of the *Annalen* contained his own lengthy report (Gilbert 1820b)—all sixty-one pages of it. From this point on, the *Annalen* featured an unusually high number of reports on electromagnetism. In the two volumes that followed, more space was devoted to this field than to any other single area of research.

Tobias Mayer, professor of physics in Göttingen, had learned of the new results from his Göttingen colleague Friedrich Stromeyer, another recipient of Ørsted's letter. Mayer, too, went straight to work on his own experiments, adding his results to a hastily drafted report on Ørsted's "highly important discovery." He mailed the report directly to Ørsted, as well as publishing it in the *Göttingschen gelehrten Anzeigen* on October 23, 1820 (Mayer 1820, which also gives the quote).[38] Among other observations, he noted, citing Schweigger, that the effect could be obtained using a simple apparatus. Having thus been announced in many journals, Ørsted's findings had soon inspired experiments, lectures, and publications in many different localities. Even a preliminary listing thereof is striking. The list includes a Dr. Buch (1821b, esp. 18–10 [timeline]) and his collaborator Christian Ernst Neeff (1821) in Frankfurt that October; Julius Konrad Yelin in Munich; Professor of Physics Georg Wilhelm Muncke, in collaboration with Leopold Gmelin and Philipp Lorenz Geiger, in Heidelberg; Professor of Physics Karl Wilhelm Böckmann, in collaboration with Baron Althaus, in Karlsruhe; and Johann Joseph Prechtel, director of the Polytechnic Institute, in Vienna.[39]

Meanwhile, in Stockholm, the chemist Jöns Jakob Berzelius received a copy of Ørsted's letter firsthand and set quickly to work, reporting to the Stockholm academy and publishing his results.[40] In the Italian-speaking world, reactions to the discovery took place in at least three places. Pictet's personal visit in November brought Ørsted's news to Florence, where on November 28 the chemistry professor Giuseppe Gazzeri and collaborator Cosimo Ridolfi reproduced the experiment in Ridolfi's private laboratory. Gazzeri (1821) published a report in the December issue of the *Bibliothèque universelle.*[41] He and Ridolfi continued their work for some time, sending their reports to the French journals.[42] The Italian translation of Ørsted's (1820a) report that soon appeared in the *Giornale di fisica, chimia e storia naturale,* printed in Pavia, also credited Pictet's copy and his French translation. The Pavia professor of physics Pietro Configliachi entered the field as well. He reproduced and extended Ørsted's experiments and presented his work to the science academies of Pavia and Milan. By year's end he was also in correspondence with his Florentine counterparts. Like them, he usually took his cues from work performed in Paris, and, beginning in early 1821, he published his reports and letters in French journals (Configliachi 1821b, 1821c).[43] Along with an account of the Geneva experiments, a very brief Italian excerpt from Ørsted's first circular also appeared in the November issue of the *Giornale arcadico di scienze, lettere, e arti,* published in Rome. A month later this journal published a translation of the second circular, supplemented by a report on work by Arago and Ampère (Ørsted 1820c, 1820m). Finally, at the Cabinet of Physical Instruments of the University of Rome, Professor Saverio Barlocci and colleagues Domenico Pino Morichini, Carpi, and Giacomo Folchi performed a series of experiments, publishing a detailed report (Barlocci 1821). In short, the Italian reception of Ørsted's findings was quite lively, though principally shaped by the mediating influence of French sources.

This collection of responses should be sufficient to show that in his expectation of evoking a response, Ørsted was not to be disappointed. His publication strategy paid off handsomely: within a few weeks, the discovery was known throughout Europe, and within a year his first circular had been translated into French, English, German, Italian, and Danish, appearing in print in at least eleven scientific journals.[44] Work on the new effect was under way at every single major research institution in Europe, and none of the big names in the field had failed to involve themselves. In electricity and galvanism, nothing since Galvani's discovery of the effects named after him and Volta's invention of the pile had met with so broad and vigorous a response. In the next section, I highlight a few characteristic features of those responses.

Surprise, Replication, Recognition

In the first few months following Ørsted's announcement, most work undertaken in response ran its course "locally": experiments were performed on the spot, and communication took place in personal conversations or lectures. Thus, when Davy presented his preliminary results to the Royal Society on November 16, 1820, though he doubtless assumed that others were engaged in similar research, he had no way of knowing what they had found; "some of these results . . . may have been discovered before, or at the same time in other parts of Europe" (Davy 1821d, 8). Real exchange of information across the borders of the various scientific cultures began only toward the end of the year, when Ampère wrote to London, Davy and Berzelius, to Paris, and so on. Only then were reports of results, apparatus, experiments, and theories exchanged more broadly. As the following chapters make clear, the phase preceding this exchange is exceptionally illuminating when it comes to understanding the local cultures of Paris and London. Nonetheless, it is remarkable that, despite all the local differences, even during this preliminary phase, reactions to the new findings exhibited very significant commonalities, to which I now turn. They serve as clear indicators of a broadly sensed epistemic situation, one transcending local culture and peculiar to the research field in question.

Three points are particularly salient. First, everywhere we look, the first reaction to Ørsted's discovery was always one of simple surprise, sometimes accompanied by doubt. This was true as much for those who welcomed the result enthusiastically as for those who dismissed it as a delusion. Even Arago, who was perfectly open to the prospect of a discovery so dramatically problematic for the Laplacean view, was frankly skeptical at first: "Arago believed not a word [of Ørsted's report] until he made the attempt himself. But we obtained such decisive results, that he could not but yield to them immediately."[45]

Given that such an effect had been sought for twenty years, only now to prove so easy to produce, the surprise is perfectly understandable. Equally understandable is the fact that everyone's first step was to ensure the reality of the effect itself by repeating the experiment. And here is my second point: replication appears to have worked everywhere. Among the many reports, we find only a very few accounts of experiments that immediately failed to succeed.[46] The existence of the effect itself was thus soon beyond doubt, though it had not been obvious to begin with, as witnessed by the fact that there were published notices consisting only in the announcement of a successful replication.[47] Pictet (1821, 284) offered a pointed observation on the resulting level of certainty: "It must be said that all of these movements [of the needle in response to the wire] are prompt, decisive, and without any possibility of mistaking their direction." Such obviously effortless replication required that the experimental finding had to be accepted regardless of which of the different theoretical traditions held sway in a particular locality. Also contributing to the swift acceptance was the discovery that the effect could be achieved with smaller galvanic devices than those Ørsted had initially employed. After the experiment had been reproduced at facilities with large devices (Geneva, Paris, Berlin, Stockholm, London, and Florence), the news that it could be performed with smaller devices, or even a single pair of plates, was treated as of great importance and spread quickly. Then the finding could be easily verified even where few resources were available. In the German states, at least, replications by small laboratories did not really get under way until this news had circulated.[48] As an aside, it might be noted that such effortless replication in so many different places and the universal recognition of the experimental result that evidently followed shed new light on recent discussions of the "experimental regress." I return to this matter in greater detail in chapter 7.

This brings us to my third point. Successful replication also displayed the effect's unusual characteristics. Arago's brief account of the three "principal results" (*résultats principaux*) is typical:

> The wire connecting the two terminals of a pile exerts a force on a magnetic needle.
>
> This effect depends on the direction in which the electrical fluid in the wire is moving relative to the needle.
>
> *When the wire is below the needle, the needle deflects in the direction opposite to that assumed when the wire is above it.* (Arago 1820b, 418, original emphasis)

The fact that Arago chose to emphasize the third point in print speaks to this most conspicuous aspect of the needle's behavior as being of particular note. It was this feature that so immediately and rudely revealed the new effect as incompatible with the familiar concept of central forces whose strength depended only on distance. This incompatibility did more than simply pose serious problems for one or another physical theory; it challenged a much more fundamental concept, one in which all scientific thought of the time (not only in France) was thoroughly steeped and whose sturdiness and adequacy had been proved by successful mathematical formalization. Its revision had seemed unthinkable.

Such characteristics also left most of the researchers with a clear sense that the finding was of enormous importance, something that, welcome or unwelcome, had to be dealt with in any case. Arago (1820a, 93) spoke of the "brilliant discovery M. Ørsted has made," while the civil engineer C. J. Lehot (1820, 407–8) described the present "moment" as one when "all of scientific Europe concerns itself with the relations between galvanism and magnetism."[49] Davy (1821d, 8), for his part, asserted in his first publication on the new phenomena, "This discovery, from its importance and unexpected nature, cannot fail to awaken strong interest in the scientific world," and William Thomas Brande (1821a, 361) began his summation in January 1821 with the words, "No discovery has, for a long time, so strongly excited the attention of the philosophical world, as that of the magnetic phenomena belonging to the Voltaic apparatus." Thomas Young congratulated Ørsted on his "marvelous discovery," which he credited with the "elevation of Your country to a rank in science which it has not held since the days of Tycho Brahe."[50] In Halle, referring to the contents of Ørsted's circular, Meinecke wrote, "This is one of the most significant of recent physical discoveries, and will prove of great consequence to science. . . . Since Galvani's experiment there has been perhaps none more important than Ørsted's for electricity and chemistry" (accompanying Ørsted 1820h, 364). Schweigger (1821c, 1), in turn, called the experiments "with respect to magnetism, the most interesting to be conducted in over a millennium." Paul Erman (1821b, 51), too, was awestruck: "Not a word on the overwhelming importance of Ørsted's discovery: if it does not speak for itself, nothing in the world will," and Seebeck (1822, 289) lauded Ørsted's discovery as "epochal in the history of magnetism."

For all the impressive unanimity on the importance of the experimental result, individual opinions on the reasons for its importance differed widely. Two examples will serve as illustrations. For Ørsted, who had made his discovery in the course of a systematic search, his success naturally served as both validation and incentive to continue following the path on which he had embarked. By contrast, for Biot, as the preeminent champion of the Laplacean program in physics, the challenge posed by the finding was positively threatening. To him, as to many others, it was immediately clear that the new effect could not easily be incorporated within the existing system of concepts and theories and as such must be thoroughly investigated and assimilated. In both of these cases, theoretical presuppositions provided strong motives for continued work, as well as reasonably well defined guidelines for how to proceed. For most others, the theoretical situation was much less clear. Their interest sprang from more general motivations. Regardless of how one felt about Laplacean physics, it remained one of the most important currents of the day. By challenging its basic concepts, the new finding immediately raised general questions regarding the understanding of physical processes. This was equally true among the heterogeneous research endeavors pursued in connection with German *Naturphilosophie*, one of which had given rise to the new discovery. In addition, the new discovery touched on the question of the connection between different domains of phenomena, an issue that had everywhere exercised scientific minds at least since the discovery of galvanism and the beginnings of electrochemistry. Every different motive for investigating the new effect brought with it a different angle of approach for future investigations. But very few researchers entered the fray with as specifically goal-directed a research program as that of Biot, for example, as I discuss in chapter 4.

A New Field of Research

Seeing past the sheer novelty of the new effect, many began to form a further impression: that Ørsted's discovery had offered a glimpse into an entirely new *field* of research. In a letter to Ørsted, Blainville, for example, spoke of "the field that you have opened up."[51] Meinecke believed that "Ørsted's discoveries on the connection between magnetism, electricity, and chemical process begin a new epoch in chemistry and physics," and he was moved, in part by this occasion, to inaugurate a new series of his journal at the beginning of 1821.[52] Despite his skepticism regarding the background to the discovery, Gilbert (1820d) expressed himself quite similarly, asserting that "the great discoveries on magnetism and magnetic electricity announce a new epoch in science."[53] And Davy (1821d, 8), at the beginning of his first work on electromagnetism, proclaimed, "This discovery . . . opens a new field of inquiry, into which many experimenters will undoubtedly enter."

The situation at the opening of the new field had several distinguishing features. With respect to the competitive environment, anyone who wanted to begin research could be certain that others, all over Europe, were doing the same. The pressure of this universal contest had a huge effect on the mood in which all further investigation was conducted.

Another striking feature was the conceptual and theoretical situation of the new field. The most advanced and coherent theories of electricity, those of Biot and Poisson, had, so it seemed, been cast into great difficulties by the new finding, for their fundamental concept—the notion of central forces—was apparently inapplicable. But on the other side, the theory of electrical conflict, on which Ørsted had drawn in explaining his effects, stood equally isolated. It resonated with many of the reflections of *Naturphilosophie,* but this camp enjoyed much less internal coherence than Laplacean physics. Consequently, even in the German-speaking world, very little attention was paid to Ørsted's theory. Yelin came the closest, but, in typical fashion, he also developed his own idiosyncratic ideas and concepts. Tobias Mayer (1820, 1711) gave voice to the widespread theoretical reticence: "Professor Ørsted has contributed his own perspective on the matter of his discovery, but given the limited pages available to us, we must content ourselves with stating the important facts. It is left to the future to consider how the difficulties facing his views, whose exposition here would take us too far afield, might be dealt with satisfactorily." In France, Ørsted's theory was barely even mentioned, let alone seriously discussed; it was presumably considered simply unworthy of discussion. But even in the less rigid English-speaking context, it was generally taken as unintelligible. For most of those concerning themselves with the new effects, Ørsted's theoretical views seem to have carried little weight and less authority, and in most places they played as little role as the Laplacean scheme in guiding the first concrete excursions into the new field.

This point becomes even sharper when we compare this situation with that of nearly three decades prior, when Galvani announced his own spectacular discovery of "galvanic" effects. Galvani had also presented a theory along with his experimental results, one clearly much more detailed than Ørsted's. Though Volta would soon vigorously dispute that theory, it was Galvani's concepts and theory, and not those of Volta, that would continue to shape experimental and theoretical efforts in the field for more than a decade.[54] By comparison, the role of Ørsted's theoretical proposal in discussions of electromagnetism was extremely marginal.

For the majority of those who rushed into the new field, the theoretical terrain thus remained wide open; there was simply no authoritative theory that might have guided their engagement with the new effects. Pictet (1821, 284) summed up this state of affairs quite clearly, and not without implicit criticism of Ørsted: "Any system one might wish to formulate regarding this isolated discovery would strike us as premature." Muncke (1820, 414) also described the new phenomena as "isolated and puzzling." Such assessments bring to light the widespread judgment that even the "empirical" rules with which Ørsted had described the needle's behavior (as opposed to his theory of electrical conflict) remained incomplete and, in places, difficult to grasp. Consequently, the first efforts of many researchers were often geared toward developing just such rules. When, several weeks after

Ørsted's announcement, both Ampère and Biot described their work as an "extension" of Ørsted's, they were referring to their attempts to develop more general empirical rules than Ørsted had managed.

Ørsted's Discovery and the Académie des Sciences

Having completed an overview of the pan-European context, I now turn in greater detail to the Paris stage, so as to more closely examine the conditions and presuppositions under which, in September 1820, Ampère developed his fundamentally new approach. Immediately after returning from Geneva, Arago reported on Ørsted's discovery at the September 4 session of the Académie des Sciences.[55] The reaction of the Parisian academicians was plainly rather reserved. As Pierre-Louis Dulong wrote to Berzelius in October, "The news at first met a very cool reception here. It was thought that this was yet another German reverie."[56] The allusion to "German reverie" (*rêverie allemande*) conveys not only the general attitude toward German-language natural philosophy but also the extent to which Ørsted (a Dane, after all) was associated with such circles. Ampère's later description expresses a similar impression: "When Arago told the *Institut* about the new phenomena, the account was dismissed, just as one had dismissed stones falling from the sky. . . . They all decided it was impossible."[57] Ampère was alluding to a resolution of the academy to no longer accept reports of stones falling from the sky—presumably real or imagined meteorites.[58] So as to verify the effect, Arago was tasked with demonstrating the experiment at the next session of the academy, a week later. His successful presentation on September 11 effectively disarmed the academicians' skepticism, and there are no further reports of anyone continuing to question the effect itself. Given the unfavorable milieu, this case provides an especially striking example of an experimental result gaining acceptance by simple replication.

The reasons for the reticent, even defensive reaction of the academicians may naturally be traced, at least in part, to the theoretical situation reviewed above. At any rate, Ampère thought he had discerned the cause of their doubts: "It resides in Coulomb's hypothesis on the nature of magnetic action. This hypothesis was believed as firmly as if it were fact, and it absolutely ruled out any idea of interaction between electricity and alleged magnetic wires."[59] An additional factor was of course the memory of Ørsted's disgrace seventeen years earlier, when he had attempted to demonstrate his friend Ritter's supposedly great discovery in Paris, an echo of which doubtless played into worries about "German reverie." In an institution as self-conscious (and occasionally complacent) as the Paris academy, such matters carried great weight.

Once the effect was successfully verified, the severity of the challenges it posed to established physics was inescapable, for it was not only the existence of electromagnetic interaction that had been demonstrated. There was also the strange spatial comportment of the compass needle, which appeared even at first blush to be beyond the reach of traditional concepts. Arago had good reasons for stressing this point as being of particular importance. There was no question but that the Laplaceans would have to respond. But developing this response would take time. Jean-Baptiste Biot, the most important (and perhaps the only) proponent of the Laplacean program in physics, was traveling—"unfortunately," as Dulong had it.[60] Had he been in Paris, he would doubtless have begun his investigation of the effect immediately, as indeed he did, and with vigor, upon his return to Paris approximately six weeks later. But by then significant activity had already taken place, in the form of unanticipated efforts by Ampère.

A clear sense of the intensity of work on this topic in the fall of 1820 may be gained from a compilation of activities reported to the Académie des Sciences (table 2.1).

Though limited to events at the Paris academy, this table strikes me as representative of events in France as a whole, for, to a much greater degree than anywhere else in Europe, the recognition of research efforts in France turned on this one institution. Whereas in England, the results of research might gain broad acceptance without ever being mentioned at the Royal Society, a Parisian analog to this prospect was scarcely conceivable.

A quick glance at the table shows how extraordinarily strong the presence of electromagnetism was at the academy throughout a four-month period. Of the nineteen sessions leading up to the middle of January 1821, there were only four at which the topic did not arise in one form or another. As Dulong observed, "At sessions of the *Institut* we no longer speak of anything but new experiences related to this subject."[61] Five researchers were especially prominent: Arago, Ampère, Jean-Pierre Boisgiraud (still an unknown young researcher and *répétiteur de physique,* or physics tutor, at the École Militaire de Saint-Cyr), Biot, and Fresnel. In the middle of November, letters also began to arrive from the outside world, to be read out at sessions. Ampère was far and away the most active of these researchers, appearing almost weekly, with a total of twelve presentations, more than all the others taken together. The first of these readings took place a mere

TABLE 2.1. Activities regarding electromagnetism at the Académie des Sciences, September 1820–April 1821

Session Date	Ampère	Others
September 4, 1820		Arago reports on Ørsted's result; Arago, Laplace, and Gay-Lussac are charged with replicating the experiments
September 11		Arago reads Ørsted's report and demonstrates the effect
September 18	First mémoire, part 1	
September 25	First mémoire, part 2; "Conclusions"	Arago reports on the attraction of iron filings and a magnetization effect
October 2		
October 9	Second mémoire, the "current balance"	Boisgiraud reads a paper; Ampère and Charles appointed as referees
October 16	Third mémoire, a "note" on Arago's experiments	Biot attends his first session after a long absence
October 23		
October 30	Fourth mémoire	Biot reads his paper and presents a law ("Biot-Savart")
November 6	Fifth mémoire	Arago reports on magnetization by conventional electricity, Fresnel on the supposed decomposition of water by magnets
November 13	Sixth mémoire	A letter from Lehot is read; Ampère delivers his report on Boisgiraud
November 20		
November 27		Receipt of a survey article by Hachette, along with Ampère's "Notes"
December 4	Seventh mémoire, part 1	Biot raises objections and announces a further mémoire
December 11	Seventh mémoire, part 2	
December 18	Submission of a "note" on mémoires to date	Biot reads a second mémoire
December 26	Eighth mémoire; submission of a special offprint of the *ACP* mémoire, part 2	

Table 2.1 (*continued*)

Session Date	Ampère	Others
January 2, 1821		
January 8	Ninth mémoire, part 1; remarks on Berzelius's letter	A letter from Berzelius to Berthollet is read; La Borne reads a "note"; Biot, Arago, and Ampère are appointed as referees
Seven sessions		
March 12	Reflections on Prechtel's letter	A letter from Prechtel is read; Biot, Arago, and Ampère are appointed as referees
March 15	Ninth mémoire, part 2	
March 19		A letter from Davy to Ampère is read
March 26		
April 2 (public session)	Ampère reads a "short notice"	Biot reads a mémoire

Source: Based on accounts in the *Proc.Verb.*, together with reports in journals.

two weeks after the announcement of Ørsted's report. In addition, he soon became the referee for contributions on electromagnetism by others. It was not just the intensity of his engagement that surprised the academicians but also the very fact that he had become active in this field in the first place, not having previously been known either for work in physics or for any other experimental research.

In January 1821, work on electromagnetism began to slow, pausing noticeably. In Ampère's case, this was in the first instance a result of his falling seriously ill at the beginning of the new year, presumably at least in part in reaction to the extremely intense work of the preceding months (Blondel 1982, 103–4). In addition, his lectures of December 4 and 11, 1820, had brought the first phase in the establishment of his theory to a close. He had devised a formula for the force law operating between elements of current-carrying wires and thereby given his theory its first mathematical outlines. The theory's wide explanatory ambitions, however, had only just begun to be satisfied, and Ampère was well aware that its real construction was still before him. But, having laid the foundations out for everyone to see, he was prepared to take more time for the painstaking details. It was not until the session of April 2, 1821, that he returned to the academy to deliver a brief report. The fact that only a few weeks later he would give a lecture not on electromagnetism but on a problem in theoretical mechanics is a clear signal that his work had entered calmer waters.[62] Something similar held for Biot, though in a rather different way. In his October and December lectures, as a result of the precision measurements he had performed together with Félix Savart, he presented a formula for a force-distance law (still known as the Biot-Savart law), and attempted to explain it by recourse to centrifugal forces, as I discuss in chapter 4. These were precisely the tasks assigned by the Laplacean program. Without further conceptual innovation, this program had run its course, in the sense that it simply posed no further tasks. At the beginning of 1821, the two most important positions in Parisian electromagnetism had been staked out by Ampère and Biot. As this overview already makes clear, the last four months of 1820 were of critical importance for the field as a whole, and for Ampère in particular.

André-Marie Ampère, from the Provinces to Paris

Before turning to Ampère's activities during this period in greater detail, I offer a brief look at his background.[63] By French standards, André-Marie Ampère's career path was atypical. Raised in a well-to-do merchant family in Lyon and at a country estate in a nearby village, his education could not have been more different from the highly formalized mathematical training of his Parisian contemporaries. He received hardly any regular primary or secondary instruction, for his father, inspired by Rousseau's pedagogical ideals, had opened his library to his son rather than sending him to school, and he provided the boy with supplementary private lessons in particular subjects as needed. Besides an inclination toward natural history, expressed in his own copious observations of nature, Ampère showed a strong early interest in mathematics, manifested in his reading of Euler and Bernoulli. But his untroubled youth came to a sudden and violent end with his father's execution by the Jacobins in 1793. Profoundly shocked, the eighteen-year-old sought solace in the deep religiosity in which his mother had raised him, in stark contrast with the Enlightenment ideals of his father. His entire life was marked by the dichotomy between a rational worldview and intense, sometimes mystically tinged piety. After several years of reclusion, in which he composed numerous poems and a play, he was called upon to assist in the task of maintaining his family, which had become challenging, and began giving private lessons in mathematics. His disposition changed dramatically upon meeting and falling in love with Julie Carron. They were married in 1799, and their son Jean-Jacques was born in August 1800. Beginning in 1802, Ampère taught physics at the École Centrale (a predecessor of the later *lycées*) in the town of Bourg, about sixty kilometers from Lyon. Julie, who had been ill since the birth of their son, remained with the baby in Lyon. By the time that Ampère was able to return to Lyon, in March 1803, to take up the hoped-for post at a lycée, Julie's health had taken a significant turn for the worse. Her death in July of that year spelled the end of a period of great personal happiness.

During his time in Bourg, Ampère had worked on his own research. Among other things, with his eyes fixed on Napoleon's sixty-thousand-franc galvanism prize, he had started work on his own peculiar theory of electricity, magnetism, and light. It never progressed beyond its speculative, programmatic beginnings. He appears not to have shared the manuscript, even in the circle of his Lyon friends.[64] He was less reticent with a more detailed treatise on probability theory, which he presented to an academic society in Bourg and also had printed (Ampère 1802). It appears to have influenced his appointment to the Lyon lycée and remained helpful when, following Julie's death, he sought to leave Lyon. Paris mathematicians having become aware of this work, in 1804 he was able to assume the post of tutor (*répétiteur*) in analysis at the École Polytechnique. With a view toward improving his financial situation, in 1808 he also became an inspector in the national school system, a post that regularly required extended summer travel. His second marriage, which began under difficult circumstances in 1806, took a disastrous course, ending in 1808. Ampère's daughter by his second wife, Albine, remained with him, and was cared for, together with Jean-Jacques, by his sister Joséphine, whom Ampère was able to bring to Paris in 1807, along with their mother.

In Paris, his scientific activity expanded to include many different fields. He published a micro-scale explanation of Boyle's law, a derivation of the laws of definite and multiple proportions from a hypothetical geometrical structure for the smallest units of matter, as well as a classification of the chemical elements. Entering yet another field, he took up optical polarization.[65] None of this involved any experimental work. Ampère was always in search of a deeper, micro-scale explanation and reduction of lawlike regularities that had already been established. While these efforts gained little recognition, his diverse mathematical studies attracted considerable attention, especially a detailed contribution to the systematic treatment of partial differential equations (Ampère 1815a). It was his mathematical work that, in 1814, secured his membership in the Académie des Sciences.[66] The support of the permanent secretary of the academy, Jean-Baptiste-Joseph Delambre, was instrumental in this regard. Delambre had known Ampère since his Lyon days and held his mathematics in high esteem.

In the course of his chemical work, Ampère had already sought contacts beyond the French borders. He was presumably familiar with Ørsted's (1813) *Recherches sur l'identité des forces chimiques et électriques* (Investigations on the identity of chemical and electrical forces), an extension and translation of his *Ansicht der chemischen Naturgesetze* (View of the natural laws of chemistry).[67] With Davy, who shared his interest in German idealism, he began corresponding, mostly on chemical issues.[68] Ampère freely informed Davy of recent chemical work in Paris. In one case at least, Davy, thus inspired, began work of his own on the same topic, claiming priority. This happened with the discovery of iodine, which Nicolas Clément had isolated in 1813 but not yet thoroughly studied. Davy learned of it during the lengthy stay in Paris he inserted into his grand tour of the Continent in the company of his assistant, Michael Faraday. Such incidents quite naturally vexed the Paris academicians, and Ampère's incautious generosity surely did not win him friends (Gardiner and Gardiner 1965, 235–38).

Ampère kept his distance from the Laplacean school. He found its assumptions regarding the constitution of the microscopic world too specific, and he shared with others the conviction that it was time to expand the horizons of scientific thought. His personal relationship with Biot, the chief exponent of this school, was also strained, given his friendship with Arago and Fresnel. Arago had tangled with Biot over a priority dispute regarding optical double refraction, while Fresnel's wave theory of light stood diametrically opposed to Biot's treatment of optical phenomena. In addition, Ampère, though formerly a proponent of the corpuscular theory of light, had in 1816 been persuaded by Fresnel of the viability of his alternative and had actively supported Fresnel ever since (Daumas 1943, 86, 102, 108–10; Hofmann 1995, 217–23). In 1822, Fresnel would move into an apartment in Ampère's house that Ampère, in financial need, had placed for rent.

When Ampère's position at the École Polytechnique improved in 1815 with his appointment to a professorship in theoretical mechanics and analysis, his mathematical focus once again became central to his daily work. In regular rotation with his colleague Augustin-Louis Cauchy, he taught the two-year introduction to analysis and mechanics. Cauchy preferred to lecture on his own new insights into the foundations of the calculus rather than catering to the needs of beginners, which led to repeated complaints by students and numerous labor-intensive revisions of the course catalog (Belhoste 1991, chap. 5). In June 1820, the *conseil d'instruction* (instruction council) of the École Polytechnique assigned Ampère and Cauchy to publish their course materials, a task Ampère put off until 1824 (Ampère 1824b). These circumstances meant that Ampère's teaching responsibilities at the École Polytechnique demanded much more time than he would have liked, and he increasingly perceived them as a burden.

Besides his interests in specific scientific disciplines, Ampère also engaged with entirely different issues. In 1814, he was substantially involved in the formation of an informal *académie psychologique* (psychological academy), a discussion group devoted to topics in epistemology, sense perception, psychology, and ethics. This circle, whose participants included brothers Georges and Frédéric Cuvier along with literati and politicians such as Joseph-Marie Degérando and Philip Albert Stapfer, met in the house of a nobleman, François Pierre Gonthier Maine de Biran, who, besides pursuing a political career during the revolution, under Napoleon, and eventually in the restored monarchy, had also long been interested in such subjects. In 1802, Maine de Biran had received a prize from the *classe des sciences morales et politiques* (disbanded after 1803) of the Institut de France, on the strength of his work entitled "The Influence of Habit on the Faculty of Thought" (Influence de l'habitude sur la faculté de penser). He would subsequently publish a whole series of works on similar topics. In 1799, Dégerando, too, had attracted the attention of the institute's moral and political science group with a piece on the question, "What has been the influence of signs in the formation of thought?" (Quelle a été l'influence des signes sur la formation de la pensée?). Ampère's interactions with Maine de Biran were extensive. In addition to their numerous meetings, they enjoyed a wide-ranging correspondence, which vividly illustrates Ampère's intellectual breadth.[69]

One important reference point in these conversations was Kant's critical epistemology, or at any rate the version of it available in France, as transformed by translation and commentary. Several of Kant's main points had been distorted in the process. For example, Ampère thought that his interpretation of Kant's conception of "noumena" (equated with "things in themselves," or *Dinge an sich*) as spatiotemporally extended material things was in full agreement with Kant, who would doubtless have taken any such view as crude backsliding into a precritical position. However, Ampère was well aware of another respect in which his views contradicted Kant's: he was convinced that important aspects of the "noumenal" world were well accessible to scientific knowledge. Reflecting on the scientific methodology necessary for such knowledge, Ampère took up Kant's discussion of the (much older) distinction between analytic and synthetic procedures, adding a second distinction between "direct" and "indirect" methods, which he used to classify arguments. "Direct" inferences set out from true premises (or premises held to be true) to establish the truth of conclusions, while "indirect" inferences were those for which the truth or falsehood of the premises could not immediately be ascertained but whose conclusions were nonetheless accepted as true. Combining the two distinctions gave him a fourfold classification of scientific argumentation: direct analysis, indirect analysis, direct synthesis, and indirect synthesis. In his 1817 course on logic at the École Normale in Paris, he discussed such methodological questions in detail.[70] The degree to which such reflections actually influenced his scientific work, however, will have to be assessed on a case-by-case basis.

The context in which Ampère pursued these reflections was strikingly different from that in which he pursued his scientific work in search of professional recognition. Among the scientists of the Paris academy, engagement with epistemology or the psychology of knowledge was considered a vain diversion, one held in very low esteem. The discussion groups were largely divorced from the life of the academy, leaving Ampère and the two Cuvier brothers among the very few who were really familiar with both contexts, moving back and forth between them as they did. In the academy's *classe mathématique*, Ampère was the only one with that breadth of interests. Arago would later eulogize his ability to inhabit "two distinctly different communities. The only characteristic they had in common was the fame of their members," thus stressing the likely abruptness of the transition between the two.[71] With such peculiarities, Ampère, though not precisely an outsider, certainly was an unusual figure in Paris. His closest personal friends remained in Lyon, not Paris. He appears in no way to have personally engaged in any of the turbulent political events of the second decade. The catastrophic conclusion to the Napoleonic Wars may have contributed somehow to his turn to Catholicism around 1816, when he began to practice much more intensely than in his earlier, more doubt-ridden years in Paris.

By 1820, Ampère's professional standing was well established. While he never stood in the limelight and his diversity of interests may have seemed a bit scattered, his mathematics was important and taken seriously. His personal affairs were well ordered, and, with his two positions, his income was adequate. To be sure, the workload associated with these positions was enormous. Ampère was thus increasingly keen to relinquish his time-consuming and personally unrewarding instructional activity at the École Polytechnique in favor of some arrangement that would allow him more time to pursue his research. Ideally, that would mean a professorship at the Collège de France. These were not merely prestigious, well-paid positions; even more importantly, they required both less teaching than the École Polytechnique and imposed less rigid controls over instructional content. Professors at the Collège de France were essentially free in choosing their own topics, students came out of interest, not obligation, and there were no examinations. Biot held the professorship in mathematical physics (strange, as this was not the field in which he had excelled), but Ampère's colleague Cauchy had substituted for him during his extended absence while on an 1817 expedition (Belhoste 1991, 48–49). So Ampère did not need to look far for a role model. However, it was perfectly clear that such positions were reserved for those who had distinguished themselves in the public eye by extraordinary contributions to one or another scientific field. In 1820, even Ampère's most rosy self-assessment must have fallen short of this standard.

Philosophy, Stroke of Genius, or Something Else?

Ampère was fascinated by the news of Ørsted's discovery from the very first instant; by his own account he could scarcely think of anything else for several weeks.[72] He quickly launched himself into feverish research. His actions during the autumn months convey a sense of breathless frenzy. There were Sunday evenings when he still didn't know exactly what he planned to say at the academy session on Monday afternoon, having already announced that he would speak. What could have moved him to such lengths that were so unexpected, and not only in the eyes of his contemporaries? Neither electricity nor experimentation had previously been within Ampère's areas of interest or experience.

Historical investigations by Ken Caneva (1980) and Christine Blondel (1982, 66) have shed some light on this matter. Ampère's opposition to established Laplacean physics was surely an important motive, for the finding was manifestly problematic for the Laplaceans and might thus be expected to furnish a starting point for possible alternatives. In addition, the finding concerned the connection between two qualitatively different domains of phenomena and thus raised questions concerning the foundations of physics, in which Ampère had always been interested. He had also reacted with great interest to Domenico Pino Morichini's 1813 "discovery" of interaction between ultraviolet radiation and magnetism (Caneva 1980, 124). Most especially, Blondel stresses, the news of electromagnetic interaction must have immediately revived Ampère's nearly twenty-year ambition to devise a theory that encompassed both electricity and magnetism. In his case, by contrast with his fellow academicians, the fact that the news came from Ørsted may have made it particularly interesting. It is difficult to see how each of these individual motives may have factored in. Caneva's strong emphasis on the anti-Laplacean motive strikes me as perhaps overstated, for though Ampère had already been persuaded of the truth of Fresnel's wave theory of light, nearly all of the clearly anti-Laplacean statements we have on record for him date to a later period, one in which the study of electromagnetism had decisively changed the landscape.

There is one factor that previous studies have tended to overlook. Because the discovery had opened up a brand-new field, one in which the greatest expertise in electricity and magnetism could offer no immediate sense of how to proceed, even a newcomer to these domains might hope to achieve some novel result. What's more, such results would be guaranteed to receive a great deal of attention. The prospects for gaining prestige were brighter than in any more established field, and, as noted above, Ampère could only have welcomed such opportunities.

The situation in which Ampère now found himself had very specific characteristics. Ørsted's news had penetrated every corner of Europe, and everyone who set to work in the new field could be sure that many others were doing the same. Even in the absence of any detailed knowledge of what was going on elsewhere, the awareness of far-flung competition was constant. This awareness was more heightened in Paris than anywhere else, for nowhere else did the peculiarities of the finding demand so immediate a response. Not only concepts and theories were at stake, but also relations of power at the academy, the École Polytechnique, and the University of Paris. Dulong's description of Biot's absence as "unfortunate" for the Laplacean school testifies to the instant force of such considerations in the minds of all the participants. For his part, Ampère was resolved to make the most of the vacuum left by Biot's departure, placing himself in the lead for the ensuing discussion. This point does much to explain the intensity, indeed the frenzy, of Ampère's work. Dulong, at least, saw it this way: "Ampère was alone for a month, and knew full well how to turn this to his advantage."[73] In all discussions that follow it should be kept in mind that to conduct research in a situation like this is something completely different from having a topic all to oneself and being able to explore without regard to competition. As contrasting examples one might cite Coulomb's work with the torsion balance in the 1780s or Galvani's investigation of twitching frog legs in the early 1790s. I have elsewhere (Steinle 1996) developed such an example for the field of electromagnetism: Faraday's discovery of electromagnetic induction.

Intellectual fascination and other motives for turning to a given field are one thing; actually gaining entry is another. As shown in table 2.1, Ampère fully succeeded, which raises the question of how that might have happened. Recall that he had never particularly concerned himself with electricity or magnetism or engaged in experimentation. So the source of his apparently sudden ability to orient himself in a field that remained puzzling even to the experts is anything but obvious. Nonetheless, despite some thorough investigations of the genesis of his electrodynamics, this question has hardly ever been posed.

Retracing Ampère's steps during his early period is particularly difficult. James Maxwell's (1873, 2:176, §528) famous quip, that Ampère had done a perfect job of covering any traces he might have left along the path to his theory, is particularly fitting when it comes to this first phase. Only in recent decades has painstaking historical work succeeded in uncovering and coherently presenting the most important lines of development.[74] I sketch and discuss these results in several sections of chapter 4. For now, it suffices to note that they have little to offer when it comes to the present question. With one exception, all of these studies remain rather reticent when it comes to the period leading up to September 25, when Ampère delivered his second address to the academy. They take this lecture as the point of departure for their analyses. To be sure, it is after this point that we can discern a reasonably well directed plan, or as Blondel (1985) so aptly puts it, Ampère's "programming of research." For the period prior to this date—the first three weeks of work—the picture remains blurry. Only L. Pearce Williams (1983) has ventured to make any claims about Ampère's very first steps in the new field.[75] But by relying entirely on published sources, he becomes entangled in the thicket of Ampère's frenzied publication strategy. His attempt to resolve open questions by recourse to far-reaching ancillary assumptions also strikes me as unsuccessful. The product of Williams's efforts is a nearly linear narration of Ampère's research, proceeding uninterrupted from the outset with a more or less clear theoretical destination in mind. Williams does not address the question of how Ampère could have found his way so quickly in the new field. The tenor of his work suggests that he took Ampère's epistemological reflections on the various strategies of argument in scientific research as particularly important. If I interpret Williams correctly, it was this epistemological preparation that, in his view, gave Ampère his point of entry into an unknown field and continued to guide the minutiae of his research.

I remain unconvinced, either of Williams's methodology or of his results. Not once in his early work on electromagnetism does Ampère allude to his epistemological and psychological reflections.[76] What's more, Ampère's epistemology was extremely general by comparison with the concrete challenges of his new research; I simply fail to see how it might have guided his concrete steps in any way, especially considering that this was Ampère's very first *experimental* venture. Recent work in the history of science has underscored the fact, evident in every laboratory experience, that the problems that arise in the course of experimental research, especially by newcomers to the field, are often far removed from the formal structures of argumentation, and soon develop their own dynamics in generating new research questions and objects of study. Our increasing reluctance to view scientific work as thoroughly determined by general epistemology strikes me as particularly appropriate in Ampère's case. At any rate, the determining role of general methodological and epistemological reflections cannot simply be assumed; for any given case, it must actually be demonstrated. For Williams, however, this assumption served simply as a scheme for ordering Ampère's scattered, incoherent statements.

In addition—and this is my second objection—Williams's strategy had to disregard several points that Ampère himself had raised, though without explaining them, in his self-presentation.[77] There is, on the one hand, the task of "reduction" to "more general facts," to which Ampère repeatedly alluded but of which Williams offers neither any interpretation nor an assessment of the importance Ampère ascribed to it. On the other hand, the status and significance of the "conclusions" with which Ampère closed his second academy lecture also remain completely obscure. These are not marginal points. They are central to the question of where Ampère might have acquired the surprising confidence he had apparently gained after so short a time in the new field. Williams's selective discussion of the few rudimentary hints in Ampère's self-presentation casts as much doubt on his reconstruction as his need to attribute reflections to Ampère for which we have no documentary evidence.[78]

Unless we are prepared to simply accept Ampère's genius, and thus elide the very question at issue, our own option is to pursue new historiographical methods. A detailed reconstruction of Ampère's research *practice* in the first three weeks of engagement with electromagnetism is needed. Williams's missteps may chiefly be traced to his nearly exclusive reliance on Ampère's own explicit assertions, as found in his publications. This forced him to contend with the problem of their scattered and sometimes incoherent character. In addition—and this point has rarely been sufficiently appreciated—the accounts researchers give of their own work are always reconstructions crafted for external readership. To draw direct inferences from such accounts to the actual practice of research can lead to gross error.[79] The reliable reconstruction of research practice requires different kinds of sources. In Ampère's case, obtaining such sources has proved to be particularly difficult and laborious, but possible. The next chapter presents and discusses my results.

CHAPTER 3

Ampère's First Studies of Electromagnetism

Entering a New Field

Ampère's early work on electromagnetism proceeded in two distinct phases of unequal duration. The first phase comprised the three weeks leading up to September 25, 1820, while the second continued through the four months that followed, ending in January 1821. Previous historical studies have given careful attention to only the second of these phases. As discussed in the preceding chapter, our understanding of the first phase remains rather hazy. In consequence, it has not even been evident that we were dealing with two well-defined phases, each with its own distinct character. It has taken the investigation relayed in this chapter to make clear that Ampère's second lecture to the Académie des Sciences, on September 25, was the decisive moment in the evolution of his work. Though in one sense it marked the beginning of his targeted research—as indeed it has been generally understood—at the same time it was also the acme and culmination of the process leading up to this point, about which we knew practically nothing but in which the guiding assumptions for the remainder of his work took shape. This holds both for Ampère's theoretical approach and for the development of his experimental techniques and his standing as researcher. The three weeks during which Ampère took his first steps into this new field were of decisive significance, a significance that Ampère consciously chose not to make public. This chapter is devoted to a scrutiny of those steps.

One of the more surprising results of this investigation is the recognition that Ampère's earliest work was by no means aimed at any single, well-defined goal. Contrary to received opinion, there was no unified theoretical framework, nor even a well-ordered conceptual scheme guiding his efforts. In their place we find a multitude of different epistemic goals and ways of working, often pursued in parallel. Of the seven distinct thematic strands I have identified in this complex, one research goal in particular, which he called the "reduction to general facts," stands out. In an effort to present a wealth of material as manageably as possible, I begin by discussing each thematic strand separately, one in each of the next seven sections. Finally, I turn to investigating their mutual overlap and eventual resolution, considering the manner in which Ampère would soon go public with a new direction for his research.

The Methodology of Reconstructing a Scientific Research Episode

Since the reconstruction of this period in Ampère's research demanded special procedures, a few remarks on my methods are required. The main cause of our persistent ignorance of Ampère's very first efforts can be traced to the extraordinarily challenging state of the sources. The few statements heretofore recorded about this period, by Ampère and others, all date to a later time, and together they present an incoherent, even contradictory picture. L. Pearce Williams's problematic reconstruction relies on such sources, "combined with a sensitive attempt to resolve their inconsistencies" (Hofmann 1995, 235–36).[1] As this chapter shows, the sense of Ampère's earliest work that Williams provides needs to be revised in several crucial respects.

Ampère kept no laboratory log, and we have no direct evidence regarding the details of his research. In addition, for this period, unlike that which followed September 25, the manuscripts of his academy lectures (of September 18 and 25) no longer exist in any cohesive form. Ampère reused portions of them, greatly modified, in later publications. No further documentary evidence was thought to survive. It is thus perfectly understandable for previous historical studies of Ampère's electrodynamics to have begun with the academy session of September 25, thus relying in large measure on Ampère's own but—as we know now—much later summary, together with his published "conclusions." But they teach us nothing about the origins of a conceptual system that, by this time, was clearly already well developed. The familiar sources yield a very incomplete portrait, one in which crucial aspects remain unexplained or are simply missing.

Only a fresh investigation, availing itself of new methods, has yielded a new perspective. Laborious puzzle-solving and detective work have allowed me to locate, collate, and reassemble the individual components of the manuscripts of Ampère's first two lectures, producing a *complete* reconstruction of the original versions of both texts.[2] This serves as a very special kind of source. Though it amounts to the reconstruction of a document that nowhere physically exists in this form, I treat it just like any complete archival text. Such a stance, of course, rests greatly on the reliability of my reconstruction. In appendix B, I present the arguments in support of its reliability in greater detail. In addition to employing this very specific type of source, I have also found a few other scattered but very important documents among Ampère's papers, most importantly manuscript AS 205(b).[3] Such items provide a few sketchy hints of his private "agenda" during this period. With these materials in hand, instead of later interpolations, we have for the first time significant documentation from the period itself. In appendix A, section A1, I provide an overview of the now greatly improved state of available sources and present several important documents.

But these new sources are all, in various ways, incomplete. Manuscript AS 205(b) was clearly intended for Ampère's own use and is written in highly abbreviated, sometimes idiosyncratic language. Given the short amount of time and the feverish atmosphere in which Ampère pursued his first research, he presumably felt no need for any further notes that were more detailed or extensive than a listing of a few keywords. In addition, the order of entries in this document can be taken only as representing the chronology of their inscription within certain limits. Reliable claims can be made about the ordering of whole blocks of text but not about the internal structure of the individual blocks. The lecture manuscripts have a very different character. Ampère composed them for public presentation and thus formulated his results in much greater detail. It is possible to discern the state of his experiments and reflections, though Ampère by no means touches on all of his activities. With respect to their chronology, the lecture manuscripts provide only a few isolated points of reference. The limitations of reconstruction and the strict constraints they impose must be taken into account. In the following sections I present the results of my reconstruction in the form of a coherent historical narrative, only occasionally pointing toward the intricacies of the reconstruction process itself.

Exploring Unknown Territory, or the Reduction to "General Facts"

Separating the Facts with the Astatic Needle

Ampère's first efforts were geared toward a closer inspection of Ørsted's effects as such. They were presumably no less puzzling to him than to most of his contemporaries. The needle's tendency to align itself at an angle to the wire contradicted all conventional paradigms of central forces depending on distance alone, all the more so in face of the fact that the needle reversed its deflection when placed below the wire as opposed to above it, but with the distance remaining the same. Even for Ampère, who, after all, taught the received conception in his day-to-day work at the École Polytechnique, this aspect posed a serious challenge, and Ørsted's report offered very little help. To be sure, Ørsted had specified several factors on which the deflection of the compass needle depended. But he had not offered a comprehensive insight on the nature of those dependencies, nor any general rule. Ørsted had suggested conceiving of the "electrical conflict" as taking place in circles surrounding the wire, though without much elaboration. For someone like Ampère, the idea of circular action was extremely alien, and the fact that Ørsted had provided only a rudimentary account of what he meant by "electrical conflict" could not have inspired trust. The impression that there was much left to be done was shared by many among Ørsted's readers. It was not least for this reason that Ampère (1822b, 309, mispaginated as 209) would later describe his first efforts as an attempt "to complete Ørsted's investigation."

From the outset Ampère criticized Ørsted for failing to separate action produced by the wire from the familiar action of terrestrial magnetism, instead investigating the combined action of the two. But for Ampère, given his familiarity with problems of mechanics, the idea of such separation was close at hand, suggesting the first step of his work: he began with a project he called *séparer les faits* (separating the facts).[4] Because terrestrial magnetism cannot be switched off or shielded, Ampère developed a setup in which the needle was arranged so that terrestrial magnetism could no longer move it: the needle's axis of rotation was aligned with the earth's magnetic inclination. In this way he eliminated the needle's freedom of motion in response to terrestrial magnetism.

In his (later) illustration of this apparatus (fig. 3.1), the magnetic needle AB is oriented orthogonally to the direction of the magnetic clip. The wire is stretched between the movable posts HG and KI and can be positioned

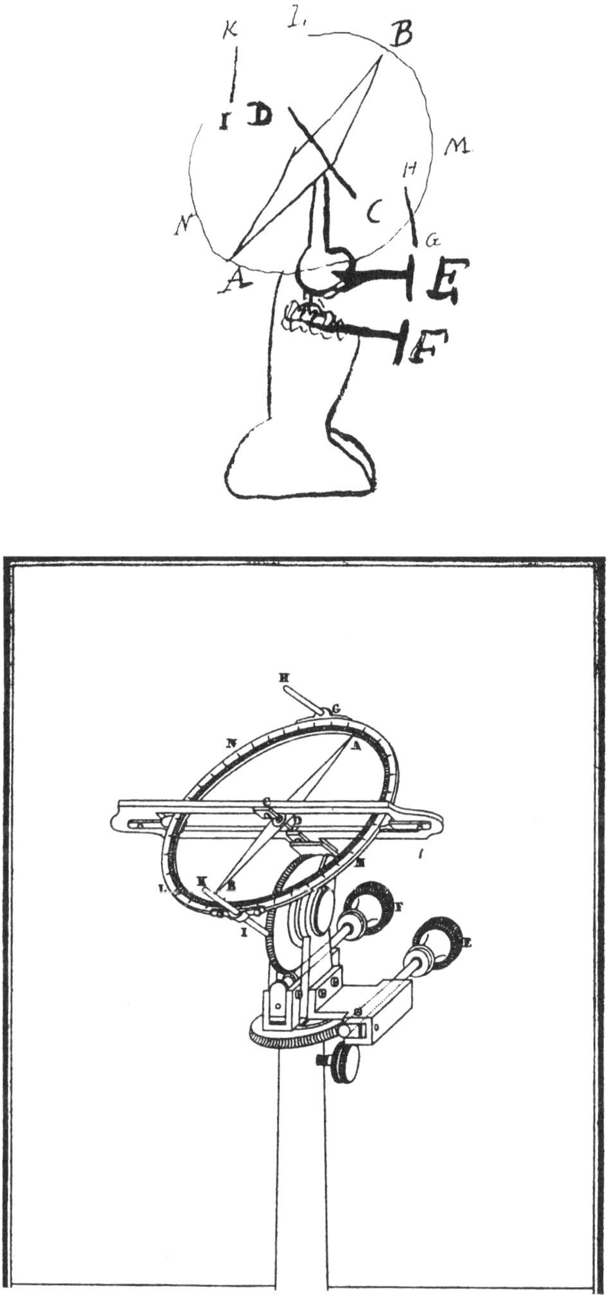

Fig. 3.1. Ampère's "astatic needle," in draft and in clean copy (AS 206bis[g], © Académie des Sciences–Institut de France; Ampère 1820j, plate 4).

arbitrarily. Ampère (1820a, 239) would compare this device with an inclination needle, suggesting that he developed it as a modification, or spatial reorientation of such a needle.[5] With this "astatic needle," the action of the wire on the needle could be investigated "without interference from terrestrial magnetism."[6] According to Ampère's report, the experiment yielded clear results: when the wire was positioned centrally, that is, over the center of the needle, it always imposed a right angle between wire and needle. This finding went well beyond Ørsted, who had reported only that the angle between needle and wire varied as a function of the strength of electrical activity (e.g., Ørsted 1820l, 260).

In the illustration, the "astatic needle" appears much more straightforward than it actually was. As Ampère's report makes clear, the illustrated device was the crowning culmination of a whole series of test versions. Presumably his first, improvised arrangements couldn't always produce clear, repeatable results. In his second lecture he went into some detail on the procedure for adjusting the needle to ensure that it was really astatic.[7] This was a necessary condition for it to serve its purpose in the investigation for which it was devised. Achieving the astatic condition took considerable trial and error. In this process, his collaboration with instrument builder Nicolas Constant Pixii, to which I return below, played an essential role. Ampère had clean drawings made of the final device, along with detailed instructions for its assembly and experimental deployment, presumably intended for distribution to interested colleagues.[8]

Exploration and an Unexpected Finding

For Ampère, establishing the right-angled deviation of the needle was only a first step; he still had to determine which of the two possible positions the north pole of the needle would assume. Moreover, with the astatic needle in hand, Ampère now had an instrument that would allow him to isolate the various factors contributing to the movement of the compass needle, investigate them individually, and, if possible, formulate general rules. These goals set the direction of his subsequent work. His experimental technique consisted in varying many different parameters of his apparatus. Chief among these was the relative position of wire and compass needle, followed by the strength and polarity of the battery.[9] In the course of these efforts, Ampère made an unexpected discovery: not only the wire connecting the battery's terminals but the battery itself was magnetically active. In all probability, this was a genuinely fortuitous find, and it is easy to envision scenarios in which it might have occurred in the course of his experimentation. There is, however, no unequivocal textual evidence on this, though it is significant that Ampère would never place this effect within an explanatory context.

The magnetic activity of the battery constituted an entirely new kind of effect. Its wholly unexpected character is made clear by the fact that in all of his experiments, Ørsted studied only the activity of the wire, never taking the battery itself into account, all without anyone's perceiving this as an oversight. Ever since the early days of galvanism it had been clear that, to produce galvanic effects, the right materials always had to be arranged in a closed circuit. In Johann Wilhelm Ritter's accounts, this point had been highlighted emphatically. But even in 1820, all of the reflections on or theories about the causes and nature of galvanism treated the underlying processes, regardless of whether they involved animal tissue, contact between metals, or chemical activity, in very different terms from those applied to the conduction or discharge in an "outer" circuit. Discussions of ordinary electricity were quite analogous in that they treated the operation of the generator completely separately from processes taking place in the discharging circuit—though here, too, the requirement of a closed circuit was taken as self-evident. We also should not take the term "circuit" as implying that, in a discharge, the electrical flow passed equally through both the generator and the outer circuit. The electrostatic theory of the pile, which enjoyed broad acceptance in France, really applied only to the open pile and thus failed to sanction any analogy between the pile itself and the wire connecting the terminals of a closed pile. So the idea of investigating potential magnetic activity on the part of the battery itself, though obvious in retrospect, was well beyond the horizon, even for Ørsted. In particular, we should recall that there was as yet no inkling of a circuit of *current,* encompassing both wire and battery.

The "Alignment Effect" as the First Rule

Discovering the magnetic activity of the battery forced Ampère to extend and ramify his investigation; in searching for a rule governing the alignment of the needle, he now also had to take the battery into account. He expended a great deal of effort developing this rule, part of which involved crafting new concepts:

> *The alignment effect:* When a magnet and a galvanic conductor act on each other while one of them is fixed, and the other can only rotate in a plane perpendicular to the shortest line between the conductor and the axis of the magnet, the mobile one will tend to move in such a way that the directions of the conductor and the axis of the magnet form a right angle, and that the pole of the magnet that habitually points north points to the left of what has been called the galvanic current, and the south pole points to the right.[10]

This rule has an "if-then" form that is typical of Ampère's approach. It attempts to express the conditions under which electromagnetic alignment occurs as precisely and comprehensively as possible. There is not the slightest hint of a more narrow or theoretical reflection on the underlying nature of electricity and magnetism.

As to Ampère's use of the expression "galvanic current" (*courant galvanique*), there is a very specific background. As Ørsted had observed, the electromagnetic alignment of the needle depends crucially on the polarity of the battery; inverting the polarity changes the effect. So as to express this dependence more concisely, without always returning to circumlocutions involving the "connecting wire" of the battery, and the poles of the battery themselves, Ampère availed himself of the notion, already in use by others, of a directional current between the poles. In marked contrast with his contemporaries, however, he wanted this understood as nothing but a convenient manner of speaking: "Finally, I am using the phrase 'galvanic current' only as a convenient term with which to define the directions at issue. This expression may be used as an expedient, so long as one does not admit the real existence of any such current."[11]

Seeking a frame of reference for establishing the direction of current, Ampère looked to the battery poles, as was customary. But he did not characterize them in the usual way by referring to the positive or negative polarity of the terminals of an open battery but by reference to the gases produced at the submerged ends of wires connected to each terminal in the decomposition of water. The fact that hydrogen and oxygen were regularly generated at the ends of wires connected to specific poles was experimentally well established. Exploiting this fact, in reverse, to designate polarity had the pragmatic advantage of enabling the swift determination of the polarity of a closed battery, without breaking the circuit. Ampère understood his stipulation of the direction of current, from the "hydrogen" pole to the "oxygen" pole, as pure convention, consistent with earlier conventions.

In his repeated reminders of the "'instrumental" character of his concepts and terminology, Ampère differed from other researchers of his day. The concept of "electrical current" had been used before, by Volta (1800c, 426) and Biot (1804, 203), for example.[12] However, they had always done so by reference to a specific understanding of the microscopic processes at work in the wire. Much the same can be said of Ørsted's notion of electrical conflict, which, despite its instrumental character at the beginning of his circular, had by the end assumed a realistically construed physical significance. In what was presumably a self-conscious attempt to distinguish his approach, Ampère initially sought to avoid committing himself. He was more aware than others of the potentially far-reaching theoretical implications of terminological choices.[13] This explains his consistent reference to "galvanic," as opposed to "electrical" current. Nor did he adopt Ørsted's term "electromagnetic," preferring to call the new effects "galvanomagnetic," for, after all, the essential identity of electrical and galvanic phenomena was still not universally accepted. Against the backdrop of such caution, the terminological changes he would later impose appear all the more dramatic.

The instrumental character of Ampère's approach to terminology may also be observed in his use of other descriptive vocabulary. In order to express the complexities of the needle's deflection in all its dependence on multiple factors, he had introduced the directions "right" and "left" of the current and explicated them in detail: "So as to make more precise the meaning that I attribute to the words 'left' and 'right' relative to the galvanic current, we must make a supposition analogous to that by means of which one defines left and right relative to [a] river. . . . [We] imagine a man immersed in the galvanic current with it flowing from his feet toward his head, and with his face turned toward the needle. His right hand will be on the side that I call the right of the current, and his left hand on the current's left."[14]

The sole function of this complex construction was to provide an unequivocal and manageable formulation of the directions relevant to electromagnetic activity. Indeed, for the challenge that arose here—to describe the relationship between three spatial directions (current, needle, and its movement)—there were simply no conceptual resources in the geometry of Ampère's time, and Ampère had to create new geometrical concepts. But again he emphasized that there was no physical significance to be deduced from them.

Ampère appended a third clarification to his rule, this one pertaining to the relative directions of current in wire and battery. His experimental results had led him to distinguish cases in a way that seems odd to contemporary readers: "So as to maintain all of the generality of which this statement is susceptible, it is necessary to distinguish between two kinds of galvanic conductors. (1) The pile itself, within which what we call the galvanic current travels from the pole that produces hydrogen in the decomposition of water to the pole that releases oxygen. (2) The metallic wire connecting the two terminals of the pile, in which we must now consider the current as moving in the opposite direction, from the oxygen-emitting pole to the pole that develops hydrogen."[15]

This stipulation accounts for the experimental finding that the magnetic activity of the battery operates in the opposite direction from that of the wire, when the direction of the current in both parts of the circuit is understood relative to the two poles of the battery. If one wishes to capture both in a single rule, formulated in terms of the concept of galvanic current, the current within the battery must be ascribed a direction opposite to that in the wire. Ampère's construction strikes us as odd only because we are accustomed to the concept of a continuous electrical circuit, of which both wire and battery are part. Ampère and his contemporaries lacked any such notion. As Ampère explicitly stressed, by distinguishing two cases he could establish the uniformity of his rule for all instances of the alignment affect. Once more, his concepts prove resistant to any physical interpretation. But that was not what he was after. He was concerned with the search for the most general possible phenomenological rules, and all of his terminological choices were motivated by that concern.[16]

The experiments that led him to these stipulations were easy to perform but difficult to grasp. Ampère's descriptions of directional relationships were complicated and required many steps. Any attempt to repeat such experiments quickly reveals how complex the relevant spatial orientations might easily become, causing errors to creep in. It was presumably with this in mind that Ampère, wanting to be really sure of his approach, undertook further experiments. In one such experiment, he placed a needle above the battery while leaving a second one in its customary position by the wire. Such arrangements, as he explained in detail, made the difference between the two cases abundantly clear and compelling.[17]

Ampère was also clearly aware of an additional problem. The need to speak of current flowing within the battery, and to stipulate the direction of flow, arose out of *electromagnetic* experiments. But his previous stipulations regarding direction of flow had all been expressed in *chemical* terms, by recourse to the poles of the battery associated with the segregation of hydrogen and oxygen. With respect to the current flowing through the wire, the two conventions could be easily accommodated. But when it came to including the current flowing through the battery itself, the choice between them was no longer a matter of convenience only; it had to be empirically established whether, for the battery as for the wire, the two conventions for establishing direction of flow really yielded the same result. Accordingly, Ampère designed experiments and placed them on his agenda. By inserting "dissolved potash between the copper and zinc" of adjacent cells in his battery, he could observe the decomposition products within the battery itself.[18] In his second academy lecture he alluded to the success of these experiments, without further elaboration. This result would prove very important to his later decision to introduce a uniform notion of an electrical circuit.

In introducing the concepts of current direction and the "right" and "left" of current, Ampère went well beyond Ørsted, who had always had to specify which end of the wire was connected to which pole of the battery. Ørsted's spatial frame of reference was defined by the four compass points and the vertical direction, as illustrated by formulations such as "the upper end received negative electricity" or "the western end received positive electricity." Thus, describing experiments required specifying the orientation of the wire with respect to the compass points, and other aspects of the geometry of the setup and any observed motions also had to be described relative to the same points of reference. Consequently, Ørsted's accounts of his experiments were necessarily long, convoluted, and difficult to grasp. By contrast, Ampère sought to develop a simpler, clearer language, liberating him from the fixed spatial frame of the compass points and allowing him to formulate his experimental results for arbitrary alignments of the wire. The new conceptual framework and terminology were framed to grasp a variety of specific experimental arrangements, eschewing any speculation on the nature of galvanism. That Ampère found Ørsted's formulations unsatisfactory in this respect may have reinforced that approach.

In retrospect we can see that Ampère was alone in these efforts. Although work on electromagnetism was taking place all over Europe, during this period no one else was devoting so much energy to the problem of the missing spatial concepts (Steinle 2011). Of course he had no way of making any such comparisons: though doubtless aware that others, elsewhere in Europe, were working on electromagnetism, he had no specific information on what they were doing or where. During the first phase of his work, his only source was Ørsted's report.

Varying Further Parameters

Following his success with the alignment rule, Ampère turned to other aspects of the experimental setup. Without further elaboration, by calling it "electro-magnetic" Ørsted had ascribed his new effect to the activity of electricity and magnetism. For his part, Ampère asked whether this description was really justified—whether the effect was genuinely *magnetic* and really evoked by *electricity*. His experiments were geared toward answering these two questions.

To begin with, he took note of his intention to build "a compass needle of wood, paper, manganese, tin, or copper and zinc."[19] All of these materials are nonmagnetic and some of them also nonconductive. The expression "compass needle" (*aiguille de boussole*) makes it clear that Ampère intended these new needles to be mobile, so as to investigate whether they would be influenced by the wire. By no means was it self-evident to him that the wire would only affect magnetic materials. A negative result in this experiment would increase the likelihood that the effect was really confined to magnets and not something more widespread. Ørsted (e.g., 1820e, 279) had also used brass, glass, and shellac needles.[20] However, Ampère was apparently unsatisfied with them and greatly expanded the range of materials. We do not know whether he actually carried out these experiments; what we do know is that he was henceforth convinced of the genuinely magnetic character of the effect. Throughout all of his subsequent terminological changes, he would continue to use the term "magnet" without further modification.

The question as to whether the effect was really being evoked by electricity posed an analogous task. Ørsted had described the activity within or outside the wire connecting the poles of a closed pile as "electrical conflict," thus bringing it under the umbrella of familiar (i.e., electrostatic) electrical phenomena. But it was not self-evident that it belonged there, for even in 1820 there was still no general consensus on the relationship between electricity and galvanism. There was agreement that the "open" pile produced electrical effects in the usual sense. The situation was more complicated with respect to the closed pile. The objection continued to be raised that all electrical effects in the strict sense—the charging of electrometers, the production of spark-gap discharges, and so on—disappeared immediately once the pile was closed. In their place, an entirely different set of phenomena appeared: the heating of the connection wire and the chemical decomposition of water or salt solutions. To be sure, one might *postulate*, as indeed many did, that the new phenomena were heretofore unknown aspects of electricity or were effects of the same force that under other circumstances produced the usual electrical phenomena. Support for this postulate came from the fact that the new phenomena were produced by the same apparatus known to cause strictly electrical phenomena. In addition, some of the new phenomena could be produced with conventional sources of electricity, such as Leyden jars or electrostatic generators.[21] But despite numerous experiments, in 1820 the question had not yet been definitively closed.

Ampère proceeded cautiously, though as we know from his draft *mémoire* of 1801, he was certainly inclined to trace disparate physical phenomena to the fewest possible underlying agencies.[22] He took great pains investigating this question experimentally, going so far as to seek laboratory access at another institution, the École de Médecine, because electrostatic generators and Leyden jars of sufficient size to perform the experiment to his satisfaction were apparently not to be had anywhere else. His experiments sought to show whether the alignment of the compass needle provoked by the connecting wire of a voltaic pile could also be achieved by the discharge of conventional electricity from a Leyden jar. When his first attempts failed to produce any detectable effect, he explained the negative result as a consequence of the duration of the discharge being too brief to cause the needle to move. He apparently returned to this experiment later, with the thought of discharging the Leyden jar using "a half conductor" so as to "prolong the galvanic current."[23] By a "half conductor" he presumably meant simply a poor conductor, such as a wire made of a weakly conductive metal or a nonmetallic material, like the traditional electrician's packthread. Once again, we do not know whether he carried out these experiments. His failure ever to report on them may have to do with the fact that even with his modified setup he was unable to set the compass needle in motion.

On the whole, the experiments involving electrostatic generators were inconclusive. Ampère continued to exercise caution in his use of terminology, referring only to "galvanic" currents. As I discuss later on, a few weeks later he would decide to speak of "electrical" currents, a decision that can be traced in part to Arago's success in demonstrating a magnetic effect using conventional electricity. However, this was an instance of magnetization, not an alignment effect of the sort Ampère had sought. By this time, in any case, Ampère's theoretical reflections presumably played a much more prominent role than any new experimental findings.

Simplifying the Regularity with the Concept of the Electrical Circuit

Ampère was probably not satisfied with the complex contours of the rule governing the alignment effect. At any rate, he continued to look for ways to formulate it with greater simplicity and clarity. A few days after he formulated the two-case version of the rule cited above, he was successful. Once again the simplification succeeded by introducing a new concept:

> The two cases [i.e., the direction of current in the wire and its opposing direction in the battery] may be captured in a single definition by understanding galvanic current as the direction in which hydrogen and the bases of salts are transported under the influence of the whole pile, and by imagining this as forming one single circuit with the conducting wire. If one interrupts that circuit at any point by water or saline solution, these will be decomposed. Finally, all that I shall say in this paragraph and the two that follow in no way presupposes that there is any real current in that direction; and one must consider the phrase I have coined, "galvanic current," only as a convenient and familiar way to indicate that direction.[24]

His new insight consisted in recognizing that a uniform formulation of the rule was possible when current was construed differently: instead of giving its direction in the usual way, by reference to the poles of the battery, the current was understood as propagating through both battery and conductor, in a continuous circle, or "circuit." The concept so familiar and self-evident to us, that of a single electrical circuit, incorporating both battery *and* conductor, appears here for the very first time.

The term "circuit" itself was not new, of course: in both conventional and voltaic electricity, the need for a closed "circle," called a *circuit voltaïque,* was well known (e.g., Pictet 1821, 281). This term was understood relative to an experimental arrangement in which a closed circle of conductive material must be maintained in order to produce certain effects. But this notion of circle, or "circuit," never had to confront the aforementioned difference between the connecting wire and the electrical source. It was this factor that motivated Ampère to introduce a new sense of "circuit," expanding the meaning of the term well beyond previous usage.

The background to this move was his attempt to devise a uniform formulation of the directionality of electromagnetic activity *for both battery and wire.* This, after all, had been his reason for introducing the notion of

Fig. 3.2. Schematic representation of Ampère's conceptual transition from battery poles to direction of current.

current in the first place. By changing his frame of reference from the poles of the battery to the direction of current in a closed circuit (fig. 3.2), he made the two-case definition superfluous. With this shift in perspective, it was no longer necessary to refer to the poles at all; instead, the idea of a closed circuit with an intrinsic direction of circulation came to the fore. This shift had its consequences for Ampère's subsequent terminology, leading him to speak increasingly of the *direction* of the current, rather than the battery poles.

In the passage cited above, Ampère explicitly alluded to experiments in which decomposition cells filled with water or some salt solution were inserted at arbitrary points in the circuit. Such points include locations *within* the battery. Toward that end, two plates of the battery that would ordinarily be in contact were separated (e.g., by placing them in separate troughs) and a decomposition cell wired between them. For Ampère, the experimental finding that in such cells the transport of hydrogen continued to follow the same direction as in the "outer" circuit was an important validation of his new concept.[25]

Like the concept of current itself, the new concept of circuit was used in an entirely instrumental way; in the quoted text above, Ampère called it a "convenient and familiar way" of speaking and took pains to disclaim any implication of physical reality. Current was defined as a *direction*, not as a physical flow or anything of the sort. It is worth highlighting that the concept of a circuit was introduced, independently of any micro-scale theory, merely as an improved device with which to formulate a rule. Not much later Ampère would indeed develop a micro-scale theory of the electrical circuit, but in its origins the term arose out of the representational dynamics of his approach to the new phenomena. In this respect, his early work differed dramatically from what would follow.[26] At this point, though, Ampère could have had no inkling that the new concept would profoundly shape all of his future work. It strikes me as remarkable that so fundamental a notion as the electrical circuit could arise in the context of the search for the simplest and most general possible formulation of an empirical rule, a context far removed from any micro-scale theory of electricity, magnetism, or the like. This case clearly illustrates the extent to which scientific innovations can arise at a very great remove from theory. I return to this issue in chapter 7.

The Rule of Attraction and Repulsion

Shortly after completing his formulation of the alignment effect, Ampère introduced a second fundamental rule, this one governing the attraction between wire and compass needle:

> *The attraction-or-repulsion effect:* This second general result consists of the fact that a galvanic current and a magnet whose axis is at a right angle with the direction of the current, conceived . . . in accordance with the preceding definitions, attract each other when the north pole is to the left of the current, i.e., when their relative position is that which the conductor and the wire tend to assume by virtue of their mutual interaction, and repel each other when the north pole of the magnet is to the right of the current, i.e., when the conductor and the magnet are (maintained) in the position opposing that which they tend mutually to assume.[27]

A later representation of the experimental setup (fig. 3.3) compellingly illustrates this principle. Magnet MN is hung vertically, suspended symmetrically opposite the horizontally stretched wire GH such that shortest-distance line between them exactly intersects the middle of the magnet. The magnet is thus either in precisely the position assumed in accordance with the alignment effect or, with the polarity reversed, in the antiparallel position. As Ampère reported, the experiment unequivocally showed that when the wire is connected to the battery, the magnet is either attracted or repelled, depending on the polarity of battery and magnet.

It is not possible to ascertain with any confidence what led Ampère to perform these experiments. His own retrospective sketch of the course of his work reads, "I will say only that after having achieved only the first general result described in Mr. Ørsted's note, I deduced the explanation whose basis I shall expound in the third portion of this mémoire; and that

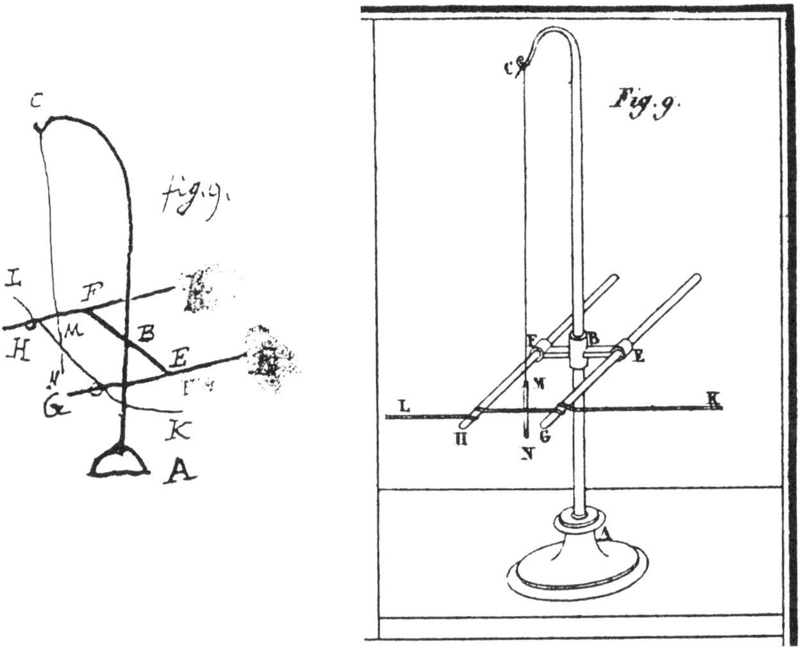

Fig. 3.3. Ampère's apparatus for the attraction-or-repulsion effect, in draft form and clean copy (AS 206bis[g], © Académie des Sciences–Institut de France; Ampère 1820j, plate 5).

this explanation led me to the second general result, and suggested an experiment for its confirmation that succeeded completely."[28]

The meaning of the passage is not immediately clear. The "explanation" in question is the idea of terrestrial circuits, to which I return below. This account presupposes his very early arrival at the concept of an electrical circuit, for the hypothesis of terrestrial circuits rests on this notion. What is less clear is how this hypothesis might have led him to the experiments just described. Conceivably, the electrical circuit hypothesis might have made electromagnetic phenomena appear closer to the more familiar phenomena of ordinary and terrestrial magnetism, suggesting possible analogies. With respect to ordinary magnetism, it was traditional to distinguish two different effects. First, two whole magnets will align themselves to each other and, in the aligned position, mutually attract. Second, if one of the two is positioned antiparallel to the aligned position, the two repel each other. Ampère's "alignment effect" (*action directrice*) was the electromagnetic analog of the first effect. The analogy to the second would be observed in an experimental arrangement of exactly the sort depicted in figure 3.3. In that setup the magnet's position is precisely that assumed in accordance with the alignment effect. But against this reconstruction we must weigh the fact that Ampère himself never mentioned any such analogy in connection with the attraction effect, not even when asserting the positive outcome of his experiment. Had the experiment served primarily as a test of the analogy, he could have claimed a decisive success—but he did not. Instead, he presented it in the entirely different context of "reduction to general facts." He may even have devised this experiment in that context in the first place, which would make the passage cited above a later interpolation based on imprecise memory; such things are not an uncommon feature of Ampère's papers. The question as to the origin of the attraction effect simply cannot be resolved with the available materials.

"General Facts" and the Explanation of Electromagnetic Phenomena

In his first academy lecture (September 18, 1820), Ampère presented the attraction effect in a very specific context. Along with the alignment effect, he described it as a "general fact" (*fait général,* or sometimes *résultat général*). Both rules played a central role in his attempt to make Ørsted's discoveries phenomenologically intelligible, that is, to understand them as special cases of general laws. In this effort he was greatly encouraged by Arago, who was also working on electromagnetism and with whom he had been interacting closely since the beginning of his electromagnetic studies. Ampère demonstrated the attraction effect at the Paris observatory, in Arago's laboratory, the day before his first academy lecture, but he had already told Arago about it: "When I informed Mr. Arago about that result, he observed, correctly, that this attraction and repulsion between a magnet and a galvanic conductor placed at right angles in the direction toward which they tend, and opposite the direction in which they repel, could make sense of the results that the author of the discovery [Ørsted] reported with respect to the case in which a horizontally mounted magnetic needle was approached with a vertically mounted galvanic conductor."[29]

Arago evidently saw in Ampère's results a potential explanation for Ørsted's findings not to appear in the original report. It is instructive to examine this more closely. In these experiments Ørsted had approached a horizontally mounted compass needle (see fig. 2.2) from various directions with a vertically mounted wire connected to a battery. In each trial the magnetic needle was deflected from its rest position (in north-south alignment) in a direction that depended on the polarity of the battery and the relative position of wire and needle—and most especially on whether the wire was closer to the needle's center or one of its ends. When the wire approached a point between pole and center, that limb of the needle would deviate to the west; when it approached the pole, it would deviate to the east. Ørsted did not explain whether in these trials the wire itself was to the west or east of the needle. Ampère and Arago found these results particularly in need of explanation. Though Ampère could not know this, researchers elsewhere in Europe, including Pfaff in Kiel and Pictet in Geneva, were also taking pains with this problem. Pictet in particular stressed the difficulty of devising a rule for these effects.[30]

In the remark cited above, Arago was said to have observed that although the alignment effect by itself was insufficient to enable him to make sense of the puzzling finding, when taken together with the recently discovered attraction-and-repulsion effect it might well succeed. What's more, the attraction effect might be derived from Ørsted's findings. Thus motivated, Ampère set about explaining the effects in question. He soon succeeded in one case, in which the vertical wire, with current flowing downward, was placed opposite a point between one pole and the center of the needle, causing the needle to deviate to the west. Ampère attributed this movement to the attraction effect: "For this movement of the needle, which as indicated [in Ørsted's report] takes place whether the conductor is either to the east or west of the needle, is in the first case an attraction, because the north pole is to the left of the current, and in the second a repulsion, because it is to the right."[31]

His idea appears to be that the specified relative position of wire and needle differs only slightly from the case described in his account of the attraction effect. Aside from the minor transformation of the geometry of the setup, the important difference between the two arrangements consisted in the needle's freedom of motion. In Ørsted's experiment it could rotate around its pivot point but not, as in Ampère's demonstration of the attraction effect, move as a whole toward or away from the wire. For a truly symmetrical configuration, Ampère's rule would predict no motion at all. However, even a tiny asymmetry, with the wire positioned ever so slightly to the right or left of the needle's center, would cause attraction and repulsion to be expressed in a rotation of the entire needle. This was precisely the behavior Ørsted had observed.

The alignment effect, on the other hand, plays no role in the explanation of this experiment; after all, the mutual orientation of wire and needle is practically what one would expect as the *outcome* of alignment, so the rule governing the alignment effect predicts no further movement. Thus, if further movement does take place, it must be accounted for by some further cause and, in particular, one that predicts attraction or repulsion. Some such reasoning presumably led Arago to his daring remark that an additional attraction effect might be deduced from Ørsted's findings.

Ampère's attempt to come to grips with Ørsted's second case proved more challenging. When (in the configuration depicted in fig. 2.2) the wire approaches the magnet not at the region between pole and center but toward the pole itself, the previously observed motion is reversed and the needle deviates not to the west but to the east. Ampère explained this as follows: "This movement only takes place because the magnetic needle adopts an orientation toward the conductor determined by the first general result, under all the conditions I included in its specification, and in particular the remark with which I concluded."[32]

His invocation of the alignment effect (his first "general result") as explanation is difficult to understand, for two reasons. First, the statement of the alignment rule presupposes an initial condition in which the needle is arranged symmetrically, that is, with its center directly opposite the wire. The configuration now under discussion is far removed from this state. But second, and even more perplexingly, the rule for the alignment effect specifies motion only within a plane perpendicular to the shortest-distance line. In this case, that plane would be oriented vertically, parallel to the wire, intersecting the needle. However, the mechanics of this setup do not allow the needle any freedom of motion in this plane. All possible and observed motion takes place perpendicular to it.

Neither Ampère's reference to a remark at the conclusion of his formulation of the first "general result" nor the passage that follows our previous citation offers any help: "The condition expressed in that remark does not appear necessarily linked to this particular determination of the direction of the needle; instead it results immediately from considerations leading me to attribute to a single cause the two general results I have just expounded."[33]

I interpret these perhaps deliberately fuzzy, somewhat cryptic sentences as pointing to Ampère's cognizance of the limitations of his attempt to explain Ørsted's result, as well as to a move to hedge claims of explanatory power. He had only partly made good on his claim to have made sense of Ørsted's findings, and the quoted formulations strike me as suggesting he was fully aware of this incompleteness of his explanatory attempt. His first academy lecture emphasized that in many cases it was difficult to keep the effects of alignment and attraction rules separate. He thus announced his intention to return to the explanation of this particular finding and also to treat any further phenomena, such as the influence of a wire on the behavior of an inclination needle or Ørsted's observations on the action of a U-shaped wire, in like manner.[34]

One critical problem with Ampère's proposed explanations, vague though they were, cannot have escaped his attention. He had accounted for Ørsted's first case, in which the wire is not far removed from the needle's center, by the attraction-and-repulsion effect alone. By contrast, for the second case, in which the wire is closer to one of the poles, he had referred only to the alignment effect. This raised the question as to the precise point at which one effect took over from or transitioned into the other. It is hard to imagine that Ampère, whose mathematical training had accustomed him to dealing with questions of continuity, would have been unaware of this one. It would have been equally clear to him that working through the question would require a lot of effort, in particular an experimental investigation of various wire positions intermediate between the two cases discussed by Ørsted.

But clearly Ampère had no desire to undertake this work. The low priority he assigned to this question is indicated by the fact that he had evidently performed no experiments of his own on such matters but had relied entirely on Ørsted's report. He spoke of it having been "indicated" (*indiqué*) that the motions of the compass needle belonged to very different cases, and he emphasized that he was simply "accepting the correctness of this [Ørsted's] observation."[35] Such language has a very different tone from that of accounts of his own experiments. His objective in this context was simply to situate these aspects of Ørsted's report in the context of his own explanatory framework, without testing their experimental accuracy or attempting any further experimental insight. For the moment, he was not inclined to expend any further effort on the matter.

Nonetheless, in his first academy lecture, Ampère appeared hopeful regarding the explanatory promise of his approach. Irrespective of the aforementioned difficulties, he saw his integration of Ørsted's two cases involving the vertically mounted wire as a success, one that demonstrated the importance of distinguishing between his two "general facts," while always referring to *both* in any explanatory context. In addition, he asserted that, taken together, the two "general facts" could explain *all* of Ørsted's findings, as well as predict new phenomena, as yet unobserved, such as the action of the wire on an inclination needle. Given the status of his previous explanations, this claim was daring and largely unsupported.[36]

Interlude on Modes of Explanation and the Roles of Experiment

Before I turn to further themes in Ampère's investigation, I would like to pause to take note of an important peculiarity of the work discussed so far. Ampère's slogan, "reduction to general facts," encompassed a specific explanatory goal and a specific explanatory procedure. The goal was that of expressing some general rules for electromagnetic action. The results of Ampère's explorations typically took the form of "if-then" rules: *if* one or another set of experimental conditions obtains, *then* a particular effect occurs or is modified in a particular way. Among such rules, those he called "general facts," or occasionally even "laws," had a special status.[37] They contained only essential conditions for the occurrence of the effect in question (e.g., alignment or attraction/repulsion of the needle), and they specified in detail the dependence of the effect on particular conditions. "General facts" presented rules or laws in "pure" form, so to speak. For both rules, Ampère also provided an instrument or experimental setup that served, in a sense, as material expression of the general law.

The fundamental character and explanatory power of the "general facts" rested on the possibility of using them to make sense (*rendre raison*) of all other electromagnetic phenomena, or as Ampère also said, to "reduce" such phenomena to the general facts. An electromagnetic effect was made intelligible by representing it as a particular instantiation of one of the "general facts" or of some combination of the two. Significantly, the general facts themselves were as much electromagnetic phenomena as are any of their instances. Their specific status, their centrality, rested on the peculiarity of containing only essential conditions for the occurrence of particular electromagnetic effects. They were core effects, reproduced in modified form in many other phenomena.

The entire explanatory model operated on a phenomenological level: the *explanans* was as much an experimentally demonstrable phenomenon as the *explanandum*. This model lent the broad field of electromagnetic phenomena its first structure, one that introduced a distinction between simple and compound effects. It is against this backdrop that we can understand Ampère's critique of Ørsted's letter for presenting very different effects in overlapping conjunction, reporting on merged experiments, and, in general, failing to provide any insight into distinct phenomena or the laws governing them.[38] For his part, Ampère sought to resolve the effects into their "pure" components. He ascribed a great deal of power to this explanatory mode, which he understood not merely as *post facto* analysis but as a tool for generating predictions on the behavior of the needle in novel, as yet unrealized conditions.

At the same time, reflections on the "nature" of electricity, galvanism, magnetism, and so on played no role in his pursuit of this explanatory goal. There was no inkling of any micro-scale theory of these phenomena. Even if Ampère had one, it is impossible to tell how it might have affected the course of his research. Indeed, such a theory would have been more hindrance than help in the quest for general phenomenological rules. To be sure, Ampère alluded to concepts from other domains, such as mechanics, as when he considered the composition of forces. But such concepts were not specific to the current field. The explanatory goal of "reduction to general facts" went hand in hand with the particular experimental style I have called "exploratory experimentation," about which I have much more to say in chapter 7 (see also Steinle 1997, 1998, 272–92). This experimental style is driven not by theoretical precepts in any specific sense but by general methodological guidelines. Exploratory experimentation is typically found in epistemic situations in which researchers seek to orient themselves for the first time in a new, wide-open field. Such was the context for Ampère's earliest work on electromagnetism.

Reciprocity

In the course of his exploratory work, Ampère also sought a "reversal" of Ørsted's effect of the sort to be expected on general mechanical grounds: if the wire causes an alignment, attraction, or repulsion on the part of the needle, then under appropriately modified mechanical conditions the opposing action should also occur: "The formulation of these results already indicates that the interaction between conductor and magnet is always reciprocal. It is that reciprocity that I set out to verify, though it seems self-evident to me. Some of the experiments I shall describe are intended to demonstrate it."[39]

Ampère wished to experimentally demonstrate the mechanical reciprocity of electromagnetic phenomena—if not of each individual phenomenon, then at least of the two "general facts." This turned out to be relatively easy in the case of the second "general fact," that is, the attraction-and-repulsion effect. By Ampère's description, the apparatus for demonstrating that effect (see fig. 3.3) could be rebuilt, with the magnetic components replaced by galvanic, and vice versa. A bar magnet would be mounted in place of the fixed, horizontally strung wire and an appropriately placed wire in place of the vertically suspended needle. Ampère called this setup an "apparatus for magneto-galvanic attraction and repulsion on a movable conductor."[40] His description does not, however, make clear whether he actually performed the experiment.

An experiment for the reversal of the astatic needle was more difficult to devise. It required that the wire be movable, raising the technical problem of mounting a galvanic conductor in such a way as to make it freely mobile while retaining best contact with its electrical source. The two goals were difficult to attain in concert, as an arrangement optimized for one would tend to undermine the other. Ampère described his first attempted solution as follows:

> For the alignment effect, the apparatus consists of two nearly semicircular wooden canals, placed on the circumference of a single circle, but separated by half an inch. The canals are filled with mercury and connected to the two ends of the pile. At the center of the circle is a pivot on which a copper needle is mounted. Each end of the needle bears a tiny platinum wheel, which dips quarter of a line [i.e., 1/32 inch] into the mercury. This galvanic needle is influenced by a magnetic bar, just as a magnetic needle is influenced by a galvanic conductor. I have called this apparatus a *mobile galvanic conductor* [*conducteur galvanique mobile*] and have performed the experiment of its alignment by a bar magnet yesterday at the Observatory.[41]

Ampère's initial experiments with this apparatus involved aligning the mobile conductor with a bar magnet, and he was evidently largely successful. No illustrations of the two devices, presumably constructed with the active participation of instrument builder Pixii, have survived.[42] Figure 3.4 offers a schematic representation of the two instruments based on Ampère's descriptions.

Encouraged by his success at reversing the alignment effect, Ampère immediately went further, attempting to achieve an analogous result by means of *terrestrial* magnetism. It's not clear exactly what moved him to take this step, which goes beyond the context of his reflections on reciprocity, but presumably his success with reciprocity encouraged him: the alignment of a wire by a magnet must have suggested the question as to

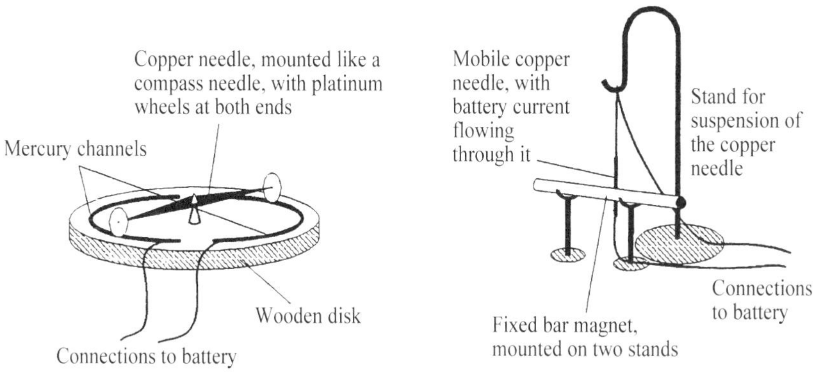

Fig. 3.4. Schematic representation of Ampère's "reciprocal" devices.

whether the classical case of magnetic alignment—alignment by terrestrial magnetism—could also be demonstrated with an electrical conductor. The expected outcome of this experiment was clear: "I hope to explore tomorrow whether the magnetism of the terrestrial globe is sufficient to align the conductor, and consequently to coax it into a position approximating perpendicular to the plain of the magnetic meridian."[43] Despite some effort, these experiments remained unsuccessful. Ampère attributed their failure to the insufficient mobility of the device, which he sought to improve, in particular by adjusting the mounting of the two platinum wheels.[44] The experimental techniques he developed to address this problem, again in close collaboration with Pixii, would prove decisive in the future evolution of electrodynamics. The famous "table d'Ampère" at the Collège de France is dominated by mobile yet current-bearing wires, central components in all of his experiments.[45] Their origins may be traced to these experimental tests of reciprocity.

To give background to these experiments, it is worth noting that Ampère was not the only one studying the mechanical reversibility of electromagnetic phenomena. Halle professor Johann Salomo Christoph Schweigger also predicted a reciprocal effect, in accordance "with mechanical laws," but he had little success with his experiments. He attempted to solve the inevitable problem of combining mobility with sufficiently solid electrical contact by using wires submerged in mercury, but he was unable to attain sufficient mobility. He also tried mounting both wire and compass needle in mobile configurations (Schweigger 1821c, 5). For a researcher like Ampère, schooled in mechanical thinking, the disadvantages of such an arrangement would have been immediately clear. By contrast, Ørsted was more successful. In his second circular, dispatched in October, he described successful experiments for both reciprocity and alignment to terrestrial magnetism. In these attempts, as noted in chapter 2, he suspended an entire galvanic battery from a thread, leaving it extremely mobile (Ørsted 1820k). In London, Humphry Davy also undertook experiments on reciprocity. He was able to demonstrate the effect using an entirely different arrangement, one consisting of a mobile wire resting atop two knife blades, each connected to one battery terminal (Davy 1821d, 17).[46]

Ampère would learn of all these efforts only much later. The fact that, despite great differences in theoretical context and local conditions, such similar questions were pursued in four different places points both toward shared mechanical convictions and toward the peculiarities of what I have been calling the "exploratory" way of working. In Ampère's case, the turn to terrestrial magnetism would have particularly far-reaching consequences. It presumably inspired him to think more carefully about a possible analogy between terrestrial magnetism and the new galvanomagnetic effect, a line of reasoning that would ultimately lead him to his hypothesis on terrestrial circuits. Given the significance of such reflections, it is all the more remarkable that Ampère hardly ever mentioned his studies of reciprocal effects in his published work. I return to this curious fact later.

The Galvanometer

Ampère first exploited his discovery of the magnetic activity of the battery in the context of a closer investigation of the alignment effect. But soon an entirely new possibility would occur to him, that of measuring the battery's activity and even making crude quantitative claims about its intensity. When he told Arago of his idea, Arago alerted him to similar experiments already under way in Geneva. In the course of the replication there of Ørsted's experiments, Charles-Gaspard de la Rive had proposed a "galvanometer" (*galvanomètre*). Witnessing these experiments, along with Arago, was the naturalist Alexandre Marcet. He described the instrument in greater detail.[47] In essence, a platinum wire was strung above a magnetic needle, which rotated above a compass rose. The wire could be made to glow strongly or weakly, depending on the strength of the battery to which it was connected. The observers noted that the deflection of the needle increased as a function of the intensity of the wire's heating. The combination of thermal with magnetic measurement—an old principle of measurement together with a new one—lent certain validation to the new principle. When Ampère learned of these details, it became clear to him that he could not claim priority either for the name of the new instrument or for its operating principle, though he had developed both independently.[48] But while de la Rive had envisioned only a very narrow range of applications, not concerning himself with the direction of current or any magnetic activity on the part of the battery, Ampère's ideas were much more ambitious:

> I mention this here only to suggest that physicists who perform experiments on the galvanic pile use a magnetic needle, so as to tell from its position at any moment whether galvanic current has been established, and what its intensity is—just as one uses an electroscope on an electrical machine [i.e., an electrostatic generator]. This little instrument, which I believe ought, by analogy, to be named *galvanoscope,* need not differ from an ordinary compass except in the use to which it is put. By making it very mobile and light, one might render this instrument capable of detecting very weak galvanic currents, for which purpose it would serve as the sole means available.[49]

Ampère could legitimately claim this proposal as a significant innovation. As I noted in chapter 1, the lack of any instrument for measuring the "strength" of voltaic piles had increasingly become a cause for complaint. Even the most sensitive electroscope could determine only the condition of an open pile; when the pile was closed, it immediately fell to zero. In its closed configuration, the activity of the pile could be measured only by the evolution of gas in a decomposition cell or by the heating of a span of wire. The strength of various batteries could thus be specified by their ability to cause a length of platinum wire of a specific gauge to glow or to easily produce particular chemical substances.[50] But such comparisons were very rough and had the decided disadvantage that the very measures employed to determine the strength of the battery would significantly alter precisely that parameter. Ampère's proposal sought to improve this situation. Accordingly, he viewed the new instrument primarily as a monitor for the condition of a battery. In a further step, he expanded its application by placing a scale beneath the needle and using it as an indicator for electrical current at any point in a circuit, whether in the battery or the wire.[51] The galvano*scope* quickly became the galvano*meter*. By "cultivating" Ørsted's effect, Ampère was able to invent a manageable instrument that permitted the measurement of electrical currents simultaneously with other (chemical or thermal) effects under investigation.

Ampère was not the only one pursuing such efforts. The need for better detection and measurement protocols was widely perceived. I have already mentioned de la Rive in Geneva and Schweigger in Halle. Schweigger (1821c, 12–15) also proposed the first "multiplier," that is, an instrument with a coil that enhanced its sensitivity. It was joined by a similar instrument, the "condenser," designed by Berlin student Johann Christian Poggendorff (1821). A little later, James Cumming (1821a) in Cambridge developed yet another electromagnetic device for the same purpose. The galvanometer, which quickly became an instrument of paramount importance, was invented more or less simultaneously and independently in five different places. This remarkable coincidence points both to the immediacy of the idea and to the great need for such a device, a need that in turn explains the high salience of potential solutions.

Interlude on Research Resources and Communication

The case of the galvanometer suggests a short digression concerning the questions of the boundary conditions for Ampère's intense research and its exceptional fruitfulness, as well as how they compare with those of other researchers. With respect to his access to laboratory resources, Ampère was in a relatively unfavorable situation. Embarking on his own investigations of electromagnetism shortly after Ampère, Biot, for example, could call on the well-equipped laboratories of the Bureau des Longitudes and the Collège de France. Ampère, by contrast, had no institutional affiliation that allowed him direct, unfettered access to a laboratory. The Académie des Sciences had neither its own laboratory nor any instruments. The physics and chemistry laboratories of the École Polytechnique were well furnished but reserved for the use of their respective professors in the preparation of their lectures.[52] As a professor of mathematics, Ampère had no guaranteed access. Since the death of the physics professor Petit in June 1820, teaching responsibilities in physics had been assumed by various substitutes, including Dulong and Arago.[53] Ampère might conceivably have gained laboratory access through Arago, through his friend Fresnel, or through César Monsuète Despretz, who rented an apartment in Ampère's house and was, like Fresnel, a tutor (*répétiteur*) at the École Polytechnique.[54] However, for specialized experiments, Ampère had to look elsewhere. When he needed conventional electrical devices, such as electrostatic generators, Leyden jars, electrometers, and the like, he could turn to the École de Médecine; other experiments he conducted at Arago's Observatoire de Paris.

But Ampère's own laboratory, the assembly of which began in great haste immediately after the announcement of Ørsted's discovery at his private house on 19 rue des Fossés-Saint-Victor, soon became far and away the most important. At considerable expense, he acquired the basic equipment required for electromagnetic experimentation: voltaic piles, wires, magnets, mounts, and so forth. But he quickly ran up against his own financial limitations. Large batteries, for example, were extremely costly. In addition, over the course of his research he developed entirely new experimental setups, impossible to achieve with standard equipment. He therefore rapidly expanded his involvement with instrument-building workshops to something far beyond the mere purchase of finished devices. His notes abound with such phrases as, "I will have . . . many devices made," "I have had an instrument built," or "a device made," all testifying to commissions for the construction of novel experimental tools. Other notes, such as "pick up Pixii's new device," tell us where the work was carried out.[55] The name of Nicolas Constant Pixii, Paris's leading instrument maker, appears frequently. We recall that Ampère, unlike Pixii, had very little experience in building instruments or performing experiments with electricity and magnetism. Accordingly, their collaboration could not have been a matter of Ampère's simply dropping off a finished design for Pixii to execute. He would have been far more likely to come to the workshop with a fairly vague notion of what he needed, arriving at a design for an actual instrument only after significant discussion in which the instrument builder's experience would have weighed heavily. However, no documents offering reliable insight into this process have survived.[56]

Research conditions were determined not only by available resources but also by opportunities for, and styles of, communication. With which other researchers did Ampère communicate in the course of his work? In this early period, as discussed earlier, international scientific communication remained sparse. Arago's oral report of events in Geneva was all that had reached Paris from the outside. Ørsted's circular would not be widely available in printed form until two weeks later.[57] Via Arago, Ampère presumably had earlier access not only to the text but to detailed information on the replication of Ørsted's experiments in Geneva, which Arago had witnessed; for a written account of them, he would also have to wait two weeks.[58] By then, the development of his own ideas was already quite advanced. Besides his collaboration with Pixii, Ampère's interactions with Arago constituted his most important opportunity for exchange. The two were good friends, and in these first weeks Arago was the only other person in Paris working on this topic. Biot was still traveling, and, even had he been in town, he and Ampère would presumably not have collaborated. From the outset, as Ampère was well aware, his relationship with Biot and the Laplacean school was one of fierce competition. His relationship with Arago was completely different. Arago's engagement with the new field had much to do with the fact that, like Ampère, he saw the challenges it posed to Laplacean physics as an opportunity to seek alternative approaches. He concerned himself with the magnetizing effects of galvanic currents, sharing his results with Ampère. Ampère's account of Arago's alerting him to broader implications of his work is representative of their collaboration, which also afforded Ampère the occasional opportunity to work in the well-appointed laboratory of Arago's Observatoire de Paris.

The Theory of Terrestrial Magnetism

Ørsted's discovery had added to the familiar alignment by terrestrial magnetism a new alignment effect, the electromagnetic one, which Ampère now had been able to express in a general rule. Ampère always saw the two effects as connected, and it was no coincidence that his device for demonstrating electromagnetic alignment, not to mention the device for showing "reciprocity," bore visible resemblance to an ordinary compass needle. Ampère quickly began asking whether the two alignment effects might be addressed in a single, uniform description. There were at least two possible ways of approaching this task. First, one might attempt to capture the newly discovered action of the wire using the familiar conceptual framework of ordinary magnetism. This was the path that Ampère's soon-to-be competitor, Biot, would follow. For his part, Ampère took a different tack, attempting to capture terrestrial magnetism within the framework of his newly devised rules for the galvano-magnetic effect. His novel concept of an electrical circuit encompassing both wire and battery had given rise to new ways of thinking and to the emergence of an unexpected dynamic. Ampère's point of departure was the recognition that the order in which two "facts" are discovered has nothing to do with the connection between them:

> We might suppose that, before knowing that a magnetized needle will always assume a north-south alignment, we had first known its prop-

Fig. 3.5. Ampère's sketch of terrestrial currents (AS 156[a], © Académie des Sciences–Institut de France).

> erty of being aligned by a galvanic current into an orientation perpendicular to the current, such that the north pole of the needle is moved to the current's left, only later to discover that the end that points to the left of the current also has the property of regularly turning to the north. To someone wishing to explain the resulting north-south orientation, wouldn't the simplest and most immediate idea be to postulate what we call a galvanic current within the earth, in such a direction that the north would be found on the left of a man lying on the earth's surface with his face turned toward the needle, and receiving that current from his feet to his head, that is, that the current would run from east to west in a direction perpendicular to the magnetic meridian?[59]

A sketch by Ampère in his archive provides the best way of visualizing these reflections and shows the man he mentioned in the text (fig. 3.5). Previous studies have missed the extent to which this new hypothesis of gigantic electrical circuits within the earth rested on Ampère's prior contemplation of the alignment effect.[60] Both the image of an observer lying in the current and the definition of the directions "right" and "left" relative to that current were adapted directly from the rule for the alignment effect. In its origins—and this strikes me as particularly worth noting—the hypothesis of circuits of terrestrial current arose entirely in the context of an attempt to expand the applicability of a rule developed for a narrow domain. This tendency toward generalization was typical of Ampère's approach during these first weeks.

Though Ampère's initial formulation of the hypothesis of terrestrial currents was very cautious, he quickly began considering whether the globe might actually harbor conditions necessary for the production of "what we call a galvanic current":

> If it exists, this current ought to be compared to that which I have demonstrated acting on the magnetic needle from within the pile, where it is directed from the hydrogen pole to the oxygen pole when a conductor is placed between them. That current would also run even if the pile formed a closed curve with a pair [of plates] like all the others making the uniting connection. For there is probably nothing in our globe that resembles a continuous, homogeneous conductor; instead, the diverse materials of which it is composed are arranged just like the elements of a voltaic pile assembled by chance, which circling back to meet itself forms a continuous belt around the whole earth.[61]

This vision of a voltaic pile curved to form a closed circle, into which stretches of metallic conductor could also be inserted, mediates between electromagnetic effects and the new hypotheses of circuits of terrestrial currents. Such currents would also have to flow both through regions that, like a battery, serve as a source of galvanic activity, as well as regions serving merely as conductors. The idea of a closed, circular battery interrupted by segments of conductor itself presupposes that of a continuous circuit of current running through both battery and conductor. To put it succinctly: no circulation without a circuit. The new hypothesis would have been literally unthinkable without the concepts previously formed in the exploratory context.

Ampère next turned to careful reflection on the generation of possible terrestrial currents, arriving at a seemingly astonishing conclusion: given all that was known about the composition of the globe and the conditions for galvanism, it would, so he claimed, be extremely unlikely if *no* galvanic effects occurred.[62] This insight opened him to new explanatory prospects for further phenomena. He asked whether the formation of heat in the depths of the earth might be caused by galvanic currents or if the observed daily and yearly fluctuations in terrestrial magnetism might result from the increasing oxidation of ore-bearing mineral formations. Ampère even attributed the brilliance of the stars to galvanic currents and finally went so far as to ascribe the earth's rotation to a kind of recoil effect, also produced by currents. Though this final speculation was confined to his private letters, he presented the others in considerable detail to the Académie des Sciences.[63] Incomplete though they might have been individually, collectively such reflections struck him as supporting his hypothesis. His initial caution soon yielded to firm conviction regarding the existence of electrical currents in the earth's interior.

The General Theory of Magnetism

This conviction about terrestrial currents in turn served as the foundation for an even broader generalization: "Now, if this [terrestrial currents] is the cause of the aligning action of the earth, this will also be the case for a magnet, whence it follows that a magnet is an assemblage of galvanic currents that run in planes perpendicular to its axis, and are directed in such a way that the south pole of the magnet, which aligns itself to the north, is to the right of these currents; for it is always to the left of a current placed outside the magnet."[64] Here Ampère formulated for the first time the hypothesis that would eventually bring him so much fame—that *all* magnetism can be traced to electrical currents within the magnetic body. Regardless of the path leading him to this proposal, it cannot have escaped his attention that such speculation was much more daring than the assumption of electrical currents in the earth's interior. Magnetic bodies had been manipulated and experimented on for centuries, without anyone's previously detecting any signs of galvanic activity. A further point posed still greater difficulties. Whereas the heterogeneous material composition of the earth could be used as argument for the existence of terrestrial currents, by analogy to the voltaic pile, no such argument could apply to homogeneous metallic bodies such as magnets. Nor did conventional magnets offer any hint of the heating effects cited in support of terrestrial currents. In short, none of the arguments Ampère had adduced for terrestrial currents appeared to support his general current hypothesis; rather, they detracted from it.

On the other hand, Ampère could point to what his hypothesis had to offer: a unified account of galvanism and magnetism or rather a reduction of the latter to the former. This prospect deeply fascinated him, and it is thus not surprising to find him engaged in extremely laborious experimental investigations. These efforts had a very specific character. In view of the aforementioned difficulties, and in sharp contrast to the case of terrestrial currents, Ampère bracketed the question of the physical causes and conditions for currents and made no attempt to directly establish the existence of the currents themselves. Instead, he sought indirect ways of making his hypothesis plausible. His overarching goal was to reproduce the imagined configurations of microscopic currents on a macroscopic scale and thereby to "imitate" or "ape" magnets.[65] Should such arrangements of conductors behave like magnets, the hypothesis would have gained empirical support. Ampère's experiments were geared toward producing quasi-magnetic analogs for every familiar magnetic phenomenon: bipolarity, the attraction and repulsion of other magnets, alignment by terrestrial magnetism, and the capacity to magnetize iron.

This effort brought about an entirely new class of experimental arrangements whose every detail was determined by the hypothesis being tested. They consisted in various configurations of circular galvanic conductors. As the analogy to an ordinary magnet suggested, Ampère began by using cylindrical coils, which he called "helixes" (*hélice*) as contrasted to planar "spirals" (*spirale*).[66] He went to some trouble with these experiments, as the numerous revisions to his lecture manuscript attest. Winding coils of conductor about cardboard or glass tubes and connecting them to a battery, he investigated whether they exerted any effect on magnets or were themselves affected by magnets or by terrestrial magnetism. The two corresponding arrangements, the "fixed galvanic helix" and the "mobile galvanic helix," represented rough analogs to the bar magnet and compass needle, respectively. In a later depiction of the "mobile galvanic helix," the analogy is completely evident (fig. 3.6). AB is the wire coil, wound continuously about glass tubes BEF and ACD. The wire ends are submerged in beakers of mercury GM and NR, which in turn are connected to the battery. The entire setup is presumably held by the experimenter at point H, whereupon magnets are brought close and any resulting movement investigated. As a freehand sketch by Ampère shows, the principle of this experimental arrangement was established right at the start. In a few steps he refined it by adjusting the supporting material, the wire gauge and quality, and the positioning of the mobile coil.[67]

The experiments conducted with these devices were largely successful in demonstrating both the influence of fixed coils on magnets and the

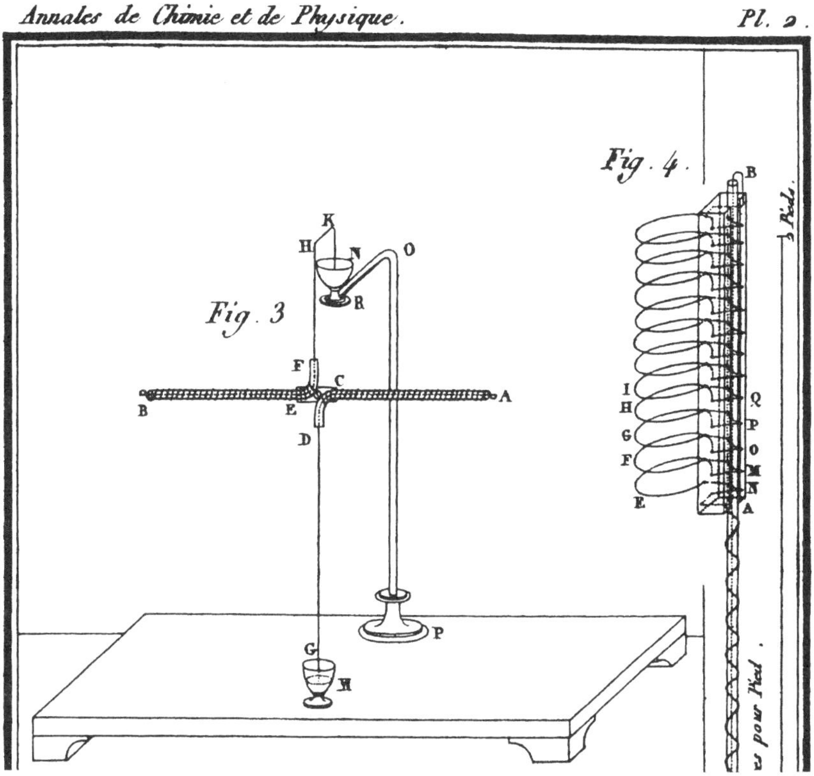

Fig. 3.6. A clean drawing of Ampère's "mobile galvanic helix" (Ampère 1820j, plate 2).

alignment of mobile coils by other magnets. They proved reliable enough for Ampère to present them to a wider audience at an evening soirée.[68] But the success was not unalloyed; it had proved extremely tough to align a mobile coil by *terrestrial* magnetism alone. The technical difficulties were truly daunting. In this case, the conflict between optimal electrical contact and mobility was exacerbated by the fact that Ampère, striving for a strict "imitation" of a magnet, had given his coils as many windings as possible. The greater weight of the resulting apparatus made it more sluggish and its positioning even trickier. But Ampère did not take this issue as grounds for questioning his hypothesis, instead blaming the failure on the inadequate sensitivity of his equipment and setting the matter aside. He would return to it a few weeks later, with better experimental techniques and a new theoretical approach.[69]

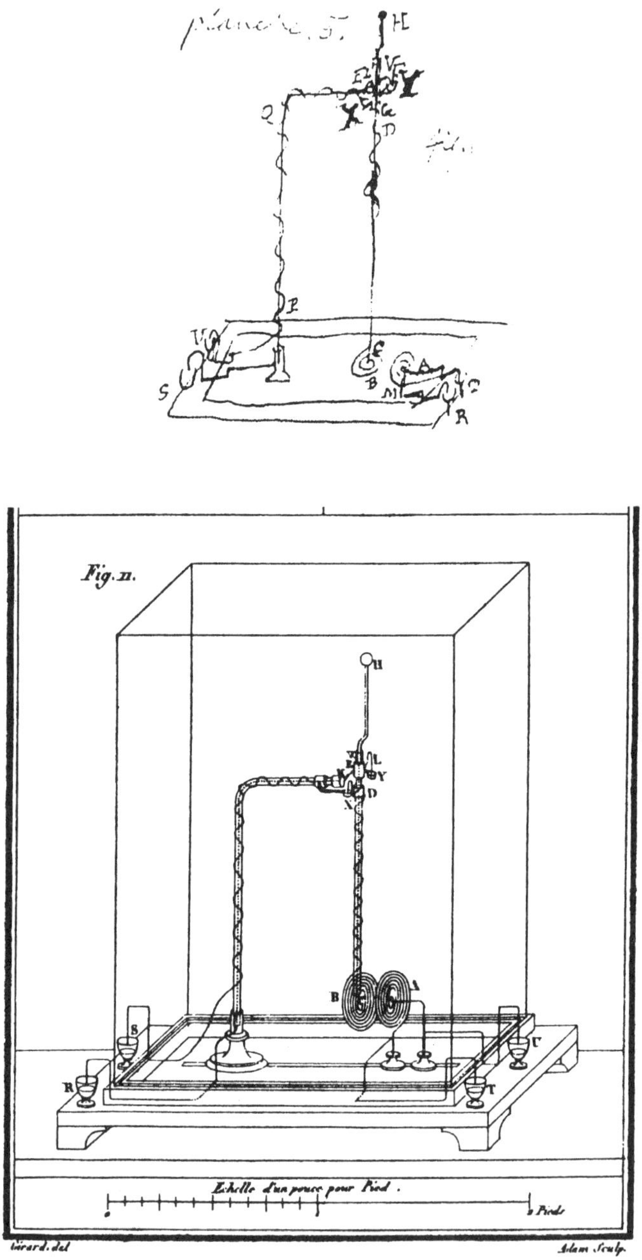

Fig. 3.7. Ampère's device for demonstrating attraction and repulsion between spirals, in rough draft and in clean drawing (AS 206bis[g], © Académie des Sciences–Institut de France; Ampère 1820j, plate 5).

The second issue on which he encountered serious experimental difficulties concerned a different problem. It had occurred to Ampère that, if his hypothesis were true, two electrical currents ought to act like magnets *on each other*, without the intervention of any actual magnet. This experiment was important enough for him to devise a new apparatus, whose principle is made clear by a rough sketch and a later clean illustration (fig. 3.7). It was in this context that he first began using flat spirals instead of cylindrical coils, thus permitting a structure with a greater number of windings to interact with a like structure in closer proximity. A and B are the two spirals, with A being mounted in fixed position while B swings on pendulum arm DB. H is a counterweight designed to balance the weight of the pendulum assembly. The entire arm HDB is supported through two mercury beakers X and Y that leave the whole assembly highly mobile and at the same time enable electric contact with a battery (not given on the figure). Thus, spiral B has full freedom of motion toward or away from A but no other degrees of freedom. By way of mercury beakers R, S and T, U the two spirals can be wired to the battery, to each other, or both.

Despite these pains, the attempt to produce the desired effect remained unsuccessful. Ampère assumed the failure to be grounded in the limited sensitivity of the device and the insufficient strength of the batteries, and he set about optimizing both parameters. He had begun by connecting each spiral to its own separate battery, in expectation of achieving a stronger effect. But lacking two batteries of sufficient strength, he contented himself with wiring them "in series" to one. When this approach failed, he set about locating the strongest battery in Paris, which he found at the workshop of the instrument builder Dumotier, who had recently completed it as a demonstration piece for the chemist Pierre-Louis Dulong.[70] Ampère was set on performing the experiment before his second lecture to the academy, already firmly scheduled for Monday, September 25. On Sunday, September 24, he received Dulong's permission to use the battery, provided he agree really to purchase it, as contrasted to temporary loan. This condition probably implied no hostility on Dulong's part; after all, voltaic piles were consumables, depreciating significantly with each use. Strikingly, Ampère was not deterred by the great expense—six hundred francs, which was more than half his monthly income.[71] He finalized the purchase on Monday morning and immediately performed his experiment right in the Dumotier workshop, finally achieving great success. The predicted effect could be observed: the two spirals attracted each other when current flowed through them in the same direction; when it flowed in opposing directions, they repelled each other.

Unsurprisingly, achieving this result after such protracted efforts left Ampère in a state of euphoria, one in which he described the observation as a "definitive proof" (*preuve définitive*) of his hypothesis. Thus bolstered, he now called it a "theory." He could not refrain from some mention of the morning's success at his afternoon address to the academy, adding a brief paragraph to the already completed manuscript. After describing his experiments on interactions between spiral and magnet, he continued, "When the magnet is replaced by a second spiral, in which the direction of current is the same, the same attractions and repulsions occur. Thus did I discover that two galvanic currents attract each other when they have the same direction, and repel each other in the opposing case."[72] He also had his equipment brought to the academy, and he then demonstrated his experiment at the end of the session.[73] As the enthusiastic letter he wrote to his son that evening indicates, any doubts he might previously have entertained regarding his general circuit hypothesis had now been banished for good: "and here we have a new theory of the magnet that actually reduces all of its phenomena to those of galvanism."[74] Ampère's ebullience may also be explained by the fact that his latest finding represented an entirely new effect, well beyond the one reported by Ørsted, and guaranteed to evoke an extraordinary response, quite independent of the hypothesis leading to the discovery.

New Perspectives in the Theory of Interaction between Galvanic Currents

It is possible to discern a fundamentally new perspective that, from the evening of September 25 on, determined much of Ampère's subsequent work. At the conclusion of the paragraph he'd appended to his lecture manuscript, he made a brief but decisive announcement: "I have undertaken to verify this [the attraction and repulsion of currents] for rectilinear currents."[75] The proposal to investigate interactions between *straight* conductors indicates the reassessment of a research interest. The interaction between galvanic currents was now a worthy object of scientific study *in its own right* and no longer only with respect to the explanation of magnetism. The two-spiral experiment offered only a first glimpse of this object, albeit one that conclusively demonstrated that currents could act on each other without magnets. But when it came to explaining *how* this interaction took place, the experiment was opaque, not to say hopelessly complicated, involving so many different currents interacting in such a geometrically complex setup. Ampère was therefore determined to find a less complex geometric arrangement of the sort afforded by rectilinear conductors, in which only *two* currents interacted. The immediacy of this effort also suggests that he saw the prospect of a mathematical treatment as being within reach.

In the following days Ampère quickly set about realizing his objectives, working with Pixii on the design and construction of an appropriate apparatus. This was no easy undertaking, for it required detecting effects orders of magnitude weaker than those produced by the spirals, whose observation had already demanded considerable effort. Nonetheless, Ampère presented his device to the academy a mere two weeks later, on October 9. It would become famous as the Ampère current balance, serving as a symbol for his entire theory.

The passages from Ampère's manuscript cited above demonstrate for the first time that Ampère deliberately looked for an attraction effect between two straight, parallel wires. Previously, questions regarding the background to Ampère's experiments could not be decided, for lack of definitive source materials. L. Pearce Williams, one of the few scholars to concern himself with these questions, has repeatedly given voice to the assumption that Ampère had discovered the interaction between rectilinear currents by accident. It was this discovery, Williams (1983, 506) claimed, that *subsequently* led him to experiment with spirals. In his recent biography, James Hofmann (1995, 238) more or less explicitly signs on to this view. But as the analysis presented here shows, the order of Ampère's investigations was precisely the opposite, and the reflections that guided them were of an entirely different tenor from those assumed by Williams.

It is this insight that reveals the shift in Ampère's perspective as a compelling example of the unplanned course and peculiar internal dynamics of experimental research. Ampère had begun his experiments with the goal of finding indirect experimental support for his hypothesis on the production of magnetism by galvanic currents. But in addition to achieving this goal, he ended with a new one, that of directly establishing the interaction between two such currents. This second goal would continue to occupy his experimental and theoretical attention for a long time, while the first receded into the background. Only over the course of experimental work could new, previously unimagined perspectives arise, in turn steering such work in an entirely new direction. In recent years, and most especially in

Hans-Jörg Rheinberger's (1997) study of microbiology and the theory of inheritance, such shifts have repeatedly been shown to be characteristic of experimental investigations (see also Rheinberger 2003). A case like the one at hand demonstrates that the course of experimentation is not fundamentally different in the physical sciences.

The Micro-Scale Theory of Galvanism and Electricity

One very brief remark: the item "a single electrical fluid" on Ampère's list of themes shows that his speculations on a microscopic fluid theory began at an early date.[76] At first he aimed at the traditional debate between one- and two-fluid theories of electricity. Even in his 1801 speculations (discussed above), he had argued for a single-fluid theory. Now this interest was deflected by a broader issue. In the course of his explanation of all magnetism by recourse to galvanic currents, it had occurred to him to attempt a corresponding reduction in the microscopic realm, that is, to abandon the idea of a separate magnetic fluid. In his "Conclusions," to be discussed below, he would formulate this thesis without explicating it in any detail. The whole issue of the fluid theory remained present, if more or less on the sidelines throughout this early period. Over the course of his later work, this tendency would become stronger as he held his speculations on an ether theory rigorously separate from his thoughts on electrodynamics, even pursuing a different publication strategy.[77]

Over the course of his reflections on the galvanic properties of the earth his attention was, however, drawn to the question of the conditions and causes of galvanism itself, especially to the relationship between galvanism and electricity. This relationship had been cast into stark relief by the difference between the effects of open and closed piles, and it was still the subject of open debate among researchers. Ampère's thoughts on the matter began with a scheme for summing the tensions of multiple voltaic sources. Going beyond Volta, he postulated galvanic effects arising between *all* manner of different materials (not only between different metals), proposing the result as a generalization of Biot's method for calculating the total tension of voltaic combinations.[78] In a further step, he attempted a unified understanding of the phenomena of open and closed piles. Though he began with the theory of magnetism, it soon became clear to him that he was dealing with a much more fundamental issue, worth treating in its own right. So he struck out his discussion of it from the draft lecture text, placing it instead in a manuscript devoted exclusively to the relationship between electricity and magnetism.[79] The special significance he attached to this text is signaled by the fact that, in a departure from his usual practice, he dated it ("20.7.bre").[80]

In all of these reflections, Ampère's new notion of an electrical circuit with a uniform direction of current through both battery and conductor played a central role. In particular, it was the aforementioned experiments on chemical decomposition at variously chosen points within the circuit that moved him to consider the micro-scale processes that might be at work.[81] One consequence was that he dropped the expressions "galvanic circuit" (*circuit galvanique*) and "voltaic circuit" (*circuit voltaïque*) in favor of "electro-motive circuit" (*circuit électromoteur*). Such a circuit was defined as "a collection of bodies conducting electricity arranged in a closed curve and in which, by virtue of the causes I shall detail in a moment, or by whatever other cause is capable of producing the same effects, an action of the sort I shall call 'electro-motive' takes place, known to us only by the effects it produces. One may therefore define it only by reference to those effects."[82] As the two essential effects of "electro-motive action," Ampère lists, first, the familiar, properly electric effects of the open pile, using the established concept of "electrical tension," and, second, the chemical effects observed when the pile is closed. By seeking to understand both kinds of effect as caused by one and the same "electro-motive action," he establishes for the first time a tight connection between the effects of open and closed piles, or between those attributable to *tension* and to *current*.[83] For the time being, the causes of this "action" were left open. However, Ampère's notion of an electrical circuit had now—and only now!—taken on an essential function in micro-scale theory.

The manuscript recording these thoughts was produced during the third week of Ampère's work on electromagnetism, that is, toward the end of the period concerning us here. It remained a fragment, not to be mentioned in either of his two academy lectures. This did not mean he considered its subject of lesser importance: indeed, he would shortly take it up more thoroughly and, most significantly, present it as a standalone text at the beginning of his first in-depth publication on electromagnetism (Ampère 1820d). In the course of these reflections, he would even take a further important step, transitioning from his initial terminology of "galvanic" current to talk of "electrical" current, and would implement that change with strict consistency, even retroactively applying it to his older manuscripts. I return to this matter in greater detail in the next chapter.

The Diversity of Ampère's Research Agenda

In the preceding sections I have canvassed the various thematic strands on which Ampère worked during his first three weeks of research. The different activities associated with each strand were conducted sometimes in sequence, sometimes in parallel. Table 3.1 provides a chronological overview.[84] In it, I assign each activity to one of the seven thematically defined cells in the right-hand column. Such assignments are inevitably rather rough, disguising subtler associations. The seven thematic strands are Ampère's own, some of them explicitly named. In a list of experimental apparatus and instrumentation intended for his own use, he divided them into four groups, evidently coinciding with his understanding of his experiments and reflections.[85] The four are as follows:

> Separate the facts
> Reciprocity
> Identification of galvanic currents
> Galvanic apparatus imitating the magnetic needle

Under the first category, Ampère listed both devices for the presentation of "general facts" and, under the second, his apparatus for demonstrating reciprocity. The third contained his galvanometer and the fourth, the "fixed" and "movable galvanic helix." The first three correspond to the first three sections of this chapter and the fourth category, to the fifth section. Though Ampère did not explicitly make the further differentiations I have undertaken in sections four, six, and seven, they are clearly discernible as separate domains of inquiry in his writings and presentations to the academy. For example, Ampère certainly concerned himself with the issue treated in my seventh section during the period in question, though he deferred its discussion from his early presentations so as to publish his insights separately. My identification of seven thematic strands is thus not far removed from the manner in which Ampère himself organized his efforts.

Two distinct tendencies may be observed in Ampère's procedures. In the first three strands—reduction to general facts, reciprocity, and the galvanometer—specific theories of electricity or magnetism played absolutely no guiding role. Instead, general methodological precepts took the lead, with the essential goal of making the new field accessible and manageable. New *concepts* were proposed and stabilized, opening the field to conceptual and experimental manipulations and at the same time imposing a preliminary structure. As the case of the concept of an electrical circuit shows, this was no mere preparatory undertaking but one that shaped all subsequent work at a fundamental level. The same holds for the experimental techniques developed in the process, as is clearly illustrated by Ampère's first apparatus with movable conductors. This device also demonstrates how closely experimental progress and conceptual progress were intertwined during this phase of research.

TABLE 3.1. Ampère's research activities, September 4–25, 1820

	Thematic strands						
Activity	GF	RE	GS	TM	GM	GC	GE
September 4: report on Ørsted's result in Paris							
Experiments on *action directrice*: the astatic needle	•						
First formulation of a rule for *action directrice*; definition of "right" and "left" of current; the "*bonhomme*"	•						
Experiments on the reciprocity of *action directrice*; unsatisfactory result with terrestrial magnetism; proposal for trial using a strong magnet		•					
Reflections and experiments on the *action attractive ou répulsive*	•						
Formulation of a rule for the *action attractive ou répulsive*	•						
Experiments on *action directrice*: needle material varied (wood, paper, manganese, tin, copper, zinc)	•						
Unsuccessful experiments at the École de Médecine on the magnetic action of conventional electricity	•						
Experiments on the magnetic action of the pile	•						

TABLE 3.1 (*continued*)

	Thematic strands						
Activity	GF	RE	GS	TM	GM	GC	GE
Second, more condensed formulation of the rule for *action directrice*; the concept of a *circuit galvanique*	•						
Experiments to determine the direction of current within the pile by means of chemical decomposition	•						
The hypothesis of terrestrial circuits				•			
Reflections toward a micro-scale theory of a single electrical fluid							•
Further reflections on the causes and effects of possible terrestrial circuits				•			
Reflection and envisaged experiments on the magnetic activity of the pile; *galvanoscope* proposal			•				
Reflections on the improvement of experiments on the magnetic action of conventional electricity	•						
Proposal of a graduated scale for the galvanoscope; further development toward the *galvanomètre*			•				
Reflections on the improvement of experiments for the reversal of *action directrice*; the *aiguille galvanique réctilinéaire*		•					
Generalization of the hypothesis of circuits; idea to "imitate" the behavior of the magnetic needle by appropriately arranged circuits					•		
Idea of aligning mobile coils (*hélice galvanique mobile*) by magnets and by terrestrial magnetism					•		
Idea of using coils (*hélice galvanique fixe*) to magnetize iron					•		
Successful experimental reversal of the *action attractive ou répulsive*		•					

TABLE 3.1 (*continued*)

Activity	Thematic strands						
	GF	RE	GS	TM	GM	GC	GE
September 17: repetition of the experiments on the *action attractive ou répulsive* with Arago in the Obsérvatoire	•						
September 17: improved experiment for the reciprocity of the *action directrice* performed successfully for the first time in the Obsérvatoire		•					
First experiments with the mobile galvanic helix					•		
Following a conversation with Arago, detailed reflections on the explanation of all of Ørsted's findings by recourse to the two "general facts"	•						
September 18, afternoon: first Académie des Sciences lecture	•	•	•	•	•		
September 19: unsuccessful experiments on the alignment of a straight conductor by terrestrial magnetism		•					
Careful consideration of the possible causes of terrestrial circuits				•			
September 20: general reflections on the conditions for galvanic effects; further development of the concept of the circuit—the *circuit electro-moteur*							•
Reflections on further effects of terrestrial circuits: magnetic variation, heat, light				•			
Further experiments (with Fresnel and Despretz) on the "imitation" of magnets by means of coils					•		
Together with Fresnel, experiments on the supposed action of two spirals on one another. Two batteries used; no discernible effect. Ampère ascribes the failure to insufficient battery strength.					•		

Table 3.1 (*continued*)

	Thematic strands						
Activity	**GF**	**RE**	**GS**	**TM**	**GM**	**GC**	**GE**
September 22: demonstration of the experiments with spirals and coils at a soirée hosted by Poisson					•		
September 24: Ampère asks Dulong for permission to purchase the pile Dulong had ordered from Dumotier					•		
September 25, morning: the pile is purchased and the two-spiral experiment performed successfully while it is still in Dumotier's workshop					•		
Ampère makes additions to his lecture manuscript and plans to conduct experiments with two straight conductors						•	
Reflection on a possible generalization: the interaction between two spirals as wholly equivalent to the interaction between two magnets					•		
First draft of the "Conclusions" composed							
September 25, afternoon: second academy lecture, topped off with a reading of the "Conclusions"; at session's end, demonstration of the two-spiral apparatus	•			•	•	•	•
Appointments made to meet with Humboldt and Laplace on September 26 and 27 to present the new theory							
September 25, evening: letter to son, Jean-Jacques: "there were no more objections—and here we have a new theory of the magnet."							

Note: Column headings are as follows: GF, reduction to general facts; RE, reciprocity; GS, galvanoscope/galvanometer; TM, terrestrial magnetism theory; GM, general magnetism theory; GC, galvanic currents interaction theory; GE, galvanism and electricity *single* electro-motive force theory.

The four remaining strands differ in characteristic ways from these exploratory domains.[86] Instead of being undergirded by the broad-based search for empirical rules, they were guided by reflections on the underlying "cause" of the phenomena in question. In consequence, they typically involved positing entities inaccessible to observation. In the language of the time (which remains current in the contemporary philosophy of science), such reflections may be called "theoretical" in a more narrow sense in contrast with the "phenomenological" pursuits of the first three strands.[87] Different epistemic interests came along with differing methodologies. In this second group of strands, the role of experiment came close to the classical understanding of experimentation as hypothesis or theory testing. Ampère's experimental work on the general electrical circuit hypothesis may serve as a typical case, and it is significant to find him referring to the notion of an experimental "proof" (*preuve*) in precisely this context. In an exploratory context it would make little sense.

Ampère's Academy Lectures and the Timeline of His Research

Ampère's activities were structured in important ways by his lectures to the academy of September 18 and 25, marking the ends, respectively, of his second and third weeks of work on electromagnetism.[88] The first two weeks were devoted primarily to exploratory work, including the discovery of regularities, the development of the tools necessary for representing such regularities, the formulation of the two "general facts," and the "reduction" of other effects to them. Ampère's work on reciprocity and his conception and implementation of the galvanometer also took place during these two weeks. At the same time, he was developing his hypothesis on terrestrial electrical circuits and attempting to use circuits as the basis for an explanation of all magnetism. He designed and commissioned the required experimental apparatus but was only able to begin his experimental work on these hypotheses shortly before his first lecture, held on September 18.

Accordingly, at the heart of this lecture were the two "general facts"; the general circuit hypothesis was mentioned only in passing. The heading for the first section of the lecture reads,

> §1. Exposition of the two general results by means of which one can make sense of the diverse experiments of Mr. Ørsted and predict the interaction between a magnet and a galvanic conductor, which has not yet been subject to observation.[89]

His second section was devoted to his various new instruments:

> §2. Description of apparatus and indication of the experiments that seem to me most appropriate for establishing the two general results expounded in the preceding section.[90]

Whereas the text of the first section stayed largely within the bounds advertised by its heading, the second went well beyond them. Ampère not only discussed his apparatus in support of the "general facts" but also presented the two reciprocal devices and improvements to them, introduced the galvanometer, and even announced the "fixed" and "movable galvanic helix" with which he would investigate his circuit hypothesis. Experiments and instruments played a central role throughout his first academy lecture.

This first lecture already marked some turning points in Ampère's research. One of them concerned Ampère's interaction with others. Where his first two weeks had found him toiling mostly on his own, in the third week he often experimented with his housemate, Fresnel, and their mutual acquaintance, Despretz, also a tutor at the École Polytechnique. His collaboration with Arago became particularly intense. Arago had begun his own study of magnetization effects using iron filings, which arranged themselves about a wire through which current was flowing.[91] A conversation with Ampère alerted him to expect much stronger effects from a coiled wire. This conversation thus took place after Ampère had developed his general circuit hypothesis, when any study by Arago of the magnetizing effects of the coil could only be welcome. His experiments were successful and led to important insights. For example, it was Arago who first used in the French context the labels "dextrorsum" and "sinistrorsum" to distinguish the two directions in which coils might be wound—concepts that originated in botanical terminology and had been introduced by Ørsted in his circular, though Arago did not acknowledge that. In the public presentation of his results, Arago openly acknowledged Ampère, just as Ampère's academy lectures repeatedly credited Arago's collaboration. While both clearly benefited greatly from working together, it is plain that Ampère brought more activities and ideas to the table than Arago, a point Arago never failed to recognize. In addition to his personal friendship with Ampère, his support could also be linked with his desire to foster any program competing with the Laplacean school. The almost symbiotic relationship between Ampère's and Arago's activities throughout this early period would soon be understood for what it was even by outsiders, to such a degree that Ampère and Arago were often mentioned in the same breath.[92]

In another, very marked shift, the focal point of Ampère's work also changed following his first lecture. Though part of his third week was also devoted to experiments on reciprocity, his "theoretical" reflections on the connection between galvanism and electricity began to carry much greater weight, and the general circuit hypothesis became the most urgent matter of all. Ampère undertook numerous experiments, refining his apparatus and stabilizing his results to the point where the effects produced could finally be presented in public. These activities culminated, toward the end of the week, in his hectic efforts to demonstrate the attraction between two spirals and then, spurred on by the final success of this experiment, in his first encounter with the idea of a theory of interaction between galvanic currents independent of magnetism.

Ampère's second academy lecture, held on September 25, was announced as a continuation of the first:

> §3. Explanation of the preceding phenomena: reduction of the two general results to a single principle by means of which one may make sense not only of these phenomena but also of others depending on the theory of magnetism.[93]

He first presented his hypothesis of circuits of terrestrial currents as an explanation of terrestrial magnetism, to which he appended extraordinarily detailed reflections on the conditions within the earth that might be causing such currents.[94] The general circuit hypothesis for magnetism, along with several experiments related to it, were treated much more briefly.[95] Finally, in one hastily inserted sentence, cited above, the lecture manuscript alluded to his planned investigation of interactions between straight wires. Discussions of experiments and instrumentation were confined to those involving the movable and fixed galvanic helixes, and, at the very end, the two-spiral apparatus. The small number of experiments presented marked a sharp contrast with the first academy lecture, in which Ampère discussed at least eight different experiments and their apparatus. This, too, is a clear signal of a shift in Ampère's research interests.

Much of the manuscript of Ampère's second academy lecture was produced over the course of his third week of work, and part of it came earlier. It cannot be made out with certainty whether he presented the complete text in this form. The success of his two-spiral experiment in Dumotier's workshop that morning (Monday, September 25) would have left him considering his work from entirely new perspectives throughout the day—perspectives he could not have incorporated into the already completed and lengthy text of his lecture. He did add two short paragraphs in which he reported the experiment's positive outcome, but, by comparison with his pages-long exposition of terrestrial circuits, this account is extremely short. Ampère made two compensatory moves. First, he had his two-spiral experimental setup transported to the academy, which, given the size of the battery and the nature of the chemicals required for its operation, was no small undertaking. But with this equipment at hand, he was able to demonstrate his experiment at the conclusion of the session. A second move was at least equally effective, underscoring the novelty of his approach. At a very late point, presumably immediately before the lecture itself, he added a stand-alone section of quite different character to the already finished text. He called this section his "Conclusions."[96]

The Formulation of a Research Program

"I summarized what I had read at this session and the preceding" (Ampère 1820h, 167). Thus would Ampère later describe this text, read at the end of his second academy lecture. This description fits the title, "Conclusions," he gave to the published version of this précis of inferences drawn from his work to that point. Though I noted in the last section of chapter 2 what a far-reaching step this text entailed for Ampère, its full significance can only be apparent now, against the backdrop of all the research activity of the previous three weeks. He articulated a series of remarkable theses:

1. Two parallel galvanic currents attract; while antiparallel currents repel.
2. From this fact the existence of an alignment effect for appropriately oriented wires may be deduced.
3. The observed attractions and repulsions are fundamentally different from those that occur with conventional electricity.
4. Under the assumption that a magnet is nothing but an arrangement of circular electrical currents, all of Ørsted's results may be attributed, by way of the two "general facts," to the law of the mutual attraction of currents.
5. When a magnet is positioned by alignment with terrestrial mag-

netism, the postulated circular electrical currents within it must circulate in a direction opposite to the apparent motion of the sun.

6. All interactions between magnets may be explained by recourse to this law.
7. The same holds for terrestrial magnetism, assuming circular electrical currents running from east to west in planes perpendicular to that of the inclination needle.
8. There is no principled difference between the two poles of a magnet; they differ only with respect to their position relative to the circular galvanic currents.
9. The interaction between currents demonstrates that the electrical fluid follows the same laws as the magnetic fluid. In addition, Ørsted's experiments show that a current exerts the same effects on a magnet as on another current and therefore that the two fluids, rather than merely obeying the same laws, are one and the same. This reasoning is wholly analogous to Volta's proof of the identity of conventional with voltaic electricity.[97]

The "Conclusions" are known to us only in their printed form, as published by Ampère shortly after his second lecture. While they doubtless received their final textual form only after the academy session, there are indications that the version he delivered at the end of his lecture was substantially the same as the later published text.[98]

A glance at the text of the "Conclusions" makes clear that, contrary to his claims, Ampère was by no means merely enumerating *inferences* drawn from his previous work but presenting a far-reaching, highly ambitious *research program.* His theoretical reach encompasses a different order of magnitude from anything proposed before. At the heart of the envisioned theory is the supposition of universal interaction between currents, from which a broad range of phenomena might be deduced—not only the specific effects of interactions between currents, such as the attraction effect and the (yet to be demonstrated) alignment effect between currents but all sorts of interactions between currents and magnets, including those that Ørsted had discovered. As for the two "general facts" to which Ampère had (at least as he claimed) reduced all of Ørsted's results, these were to be derived in turn from interactions between currents, under the assumption that such currents circulate within all magnets. But this broached the prospect of viewing all interactions between magnets, and even the effects of terrestrial magnetism, as the products of interactions among currents.

With this proposal, to borrow Rheinberger's phrase, a new "epistemic thing" had come to occupy the center around which subsequent research had to revolve.[99]

This was an extraordinarily bold proposal, involving truly breathtaking claims and implications. After all, after centuries-long investigations there was by this time a vast store of knowledge regarding magnetic effects, especially those of terrestrial magnetism. Numerous empirical and theoretical contributions had been made to this field in the first two decades of the nineteenth century, especially in Paris. In this context, the claim that all pertinent phenomena could be derived from an interaction between currents discovered only a few days earlier, and about which most details remained sketchy, would have to be taken as extremely audacious—all the more so given its limited empirical grounding. Its whole foundation—the attraction and repulsion between currents—rested on a single rather involved experiment, hardly transparent. The alignment effect between rectilinear currents that Ampère cited in thesis number 2 remained a purely theoretical prediction, he having yet to perform any experiment to verify it. To be sure, his experiments made his far-reaching claims on the reduction of all magnetism to the interactions among currents at least plausible, but the mathematical treatment of this reduction he clearly envisioned had not yet even begun. As of this moment, Ampère thus had no real way of judging the soundness and eventual success of his thesis.[100]

With their grand speculative consequences, the "Conclusions" were clearly geared toward attracting broader attention. By comparison, the reduction Ampère had sketched, a scant week earlier, of all of Ørsted's findings to two "general facts," seems positively modest. To be sure, this reduction had not entirely lost its value, though it had been absorbed into a theoretical undertaking of an entirely different order. Significantly, the "Conclusions" mentioned it only in passing.

We have no direct accounts of the reactions of the academicians to Ampère's second lecture and demonstration. To judge by Ampère's own assertions, in contrast with the response provoked by Ørsted's report three weeks earlier, this time there were no decisive objections. In a letter to his son written that very evening, following the lecture, he emphasized this point: "There were no more objections—and here we have a new theory of the magnet."[101] That the reaction to the second lecture was surely more positive than to the first is also clear from the fact that immediately after the session closed he agreed to appointments with Laplace and Humboldt, at which he would continue the presentation of his theory.[102] Regardless of who might have taken the initiative here, Ampère could view the very fact of these appointments as a clear sign of recognition, for both savants remained among the most important and prominent people on the Parisian academic scene.

A further circumstance contributed to this success: Ampère was not the only one to present on electromagnetism at this session. Another was Arago, who reported on his magnetization experiments, both those with straight wires and iron filings, as well as those with coils and steel needles. Arago repeatedly acknowledged having devised his experimental setups against the background of Ampère's circular current theory, making the success of his experiments into a success for the theory. In this situation, Ampère could hardly have wished for better support for his hypothesis.

The events of September 25, from the positive outcome of the experiment in Dumotier's workshop to his successful presentation to the academy and Arago's public support, spurred Ampère on to further strides. Considering his general orientation to have been vindicated, he made a clear decision as to the direction of his future research. Henceforth, his scientific inquiries would be geared toward pursuit of the program formulated in his "Conclusions." All of the uncertainty and tentative grasping of the first three weeks now vanished. The firmness of his conviction is plainly indicated by his effort, shortly following his lecture, to get his programmatic text, the "Conclusions," into circulation. Far from contenting himself with its oral presentation to the academy, he completed his text, commissioned a special printing, placed copies in the hands of other academicians, and circulated them internationally, for example, to the Royal Society in London.[103] He also pressed for its rapid publication in the periodical press. Its first appearance in a scientific journal came within a few weeks (Ampère also ordered reprints), and by the end of the year it would appear in two more.[104] Ampère's self-publication of his "Conclusions"—initially without any accompanying text—constituted his first publication on electromagnetism. This proactive publication strategy not only made his program widely known but also placed him under great pressure, for he could no longer deviate significantly from the advertised research program without undermining his own credibility. Its pursuit would continue to occupy him for the next six years. As for his other lines of inquiry, and in particular the more exploratory among them, he set them aside, never to pursue them again. The events of September 25 marked a definitive end to the first phase of Ampère's work on electromagnetism—the phase of tentative experiments, exploration, tinkering, and the development and stabilization of concepts.

September 25 was also a decisive moment for Ampère's standing in Paris. The positive reaction to his lecture and the scheduling of appointments with Laplace and Humboldt served as clear indications that he had entered the limelight, establishing himself as a researcher to be taken seriously in this new research field. He had become the leading figure in electromagnetism and would be called upon as referee (*rapporteur*) for all subsequent contributions to the field. In this respect, too, the bold programmatic step he took in his "Conclusions" must be judged a success.

A Formative Period

Two previously unknown points emerge with particular salience from the account constructed in this chapter of the first three weeks of Ampère's work on electromagnetism—a research period about which scarcely anything was previously known. First, it is clear that Ampère did not begin his work with a very specific goal in mind, instead pursuing numerous diverse efforts simultaneously, all of which aimed at giving the new field a first, perhaps preliminary structure. Only toward the end of this period did he select one of them as his future research program. This choice rested in important ways on the various directions his work had taken to date. The first phase constituted a formative period, one in which a clearly defined research direction emerged out of vague, uncertain beginnings, and finally stabilized. My second point concerns the peculiar character of the efforts that occupied the largest part of Ampère's attention throughout the formative period. Ampère's experiments ranged widely as he undertook a targeted, self-conscious exploration of the new field, refining suitable concepts, means of representation, and experimental procedures. The articulation of a well-defined theory, and the sharply focused experimentation that came with it, began only toward the end of this period.

The received historical picture looks rather different. Ampère was, and is, often characterized as a scientist exclusively oriented toward the construction of mathematical theory and never undertaking broad experimentation. The received view may well be applicable to his later work, but it is usually also imposed, more or less explicitly, on the early period, from which we had previously had no reliable sources. The few attempts to shed light on this period have not altered this perception.

The status that historical studies have ascribed to the two "general facts" serves as a good illustration. L. Pearce Williams (1985a, 94), whose study had been the most detailed to date, misses the significance of the "general facts" entirely, not mentioning them once in his overview of Ampère's early activities. Christine Blondel (1982, 69–72), by contrast, does present the two "facts" but does not situate them within the explanatory framework in which Ampère saw them. Such accounts precisely reflect the picture Ampère himself would craft for public presentation, anticipated as early as the "Conclusions," which I discuss in the following chapter. The "facts" also appear, similarly isolated, in the comprehensive study by James Hofmann. While claiming to be interested in Ampère's "early experimental methods," Hofmann (1995, 229) confines his attention to the method of equilibrium experiments, which Ampère used for the first time toward the end of 1820. The earlier period, in which Ampère was more intensively engaged in experimentation than he would ever be again, is not taken into account. In consequence, Hofmann overlooks the fact that Ampère used experimental procedures entirely different from the method of equilibrium experiments. Hofmann's (1995, 230) assertion that, "with rare but important exceptions, Ampère did not experiment with an exploratory mentality" thus has to be significantly qualified, for the most important of these "exceptions"—the entire formative period—is absent from Hofmann's discussion. Given the fact that it was during this period that Ampère's research program was first articulated, characterizing it as an "exception" strikes me as missing its true significance. To be sure, such misinterpretations may be traced back squarely to Ampère's self-presentation, thereby underscoring the necessity of seeing past such self-presentations by turning to actual research practice.

We must also reconsider the view, advocated by Williams and Hofmann, that Ampère's work on electromagnetism was, from the very beginning, essentially shaped by his foundational reflections on metaphysics and epistemology. The fact that, in a more general context, Ampère did indeed engage in deep epistemological ruminations does not by itself mean that he was led by them in conducting his more concrete research activities. As is well known, much the same holds for many other scientific episodes, both historical and recent. With regard to the formative period of Ampère's work, there is no trace of these ruminations or their consequences to be found. With certain qualifications, this is also true of the period through January 1821, to which I turn in the next chapter. Ampère's exploratory methods were suggested by the particular given epistemic situation, rather than by specific philosophical convictions. It is equally significant that, unlike many others, he did *not* consider himself constrained by such metaphysical or epistemological convictions and thus enjoyed greater freedom in analyzing the new phenomena.

It is a highly significant feature of the epistemic situation that exploratory work was a common response to Ørsted's report all over Europe. With varying intensity and duration, Pictet and de la Rive in Geneva, Davy in London, Schweigger in Halle, and Erman in Berlin all engaged in highly innovative exploratory work. But although in the broader European context (of which, it must be stressed, Ampère knew nothing during the early period) such exploratory approaches were not particularly unusual, the situation in Paris was decidedly different, for there, more than anywhere else in Europe, the expectations of the scientific public had been shaped by the ideal of mathematical theory construction. In such an environment, the decision to proceed in an entirely exploratory mode and, above all, at least initially, in a qualitative mode was anything but self-evident. There were other options, as the case of Biot, yet to be discussed, clearly demonstrates. It is hardly chance that Ampère of all people, with his unusual background, was able to resist these expectations, at least for a time, treading paths no properly trained Parisian physicist could easily have discerned.

Ampère differed from many other European researchers in the extraordinary originality of his ideas—ideas that would eventually lead him away from exploratory work. But also in another important respect he conducted himself differently. He was much more vehement and proactive than most in bringing his results to the public as quickly as possible. The fact that after three weeks of effort he not only had remarkable new results and ideas but was also able to publicize them immediately and effectively demonstrates the urgency with which he sought to make his mark. Ampère's feverish publication strategy might well be compared with Ørsted's actions a few months earlier. But while Ørsted indeed opened a new field and was well aware of that, Ampère's situation was different. Probably the strongest incentive for his zeal for being first to press was the local Paris milieu: he knew that the new field posed a most serious challenge to the leading Laplacean program, and he could take for granted that on his return Biot would immediately jump into research. So he had only a short time in which to gain a head start against his much better equipped Laplacean competitor and to establish himself as the expert in the new field. In this regard, the contrast with Davy is particularly noteworthy. Though Davy conducted his own innovative work in October 1820, his results were not in print until June 1821, by which time they were old news. Schweigger's first lecture on electromagnetism to the Halle academy (September 16, 1820) came two days before Ampère's, but it was also not published until the following spring. Neither seems to have judged the new effects to be as dramatically significant as Ampère did. To be sure, this drama was nowhere more potently felt than in Paris.

To penetrate so far into a new field, propose new instruments, demonstrate new experiments, devise and stabilize new concepts, present a fully formed theory, and establish himself as the leading authority, all within three weeks, was certainly most unusual for a beginner. The existence of well-established academic structures such as those in Paris made it even more challenging. That it was possible at all has to do with a very particular aspect of the Parisian situation. Ampère's meteoric rise would have been difficult to conceive had he not already been a well-established member of the academy—albeit from another field—and, by virtue of his work at the École Polytechnique, a fixture in Parisian academic life. Biot's absence from Paris and the resulting delay in the response of the Laplacean physicists were also decisive points of which Ampère was well aware and which he sought to exploit. Arago's support was both encouraging and important, as was the fact that Delambre, permanent secretary to the academy, was favorably disposed toward Ampère and presumably played a role in allowing him to bring his results before the academy so quickly and spontaneously.

To all this we must add Ampère's ability and willingness to make a vast investment of an entirely different sort. His achievements required not only careful thought and numerous experiments but also a great deal of time and money. A very specific set of circumstances had brought this immense investment within his reach. His financial resources were considerable, and he strained them to the very limit in the procurement of costly new instruments.[105] Given his two positions, the time and effort needed for all this research could not be taken for granted. Here, too, he was fortunate, for he had already discharged his obligations as traveling inspector of schools for the year, and classes at the École Polytechnique had not yet begun. Ampère was entirely aware of this happy accident of timing, and the extraordinary haste in which he worked reflected not only his desire to present his ideas earlier than Biot but also that class instruction would commence in the middle of November and then significantly curtail the time available to him for electromagnetic research.[106] The extent of

Ampère's success during the opening phase can be imagined only against the backdrop of this configuration of circumstances. Ampère was, without a doubt, one of the most original and innovative researchers to enter the new field. But as this episode makes clear, such talent can really flower only when graced by favorable conditions. Though in general such claims sound nearly trivial, when supported by detailed studies of research practice they acquire definite, instructive concreteness.

CHAPTER 4

Competing Pursuits in Paris, 1820–1821

This chapter turns to Ampère's further work through January 1821, that is, to the point when a lengthy pause in his efforts set in. My objective is to present Ampère's various activities in such a way as to shed light on the connections among them, as well as between them and aspects of his situation in Paris. The changes in his experimental practice, his energetic attempts to perform a measurement and the consequences of their failure, his fraught relationship with Biot, his numerous publicity efforts, and his deliberately crafted self-presentation are all more tightly interwoven than has previously been appreciated. I begin by retracing the steps he took to stabilize his research program and by examining the mathematical treatment he intended. The subsequent sections focus on work by Biot and Savart, who gave Ampère his first serious competition. In the final sections of this chapter I turn to Ampère's public relations and the manner of his self-presentation. The fresh perspective on all of his early research that ultimately emerges allows us to clearly distinguish the characteristic differences between its two phases, as treated in this and the preceding chapter.

The state of available sources is much more conducive to such reconstruction for the period discussed in this chapter—the period following September 25, 1820—than for the initial phase. Ampère's publications now offer clear indications of his trajectory, but, even more important, from this point forward most of his lecture manuscripts have been preserved either intact or in easily reconstructable form, licensing a chronological inventory of his activities. It is thus no coincidence that it is this period for which detailed studies are already available. In the first such, which remains the most thorough, Christine Blondel (1982) has offered a superb analysis of the contents and chronology of Ampère's evolving research. Following the available source materials, she structures her account in accordance with the weekly rhythm of Ampère's Académie des Sciences lectures. L. Pearce Williams (1983) has attempted a rather different kind of reconstruction. Focusing on theoretical developments, his study tends to overemphasize their significance, suggesting more consistency than the sources support. In consequence, some of the discontinuities in Ampère's reports escape his attention. While James Hofmann (1982, 1987a, 1987b, 1995, chap. 5) endorses Williams's analysis in its essentials, he also makes a significant, novel contribution, offering a detailed investigation of the background to the method of equilibrium experiments, which had become so important to Ampère. But as I discuss in greater detail below, in my view an overemphasis on Ampère's philosophical background has disguised the complexity of the research process.

Stabilization and Public Presentation

Even after the success of his second academy lecture, Ampère continued to pursue various activities in parallel. They were, however, all subordinated to one well-defined goal: to flesh out the research program sketched in his "Conclusions." In addition, his efforts to communicate his results were of great importance to him. We know nothing of the outcome of the appointment with Humboldt arranged following the academy session of September 25. His meeting with Laplace, however, appears to have been conducted in a constructive atmosphere and accompanied by an experimental demonstration (Grattan-Guinness 1990, 2:920). At Laplace's suggestion, Ampère performed experiments aimed at determining whether electromagnetic effects could also be obtained using very long wires. The background question was whether what was being called a current ought really to be understood as propagating down a wire or, instead, as a kind of polarization effect (Blondel 1982, 81; Ampère 1820a, 247).[1] The success of this experiment not only clearly supported the first option (toward which Ampère was already inclined) but also spurred Ampère toward further reflections; for example, technically, it ought to be possible to use this kind of propagation across greater distances for transmitting signals, with the magnetic needle serving as their detector. The natural pride he took in this proposal of an electromagnetic telegraph was somewhat dampened when, while editing the paper in which he planned to announce this device, he was made aware (presumably by Arago) of Samuel Thomas von Soemmering's much older proposal for an "electrochemical" telegraph (Ampère 1820d, 73).[2] As in the case of the galvanometer, though he had developed it independently, Ampère could make no claim for the basic originality of his idea, only for the specific way of realizing the apparatus. Indeed, he would take no more interest in developing the telegraph than he did in the galvanometer. The improvement of galvanometers was quickly taken in hand by instrument builders, who pursued it successfully. The idea of an electromagnetic telegraph, by contrast, remained on the shelf until Carl Friedrich Gauss and Wilhelm Eduard Weber returned to it in 1833.

The Current Balance

One of Ampère's core interests was to firmly establish the central effect of his theory, that of current-to-current interaction. As promised in his lecture of September 25, he set about studying interactions between straight wires, working with Pixii to develop an appropriate device. This turned out to be rather challenging, since the strength of the effect was orders of magnitude less than could be observed with spirals.[3] It was thus no surprise that it took Pixii two weeks to build an instrument sensitive enough to display it (fig. 4.1). From a drawing preserved in the archive (fig. 4.2), it is clear that, with regard to the placement of movable parts and the maintenance of good electrical contacts, significant developments had occurred within a relatively short period of time.[4] The two conductor segments AB and CD (letters refer to the published drawing, fig. 4.1) were the central components of the assembly; it was their interaction that must be detected. AB was mounted on fixed supports, whereas CD, as a segment of wire loop ECDF, could move toward or away from AB. The two conductor segments were linked with mercury beakers R and T (via supports MA and NB), and with beakers S and U (via supports IPX and JQY). Through them they could be attached to a battery in either polarity. By means of tracks MI and NJ, the distance between the two conductor segments in their rest position could be adjusted. Counterweight GVH was attached in the middle of the (non-conducting) truss EOGZF, giving the wire loop ECDF greater mobility by lifting its center of mass, fine-tuned by means of flexible wire VH. The entire device was placed within a glass case to isolate it from any air currents.

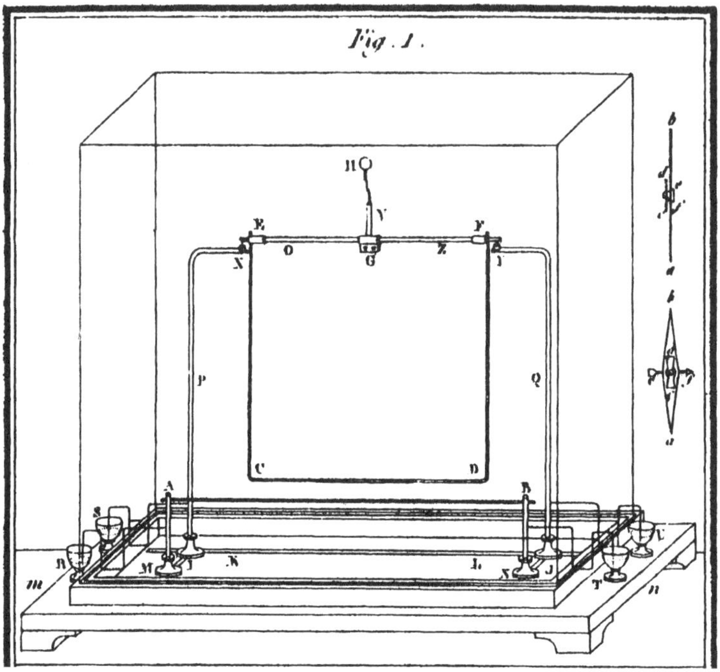

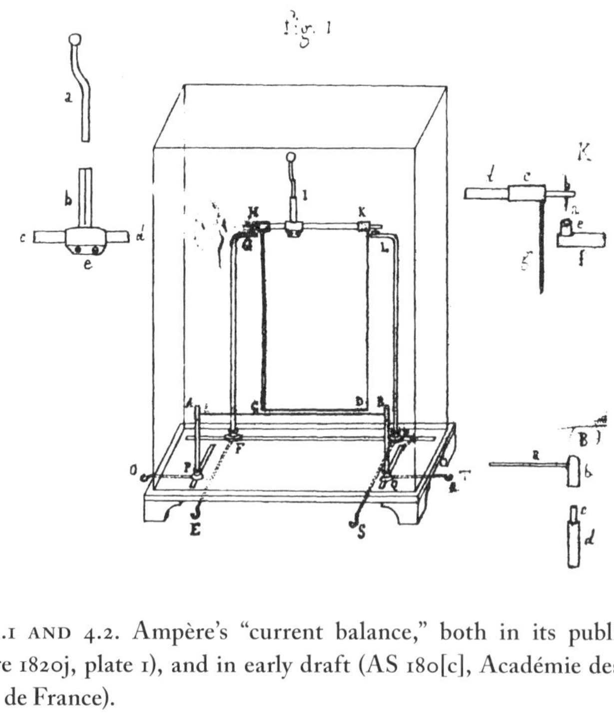

Figs. 4.1 and 4.2. Ampère's "current balance," both in its published form (Ampère 1820j, plate 1), and in early draft (AS 180[c], Académie des Sciences–Institut de France).

As the experiment showed, when the two conductor segments were subjected to parallel currents, there was an attraction between AB and CD and a repulsion in the case of antiparallel currents. Following his successful demonstration of this experiment at the academy session of October 9, Ampère had reason to feel vindicated. The central effect of his theory had been demonstrated in a much clearer and more compelling way than two weeks prior, when he used the two-spiral device depicted in figure 3.7. It must be noted, however, that while the current balance *demonstrated* the expected effect as such, it was not designed for *measuring* it and was never used for that purpose.

First Publications

This success encouraged Ampère in his efforts to publish. After his October 9 lecture, he strove to get into print with all due haste, so as to spread news of his work beyond Paris and secure his priority claims. His publication strategy was carefully considered. The proceedings of the academy were out of the question, for they took years to appear in print, which was far too slow for establishing priority.[5] The scientific journals seemed a much better choice, with Arago's *Annales de chimie et de physique* chief among them. But even they would take quite a few weeks—more than Ampère could have wished. He did indeed get such an article in the pipeline, but he also preempted it by other means. As discussed above, he drafted his "Conclusions" in concise form and ordered a special printing, passing out copies and mailing them far and wide—a strategy quite similar to what he had seen Ørsted use. His success with the current balance must have given him the confidence necessary for such a daring move. This aggressive distribution of his first print publication on electromagnetism was a success. By the end of October, Berzelius, for example, had learned of Ampère's work via a copy of his offprint sent to him by Dulong. The "Conclusions" piece was also Ørsted's first news of Ampère's results, and it presumably reached him before the end of October as well.[6] Though Ampère had no way of knowing exactly whom he would catch in this wide net, this sort of snowball effect was exactly what he had in mind. As a further step toward publication he made the manuscript of his first two academy lectures available to a correspondent of the Brussels-based *Annales générales des sciences physiques* (*AGSP*).[7]

The "Conclusions" as such merely presented his program, however. Ampère immediately went to work on a more thorough exposition, of which the first piece of text was presumably completed in time for his academy lecture of October 9 (Ampère 1820d).[8] It is highly instructive to analyze this first in-depth publication on electromagnetism in detail. Ampère's choice of focus is already highly significant: he had, after all, worked on a wide range of different thematic strands, at least the seven enumerated in chapter 3. Now he placed greatest importance on a matter that, while he had worked on it before, had not made it into his first academy lectures: a general theory of the relationship between galvanism and electricity. This he laid out systematically. His starting point (Ampère 1820d, 59–61) was a "phenomenological" comparison between the two familiar kinds of electrical effects: on the one hand, attractive and repulsive forces, sparks, shocks, crackling, and the like and, on the other, chemical decomposition, the production of heat, and magnetic deflection. He characterized the two as effects of "tension" and of "current," respectively, supporting this distinction by reference to the different sorts of equipment by which each was evoked. From the outset, he ascribed both domains of phenomena to a single "electro-motive action" (*action électromotrice*). Underpinning this claim, on which there was still no general consensus, we find, in the second part of his essay (Ampère 1820d, 61–66) a micro-scale theory. In very rough terms, his idea was this: in the tension condition, the two electricities were understood as dammed up and accumulating at either end of a charged body. In a voltaic pile, the tension condition was maintained by the constant operation of an "electro-motive force." By contrast, in insulating bodies made of such materials as lacquer, wax, or glass, the nonconductive property of the materials kept the electricities from resolving. Ampère explicitly left open the question of the origins of the electro-motive force. According to his theory, with the addition of a conductive connector the tension was immediately resolved. In the case of the closed pile, current effects result, and since the closure of a conductive connection was the only change made in the setup, we must, Ampère argues, attribute these effects to the very same electro-motive force as before (Ampère 1820d, 63). The two electricities must now be imagined as flowing through the conductor and through the pile itself, assuming two currents in opposite directions circulating through the entire circuit.

Ampère's theory cannot be assigned to any specific established theoretical tradition. It contained elements of Volta's contact theory but also some borrowed from the refinement, expansion, and mutation of that theory in Biot's electrostatic schema, as well as from the more critical efforts of the chemists Gay-Lussac and Louis Jacques Thénard. For example, the sharp distinction between current effects and tension effects, wholly alien to Biot's theory, constituted an important element of Gay-Lussac and Thénard's approach. And unlike Biot, Ampère refrained from any assertions on the causes of tension or of the electro-motive force in general.[9] The idea of two currents flowing in opposite directions had also been proposed before.[10] What was new and unfamiliar, however, was the suggestion that these currents flow through both connecting wire and battery in one and the same circuit. This was the real novelty of Ampère's theory, and a great enhancement to the uniformity of the traditional view. While working on this essay, sometime between September 25 and October 9, Ampère introduced the aforementioned fundamental terminological transition, and he would henceforth speak no longer of "galvanic currents" but of "electrical currents." Well aware that such a step would ensconce theoretical insights at a deep terminological level, thereby holding them relatively fixed and insulating them from any possible revision, he now felt confident enough to undertake it. His lecture of October 9 used the new terminology throughout.[11]

It should be stressed that the status of the concept of the electrical circuit had now significantly changed. Recall that Ampère had originally developed it in the context of his attempt to capture the magnetic effects of both wire and pile in a single rule, without reference to any micro-scale theory. Now, however, the concept was presented wholly within the framework of a micro-scale theory for which it promised significant gains, and not simply as an afterthought. The fact that it had been developed elsewhere and was only later inserted into this context, and may even have lent the initial impetus toward the theoretical reflections taking center stage, has been completely disguised. Ampère was interested in a systematic account, not a historical one. This should serve as a warning to historical researchers not to be too quick to take the connections made in representations crafted for publication, no matter how clear and compelling they might seem, for those that guided the actual course of research.

In the third part of his essay (Ampère 1820d, 66–68), Ampère concerned himself with instruments devised for the detection and measurement of the various phenomena. Now placing stronger emphasis on parallels with the electrometer, he presented the galvanometer, together with a rule for determining the *direction* of current. It was here (Ampère 1820d, 67) that, among other things, he formulated the mnemonic for the direction of magnetic action long known as "Ampère's swimmer rule." Not until the fourth and longest part (Ampère 1820d, 68–76) did he finally offer a wide-ranging discussion of his experimental discoveries. At its heart he placed the attraction-and-repulsion effects between current-bearing conductors, laying particular stress on the current balance (Ampère 1820d, 71–73) and also mentioning the experiment with long wires proposed by Laplace, along with his idea for an electromagnetic telegraph. One important point of discussion was the question of the difference between the new attraction effects and those familiar from conventional electricity. The objection that the attraction and repulsion between conductors might well be just a variation on the well-known phenomenon of electrical attraction and repulsion had been brought to his attention by members of the academy. So as to forestall it, he carefully expounded the different properties of the two classes of phenomena (Ampère 1820d, 70). Next, predicting an alignment effect between two straight conductors, he broached a further topic: the attraction and repulsion between current segments should also explain the exertion of electromagnetic torque (Ampère 1820d, 74). Furthermore, the distinction between attractive and alignment effects so important to all magnetic and electromagnetic phenomena would lose its importance when a magnet was understood as itself a collection of currents circulating in planes perpendicular to the north-south axis of the magnet. For further explanations of magnetism and electromagnetism, Ampère referred his readers to the second section of his *mémoire,* to be published soon.

A most important goal of this first essay was to draw attention to the central importance of the newly discovered interaction between currents. This discovery led to a noticeable shift in emphasis. The claim that all magnetism could be explained by recourse to such interaction now took second place, though, in chronological terms, it had been Ampère's starting point. Pride of place was given instead to the much older problem of the relationship between electricity and galvanism, which Biot's electrostatic theory of the pile had already addressed, and without much success. By stressing the effects of current proper, Ampère established the independent significance of his new attraction-and-repulsion effect. Rather than being derived from electromagnetism, it served as the basis for explaining all electromagnetic and magnetic interactions. Emphasizing the independence of his new effect also had a great deal to do with Ampère's desire to step out from under Ørsted's shadow. After all, he had faced the objection that current-current interaction could be deduced from Ørsted's findings alone. By presenting the corresponding device, the current balance, in all its detail, Ampère underscored the independent meaning of the new effect. It is no coincidence that this device would henceforth become emblematic of Ampère's theory.

Further Research

Theory of Magnetism and the Force Law

As Ampère (1820h) himself reported, he continued to work on stabilizing and refining the experiments geared toward simulating magnets by means of spirals and helixes, deriving evidence in support of his theory of magnetism from them. However, the effects predicted by this theory that had not yet been experimentally demonstrated were of greater urgency. Of particular importance was the aforementioned rotation effect between electrical currents, as announced both in the "Conclusions" (thesis 2) and the first essay. Demonstrating it would show that attractive and repulsive forces alone could bring about alignment effects. Ampère's theory directly predicted such an effect, for between two current-carrying wires free to rotate in parallel planes about a common axis (fig. 4.3) the forces of attraction and repulsion ought indeed to induce rotation. The wires would continue to rotate until they reached the position at which only attractive forces remained—when their currents ran in strictly parallel directions.

The experiment proved to be very challenging. Despite the considerable effort Ampère and Pixii expended on the apparatus, they initially met with no success in demonstrating the effect. In his lecture of October 9, Ampère promised to press on (Ampère 1820a, 245; Blondel 1982, 80). Only three weeks later, on October 30, was he ready to report success, presenting the completed device (fig. 4.4). Once again, two conductor segments (AB and MN), whose mutual alignment was to be observed, formed the core of the assembly. AB was a component of the fixed frame CABD, while MN

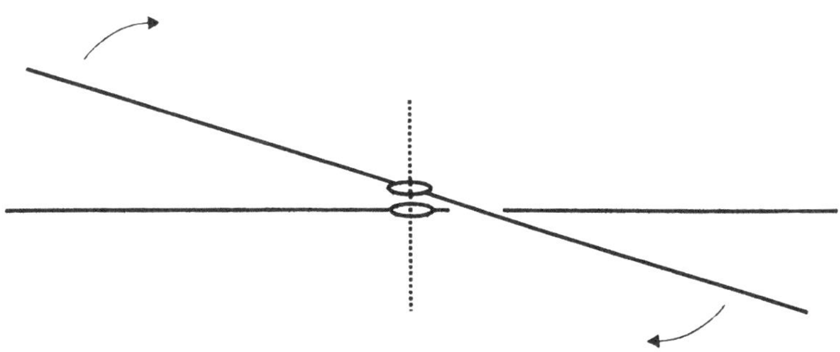

FIG. 4.3. Schematic illustration of the alignment effect between two straight wires.

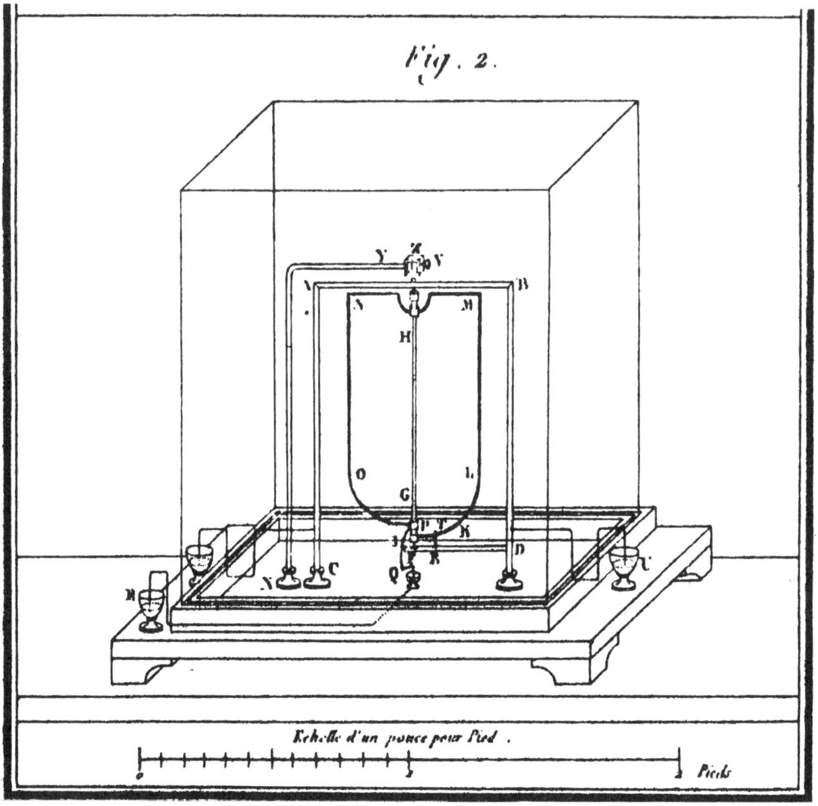

FIG. 4.4. Ampère's apparatus for demonstrating alignment between currents (Ampère 1820j, plate 1).

belonged to the wire loop KLMNOP, which could freely rotate around axle FPHZ. To facilitate rotation, the axle was suspended by a wire from stand XYZ at point Z. Its lower end F was submerged in a beaker of mercury. Both conductor segments formed part of a single electrical circuit, which ran from the contact at the mercury beaker to the left through frame CABD, then to the lower end F of the axle, through wire loop KLMNOP, back to beaker Q, and from there to external contact R. In the position shown in figure 4.4, the current thus passed through MN and AB in opposing directions. Ampère predicted that the wire loop would rotate, as indeed it did when the experiment was run. To be sure, this rotation effect might also be interpreted as a consequence of the repulsion between the vertical conductor segments CA and DB and the movable segments

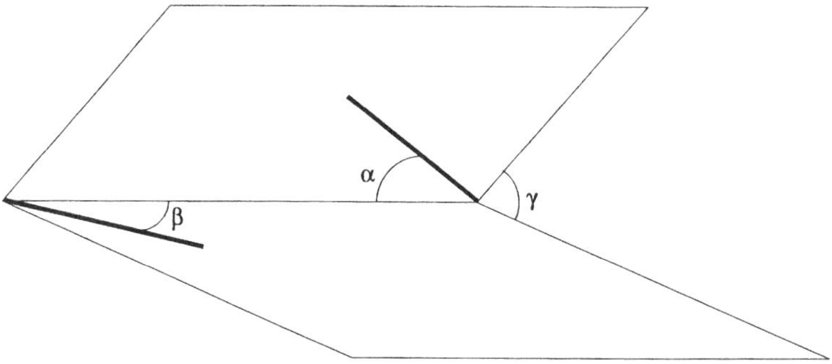

FIG. 4.5. Diagram of the angles between two current segments.

NO and LM. But Ampère's gaze was focused on the interaction between the horizontal segments, and he seems not to have taken the effects of the vertical segments into account. He took the effect obtained with this instrument as confirmation of his theoretical expectations.

Another important question was the mathematical form of the law governing the force between current-carrying conductors. Ampère's reflections on the possible contours of such a law began quite early.[12] In a thoroughly Laplacean spirit, the ultimate cause, the *fait primitif* on which everything had to rest, could only be a central force operating between pointlike centers. Ampère assumed current segments of infinitesimal length as his centers of force. In general, the force had to depend on distance, and, without any argument, in a kind of a priori assumption, Ampère postulated an inverse-square law of the sort accepted for all classical interactions, whether gravitational, electrical, or magnetic. Equally without justification, he assumed the proportionality of the force to the "intensity" of the current. As the intensity of a current was at least as vague a notion as that of current itself, and in the absence of any means of measuring current intensity, this assumption had to be provisional, as well as less specific than the inverse-square law. Finally, so as to account for the dependence of macroscopic effects on spatial orientation, Ampère introduced angular variables. For the interaction between two rectilinear current segments, three angles had to be considered (fig. 4.5). For the force law in its most general form, this yielded the formula,

$$\frac{g \cdot h \cdot F(\alpha, \beta, \gamma)}{r^2}$$

where *g* and *h* stand for the current intensities, *r* for the distance between current segments, and *F* for a yet-to-be-determined function on angles α, β, and γ.[13] This force law applied to infinitesimal current segments, so that the effect between extended conductors could be determined only by integration.

Here, at the very latest, it becomes clear to what extent Ampère's thinking remained firmly within the Laplacean tradition. Despite the fact that he was studying the interaction between linear segments, he confined himself to central forces of attraction or repulsion exerted along straight lines connecting centers of force. He gave no thought to torque between centers of force, though for macroscopic configurations this might be derived from a superposition of central forces. It was for this reason that the aforementioned experimental demonstration of rotational alignment between currents was of such paramount importance. The fact that Ampère never considered any other conceptual structure had to do with the available mathematical tools, for at this time there was not a single example of the successful mathematical formalization of forces and motion that did *not* rest on central forces. Mechanics, the subject Ampère taught in his day-to-day life, remained the classical domain par excellence of central force thinking.

In one decisive respect, however, Ampère broke sharply with tradition. He conceived the centers of force as also having spatial directions and, in a dramatic departure from the traditional understanding, conceived the force itself as depending on the specific constellation of those directions, expressed in the three angles α, β, and γ—a move that was totally unthinkable in a strict Laplacean framework. One consequence of this departure was that most parts of the highly developed mathematical repertoire of the tradition, with which Ampère's teaching experience had left him intimately acquainted, could not be applied. This held especially for the procedures perfected by Siméon-Denis Poisson, later known as potential theory. By comparison with that sophisticated tool kit, the mathematical devices Ampère sought to use instead—trigonometric functions—looked absolutely elementary.

Ampère's efforts to further specify the function *F* proceeded from general reflections extracted from his experiments to date. Attraction occurred between two straight conductor segments when they carried parallel currents and repulsion, when the currents were antiparallel. By interpolation, Ampère felt certain that the effect would disappear when the two segments were arranged perpendicular to each other. At a fairly early stage, this led him to suppose that F must be an odd function on the cosine of angle γ. As Ampère stressed, the simplest such function, or at any rate the first in a series of approximations, was the cosine function itself. By way of similar, albeit less explicitly worked out reflections, Ampère came to assume that F depended on the sines of the other two angles. Now his law had a provisional form that could also be used for integration:

$$\frac{g \cdot h \cdot \cos\gamma \cdot \sin\alpha \cdot \sin\beta}{r^2}$$

Ampère presented this law more or less explicitly in his academy lecture of October 30, while stressing its provisional, speculative character.[14]

"Primitive Fact" and "General Fact"

In characterizing these endeavors, it is fruitful to consider Ampère's epistemological terminology and its relationship to his earlier talk of "reduction to general facts." These general facts had served as a starting point in the explanation of his electromagnetic findings, and in this respect their pivotal role was similar to that now ascribed by Ampère to the notion of a central force between two infinitesimal current segments. To be sure, the earlier "reduction" had had a very specific character. The "general facts" in question were lawlike statements, as well as phenomena every bit as experimentally demonstrable as the more complex effects in whose explanation they aided. They could be characterized by reference to a minimal set of experimental conditions. Complex phenomena could be taken as having been "reduced" when they were understood as special cases or superpositions of several "general facts." In stark contrast, the mathematical derivation Ampère was now pursuing worked very differently and was described in different terms. At its center stood what Ampère would soon begin to call the "primitive fact" (fait primitif): the central force between two current segments. By its very nature, this fact could not be directly established by experiment. It was rather an abstract construct, devised first and foremost for mathematical convenience. When he was called upon a year later, in responding to Faraday, to comment on the status of his explanatory approach, he would explicitly stress this point. Then he would insist that not even attraction and repulsion between straight wires could be considered "simple facts" (*faits simples*): "It strikes us that this name can only be given to such laws of mutual action as must be admitted between two points . . . ; therefore simple facts cannot be observed directly, but only inferred from observations by means of calculation."[15] If the effects observed between straight wires could not be taken as "simple," neither could the rotation effects discovered by Faraday, despite the latter's claims (see chap. 6). They were ill suited to "submitting phenomena to calculation."

Such remarks drive home the extent to which notions of "simple fact" or "elementary action" depend on a choice of specific explanatory goal. With the goal of mathematical deduction in mind, Ampère could conceive no other way of grasping the observed effect between extended segments than to treat them as the result of integration over elementary effects, which in turn had to be understood as effects between single-point centers of force. This was the only suitable point of attack, given the mathematical tools at his disposal. His previous program of "reduction to general facts" had not been subject to any such constraints. With "derivations" proposed at a phenomenological level, the "basic" explanatory ground or "general fact" also had to be sought at that level. I return to such a program in chapter 6, when I consider Faraday's attempt at a "reduction" to "simple cases."

Attempting Mathematization by Quantification

Ampère was well aware that his reflections were in need of empirical support, especially in light of his partial break with tradition. With respect to the precise formulation of the force law, he noted, "This function can only be determined empirically."[16] However, a force law that rested on infinitesimal current segments could not in principle be directly established empirically. Any description of experimental, that is, macroscopic conditions would require integrating over that basic law, so that an unambiguous inference from experiment to the force law was therefore impossible. Still, one might strive to reduce the degree of uncertainty, and increase the plausibility of the law, by means of experiment. This was presumably Ampère's intention when he and Pixii set about devising instruments for the empirical determination of the dependence of the force between straight wires on the angles between them.

The first step consisted of a modification of the current balance, introducing a new parameter by changing the mounting of the fixed conductor. This would now be placed upon an assembly that allowed it to be rotated about its geometric center in the vertical plane and to be fixed in a given orientation.[17] Ampère hoped that this device would permit the qualitative demonstration of the effect in question, but as he must have soon dis-

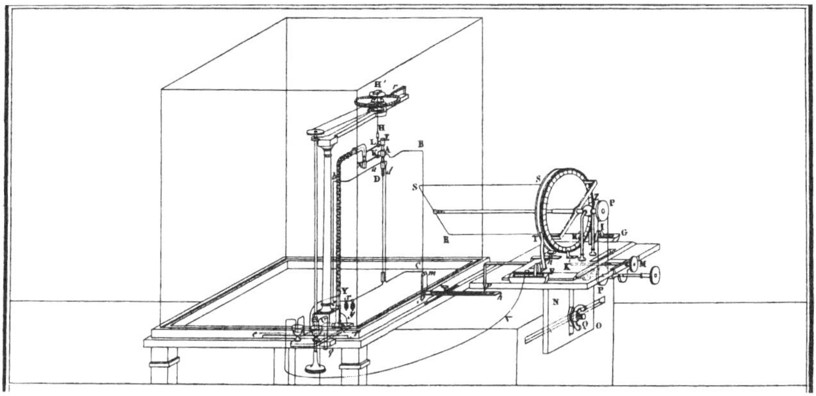

FIG. 4.6. Ampère's device for measuring angle dependence (Ampère 1820j, plate 3).

covered, it was as ill suited to exact measurement as the original current balance. And this was his true goal, for mathematical formalization could only follow successful quantification. Ampère and Pixii therefore turned to the construction of a second device "for making exact measurements" (Ampère 1820j, 182).

Once again (fig 4.6), the effect at issue was the attraction and repulsion between two conductor segments: mobile segment BC and fixed segment RS. By means of the rather complex assembly on the right side of the device, the orientation of RS in the vertical plane could be varied. Wheel R′ST was designed both to allow for the precise adjustment of the segment's angle and to close the electrical connection between wire loop SSRT and mercury beakers G and F, which in turn were connected to the battery. Mobile segment BC comprised part of the larger loop ABCDabcdE, the whole of which was suspended from torsion wire H′H and free to oscillate about a vertical axis (the extension of H′H). Mercury beakers K and L provided electrical connectivity to the wire loop via the two battery leads passing through stand XY. The incorporation of the mobile loop in the symmetrical *double* loop ABCDabcd was designed to neutralize the effects of terrestrial magnetism, with small weights xy there to balance the loops as needed. The starting distance between mobile and fixed conductors could also be adjusted to various values by means of track gph. Furthermore, by displacing the entire table onto its right side, using screw and groove PQ, the height and horizontal position of fixed segment RS could be adjusted. Measurements using this rather complex device were to begin with calibration in its inactive state, ensuring the vertical alignment of the mobile conductor above point m. When the current was switched on, it would move from this default position, toward or away from RS, and could then be returned to it by twisting the torsion wire. The degree of twisting necessary for this purpose could be read off scale r, yielding the magnitude of the electrical force now exerted between the two wire segments (Ampère 1820j, 184–86).

It is worth noting, and has thus far escaped the attention of historical scholarship, that Ampère here took recourse to a thoroughly classical principle of measurement. The force exerted between the two wire segments in a given, preset configuration was to be equalized by torsion forces and its magnitude determined thereby. By varying the parameters on which this force was assumed to depend (distance, direction of current, and angle), he thought to determine that dependence in exact, quantitative terms. The classical paradigm for this sort of measurement was, of course, Coulomb's torsion balance. Even Ampère's measurement protocol was a direct adoption of Coulomb's procedures. What's more, the torsion balance was emblematic of the program of precision measurement as such, as well as of the very idea of an exact natural science in the Laplacean spirit. It is highly significant that the very first measurement instrument Ampère ever devised was an adaptation of the torsion balance to the newly discovered interaction between currents. Despite all Ampère's opposition to Laplacean physics, he assimilated some of its most important principles—a fact that has thus far escaped scholarly attention.

Putting this device to work posed serious problems, however. In hindsight, this is no surprise since using the original torsion balance had also been an extremely delicate undertaking (Blondel and Dörries 1994; Heering 1998). Ampère's attempt to measure the forces between currents also faced additional difficulties. To begin with, he was dealing with an entirely new kind of force, one exerted only when current flowed through both wires. In addition to the notoriously difficult problem he had encountered in his first electromagnetic experiments, that of maintaining the best possible electrical contact without sacrificing mobility, he now had to contend with the fact that the strength of voltaic batteries did not remain constant over time but decreased constantly and rapidly. Whereas the first of these problems could at least be mitigated by submerging the wire ends in beakers of mercury, there was no technical way of eliminating the latter. Instead, he was forced to consider measurement and interpolation protocols by which the effect could be calculated from the experimental results. In later physics, one would speak of data reduction, a concept unknown in early nineteenth-century physics. In a series of measurements with incrementally varying angles between the two conductor segments, every measurement for a given angle setting had to be preceded by a calibration step in which the two segments were realigned in parallel. This yielded a curve for the decrease in battery strength over time, by which measurements could be renormalized to constant strength.[18] Guided by his previous conceptual analysis, Ampère intended to evaluate the results of these measurements with a view toward the question "by which cosine function on the relevant angle the attraction between the two currents varied."[19]

Puzzling Effects, New Insights, New Actors

Work with the device likely began shortly after October 9 but did not initially yield any intelligible data. For three weeks, Ampère had no new results to present. In the interim (on October 16) he did read a short notice in which he showed how Arago's magnetization experiments could be taken as supporting his theory of magnetism. And his work with the apparatus in figure 4.6 did offer unexpected hints toward the solution of other problems. Ampère became aware that, in these experiments, terrestrial magnetism was evidently having a strong, if unwelcome effect. Over the course of his attempts to investigate this source of interference (which eventually led to his introduction of the symmetrical double loop, designed to exclude it), he observed that it was particularly strong when the wire loops enclosed a large area. Though he had no explanation, he took this observation as a welcome hint toward addressing an as yet unsolved problem: while a current-bearing wire would align itself to a conventional magnet, it could not be made to align itself to terrestrial magnetism. Thus far Ampère had been guided by the idea of "naturalistically" simulating the circulation of currents within magnets, as he envisioned it, with dense arrangements of electrical circuits. This had led to small-scale, but densely packed configurations, such as that in figure 3.6. This principle was now replaced by a new approach, with wire loops enclosing the largest possible areas and, given the difficulty of supporting them, with the lowest possible weight (Ampère 1820j, 172). With such equipment in hand, Ampère finally succeeded in demonstrating a terrestrial alignment effect.

At his lecture of October 30, he presented two different devices. In the apparatus depicted in figure 4.7, wire loop ABCD, with a diameter larger than 40 centimeters, was mounted on points E and F in such a way as

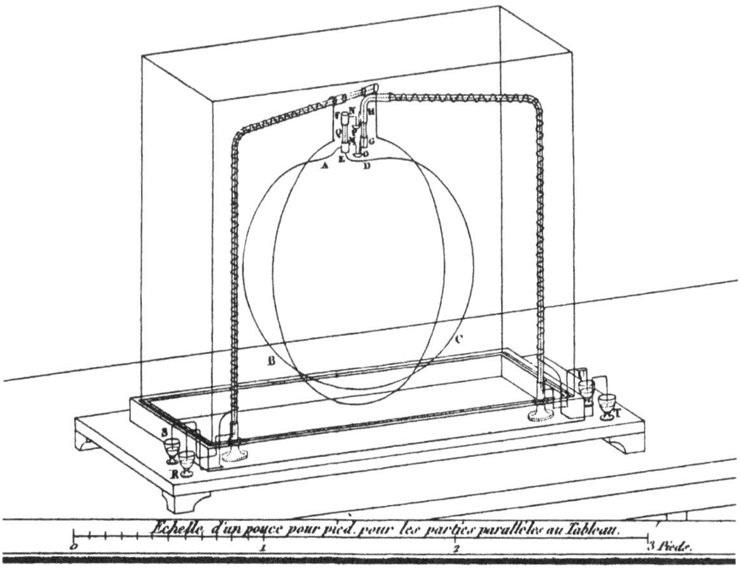

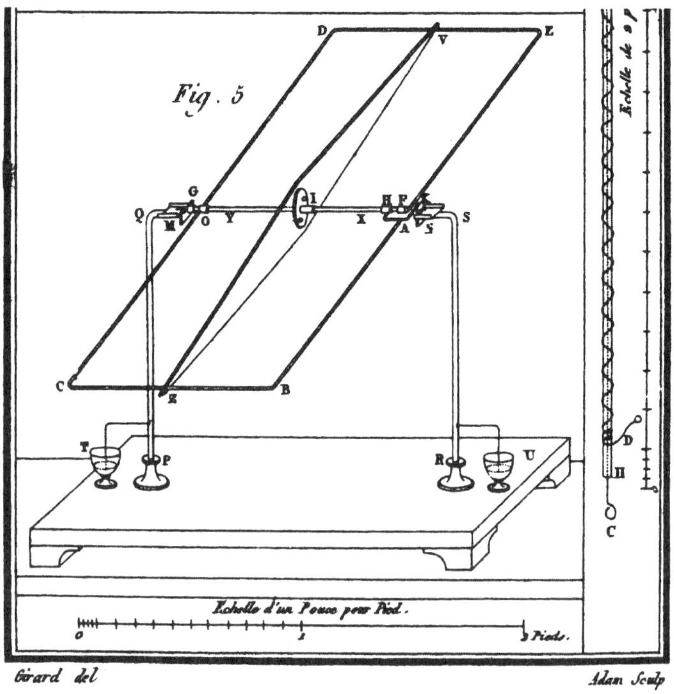

FIGS. 4.7 AND 4.8. Ampère's devices for the terrestrial alignment effect (Ampère 1820j, plates 3 and 2).

to allow rotation about the vertical axis. When connected to the battery, it immediately oriented itself east-west, perpendicular to the magnetic meridian. Ampère incorporated the second, fixed loop so as to allow him to also demonstrate alignment between current segments (previously observed with the equipment in fig. 4.4) using the same device. While this device provided an analog to the compass needle (i.e., to the declination needle), another (fig 4.8) was analogous to the inclination needle. Its large conductor loop ABCDE, about 30 centimeters wide and 60 centimeters long, could rotate about horizontally mounted axle GIK. Mountings G and K also provided connectivity for the electrical current. Once again, when the loop was connected to the battery, it assumed a position at least approximately perpendicular to the direction of terrestrial magnetic inclination or magnetic dip. Ampère took both results as strong confirmation of his theory.

My brief account might give the impression that, once Pixii had built these devices and they were placed in service, all of them worked perfectly, displaying the expected effects. But in the case of the inclination instrument, Ampère described a whole series of difficulties that could not be fully solved, for example, those confronting the mechanical balancing or placement of the loop. In consequence, even in its final version, the device failed to yield consistent and stable results (Ampère 1820j, 193–96). It is also easy to be misled by the sense of perfection Ampère's illustrations evoke. These illustrations were produced weeks after the fact, representing a degree of technical maturity most unlikely to have been reached by the time of Ampère's public demonstrations. Furthermore, though he never even mentioned it in print, Ampère's collaboration with Pixii was an absolutely decisive factor in all of these technical developments. Without the resources in Pixii's workshop—not only manual skill but extensive knowledge of techniques, materials, and feasibilities—it is inconceivable that anyone as inexperienced in experimental work could have managed such rapid and innovative developments in experimentation and the necessary equipment. We can assume that many of the devices were assembled in the workshop, placed in service, and fine-tuned under Pixii's supervision. No doubt such efforts entailed significant expenditures by Ampère.

The two thematic strands Ampère pursued in the period following September 25—the theory of magnetism and the force law—were distinguished by characteristic procedural differences. All investigations of the theory of magnetism remained within the bounds of qualitative research. The reflections by which he designed and evaluated his numerous experiments typically took the form of analogies. By contrast, his studies of the force law had a mathematical character, and the role of experiments was completely different: there was essentially one single experiment whose goal was exact measurement. Analogies, as used for his theory of magnetism, played absolutely no role here. Though both strands were pursued in parallel, in the weeks leading up to October 30 the emphasis was decidedly on the theory of magnetism. That this would change thereafter had to do with events in Ampère's academic environment.

Ampère was not alone in the field for long. At the October 9 session, at which he presented his current balance, there was another lecture on electromagnetism. It was delivered by the physicist Jean-Pierre Boisgiraud, physics tutor at the Saint Cyr military school near Paris (Arago 1820a, 100). Boisgiraud was clearly in contact with Arago, who may also have sparked his interest in the new effects. But he also carried out some of his experiments at the request of Poisson, who for a short time appears also to have been interested in electromagnetism (Boisgiraud 1820, 281).[20] In the experiments reported on October 9, Boisgiraud had begun with Ørsted's setup and attempted to increase the mobility of the needle by adjusting its position in various ways. In addition to the traditional compass mount, Boisgiraud tried suspending it from a thread and finally floating it, embedded in a great deal of fat, in water. In this last configuration he studied the needle's response to various placements and orientations of the wire: horizontal, above or below the water's surface, or vertical, to the right or left of the needle. Without attempting any systematic expression of his results in the form of a general rule, he drew the conclusion that electromagnetic effects were not confined to the poles of the magnetic needle but acted on it in its entirety. Not surprisingly, Ampère showed great interest. In his official capacity as reporter, to which he had been appointed along with the much older physicist Jacques Alexandre César Charles, he asked to have the experiments demonstrated.[21] He then investigated them thoroughly. In his November 13 report to the academy, Ampère singled Boisgiraud out for praise, though he could not refrain from noting that Boisgiraud's observations of the motion of the floating needle could all be more easily deduced from the general law he himself had proposed.[22] Like Arago's results before them, Ampère could take Boisgiraud's as welcome support for his theory.

For the sake of completeness, and in anticipation of later events, it should be noted that at the beginning of November Ampère's friend Fresnel also entered the field. In a short notice, he reported on experiments in which he had apparently managed to decompose water by magnetism.

Arago's experiments on magnetization had led Fresnel to consider their reversibility: he surmised that inserting a magnet into a coil should produce an electrical current, which in turn could be detected by the decomposition of water. Several experiments appeared to confirm this expectation. Following Fresnel's report, Ampère began his own experimental search for this effect, but, before he attained any result, Fresnel retracted his claim, cautioning that the experimental finding was by no means as unequivocal as it had first appeared. Fresnel's experiments constituted the very first attempt to produce electromagnetic induction, and this episode opened new perspectives to Ampère, though he did not pursue them at this time. His constructive, friendly collaboration with Fresnel would continue for some time to come (Fresnel 1820).[23]

For Ampère, the entrance of another actor on the stage had much higher significance. Toward the middle of October, Biot returned to Paris, attending academy sessions from October 16 on. With Biot's appearance, Ampère's situation changed profoundly. Though competition was to be expected, it had remained a remote threat so far. Now it had arrived and was present right at the academy. Ampère (1820j, 182) took the initiative by presenting his first device for demonstrating angular dependence to Biot and Gay-Lussac. We know nothing of Biot's immediate reaction, except that it is clear that he immediately set to work.

The Laplacean Response of Biot and Savart

Biot's Precision Measurements

Ampère's demonstration at the academy was surely not Biot's first taste of the new effect since his return to Paris. He recognized the challenge immediately and set to work with an intensity no less than Ampère's. "It is said that Biot, who has only just arrived, works day and night on this matter"—so wrote Dulong about the situation.[24] Biot was supported in these efforts by Félix Savart, a technician at the physical cabinet of the Collège de France, where Biot held the professorship in physics. Savart had first come to Biot's attention by virtue of his work on acoustic figures. The pair took an approach completely different from Ampère's. Biot (1821c, 228) later formulated his initial goal as follows: "The first thing that had to be discovered was the law by which the force emanating from the connecting wire decreased at various distances from its axis."[25] Both the goal, and the experimental arrangement with which the two researchers sought to attain it (figs. 4.9 and 4.10), were entirely in the spirit of the Laplacean program. A small bar magnet (on the left in fig. 4.9) was suspended from a torsion thread in such a way as to allow it to oscillate in a horizontal plane, about a vertical axis. Close by, long vertical wire MC was mounted such that, when the magnet was at rest, a shortest-distance line from the wire to the center of the magnet would be perpendicular to its north-south axis. The wire would be connected to a battery. In some of the experiments, the influence of terrestrial magnetism would be neutralized by large magnet AB. What was measured was the oscillation of the small magnet at various distances from the wire. By determining the relationship between oscillation period and distance from the wire, the distance function of the force exerted by the wire on the magnet could be established. This was precisely the goal Biot had set himself.

The measurement method used in these experiments was well known. Coulomb had established it in the 1780s, performing measurements to establish the distance function of electrical attraction and developing the torsion compass (see chap. 1). This means of determining the intensity of a magnetic force was used in the wide-ranging campaigns to measure ter-

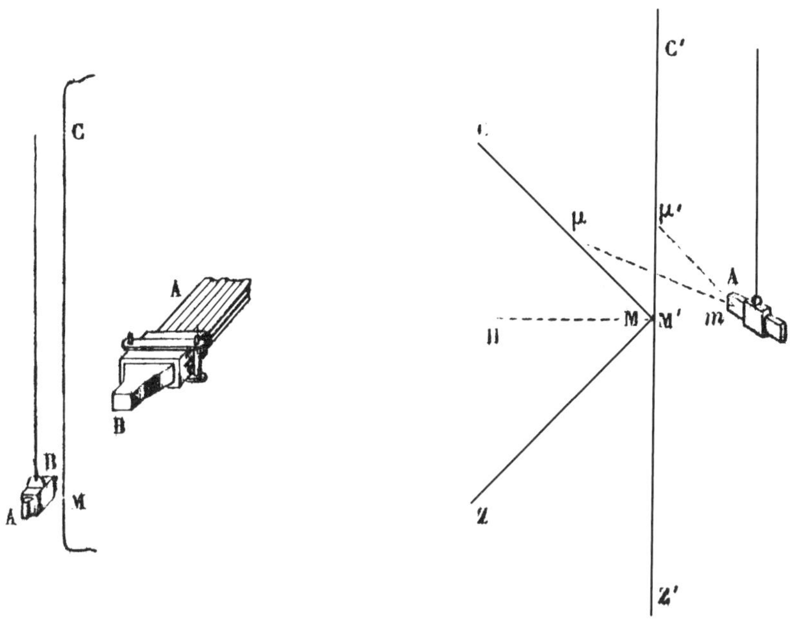

Figs. 4.9 and 4.10. Biot and Savart's measurement setups (Biot 1824, repr. Joubert 1885, 84, 93).

restrial magnetism. Shortly after the turn of the century, in collaboration with Alexander von Humboldt, Biot himself had performed such measurements (Biot and Humboldt 1804), and so he was intimately familiar with the procedure. There was, however, one new component in the latest experiments: the voltaic pile. Its use forced the experimenters to confront some of the same problems Ampère was encountering with the instrument depicted in figure 4.6 at roughly the same time. Particularly difficult was the fact that the intensity of a working battery quickly diminished, so that the force to be measured did not remain constant over time and could not be easily reproduced. As a series of measurements at various distances between wire and magnet took some time to perform, Biot and Savart also had to adopt special procedures for the interpolation and numerical reduction of their experimental data. But they were not confronted with some of the other problems with which Ampère contended. Because they were measuring the interaction between current and magnet, and not between currents, they had no need of movable conductors, with all the challenges those entailed. This may be why Biot and Savart quickly surmounted all experimental difficulties. On October 30, after only two weeks of work, Biot presented his results to the academy.[26]

The first result was the law that the force acting on a small particle within the magnet (a *molécule de magnétisme*) was always aligned so that it was perpendicular to the plane defined by the wire and the shortest-distance line between the wire and the magnet, with an intensity inversely proportional to the square of the distance between magnet and wire. Biot next suggested that the reader conceive of the "nature" of this force by imagining that the electrical current induced tiny magnets down the length of the wire, with their axes perpendicular to the wire. This presumably took place around the whole circumference of the wire, such that when facing the vertically mounted wire from any direction, one would be gazing at tiny, horizontal magnetic dipoles (Biot and Savart 1820, 223). Of the two parts of this report—the geometric, quantitative determination of the force and the explanation of its underlying character—I now turn to the first.

The Determination of a Force Law and Its Limitations

Biot's assertions regarding the direction of the attractive force can only partly be taken as the product of observation. After all, the experimental arrangement itself left the magnet with no other degree of freedom of movement except oscillation in the horizontal plane, perpendicular to the wire. The force required to produce a deviation from the horizontal plane was many times greater than that needed to cause oscillations within the plane. We cannot decide whether Biot was simply unaware of this limitation or rather sought to disguise it by the same rhetorical means with which he emphasized the soundness of his empirical grounding throughout. Though Biot's distance law was explicitly formulated for (elemental) magnetic particles, he had only actually measured the motion of an *entire* magnet. The immediacy with which he deployed his mathematical tools in an inference to microscopic particles and the fact that he presented the resulting law, and only this law, as "empirical" are thoroughly typical of his Laplacean mentality.

A further step, taken soon thereafter, also fits this picture. Citing a study by Laplace, Biot claimed that a $1/r^2$ law for the action on an infinitesimal segment of the wire could be deduced from a $1/r$ law for the action on the *whole* wire (assumed to have infinite length). However, the details of this consideration, and especially of any argument for ruling out other force laws that, by integration, might also have yielded $1/r$ action on the entire wire, were never explicated. One such alternative, for example, would have been a $1/r^3$ law. Much later, Ampère (1827) offered some intimation that the argument may have had to do with dimensions. In that same context (1827, 374–75), he also hinted that Laplace might have suggested the possible dependence of the force on the angle joining the wire with the least-distance line between magnet and wire segment.[27]

It was presumably this suggestion of a possible angular dependence that led Biot to undertake another series of measurements, despite the fact that the first had left him completely convinced that his law allowed for the calculation and prediction of "all of the movements imposed by the connecting wire on magnetic needles" (Biot and Savart 1820, 223). The experimental configuration required for this second series was a bit more complicated. Now, one wire would be bent to a specific angle at the height of the magnet (see fig. 4.10) and the action of the bent wire (CMZ) compared with that of a straight wire (C′M′Z′). Otherwise, the measurement protocol remained unchanged from the first series. On December 18, Biot presented his latest results to the academy. His measurements, so he reported, had shown that the force exerted by the whole wire was proportional to angle CMH (which he called *i*). For the force most of interest to him, that exerted by an infinitesimal wire *segment*, he inferred the complex formula,

$$1 / r^2 \times \sin \omega$$

where ω is the angle between the wire and the line between the magnet and the given infinitesimal wire segment. However, this inference contained an error: integrating Biot's microscopic force law over the entire wire yields a force proportional to the tangent of $i/2$ and not, as Biot had measured, to i itself. As Blondel has aptly observed, this calculation would have constituted no more than a practice exercise for any student at the École Polytechnique. It is hard to imagine that Biot would have been unaware of the discrepancy. It may be that he contented himself with the reflection that, for small values of angle i, the two functions would be too similar to clearly distinguish by measurement.[28] In any case, it is clear that the above formula, now known as the Biot-Savart law and expressed in modern notation as,

$$d\mathbf{B} \sim (d\mathbf{l} \times \mathbf{r}) \,/\, r^3$$

was by no means grounded directly in measurements but instead derived from general considerations and then found to be sufficiently consistent with the data. In early nineteenth-century efforts toward mathematical formalization and quantification, such a procedure was hardly unusual, though for a long time its post hoc character was not perceived as such, thanks to empiricist rhetoric and self-presentation in published papers.

With his two series of measurements and the improved force law in hand, Biot evidently saw the task of quantitative determination as essentially completed. We have no evidence that he saw any need for further research. This is noteworthy in that he had only dealt with rather specific geometric configurations. In all of his experiments, the magnet had been in a symmetrical position, never an eccentric position relative to the wire. Nor did Biot ever depart from the placement of magnet and wire at a right angle to each other. Other researchers, by contrast, had allowed the geometrical parameters to vary much more widely, so as to obtain a more comprehensive sense of the new effect. They had explored eccentric positioning, or configurations in which wire and needle lay in the same plane. This was especially true for Ampère, whose work Biot must have followed very carefully. It is not that Biot was unaware of the more complex phenomena, for in a report in April 1821 he mentioned several of Ørsted's more complex arrangements, for example, one in which both magnet and wire were positioned in a horizontal plane (Biot 1821c, 225–26).[29] But rather than pursuing more intricate experiments, he confined himself exclusively to the two setups discussed above.

The reason for this restriction was presumably that the more complex arrangements were inaccessible to any immediate mathematical analysis. The fundamental problem emerges quite blatantly in Biot's description of Ørsted's results: in order even to describe the various configurations of magnet and wire, Biot, like Ørsted before him, had to fall back on cumbersome reference to the compass points. This was the only way to unambiguously characterize the relations between needle and wire, wire and battery poles, and between all of this equipment and the observer. Consequently, in describing only a single experiment, Biot needed more than half a printed page to account for such spatial relations. The outcome of the experiment itself, however, required only half a sentence. This case clearly illustrates the conceptual situation of early electromagnetism.[30] There was simply no adequate conceptual scheme, no language, for the description of such spatial configurations. For all its mathematical sophistication, the Laplacean physics of central forces had no need of such concepts and consequently never developed any. This problem plagued everyone who dealt with the new effects. Earlier I discussed how much effort Ampère expended simply devising the concepts necessary to permit adequate description. Biot, by contrast, set the problem aside and contented himself with the few experiments he could describe using traditional concepts. He never looked for new ones. It was this narrow focus that allowed him to attain his mathematical treatment so directly and quickly. The success of his quantitative, mathematical program was purchased at the cost of considerable loss in experimental breadth.

Qualitative Diversity versus Mathematical Formalization

In addition to developing a mathematical force law, Biot also made qualitative claims regarding the "nature" of the new effects. From the beginning, he treated them as strictly magnetic, and he retained this conception unmodified throughout his studies of electromagnetism.[31] It required that he ascribe to battery currents the capacity to induce a special kind of magnetism in the wire, a process whose precise operation Biot never discussed. This view also stood in potential conflict with the Laplacean physics of imponderable fluids, on which the several fluids (electrical, magnetic, and caloric) were always conceived as *not* interacting. Biot disregarded this difficulty. His understanding of electromagnetic attraction as essentially magnetic would have far-reaching consequences for his subsequent work, which would henceforth be guided by magnetic conceptions of poles and bipolarity. Furthermore, the experimental setup at the heart of his research relied on magnetic measurements. But in addition to the mystery of the supposed "induction" process, this view led also to other difficulties. They became especially evident in his image of the current as inducing tiny magnets down the whole length of the wire, oriented perpendicular to it. One had to imagine an arbitrary number of such magnets encircling the whole circumference of any cross section of the wire but without any clear notion of whether and how these magnets might superpose or interact with each other. In light of such difficulties, it is worth asking why Biot was so completely convinced of the magnetic nature of the effects at issue, and held to this conviction so doggedly. Historical research to date has paid little attention to this question.

To begin with, by taking the phenomena as essentially magnetic, Biot self-consciously distinguished his own program from that of Ampère, who, by treating them as essentially electric, was pulling in precisely the opposite direction. But more significantly, Biot's approach points toward a background that, once taken into account, removes much of the strangeness from his peculiar characterization of the force. Twenty years prior, in collaboration with Humboldt, Biot had studied another case of interaction between magnets of very different size: the earth and the compass needle. One goal of that study had been to explain the crooked lines inscribed on a global scale by the alignment (including the magnetic dip) of magnetic needles. Biot had developed a mathematical procedure for treating these lines as superpositions of $1/r^2$ effects exerted by the two terrestrial poles (Biot and Humboldt 1804).[32] To be sure, this case could not be applied directly to the cluster of issues raised by electromagnetism. However, Ørsted's proposal of a circular action could well evoke the crooked lines of magnetic alignment. And indeed, though this fact has not been taken into account in previous historical studies, Biot's publications repeatedly referred back to Ørsted's suggestion in which, as he emphasized, the Dane had spoken "with reason" (*avec raison*). Given the rather reluctant initial reception of Ørsted's report in Paris, this endorsement was rather striking.[33]

However, Biot saw circular action as merely the epiphenomenal product of a cluster of active forces, and of course all of these, in turn, were understood to be operating as central forces. This held for his view of terrestrial magnetism as much as for his view of electromagnetism. The earlier studies had clearly shown him how an explanation for such epiphenomena might follow from underlying central forces and how such an explanation might shed light on the particular clusters of centers of force required to bring about specific phenomena. In following up on Ørsted's results, he thus had every hope of finding a similar solution. That Biot so decisively embraced the view that electromagnetic phenomena were ultimately magnetic had a great deal to do with the fact that this view, and only this view, afforded some prospect of overcoming the difficulties that made Ørsted's findings appear so anomalous.

Against this backdrop, the peculiar circumstance of Biot and Savart's direct, unwavering march to measurement in a field in which all other researchers first had to orient themselves becomes more intelligible. That they pursued this goal in the first place followed from Biot's firm commitment to the Laplacean program. But more was needed, for measurement requires at least the outlines of a conception of that which is to be measured, that is, of the character and direction of the underlying force. Biot and Savart's access to well-developed concepts from the outset was a consequence of Biot's personal career trajectory and earlier work. This, in turn, explains why Biot, unlike Ampère and most other European scientists, had no exploratory phase. For others, these were phases in which the basic concepts and categories for dealing with the new phenomena were being proposed and questioned. In Biot's case, so far as we can tell, there is no comparable period.[34] Coming to the table with a clear notion of what the operant basic concepts had to be, he had no need of any such phase. On the one hand, this saved him a significant amount of labor and uncertainty. On the other, it left him with significant constraints, especially with regard to the breadth of the phenomena he could take into account. I return to this characteristic trade-off in my general discussion of exploratory work.

It should also be noted that, unlike Ampère six weeks earlier, Biot was not plowing previously unbroken ground. Not only Ørsted but also Ampère and Arago had forged concepts that could serve him as points of reference or contention, even when he did not accept them. After all, in his investigation of angular dependence, Biot too broke with the traditional assumption of isotropy. Even though he most likely did this at the suggestion of Laplace himself, such a step was doubtless made easier by having previously been taken and explained in a prominent venue—by Ampère.

A quick glance at the broader context of European research may reinforce these remarks. In Paris, Biot's emphasis on precision measurement and mathematical formalization represented the dominant tendency, to which Ampère's approach was the exception. That Ampère took an exploratory approach in such an environment, even briefly, was certainly due to the fact that he did not feel any deep commitment to the Laplacean ideal of knowledge. Only in this way could the new findings spur him toward the revision of established conceptions and also give him the freedom to develop new and unorthodox concepts. Viewed from a broader European perspective, however, events in Paris take on a different guise. In the world at large, the path initially taken by Ampère—exploration and the quest for concepts—was far more typical. When compared with the efforts of Schweigger, Davy, Seebeck, Ridolfi, and so forth, it is not Ampère's early studies that seem unusual but Biot's. No one else in Europe set to work with such firm concepts or such definite research objectives. Most of all, no one else immediately tried to use measurements to bring the phenomena to heel. It was the perplexing qualitative aspects of the new effects that took center stage. It is no coincidence that Biot and Savart were the only ones to undertake measurements and those who most doggedly ignored the qualitative complexity. The mere fact that they confined their experiments to only two arrangements stands in sharp contrast to the large number of experimental setups and clusters of setups devised and investigated elsewhere—or even right there in Paris, by Ampère.

A clear parallel to Biot's approach during his investigation of the voltaic pile twenty years previously, as discussed in chapter 1, should now be apparent. Then, too, his proclivity toward mathematical formalization and measurement quickly narrowed his focus to a limited range of phenomena—the effects of the open pile—and a small number of instrument and detection protocols. Consequently, his electrostatic theory of the pile was never suited to anything but a narrow subdomain. Biot himself never explicitly acknowledged this fact. To the contrary, he saw his electromagnetic research as valid and foundational for the entire domain, though he had studied only a very few, specific configurations. Then as now, the price of concentrating so determinedly on quantitative mathematical treatment was a great sacrifice in breadth, openness, and innovative potential.

The Competition with Biot and Mathematization without Measurement

An Experimental Dead End

Biot's lecture brought his competition with Ampère clearly out in the open. Ampère certainly felt immediately obliged to defend his acknowledged position as leader in the new field. He evidently strived to keep the contest on the level of matters of fact, repeatedly and appropriately citing Biot in his lectures and publications. On Biot's side, by contrast, the competition quickly took on an embittered, personal tone, stunningly unveiled in a short report on electromagnetism composed in November. It appeared in the *Bulletin des sciences* published by the Société Philomatique, a scientific group of great importance to Paris academic life and among whose members almost every important name in Parisian natural sciences could be found. Ampère had been a member since 1807.[35] In 1820, the *Bulletin*'s editorial board was dominated by adherents of the traditional Laplacean view, with Poisson in charge of mathematics and Biot, physics and astronomy. His freedom to publish reports was thus completely unchecked. In the August issue, finalized at the beginning of November and appearing toward the end of the month, he made use of this freedom in his own cause.[36] Biot's half-page report was the first on electromagnetism to appear in the *Bulletin.* In it he sketched Ørsted's discovery, stressed its extraordinary character, asked what laws it might obey, and replied by referring to the mémoire he and Savart had delivered on October 30 (Biot 1820).[37] He also took note of the finding that the pile itself exerted electromagnetic effects. Ampère's name, however, was strikingly missing from the report. Combined with Biot's emphasis on his own contributions, this was an unmistakable affront. Not only was it Ampère who had discovered the electromagnetic effects of the pile but by this time he had given many academy lectures in which he had gone well beyond the effects and regularities presented by Ørsted. Ampère understood the message. In the very next issue of the *Bulletin,* he published his own contribution (Ampère 1820f), in which he mentioned his own name and those of Ørsted and Arago but not that of Biot.[38]

This new situation had a noteworthy effect on Ampère's research interests. More than ever before, he now focused on the formulation of a force law. From October 30 on it was clear that Biot had set himself this very goal and was in pursuit of a promising lead. It even seemed possible he might present a fully fleshed-out theory before Ampère did. Even if this theory were to have a much narrower scope, it would be based on exact measurements. Given the weight such measurements carried in Paris, it is understandable that Ampère's new priority was to make progress in this direction.

But at the same time, during those first weeks of November, his attempts to detect or even directly measure angular dependence with the device depicted in figure 4.6 were appearing less and less likely to succeed. We can only speculate on the precise reasons for these difficulties. In this device, to a much greater degree than in the current balance, all of the usual problems of suspension and the maintenance of electrical connectivity in the moving parts were exceedingly significant. After all, the goal was not merely the qualitative observation of a particular movement but precisely repeatable observations of the same movement under the same parameters. In all probability, it was the requirement that results be regularly reproducible that proved the rock on which Ampère's efforts foundered. We do not know whether the instrument ever worked properly. Ampère did not publish any measurements, nor did he ever even try to justify his force law by recourse to measurement. Despite all the labor he had invested, this first attempt at a genuinely electro*dynamic* measurement was a complete failure. This experimental disappointment stymied Ampère's intentions at a critical point. How, after all, could the mathematical formalization of a domain of phenomena be undertaken without quantitative data? The French tradition since Coulomb had especially prided itself on using precise measurements to ground all mathematical treatments on what was taken as the most solid foundation possible. This was the goal Ampère had set himself. When it became experimentally unsustainable, his proposal for mathematical formalization had reached a dead end.

The "Addition Law" and a New Formula

An unexpected hint finally pointed Ampère toward another possibility. As noted above, Ampère had undertaken numerous experiments in search of evidence for his theory of magnetism. In one of these, he tested his prediction that when current was flowing in the same direction through two cylindrical coils suspended parallel to one another, they ought to repel, just like magnets suspended in parallel, and in the same polarity. The experiment involved a variation on the current balance. Conductor segments AB and CD in figure 4.1 were replaced by horizontally positioned coils. One of these was fixed, while the other was suspended from the highly mobile "swing." To Ampère's surprise, and presumably disappointment, the coils displayed just the opposite of the expected motion. When current flowed through them in the same direction, the coils attracted, just as straight wires had.

This experimental result might well have led Ampère to question some of the basic assumptions of his theory of magnetism. He reacted differently, however. The discovery that coils behaved just like straight wires spurred him toward further reflection. Perhaps, he thought, the action of a "diagonal" coil winding could be understood as a superposition of two constituent actions: that of an electrified wire segment precisely perpendicular to the axis of the coil and that of another short wire segment, its length equal to the diameter of the winding, arranged parallel to the coil axis. The action of the whole coil could then be understood as the superposition of that of a (hypothetical) coil in which current ran only perpendicular to the coil axis and that of a straight wire running the length of the coil. The latter, longitudinal action could be experimentally compensated by means of a straight wire with current flowing in the opposite direction. The resulting "compensated" coil would then be precisely analogous to the arrangement of circuits in a magnet and should thus behave like a magnet. Without realizing it, Ampère had used such compensated coils much earlier, in his first experimental work on the circuit hypothesis (see fig. 3.6), for in those coils, for practical reasons, the circuit had always been closed by a length of wire returning down the coil axis. This was not the case for the modified current balance, but now that this device was retrofitted with current returning through a wire running down the center of each coil, the coils finally behaved as expected. Under parallel current they repelled, while under antiparallel current they attracted, just like magnets with the same and opposing polarity, respectively. Thus modified, the experiment could now be taken as a further success in support of Ampère's theory of magnetism. He presented it to the academy on November 6.

More importantly, however, this result also provided him with an urgently needed hint in his search for the force law, for it was obviously possible to treat the action of a current segment arrayed in any arbitrary direction as the superposition of two actions arrayed in specific directions. Though Ampère had attained this insight while working with a specialized device, he now formulated his result as the general "law" that "currents, like forces, may be decomposed." More thoroughly, "the law at issue is that a current running in the direction yielded by summing two component directions exerts . . . an attractive or repulsive force equal to the sum of the actions that would result from two currents running in the directions of the components."[39] Well aware that so general a statement of the principle had little empirical support, he confined its applicability to short current segments. For his purposes, this was sufficient. In modern language, we would formulate this "addition law," as Ampère would later call it, by asserting that two current segments together exert the same action as their "vector" sum, that is, as a hypothetical current segment formed by vectorial addition of the two real current segments. As will become clear in the following section, Ampère was fully cognizant that this was not the familiar principle of the vector addition of forces in mechanics. The law determined the composition not of forces—for the nature of the forces involved had yet to be established—but of the current segments thought to produce those forces. As such, it made a specific physical claim about electrical effects.

The addition law made possible a novel approach to mathematical treatment, one very different from direct measurement. Ampère's idea was to use it to decompose the interaction between any two arbitrarily positioned currents into whatever components he chose. For certain configurations—those involving parallel, antiparallel, or perpendicular currents—he believed his qualitative experiments had already yielded some firm results; now he attempted to deduce the workings of every complex arrangement from these simple ones. A brief illustration will suffice to show how his procedure worked.[40] To begin with, he considered the special case of the interaction between two currents, AG and BH, both oriented vertically with respect to line AB (fig. 4.11).[41] Each of these he split into two components at right angles to one another: AG into AN and AM, and BH into BQ and BP. An attractive force could be assumed to operate between the parallel components AM and BP, as well as between AN and BQ. At the same time, the force between perpendicular components AM and BQ (and AN and BP) could be assumed to be nil. Ampère took both assumptions as sufficiently plausible, given his macroscopic, qualitative experimental results. This inference was anything but well founded, especially considering that the negligibility of attractive force for the perpendicular orientation was warranted only by interpolation and never experimentally established. But leaving this difficulty aside, Ampère's reasoning yielded a total force proportional to

$$(AM \cdot BP + AN \cdot BQ)$$

or alternatively, by an addition theorem, to cos γ, where γ was the angle between the two currents (or between the projections of the two currents onto a common plane). The fact that this procedure yielded the very same result he had come to expect in his preliminary speculations can only have increased his confidence in it. Ampère's next step was to generalize this conjecture to encompass any two arbitrarily positioned currents AG and BH. In this configuration, all three angles were once again recognizable (see fig. 4.12): GAN = α, HBQ = β, and DBP = γ. Two of the "components," AM (= AG · sin α) and BP (= BH · sin β), reduced to the special case considered above. These, in turn, were perpendicular to components AN and BY, and Ampère could assume that the corresponding perpendicular pairs contributed no force. That left only the force between the collinear components AN (= AG · cos α) and BQ (= BH · cos β) to be considered. Ampère suspected that this force was also nil, but in the absence of any persuasive argument the matter had to be left open, introducing an additional term into the formula. Assuming (without any discussion) that the total force must be proportional to the product of the two intensities, as well as to the inverse square of the distance between the currents, it could be rendered as

$$g \cdot h \cdot 1 / r^2 \cdot (\sin\alpha \cdot \sin\beta \cdot \cos\gamma + n/m \cdot \cos\alpha \cdot \cos\beta)$$

where *n*/*m* was the as yet unknown coefficient for the contribution of the two collinear components. Ampère thought it likely to prove equal to zero, in part because the resulting formula, lacking the second term in the sum, would then be identical to the formula he had proposed in November.

I consider it quite remarkable that in this way Ampère deduced—validly, in his view—a mathematical formula from so few empirical data points. In the success of his argument, in which the addition law played a central role, Ampère saw a promising way toward *mathematical treatment without measurement.* This procedure, the exploitation of a small number of basic assumptions by means of mathematical analysis, would become characteristic of the whole of Ampère's electrodynamics.

In his lecture of December 4, before presenting the force law itself, he paved the way with another step. He showed that his force law yielded a 1/r distance function for the force between two straight, infinitely long current-bearing conductors and that the same applied to the force between such a conductor and a magnetic pole.[42] This constituted his first real effort to calculate an electromagnetic effect, going beyond his earlier thoughts on current-current interactions. Of course in presenting this calculation at precisely this time he was making a very specific point. He was deliberate-

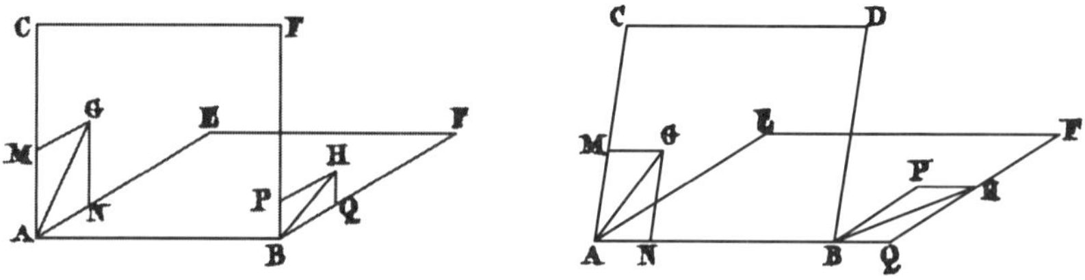

FIGS. 4.11 AND 4.12. Ampère's analysis of the force between two current segments (Ampère 1822b, plate 6).

ly recalling Biot's experimental measurements, not to mention targeting the 1/r law Biot had deduced from them and presented five weeks earlier. On the one hand, this showed that Biot's empirical finding could also be articulated and understood by means of different conceptual frameworks than Biot's. On the other, Ampère, who had no empirical measurements of his own, could now appropriate Biot's results both as welcome empirical support for the fundamental notion of reducing all electromagnetic effects to current-current interactions and as an impressive display of its explanatory power. In the bitter contest between the two rivals, this was a clever, and successful, move. We can well imagine Biot's displeasure at this arrogation. He intervened immediately, promising new measurements and a more refined force law at the very same session (Hofmann 1995, 269). When he delivered on this promise in his lecture to the academy two weeks later (December 18; see above), he was once again on an even footing with Ampère.

Avoiding Measurement with an Equilibrium Experiment

Ampère's force law left important questions unanswered, beginning with the status of the unknown factor n/m. Ampère considered this point carefully and drew on various empirical results in an attempt to show that $n/m = 0$.[43] Once again, while his arguments provided a certain plausibility, they would not withstand careful scrutiny. Ampère himself seems to have sensed this, notwithstanding his occasionally more confident, though generally rather diffuse rhetoric. In any case, he would return not much later to consider the possibility of nonzero values for n/m. Another, more significant point concerned the status of the addition principle itself (which Ampère also called a "law" and on which his argument heavily depended). Ampère was well aware that it had to be empirically justified and that he had not yet done enough in this regard: "Because it is from this law . . . that he [Ampère] deduced his analytic expression for the mutual action of two infinitely small segments of electrical current . . . one feels that he ought to have sought the means of verifying it by direct, precise experiment."[44] But here, as with the force law itself, he was confronted with the problem of having to make claims about the relationships among infinitesimal entities, which were in principle inaccessible to any direct empirical confirmation. Experiments could at best increase their plausibility. Ampère's guiding intuition was that the action of a straight wire ought to be the same as that of a wire running in a large number of very small, irregular windings but with the same starting and ending points. In presupposing the transverse dimension of the arrangement much smaller than the longitudinal, and the form of the convolutions very irregular, he was presumably thinking that such an arrangement could demonstrate the mutual compensation of a large number of small-scale windings twisted in every imaginable direction.[45]

To realize the idea experimentally, Ampère envisaged new measurements, though not of the current-current interactions in which he was ultimately most interested. Since such measurements had proved unattainable, he contented himself, as a sort of substitute, with measuring the action of currents on magnets, measurements whose practicality Biot and Savart had demonstrated so compellingly. Accordingly, Ampère designed his first instrument very much in the spirit of Biot.[46] A horizontally suspended magnet would oscillate under the influence of variously placed conductors. By measuring the duration of oscillations, it should be possible to determine whether the effect was any different with a straight wire as opposed to one coiled in the manner described. This attempt at measurement also yielded no results, or, at any rate, Ampère nowhere mentioned any data. Unlike Biot and Savart, he was evidently unable to overcome the experimental difficulties involved.

Once again, Ampère took the failure of an attempted measurement as an incentive to more carefully consider his goal and the means chosen to pursue it. What he really wanted to know was whether or not two differently configured conductors would exert the same effect. This was in the end not a matter of numerical values but a simple yes-or-no question of equality or inequality. In deciding such a question there were other options besides the method of measuring and comparing the two effects numerically. Instead, one might work to produce both effects simultaneously in an arrangement in which, should they turn out to be equal, they would neutralize each other and, if not, exhibit some residual action whose magnitude did not have to be *measured* because its mere presence would be enough to demonstrate the inequality of the two effects. The comparison of the two effects could thus be built into the experimental apparatus itself. Ampère was already familiar with experimental setups in which two composing effects compensated for one another, having just recognized the device with compensated coils (see fig. 3.6) as one such. Avoiding measurement in this way had the further advantage of allowing him to directly examine the true object of his interest: current-current interactions. In addition, he did not believe the rapid decline in battery strength would be a problem in this kind of experiment. Such reflections may well have inspired the design for

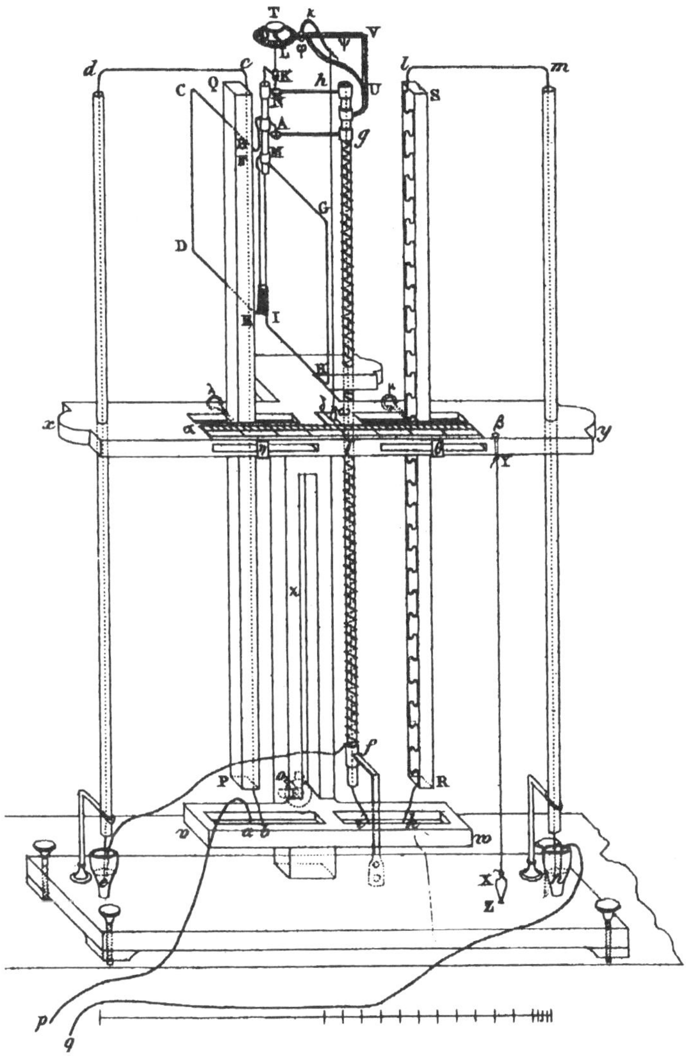

FIG. 4.13. Ampère's apparatus to prove the addition principle (Ampère 1827, plate 1).

his second apparatus for demonstrating the addition law.

In this device (fig. 4.13), mobile conductor segment GH was mounted on freely rotating axle FI, with symmetrically mounted segment CD neutralizing terrestrial magnetism, while current arrived and departed through glass tube gf. GH interacted with two other conductors—straight wire PQ on the right and coil RS, composed of many small, irregular windings, on the left. If the addition law was valid, the effect of RS should not differ from that of the straight wire (PQ) placed mirror-symmetrically across from RS. With both connected to the battery such that current flowed in the same direction, their actions on mobile segment GH ought to precisely neutralize each other. We do not know whether Ampère actually had this device built or worked with it. What we do know is that he no longer questioned his addition law.

This apparatus served as the prototype for an entire series of further devices, the cornerstone of an experimental method Ampère would pursue with great success in the years that followed and that would eventually become the foundation of all of his reasoning: the method of equilibrium experiments. One major incentive for introducing it had been the discovery that direct measurements of current-current interactions could not be realized. The fact that he persisted with this method reflected his continuing tendency to *avoid measurement.* The mathematical treatment he would develop so thoroughly over the coming years made contact with empirical evidence almost exclusively through experiments of this kind. In such experiments, the question was always *whether or not* a given motion took place, and never *to what degree* or *how strong.* Ampère's equilibrium experiments were always geared toward determining whether or not two actions mutually compensated, that is, toward answering yes-no questions. Numbers and numerical data played no role. Ampère felt that in this way he could avoid all of the difficulties associated with actual measurement.[47]

It is a truly remarkable fact, mostly neglected in previous historiography, that although Ampère's electrodynamics was a highly sophisticated mathematical theory, it did not rest on even a single measurement. The method of equilibrium experiments had shown him the way toward mathematical treatment without measurement. The result was not free-floating speculation, unfettered by any empirical findings: it did have an empirical basis, though one of very specific character. This peculiar cluster is striking, and it doubtless represents a very unusual episode in the history of science. How can a mathematical theory be grounded in empirical findings but without measurement? In Ampère's case, the answer was, by equilibrium experiments. Quantitative analysis took place on paper, prior to experiment, and guided the design of experimental apparatus. The method went so far as to sustain a complete and extremely successful mathematical treatment on the basis of yes-no answers alone.

There is a further point that has escaped attention. Ampère was not the only one who avoided measurement so systematically and with such consummate skill. No one else undertook such measurements either. The next attempt at an electro*dynamic* measurement would not take place until Wilhelm Weber's work in Göttingen, more than twenty years later. In order to attain truly stable results, Weber had to design entirely new procedures around his bifilar suspension.

The central role of equilibrium experiments in Ampère's work on electrodynamics becomes very clear against that background. In several publications, the historian of science James Hofmann (1987a, 1987b, 1995, chaps. 7–9) quite rightly stresses the significance of the first of these experiments. However, his analysis of the context in which Ampère chose this path strikes me as one-sided and insufficient. Hofmann (1995, 260) cites Ampère's "longstanding preference for the direct analysis of simple theoretical conclusions from more complex observations" as one of his chief motivations, in virtue of which he distinguished himself from "Biot and other more inductively minded Laplacean colleagues."[48] In this analysis, Hofmann is informed by Ampère's reflections on four modes of scientific argumentation, discussed in chapter 2. For the fall and winter of 1820–21, however, I find no indication that these epistemological considerations played such an important role, leaving aside the fact that Ampère's epistemology does not seem to entail any special preference for equilibrium experiments. Furthermore, Hofmann overlooks the fact that Ampère's search for a mathematical treatment indeed began with the attempt to perform actual measurement series, just as Biot had done, so as to use them as empirical justification for his law.[49] Only after the failure of all such attempts at measurement did he find himself, in the midst of a fiercely competitive situation, compelled to look for other methods. One of the constraints in this search would have to be the avoidance of measurement. It was in this context that his prior experience with experimental arrangements geared toward mutually compensating effects came into play. Such perspectives are either missing from Hofmann's discussion or not granted the importance they must have had for Ampère. In this context, I find no trace of Ampère's epistemology. It played a much smaller role than Hofmann claims.

Competing Theories

Throughout December, Ampère's work was profoundly shaped by his competition with Biot. By the time of Biot's presentation of December 18, if not earlier, a remarkable situation had arisen: in the same place and at more or less the same time, two different theoretical approaches to the field of electromagnetism had been given mathematical expression. The two had shared important features. Both assumed central forces operating between infinitesimal segments, and both broke with the tradition by introducing angular dependence. For the Parisian public, however, the differences between them were doubtless more salient. Ampère's force law governed a force operating between electrical currents, while Biot's applied to forces between currents and magnets. Furthermore, Biot had made it quite explicit that he viewed the force as purely magnetic, with the current serving only as the cause of its appearance. The applicability of his law was thus confined exclusively to the domain of electromagnetic phenomena, which Ampère viewed only as a subdomain, and not even the most important one. With his own focus on current-current interactions and the theory of magnetism, Ampère had targeted a much broader domain.

The fact that Ampère was much more worried by the coexistence of the two theories than Biot had to do with the relative standing of the rivals in the scientific community. As a physicist, Biot was internationally well established. His widely distributed textbook alone ensured both his Parisian support base and pan-European attention to and sympathy for his approach. Ampère's situation was completely different. As a newcomer to the field, he had to take the initiative, struggling actively for the dissemination and recognition of his work. He knew these efforts would quickly come to naught if his theory was absorbed by the powerful traditional school or in some other way proved inferior. The different publication strategies adopted by Ampère and Biot make clear just how stark these differences were. I will shortly consider them further.

Against this backdrop it is no wonder that Ampère viewed Biot's December 4 announcement, not to mention his December 18 presentation, as a challenge and thus immediately began considering what Biot's results meant to his own approach—not to his actual force law, which governed forces between infinitesimal segments and was thus not open to direct critique, but to the "empirical laws" justified by measurements. He had already successfully prosecuted this strategy once, with reference to Biot's first law. Laplace, who wanted to keep his distance from the conflict, as it appears, provided a further strong incentive by posing the question pointedly.[50] Thus prompted, Ampère further pursued his efforts in the field of electromagnetism—not the proper domain of his theory at all—by attempting to use his account of current-current interaction to calculate the force exerted by a current-carrying wire with a V-bend on a small magnet. The result, which he presented to the academy on January 8, 1821, was a more complicated force law than Biot's (Blondel 1982, 98–99; Grattan-Guinness 1990, 2:925–26).[51]

With this new law in hand, Ampère looked for an experimental means of deciding between his own theory and Biot's. One case on which the two made different predictions involved an arrangement in which a suspended, oscillating magnet was acted upon by two wires, one horizontal and the other vertical. On January 20, 1821, he and his tenant Despretz undertook the experiment, obtaining an unambiguous result. Unfortunately, it was precisely the result expected according to Biot's theory. Ampère's initial reaction to what was doubtless a disappointing finding was to bide his time. Though he had announced the pending experiment at his academy lecture of January 15, he kept its outcome to himself, publicly acknowledging it only two years later. By that time he could tout it as a success for his own theory, for in the meantime he had ascribed a new numerical value to the unknown factor n/m in his force law (n/m = 1/2), superseding the assumption n/m = 0, on the basis of which he had calculated the failed prediction. As James Hofmann correctly observes, this sort of reaction to negative experimental results—wait and see—was quite typical of Ampère and remained so for years to come.[52] In January 1821, however, the negative result left him with no basis for arguing the superiority of his theory over Biot's, leaving the two theories coexisting in public discourse.[53]

For the sake of completeness, it should also be noted that, in his January lectures, Ampère offered further reflections on the theory of magnetism and, for the first time, alluded to the possibility—suggested by Fresnel and by Arago's magnetization results—that the electrical circuits within magnets need not be imagined as concentric but could also be conceived as molecular—evenly distributed throughout the entire magnet body. These thoughts were not worked out in any detail until later.

Ampère's Efforts on Disseminating His Findings

Recall that as early as October Ampère had taken great pains to have his "Conclusions" printed and widely circulated. He continued to devote much effort to publicity. As a counterpart to the list of academy activities in table 2.1, table 4.1 lists all of the publications on electromagnetism circulating in Paris through March 1821. The left-hand column specifies the date on which a given publication was actually available to the public, based, for example, on when the receipt of the journal issue in which it appeared was logged at the academy or recorded in the *Bibliographie de France*.[54] This list also includes specially commissioned offprints circulated by their authors but otherwise not generally available.

The table shows that in publishing, as in research, Ampère was more active and quicker off the mark than any other author in the French-speaking world. His large number of publications may be partly attributed to the fact that, in some cases, a text once written would be printed several times, sometimes with slight variations. I have treated such "parallel" printings as separate articles, marking them as such in the table. Though such strategies were hardly unusual in early nineteenth-century Paris, Ampère pursued them more avidly than other authors. But even disregarding the parallel printings, Ampère still penned a remarkably large number of different texts during these months.

Though the "Conclusions," Ampère's first publication on electromagnetism, did not appear in a journal until November (and in three other journals shortly thereafter), his private print run had made it available much earlier, in the second week of October. Via the extensive distribution of this offprint, Ampère had ensured that by the end of that month his program would be widely known. In October he also penned the first section of the detailed mémoire discussed above, and by the middle of November this text had been printed in the *Annales de chimie et de physique* (*ACP*). Ampère was thus the first to press with an in-depth presentation, two weeks ahead of Biot. Recall that he used this piece to systematically set out the conceptual foundations and central experiment of his new theory. The full range of his experimental efforts, however, remained largely unaddressed. He had consciously deferred it to the promised second section of his paper.

However, the publication of this second half was significantly delayed, thanks to the turbulent progress of Ampère's work and Biot's entrance on the scene. Ampère was strongly motivated, especially given the stiff competition, to back his claims with as much empirical support as possible. As not all of his experiments went as expected, he continued to delay publishing, until in its final form the essay encompassed the material from his lecture of November 6. By this time, Biot's lecture had heated the competition to such a point that Ampère felt he had no time to lose. At the same time, with instruction beginning at the École Polytechnique on November 10, he began to face serious limitations in the time available to him for such efforts. Working from the drafts of his academy lectures, he assembled a detailed text in "extreme haste," as he described the process to his friends.[55] Some passages from the lecture manuscripts were used with-

TABLE 4.1. French-language publications on electromagnetism through March 1821

Date	Ampère	Other authors
September 18, 1820		Ørsted 1820j (*Bibl.Univ.*) Ørsted 1820l (*AGSP*)†
October 9	Offprint, "Conclusions"*	Ørsted 1820f (*ACP*)†
November 11	Ampère 1820b ("Conclusions," *J.Phys.*)†	Ørsted 1820b (*J.Phys.*)† Ørsted 1820k (2nd circular, *J.Phys.*)
November 13	Ampère 1820d (*ACP* mémoire, part 1)	Arago 1820a (*ACP*)
November 25		Biot 1820 (*Bulletin*)
November 27	Ampère 1820c ("Conclusions," *AGSP*)	
December 2	Ampère 1820i (offprint of "Notes")*	Hachette 1820b (offprint, survey)† Blainville 1820 (reported in *J.Phys.*)
December 18	Ampère 1820j (*ACP* mémoire, part 2)	Biot and Savart 1820 (*ACP*) Fresnel 1820 (decomposition of water, *ACP*)
December 26	Ampère 1820e (offprint of the whole *ACP* mémoire)*	
December 30	Ampère 1820h ("Notes," *J.Phys.*)† Ampère 1820g (*J.Phys.* report on December 4)	Hachette 1820a (survey, *J.Phys.*)
January 2, 1821	Ampère 1820a (*AGSP*) Ampère 1820f (report in the *Bulletin*)	Davy 1820b (*AGSP*)
January 22		Boisgiraud 1820 (*ACP*) Gazzeri 1821 (*Bibl.Univ.*)
January 29		Doebereiner 1820 (*AGSP*)
February 17		Davy 1820a (*J.Phys.*)†
February 19		Configliachi 1821c (*Bibl.Univ.*) Ridolfi 1821 (*Bibl.Univ.*)
March 23		
March 26	Ampère 1821f (continuation, *AGSP*) Ampère and Laumont 1821 (*Annales des Mines*)	Configliachi 1821b (*AGSP*) Gazzeri, Ridolfi, and Antinori 1821 (*Bibl.Univ.*) Buch 1821a (*Bibl.Univ.*)

Note: Offprints are signalled by an asterisk (*) and parallel publications by a dagger (†).

out modification, while others were completely reworked in light of his maturing theoretical insights. Much more work would have been required to integrate the various pieces from different phases of his research into a coherent text. Faced with a choice between publishing this piecemeal assemblage quickly and taking the time needed to produce a more complete presentation, he opted for the former, delivering his text to the printer over the course of November. It was important to him to underscore the experimental side of his work with as many illustrations as could be managed, and so a large number of drawings were made and plates engraved. Text and plates alike were published in the *ACP* toward the middle of December. Moreover, Ampère quickly had a small volume printed that combined a slightly modified version of the essay with the previously published first part, the volume thus comprising sixty-eight pages of text and five plates, continuously paginated (Ampère 1820e).[56] Despite the work's deficits, of which Ampère was well aware, he once again pursued a strategy of swift, wide-ranging distribution. On December 26 he presented a copy to the Paris academy, and the receipt of another copy was recorded at the Royal Society in London at the beginning of January.[57] The heterogeneous character of the presentation was evident to its readers. When, in the spring of 1821, the Leipzig physicist and editor Ludwig Gilbert published a German translation, his foreword characterized the essay as "somewhat disrupted and unformed" (Ampère 1821g, 114).

November also saw Ampère at work on a different sort of text. At the request of Jean Nicolas Pierre Hachette, an older and well-respected academician, he put together a brief chronological overview of all of his academy lectures. The fact that Hachette was already working on a survey article on electromagnetism is remarkable, considering that work in the field had begun only two months earlier. Hachette was doubtless motivated not only by the brisk pace of developments but also by the clearly discernible competition under way in Paris. In this context, it was to be feared that future reports would be more strongly shaped by the conflict than an impartial observer might wish. That Hachette would ask Ampère to supplement his own survey by providing a summary of his activities also points in this direction. The goal of such a survey would not be the discussion of experiments or theories but the documentation of their historical sequence. Ampère can only have welcomed this opportunity to display the breadth of his contributions and, not least, to underscore his priority claims. In drafting his overview, he drew on his lecture manuscripts. Leaving aside a slight tendency to pre-date certain events, the result is an accurate if very brief sketch of historical developments. Together with Hachette's (1821) article, it would be published toward the end of December in the *Journal de physique.* However, four weeks earlier, an offprint of it had already appeared, which Ampère also exploited as a publicity vehicle. And by early February 1821, an English translation of it would appear in London in the *Philosophical Magazine* (Ampère 1821c).

I have already discussed Biot's first report of his own work on electromagnetism. As one of the editors of the *Bulletin,* he could publish his work without further review, and quickly at that. His report was printed on November 25, a mere four weeks after his lecture. The calculated insult of discussing electromagnetism in general terms without mentioning Ampère's name demanded a response. As discussed above, Ampère set to work on his own report for the *Bulletin,* which appeared in the next issue, printed toward the end of the year. The tenor of this report was molded by the context of its composition and placement. Here, by contrast with his larger mémoire, he took pains to proceed very systematically. Toward this end, significantly, he introduced a new organizational principle for the very first time: "M. Ampère's work divides naturally into two clearly distinct parts: one of them is purely experimental . . . the other is composed of reflections on phenomena . . . on the basis of which he concludes that these phenomena . . . are due solely to electricity acting on magnetized bodies."[58] Such explicit segregation of experimental facts from theory had been completely absent in his earlier research activities and publications. In previous presentations he had frequently alternated between the two as need or inclination dictated. The sharp distinction he now introduced between them, and its description as "natural," was a conscious rhetorical strategy. Among other consequences, it caused the presentation of his results to deviate from their historical development even further. It is no coincidence that the strategy first appeared in the *Bulletin,* addressed to a society dominated by traditionalists. The Laplacean tradition held the distinction between theory and experiment in high esteem, treating it as a kind of epistemological baseline. With his rhetorical conformation to these categories Ampère sought both to present his results as valid by traditional standards of argument and to provide the strongest possible response to Biot. Used in this context for the first time, this structure would be retained in subsequent publications.

The rivalry with Biot also serves as backdrop to Ampère's rush to print the text of his December 4 lecture, or at least that portion of it devoted to the presentation of the mathematical force law. He was evidently able to find a place for it in the next issue of Blainville's *Journal de physique,* which appeared in less than four weeks (Ampère 1820g). His mathematical force law and its derivation were thus in print well before Biot's.

Much of Ampère's material found its way quickly and directly into the hands of a correspondent of the Brussels-based *Annales générales des sciences physiques*. The "Conclusions" were published there in October (Ampère 1820c). *AGSP* also published a larger, very heterogeneous essay consisting of important extracts from the lectures of September 18 and 25 and October 9 and 30, together with the text prepared for the *Bulletin* (Ampère 1820a). We do not know whether this piece was assembled and edited by Ampère or someone else, though it is clear that it drew directly on material in his lecture manuscripts that was not published anywhere else. However and through which individuals this communication took place, it is certain that Ampère actively supported it.

Other media also provided opportunities for publicity. For example, the daily newspaper *Le Moniteur universel* sometimes offered brief accounts of Ampère's academy lectures, including those of October 30, November 6, and December 4.[59] Such notices are of interest, because *Le Moniteur universel* was primarily a sort of unofficial outlet for government decrees and the like, as well as for reports of domestic and foreign political events. A column with the title "Académie Française" appeared very occasionally and another, "Institut Royal de France," even more irregularly. Advertisements for scientific books were equally rare. The short reports on electromagnetism were the first on this topic. That they mentioned Ampère, Arago, and Fresnel but not Biot makes it likely that Ampère or Arago took the initiative in encouraging their publication. This conjecture is supported by the fact that manuscripts of these reports are among Ampère's papers; while they are not in his hand, their author must have been closely connected to Ampère himself.[60] His mention in such outlets may have been unusual but still useful for establishing a new name on the scientific landscape. Such notices also brought news of electromagnetism to foreign researchers (Yelin 1820).[61] In sum, it is plain that Ampère's publication activities cast a very wide net and were very labor intensive. This is all the more significant since he pursued them in parallel to his no less labor-intensive research activity, on top of the demands of teaching at the École Polytechnique from the middle of November on. That he nonetheless found time for such an intense drive toward publication underscores the great importance he placed not only on producing new results and presenting them to the Parisian public but also in reaching a broader audience and persuading it of the significance of his work.

By comparison with Ampère, other Parisian authors adopted much less intensive publication strategies. Even acknowledging, as Arago would later explain, that in Paris it was standard procedure to expend great effort in discussing and publicly resolving questions of originality and priority, the degree of attention Ampère devoted to this task was certainly unusual.[62] By contrast, Biot published only two short notices in the fall of 1820. A longer text would not appear until the early summer of 1821 (Biot 1821c), and a truly thorough presentation would have to wait until the publication of the third edition of his textbook in 1824 (Biot 1824, vol. 2).[63] Biot could be certain of gaining public attention without the need for such a fevered publication strategy. Arago was relatively quick off the mark, placing an article on his findings in his *ACP* and also publishing Fresnel's notice. Broader cognizance of various international contributions did not set in until 1821, by which time Ampère's important publications were in print and distributed throughout Europe. His aggressive strategy stood out not only in the Parisian context. In all of Europe there was not a single researcher who worked so intensely and with such success to see his results published quickly and widely. To be sure, hardly any other researcher so strongly tied his studies in the new field to the goal of crafting a new personal profile.

Two Phases

After the January 15 lecture and the disappointing experiment that followed five days later, Ampère's research on electromagnetism slowed down considerably. In the months that followed he engaged in some international correspondence on experimental minutiae that had emerged over the course of replication efforts by others. In that process he modified and tightened the form in which he presented his theory. The long break was interrupted by a rather brief lecture at a public session of the academy in April 1821, in which the rivalry with Biot also weighed heavily.[64] Not until October 1821 would a report on a new, potentially problematic experiment performed in London startle Ampère into a new phase of extremely intensive research. The report on an electromagnetic rotation effect came from a newcomer to the field, the unknown laboratory assistant Michael Faraday. But that was still far in the future.

The decline in Ampère's research activity that began in January 1821 had various causes. To begin with, toward the beginning of the year he fell seriously ill (Blondel 1982, 103; Hofmann 1995, 274).[65] This was doubtless at least in part a consequence of the great pressure of the preceding months.

There were also significant reasons that the break lasted so long, however. In his December and January lectures Ampère had strengthened his theory and presented its first mathematical treatment. Its foundations had gained experimental support, a mathematical framework, and broad publicity. Though the competition with Biot remained, unresolved and unsettling, Ampère could afford a somewhat more tranquil pace.

Historians agree that the phase that ended in January 1821 was absolutely decisive in the development of Ampère's research program. In her monograph, Blondel (1982, chap. 3) titles the corresponding chapter "Les fondements de la théorie électrodynamique" (The foundations of the electrodynamic theory). But she treats the period more or less as a single unit, as do all other historical studies. According to this approach, the goals Ampère adopted later in the period are also taken as representative of the first few weeks, about which very little was known. This picture must now be fundamentally revised. To a much greater degree than previously recognized, September 25 marked an important turning point in Ampère's research. He made a crucial and far-reaching decision, after which the direction and character of his research changed markedly. It would not be an overstatement to speak of *two* research phases, one comprising the three weeks from September 4 through 25, 1820, and the second, nearly four months, from September 25 through the middle of January 1821.

Of the seven thematic strands discernible in the first period, several were no longer pursued in the second. Significantly, these were the exploratory strands. After proposing the galvanometer, Ampère took no further interest in it. He took the reciprocity of effects as unproblematic, posing no significant new questions. Matters stood differently when it came to the third of the abandoned strands. The program of "reduction to general facts" was far from having reached any satisfactory conclusion. Ampère was aware that there remained interesting open questions. He was also aware, however, that their pursuit would have required a great deal of work, might have led him in an entirely different direction, and above all, would surely have demanded that he retain his exploratory methodology, continuing to leave his basic concepts open to revision. This was not what Ampère wanted, and, given the time pressure, it would have proved counterproductive. So he decided not to pursue these questions. The concepts presented in the "Conclusions" would henceforth no longer be questioned but would serve as the basis for all of his future research. On this foundation he based his reflections on the connection between electricity and galvanism (especially as presented in his first article), his theory of magnetism, including terrestrial magnetism, and his theory of current-current interaction in the form of a mathematical force law. It is no oversimplification to say that all of Ampère's work in the second period concerned either the theory of magnetism or the mathematical force law.

Looking ahead, it is worth noting that one of the most important questions Ampère set aside at the end of his exploratory phase concerned the explanation of Ørsted's experiments with vertically aligned wires. He was well aware that there were open questions remaining. It is somewhat ironic that, a year later, when Faraday investigated precisely these experiments in much greater detail in his own exploratory work (described in chap. 6), they led him to the discovery of electromagnetic rotation, which would prove so challenging to Ampère.

The two phases differed not only thematically but in the whole character of Ampère's work. The first might be described as a period of uncertainty, (systematic) trial and error, or, to use Hans-Jörg Rheinberger's apt term, "tinkering." Ampère's experimental work throughout the first phase was primarily exploratory. It remained open, on the most basic level, as to which concepts, categories of thought, and means of representation were best suited to the field. The direction of future research was not yet clear, its most important research questions still undetermined. Only toward the end of this period did stabilization and the selection of a particular research direction take place, as documented in the "Conclusions." By contrast, the second period found Ampère's activities much more sharply focused. There was no trace of uncertainty or vacillation in the direction of research, and the conceptual scaffolding was no longer flexible. All new developments involved the fleshing out and extension of structures laid down in the "Conclusions." The theory of magnetism and the force law constrained the scope of research questions. The character of Ampère's research practice shifted accordingly. Experiments were now all rigorously integrated with theory, being designed and evaluated in theoretical context. There was no sign of exploration or deep-seated insecurity in the second period. Instead, experiments were geared toward building on and, occasionally, defending the foundations established in the first period. The second period, far more than the first, was shaped by the immediate exigencies of the rivalry with Biot and by Ampère's efforts to position and assert himself in this competitive climate.

The insight that two distinct phases in Ampère's electromagnetic research may be clearly distinguished sheds new light on his entrance into an entirely new field. In place of the received view of him as choosing his research direction almost from the outset, with all the problems that entails, we find a much more nuanced view. Only *after* September 25 did his work head in a clearly defined direction, the direction so pointedly documented in the aptly titled "Conclusions." But this date was preceded by a highly intensive period of work, one in which the basis for selecting a research direction was first developed from a variety of very diverse possibilities. This first period is a remarkable example of a "formative phase," as described in my introduction. One of its characteristics should be emphasized again: Ampère had no way of foreseeing whether the research program set out in the "Conclusions" would prove amenable to successful consolidation. Indeed, he would later claim that a kind of soothsaying had been at work.[66] One might also call it a lucky toss of the dice. But this was not what motivated his decision of September 25. Rather, it was driven by diverse contributing factors, such as experimental findings, theoretical proclivities, and the social situation in the Paris academic scene. Ampère's case strikingly illustrates the claim made in my introduction—that a formative phase reveals the complexity of research with particular clarity.

Ampère's Self-Presentation as a Researcher in Paris

Crafting an image of the researcher is part and parcel of research. The impression Ampère himself would convey of the evolution of his work during these first two phases is highly instructive. To be sure, he never provided a systematic historical account of his progress with any pretense of completeness. The closest he came were the remarks on his research trajectory in the second part of his *ACP* mémoire, in which he claimed to offer a "fair idea" (*idée juste*), if not of the whole development, then at least of several episodes (Ampère 1820j, 196). At several places in this text he refers to datable events, especially his lectures to the academy. Occasional glimpses into specific moments of his research made it clear for the reader that this was no purely systematic account, cleansed of all historical narrative, any more than it was a strictly historical report. Still, the episodes he mentioned might leave the reader with a sense of having learned something about his process.

The "Conclusions" offer a representative sample of the kind of self-presentation that would become typical of Ampère. Pride of place was given to the theory of current-current interaction and the explanation of magnetism, while the project of reduction to general facts was barely mentioned. This tendency became more pronounced in the aforementioned *ACP* mémoire, in which the two periods were weighted very differently.

The second period was described in detail, with some dates, carefully articulated background assumptions, and explicit acknowledgment of his two primary targets: the theory of magnetism and the force law. By contrast, events of the first period were discussed more rarely, as well as much less thoroughly. Most importantly, they were not placed into a context in which they might have made sense. This different weighting was no accident and had only little to do with the fact that the second episode was much longer.

In composing the *ACP* mémoire, Ampère could draw directly on the manuscripts of his academy lectures. He had the full scope of his work to date at hand, and, as the documents paraded before him, he made choices in preparing what to present. He rearranged various text passages, changed formulations, and added new paragraphs. But, above all, he drew very selectively on the materials at his disposal, leaving out a great deal.[67] A glance at the manuscript shows that he treated his records of the two periods quite differently. For the second period, he incorporated whole sections of his lecture manuscript into his text, including long passages in which he made no changes at all. The few rearrangements and changes he did make were confined to short passages. By contrast, the way in which he used documents from the first period was fundamentally different. Not only did he leave out a great deal, but he also extensively reworked the portions he retained. In the surviving manuscript of the mémoire, there is hardly a page on which we find no sentence corrected or inserted, no paragraph struck out or pasted over with another.[68] Ampère clearly had a lot invested in presenting this period to the outside world in a very specific way, one that necessitated substantial reworking of the original contours of his work.

In order to illustrate the consequences of this revision effort, I have reproduced in table 4.2 the overview of Ampère's first-phase activities given in table 3.1, this time highlighting the activities elided from his self-representation—those about which his reader could learn nothing. Without canvassing all of them, the overview makes it clear that his work before the first academy lecture, and especially the more exploratory efforts, was the target of his most drastic revisions. Only the galvanometer plays the same role in his self-presentation as it had in the actual research. While Ampère introduced the instrument in the first part of his essay and stressed its importance, he never returned to it—thus precisely mirroring the course of his work. The situation is completely different when it comes to the "reciprocal" experiments, on which he took a radical step: there was absolutely no mention of these studies in his article. This is of interest insofar as he had taken considerable pains with those experiments and learned key experimental techniques in the process. His deletion of all passages related to this work reveals an important tendency in Ampère's self-presentation. He expected that in the eyes of the Parisian audience to whom his text was primarily addressed, these experiments were simply not that interesting. From a theoretical point of view, reciprocity was naturally to be expected but was not particularly exciting. How, then, could anything new be learned from its experimental confirmation? In Paris this question was more pointed than elsewhere, and it is no surprise that no one besides Ampère made efforts to demonstrate reciprocity. The fact that Ampère had charged himself with the task may be understood only against the backdrop of his exploratory work, in the context of which such experiments made perfect sense. However, he could not see how such experiments could improve his Parisian reputation as a scientist, and he may even have feared they would become a liability.

The same holds, finally, for the topic on which he had expended far and away the most effort during the first two weeks: the search for empirical rules and the "general facts" among the phenomena to which a special role

TABLE 4.2. Ampère's first-phase activities as selected or elided for his self-presentation to the public

	Thematic strands						
Activity	GF	RE	GS	TM	GM	GC	GE
September 4: Report on Ørsted's result in Paris							
Experiments on *action directrice*: the astatic needle	•						
First formulation of a rule for *action directrice*; definition of "right" and "left" of current; the "*bonhomme*"	•						
Experiments on the reciprocity of *action directrice*; unsatisfactory result with terrestrial magnetism; proposal for trial using a strong magnet		•					
Reflections and experiments on the *action attractive ou répulsive*	•						
Formulation of a rule for the *action attractive ou répulsive*	•						
Experiments on *action directrice*: needle material varied (wood, paper, manganese, tin, copper, zinc)	•						

TABLE 4.2 (*continued*)

Activity	Thematic strands						
	GF	RE	GS	TM	GM	GC	GE
Unsuccessful experiments at the École de Médecine on the magnetic action of conventional electricity	•						
Experiments on the magnetic action of the pile	•						
Second, more condensed formulation of the rule for *action directrice*; the concept of a *circuit galvanique*	•						
Experiments to determine the direction of current within the pile by means of chemical decomposition	•						
The hypothesis of terrestrial current circuits				•			
Reflections toward a micro-scale theory of a single electrical fluid							•
Further reflections on the causes and effects of possible terrestrial current circuits				•			
Reflection and envisaged experiments on the magnetic activity of the pile; *galvanoscope* proposal			•				
Reflections on the improvement of experiments on the magnetic action of conventional electricity	•						
Proposal of a graduated scale for the galvanoscope; further development toward the *galvanomètre*			•				
Reflections on the improvement of experiments for the reversal of *action directrice*; the *aiguille galvanique réctilinéaire*		•					
Generalization of the hypothesis of current circuits; idea to "imitate" the behavior of the magnetic needle by appropriately arranged circuits					•		
Idea of aligning mobile coils (*hélice galvanique mobile*) by magnets and by terrestrial magnetism					•		

Table 4.2 (*continued*)

Activity	Thematic strands						
	GF	RE	GS	TM	GM	GC	GE
Idea of using coils (*hélice galvanique fixe*) to magnetize iron					•		
Successful experimental reversal of the *action attractive ou répulsive*		•					
September 17: Repetition of the experiments on the *action attractive ou répulsive* with Arago in the Obsérvatoire	•						
September 17: Improved experiment for the reciprocity of the *action directrice* performed successfully for the first time in the Obsérvatoire		•					
First experiments with the mobile galvanic helix					•		
Following a conversation with Arago, detailed reflections on the explanation of all of Ørsted's findings by recourse to the two "general facts"	•						
September 18, afternoon: first Académie des Sciences lecture	•	•	•	•	•		
September 19: unsuccessful experiments on the alignment of a straight conductor by terrestrial magnetism		•					
Careful consideration of the possible causes of terrestrial current circuits				•			
September 20: general reflections on the conditions for galvanic effects; further development of the concept of the circuit—the *circuit electro-moteur*							•
Reflections on further effects of terrestrial current circuits: magnetic variation, heat, light				•			
Further experiments (with Fresnel and Despretz) on the "imitation" of magnets by means of coils					•		
Together with Fresnel, experiments on the supposed action of two spirals on one another. Two batteries used; no discernible effect. Ampère ascribes the failure to insufficient battery strength.					•		

TABLE 4.2 (*continued*)

	Thematic strands						
Activity	GF	RE	GS	TM	GM	GC	GE
September 22: demonstration of the experiments with spirals and coils at a soirée hosted by Poisson					•		
September 24: Ampère asks Dulong for permission to purchase the pile Dulong had ordered from Dumotier					•		
September 25, morning: the pile is purchased, and the two-spiral experiment performed successfully while it is still in Dumotier's workshop					•		
Ampère makes additions to his lecture manuscript and plans to conduct experiments with two straight conductors						•	
Reflection on a possible generalization: the interaction between two spirals as wholly equivalent to the interaction between two magnets					•		
First draft of the "Conclusions" composed							
September 25, afternoon: second academy lecture, topped off with a reading of the "Conclusions"; at session's end, demonstration of the two-spiral apparatus	•			•	•	•	•
Appointments made to meet with Humboldt and Laplace on September 26 and 27 to present the new theory							
September 25, evening: letter to son, Jean-Jacques: "there were no more objections—and here we have a new theory of the magnet."							

Note: Those activities that Ampère did not mention in his publications are highlighted in gray. As in table 3.1, column headings are as follows: GF, reduction to general facts; RE, reciprocity; GS, galvanoscope/galvanometer; TM, terrestrial magnetism theory; GM, general magnetism theory; GC, galvanic currents interaction theory; GE, galvanism and electricity *single* electro-motive force theory.

might be ascribed. In his printed text the "general facts" were described as "laws." But he nowhere explicitly discussed the process of "reduction," in which context they arose in the first place, and thus their importance now appeared somewhat obscure. That Ampère had intended to ascribe them any special role, or had an entire explanatory project in mind, was no longer even discernible. His text disclosed neither the extent to which this undertaking had guided his whole research endeavor nor the considerable effort expended in its pursuit. Especially hidden was the fact that what would henceforth become one of his most important concepts had arisen in this context: the concept of the full current circuit, including both battery and wire. As discussed above, in the first part of his essay Ampère introduced this concept in connection with his micro-scale theory of electricity and galvanism, leaving the reader with the sense that it had arisen in this context in the first place and only later, almost retrospectively, proved suited to formulating the general rule for the alignment effect. Not even the most attentive reader could have discerned that this presentation precisely reversed the actual chronological order.

By leaving the program of "reduction to general facts" so vague, as well as pushing it so determinedly into the background, Ampère decisively steered his presentation away from his exploratory work, which receded completely. The first phase of his work appeared in a completely different light. His article conveyed the sense that the very same topics that would dominate his later studies, especially the development of a theory of magnetism, had also guided his earlier work. The difference between the two phases was blurred and, with it, the fact that there had been two phases of such distinct character in the first place. This tendency was further reinforced by another editorial change, which Ampère implemented more consistently than any other: he replaced all instances of the expression "galvanic current" (*courant galvanique*), commonly used in his early lecture manuscripts, with the expression "electrical current" (*courant électrique*). With this change, he shifted a theoretical insight gained only in the second phase to the very beginning of his work. This move left the impression of a single, uniform period of more or less goal-oriented research. The fact that there had ever been a phase of uncertainty, in which his conceptual scaffolding was still flexible, unstable, and under development and in which he had pursued an epistemic goal entirely different from that which dominated his later research while crafting and improving new concepts through qualitative experimentation, remained completely hidden from his contemporaries. Such features would not have enhanced the image of a Parisian scientist.[69]

However, through the concepts that had formed and stabilized during the exploratory phase, the contributions of that period remained. This logic applies not only to the development of the electrical circuit concept but to talk of "galvano-magnetic" effects in general. We recall that Ampère had undertaken an entire series of experiments—testing needles made of wood, paper, manganese, tin, copper, and zinc—to assure himself that the effects in question were really magnetic. Only after these tests did the expression "galvano-magnetic" appear justified and become canonical. The care taken in adopting that concept is a clear signal that Ampère was well aware of how many of its connotations could only be cashed out empirically. But in none of his many publications did he ever mention these experiments; he simply used the expression without further comment.[70] That there had ever been a potential problem here might simply escape the reader's attention. This case clearly illustrates a sense in which an entire series of experiments can be *absorbed* into concepts and thereby become invisible.

The discernible disappearing (to use a paradoxical phrase) of exploratory work from Ampère's self-presentation would persist for some time to come. Even historical scholarship has not, to date, identified the exploratory period as such.[71] A significant reason for this failure to acknowledge the exploratory phase is the fact that, in composing his article, Ampère dissected, reorganized, modified, and partially discarded not only the content but also the physical paper of his first two lecture manuscripts—the most important documents from this period. Until now, the article manuscript was all that was left of them. Only a reconstruction of the source materials made it possible to arrive at the new understanding of his research sketched here. This case amply illustrates the disappearance of exploratory phases from historical investigations as well as from self-presentations, driving home the moral that an accurate picture of research *practice* may be gained only by consciously going *behind* self-presentations.

In addition to the elision of exploratory work, Ampère's self-presentation was significantly shaped by his tendency to adapt his research to the epistemological standards of the tradition of mathematical formalization. He thus emphasized that his work "naturally" divides into two parts, one devoted to experimental facts and the other, to theory. This emphasis, as discussed above, was certainly a matter of the retrospective rhetorical adaptation of his work to a preexisting understanding of scientific process. There is no trace of such a division in the frantic first weeks and months of his work, not even in his earliest publications. Much the same holds for Ampère's epistemological reflections on the four structures of scientific argumentation. Despite the fact that these reflections arose in an entirely different context, historians have taken them as a strong influence on Ampère's electromagnetic research.[72] Though this may indeed have been the case for Ampère's work after 1822, in my investigation of the research practice of the early phase I find no evidence for it. His few allusions also strike me as retrospective interpolations of reflections developed in a completely different rhetorical and scholarly context. Unfettered by such reflections, his early research practice was driven instead by the mostly practical considerations of day-to-day science.

To be sure, in his publications Ampère soon began to take pains to construct a very different, more rigorous-seeming picture of his efforts. Looking ahead, we find the remarkable culmination of this construction in his comprehensive retrospective of 1826:

> To observe the facts, varying the circumstances as much as possible; to supplement this preliminary work with precise measurements so as to deduce from them general laws based only on experience; to deduce from these laws, independent of any hypothesis on the nature of the forces that produce these phenomena, the mathematical expression of the forces or the formula governing them: this is the path followed by Newton. In France it has been generally adopted by the savants among whom physics has made such immense progress in recent times, and it is the path I have followed in all of my research on electrodynamic phenomena. (Ampère 1827, 176–77)

But this was the methodological ideal of the French tradition of mathematical formalization, especially as embodied in Laplacean physics. By comparison, we may recall how Biot, the paradigmatic representative of this tradition, had described the parameters of his work on electromagnetism six years earlier: "The first thing that had to be discovered was the law by which the force emanating from the connecting wire decreased at various distances from its axis" (Biot 1821c, 228).[73] Just prior, Biot had emphasized the confusing character of Ørsted's discovery, asserting that only a research effort geared toward capturing the force law could protect the scientist from losing his way among the multiplicity of phenomena and performing unnecessary experimental repetitions and variations. Without mentioning Ampère by name, Biot was presumably targeting the latter's much more broad-based experimental enterprise. Ampère, who unlike Biot was forced to expend considerable effort to establish himself as a physicist and be taken seriously, appears to have taken the hint. As early as his article in the *Bulletin* published in the winter of 1820 he had begun adapting his presentation of his work to the accepted methodological ideal. The 1826 passage quoted above would take this tendency to its ultimate conclusion.

Whereas in the winter of 1820 some readers might have been inclined to take Ampère's account of his research practice at face value, the 1826 version was so stylized and overblown that even his contemporaries might have become aware that they were seeing a retroactive construction, disguising an entirely different research process. As James Clerk Maxwell so aptly asserted, with reference to this very work of 1826, Ampère had done a perfect job of hiding the path that had led him to his theory, whatever his methodological proclamations. In calling Ampère the "Newton of electricity," Maxwell (1873, vol. 2, §528) may have been alluding not only to the similarity between the two figures but to the fact that both of them so diligently redacted their research trajectories to conform to methodological guidelines prevalent in their time and place.

CHAPTER 5

Electromagnetism in London

Scientific Research in London, circa 1820

Scientific research in early nineteenth-century England was conducted within institutional structures completely different from those in France. These structures had emerged as the product of historical tradition, not deliberate planning. In sharp contrast to France, in England, as in other European states, there was nothing resembling an integrated, planned, or rigorously organized system of research and instruction (e.g., Cardwell 1957).[1]

The center for high-level research and international exchange remained the Royal Society of London (Hall 1984). But there were stirrings of discontent regarding its state. During his more than forty-year tenure as president, which began in 1778, Sir Joseph Banks had advanced the society in many ways and brought to it numerous powerful sponsors. But at the same time, many new members had been elected for whom the prestige of society membership outweighed scientific interest. With Banks strongly protecting his own disciplines of natural history and agriculture, the society's openness to new fields of research had suffered markedly. Fields emerging from within less traditional disciplines were not adequately represented, and so those interested in them turned to independent initiatives.

In 1807, a group of geologists founded the Geological Society. A group of mostly applied mathematicians (including Charles Hutton, Olinthus Gilbert Gregory, Peter Barlow, and Samuel Hunter Christie), centered on the Royal Military Academy of Woolwich, London, also went its own way, as did the Analytical Society, founded in 1811 by Cambridge mathematicians and astronomers (including Charles Babbage, John Herschel, and William Whewell) with the aim of promoting the mathematical formalization of the physical sciences in England.[2] The Royal Society's status as the fulcrum of scientific research in Britain had begun to erode. When Humphry Davy took over the presidency in the fall of 1820, following Banks's death and a brief interregnum under William Hyde Wollaston, he faced the daunting task of bringing together the diverse interests of these groups, all of which hoped for far-reaching changes. In the long term, the Royal Society could not keep pace with the rapid proliferation of new scientific disciplines. In the 1830s, this slackening progress finally led to the establishment of the British Association for the Advancement of Science (Morrell and Thackray 1981).

The Royal Society was a node for exchange and communication, but it had no laboratory of its own. Experimental research was undertaken either in private laboratories or, in the case of certain specialties, at publicly financed institutions devoted to them, such as the Greenwich Observatory or the Royal Military Academy, where in addition to questions of navigation, problems in materials science and construction theory were addressed. At the beginning of the nineteenth century, as discussed in chapter 1, the private Royal Institution also entered the scene. Buoyed by the initiative and success of Humphry Davy, it expanded its laboratory to such an unusual degree that it soon rivaled and surpassed those of older institutions. This important development in the history of laboratory research is indicative of the great extent to which the laboratory itself evolved in response to the demands of *chemical* research: specifically *physical* laboratories emerged only later, while as yet too little is known about the development of experimental research in the observatories.[3] However, such disciplinary boundaries had been firmed up earlier and more sharply in Paris than in London, where they played a much less significant role for Davy and his successor, Faraday. This difference had significant consequences for the evolution of the laboratory sciences, as the cases of Ampère and Faraday illustrate. The less strict disciplinary character of the Royal Institution laboratory may well have been an important factor in its extraordinary productivity.

The growing demand for scientific instruction and training gave rise to numerous initiatives. The Royal Institution, for example, was originally founded to provide instruction for craftsmen, though it quickly became a meeting point for the affluent, and under Davy's direction its ties to the "upper-class" science of the Royal Society were nurtured and extended. By contrast, the London Institution, modeled on the Royal Institution and founded in 1805, performed very little experimental research, instead sponsoring lectures by prominent speakers on applied and popular scientific themes.[4] The London Institution drew crowds, and its rapidly growing wealth made clear that scientific instruction and entertainment could also be a moneymaker.[5] But even London Institution events remained inaccessible to many. In 1801, the City Philosophical Society was established as a more informal mutual aid group for young amateurs who worked on scientific themes and presented their results to each other (e.g., Williams 1965, 15–20). They met at the home of the silversmith John Tatum, who was well read in many fields and gave regular series of lectures. Some of those to whom access to the flashier institutions was barred were thus afforded an opportunity for intellectual exchange, one in which they shared their autodidactic activities with the like-minded.

This broadening of the social spectrum of science found its reflection in the media of scientific communication. A number of new periodicals began publication, spreading scientific news to a far less exclusive audience than that of the *Philosophical Transactions.* William Nicholson had founded his *Journal of Natural Philosophy, Chemistry and the Arts* in 1797, and it was followed a year later by Alexander Tilloch with his similarly oriented *Philosophical Magazine.* The trend continued with the establishment of the *Annals of Philosophy* and the *Quarterly Journal of Science,* often viewed as the unofficial in-house journal of the Royal Institution, where its founder and publisher, William Thomas Brande, was employed as professor of chemistry. The success of all of these periodicals clearly demonstrates the demand for scientific media "beneath" the level of the prestigious, exclusive, and rather slow *Philosophical Transactions.* Though the new journals often simply reprinted articles originally published elsewhere, they increasingly featured original material. For example, the entire debate on electrolysis took place primarily in the pages of Nicholson's journal. Even the established researchers of the Royal Society consulted these journals, if sometimes reluctantly. Also, these journals were not merely local in distribution. They were widely read abroad, posing a long-term challenge to the leading role of the *Philosophical Transactions.*[6]

Michael Faraday, Chemical Assistant in the Royal Institution

In early nineteenth-century England, there was no standard way to become a natural philosopher or to make a living as a professional researcher (e.g., Reader 1966). By comparison with those of their French counterparts, biographies of English researchers contain far fewer of such common themes as attendance at a particular school. But even in all this diversity, Michael Faraday's path stands out.[7] His autodidactic activities began very early, encouraged by *The Improvement of the Mind* (1809), a widely read book by cleric Isaac Watts. The rules he learned from it would shape his scientific work for his entire life. He attended many of the aforementioned lectures by John Tatum and, following the conclusion of his apprenticeship as bookbinder, took the highly unusual step of soliciting any open position at the Royal Society and Royal Institution. Thanks to Davy, his inquiry at the Royal Institution eventually bore fruit. Faraday had attended Davy's lectures (with tickets given to him—he could not have afforded them on his own), transcribed them, and sent his transcription with his inquiry. The year and a half he spent traveling the European continent as Davy's assistant (and, as Faraday would soon come to regret, his personal valet), from October 1813 through April 1815, would have a profound impact. In addition to his having gained the acquaintance of many continental researchers, witnessed stimulating scientific research, and curiously taken in various other continental ways of life, the legacy of this trip was strongly shaped by disillusionment regarding the moral character of scientists.[8] On his return he resumed his former position as laboratory assistant. He would remain at the Royal Institution for the next forty years.

Faraday's duties as an assistant included preparing experiments for lectures, assisting with their demonstration, and helping Davy in his laboratory research. In addition to chemical investigations, they dealt with such technological problems as the design of a safety lamp for mines and the copper sheathing of ship hulls. Faraday also worked on Brande's *Quarterly Journal of Science* and was responsible for the "Miscellaneous Intelligence" section. Here he had increasing freedom in accepting smaller articles and notices for publication. This task not only allowed him to familiarize himself with the most diverse scientific fields but also gave him his first position of responsibility in the scientific enterprise.

Such assignments were rather unusual for a laboratory assistant. That they were entrusted to Faraday had much to do with his efforts to become a scientist in his own right. For years he spent his free time autodidactically reviewing the entire field of chemistry—not only its laboratory procedures, in which he assisted daily, but also its concepts, categories, and theories. Brande's textbook *Manual of Chemistry* served as his introduction. He also regularly consulted a large number of journals, taking note of recent findings in the margins of his copy of Brande's *Manual*. The list he himself compiled of these journals included fifteen titles, clearly indicating the huge effort he expended. L. Pearce Williams (1965, 43–44) rightly calls it a "Herculean task." Beginning in 1816, Faraday began to publish short notices and articles on chemical topics, usually in the *Quarterly Journal of Science*. He became widely enough known to attract commercial commissions for chemical analyses, though Davy's reputation doubtless played an important role here, too. He eagerly accepted such assignments, not least to improve his less-than-prosperous financial situation.[9]

The biennium of 1820–21 found Faraday intensely involved in an ambitious project for the improvement of steel and serving as an expert witness in legal proceedings.[10] He was also challenged from another quarter. His acquaintanceship with Sarah Barnard had developed into a love interest, and he had some difficulty learning to cope with unaccustomed fluctuations in his emotional life. All of these demands may explain why, when he first learned of electromagnetism from Davy in 1820, he initially took no interest.[11] Through October 1821, he reported nothing on the topic in "his" column in the *Quarterly Journal of Science*. When he assisted in Davy's experiments on the magnetic deviation of an electric arc in May 1821, he did make some notes in his laboratory journal (Martin 1932, 1:45–46, entry for May 21, 1821).[12] This activity would tend to indicate some degree of personal interest, for this journal was otherwise reserved for his own research. Still, compared with other periodicals, the *Quarterly Journal* continued to publish very little on electromagnetism.[13] Had Faraday been more taken with the subject, it would probably not have proved difficult to persuade Brande to devote more attention to the field.

In 1821, Faraday became the Royal Institution's "Superintendent of the House," with responsibility for facility maintenance.[14] The new position carried the same salary, though it also came with a larger apartment in the Royal Institution building. Faraday had pressed for this. He married Sarah Barnard in June 1821, and they moved into the apartment together. Shortly thereafter he declared his faith in the Sandemanian Church, becoming a full member of the community, like Sarah before him. It would be difficult to overstate the importance of these steps. For Faraday, they meant the abiding stabilization and consolidation of his inner constitution.[15] As

Faraday acknowledged, Sarah became an essential pillar of his life. The community life of the Sandemanian Church was extremely intensive, comprising an essential part of daily life for both of them. Faraday's duties as a member of that community would remain at least as important to him, his entire life, as his scientific labors. His decisions of the summer of 1821 were profoundly life altering and stabilizing, and it is no accident that this very moment also found him prepared to embark on a grand new scientific enterprise, choosing as his target the most pressing issue of the day: electromagnetism.

Electromagnetism in London, 1820–1821

The opening phase of Faraday's engagement with electromagnetism can be properly understood only against the backdrop of the situation in London, where the reaction to Ørsted's report took a significantly different track from that in Paris. As the overview in table 5.1 shows, the news spread slowly.[16] Wollaston received Ørsted's report quite early and conducted his own investigation, but he took no steps toward disseminating the findings any further (Faraday 1823, 289). Ørsted's circular did not reach a wider audience until the appearance of an English translation in the *Annals of Philosophy* at the beginning of October. Davy learned of the discovery from this source (Ørsted 1820g) in the October issue of the *Annals*.[17] He was immediately aware of its importance and, assisted by Faraday, set straight to work on his own experiments.[18]

Table 5.1. Electromagnetism in London, August 1820 through August 1821

Date	Royal Society	Publications	Wollaston, Davy, Faraday
Aug. 1820			Wollaston learns of Ørsted's finding, conducts his own experiments[1]
Oct. 1		*Ann.Phil.* (Oct.): English translation of Ørsted's (1820g) first circular	Davy experiments, with Faraday's assistance[2]
Nov. 1820		*Ann.Phil.* (Nov.): English translation of Ørsted's (1820i) second circular	Davy experiments together with Wollaston[3]

TABLE 5.1 (*continued*)

Date	Royal Society	Publications	Wollaston, Davy, Faraday
Nov. 9	First session after summer break, Ampère's letter is read, his "Conclusions" received[4]		
Nov. 16	Davy's first article, dated Nov. 17, is read[5]		
Nov. 30	Davy elected president, Copley Medal awarded to Ørsted, Wollaston reports on electromagnetism[6]		
Dec. 1820		*Phil.Mag.* (Nov. 1820): report on Davy's Royal Society lecture on electromagnetism and on Ørsted's additional findings	
Dec. 7	Offprint of Hachette's "Sur les expériences" received, together with Ampère's "Notes"[7]		
Jan. 1821		*QJS* (Jan. 1821): Brande provides an overview of electromagnetism, mentioning Wollaston (Brande 1821a)[8] *Edin.Phil.J.* (Jan. 1821): Brewster, survey of electromagnetism work by Davy, Ampère, Biot (Brewster 1821a)	
Jan. 11	Offprint of Ampère's *ACP* mémoire received[9]		

TABLE 5.1 (*continued*)

Date	Royal Society	Publications	Wollaston, Davy, Faraday
Feb. 1821		*Ann.Phil.* (Feb. 1821): "Anonymous" article ("J.C." 1821) *Phil.Mag.* (Jan. 1821): translation of Hachette's survey (Hachette 1821); Ørsted's (1821a) second circular and Ampère's (1821c) "Notes" appear in English	
Feb. 10			Davy writes to Ampère expressing objections on the theory of circular currents[10]
Apr. 1821		*Phil.Mag.* (Mar. 1821): article by Boisgiraud (1821a) appears and is continued in the Apr. issue (1821b) *Edin.Phil.J.* (Apr. 1821): Brewster (1821d) on Ampère's experiments as reported in the *ACP* mémoire, on Wollaston's ideas on electromagnetism (1821e), and on Ampère's telegraph (1821c)	Davy and Wollaston perform experiments in the Royal Institution, searching for rotation of wire about its own axis, in Faraday's presence[11]
May 1821		*Ann.Phil.* (May 1821): Hare (1821a) on a new galvanic device *Phil.Mag.* (Apr. 1821): Hare (1821b) on a new galvanic device	
May 21			Davy, with Faraday's assistance, experiments on the magnetic deviation of the electric arc[12]

Table 5.1 (*continued*)

Date	Royal Society	Publications	Wollaston, Davy, Faraday
June 1821		*Phil.Trans.* (part 1): Davy's (1821d) first article	
July 1821		*QJS* (July 1821): English translation of Biot's (1821a) Apr. 2 lecture to the Académie des Sciences *Phil.Mag.* (June 1821): Tatum (1821) on electromagnetic experiments *Edin.Phil.J.* (June 1821): Brewster (1821b) on Poggendorff's "Condenser"	Faraday begins work on his "Historical Sketch," including the interpretation of Ampère's *ACP* mémoire[13]
July 5	Davy presents "On the Magnetic Phenomena Produced by Electricity, and their Relation to Heat occasioned by the same Agent"[14]		
Aug. 1821	*Phil.Mag.* (July 1821): reprint of Davy's (1821b) first *Phil. Trans.* article *Ann.Phil.* (Aug. 1821): reprint of Davy's (1821c) first *Phil.Trans.* article		

Notes: 1. Faraday (1823, 289). | 2. Faraday (1836, 521). Davy (1821d, 18) describes Faraday's assistance. | 3. Davy (1821d, 18). | 4. *Phil.Trans.* (1821): 443; see also *Ann.Phil.* 6 (December 1820): 453 or Brande (1821b, 378). | 5. Published as Davy (1821d). | 6. Brande (1821b, 379–82). | 7. *Phil.Trans.* (1821): 444. | 8. See Faraday's (1836, 522) comment regarding authorship. | 9. *Phil.Trans.* (1821): 444. | 10. AC 2:563. | 11. Faraday (1823, p. 289); see also Faraday's own chronology, in which Wollaston himself inserted the date "April 1821": RI F3B, p. 152.| 12. Faraday's Diary (Martin 1932, 1:45–46). | 13. See FC letter 471; Faraday (1823, 291); and Faraday's own chronology, RI F3B, p. 152. | 14. Reported in *Ann.Phil.* 2 (August 1821): 149–50.

The first session of the Royal Society following its summer break (July through October) did not take place until November 9.[19] By this time it had received additional documents on electromagnetism: a letter from Ampère, together with two offprints of his "Conclusions." Ørsted's second circular was also published in the *Annals of Philosophy* at the beginning of November. By this point, if not earlier, members of the society would have been aware that a spectacular new research field had been opened. Davy, who in those weeks was busy prosecuting his candidacy for the presidency of the Royal Society, did everything possible to underscore its significance, including reading his own paper at the November 16 session. Well aware that the field was undergoing rapid development, he stressed that some of his results "may have been discovered before, or at the same time in other parts of Europe."[20] Two weeks later, at the very session at which the society chose Davy as its president (Saint Andrew's Day, November 30), Ørsted was publicly proclaimed the recipient of the Copley Medal. In an homage to Ørsted's discovery, Wollaston took pains to underscore its importance and the expectation of its long-term influence on the progress of science.[21] After another week (on December 7), the Royal Society had received Hachette's (1820b) report, "Sur les experiences . . . ," along with Ampère's (1820i) "Notes." The prominent attention the new field had drawn in Paris was now plain. Finally, early in the new year, on January 11, 1821, the Royal Society received an offprint of Ampère's detailed *mémoire* in the *Annales de chimie et de physique* (*ACP*). Not long after, three survey articles appeared in England. Brande (1821a) wrote one (anonymously) for the *Quarterly Review of Science*.[22] Brewster (1821a) contributed another to his *Edinburgh Philosophical Journal*, and soon thereafter the *Philosophical Magazine* published an English translation of Hachette's (1821) survey.[23]

Electromagnetism was now in England to stay. In the following months, various periodicals reported on developments in France and, to a significantly lesser degree, in the German-speaking world. Initially, there was little original English research. Besides Davy's contributions we find one short report ("J.C." 1821) by an anonymous author ("J.C." most likely being the Cambridge chemistry professor James Cumming) in February and an only slightly more detailed piece by Tatum (1821) in July.[24] Davy was the only one to concern himself deeply and persistently. In April 1821, he and Wollaston undertook a new series of experiments. Among other things, they were following up on Wollaston's idea of achieving the rotation of a current-carrying wire about its own axis. These efforts, in which Faraday also participated for a time, remained unsuccessful.[25] Davy presented his results to the Royal Society on July 5.[26] His research on electromagnetism had thus concluded, at least provisionally.

All told, London felt little of the frantic, feverish atmosphere that had gripped Paris. For one thing, few prominent actors had entered the field, with Davy the only one of them to lecture on the topic to the assembled luminaries in the Royal Society auditorium. His studies were far and away the most interesting and innovative. The competitive environment and time pressure were much less urgent than in Paris, where the topic posed a greater theoretical threat, and had a greater impact on scientific politics.

Davy's Work on Electromagnetism

Nothing would shape Faraday's later work so profoundly as Davy's investigations, whose features thus deserve some attention. They were geared toward broad-based exploration. Both at the Royal Institution and at the London Institution (which possessed a large battery), they targeted a wide range of questions. Like Arago, of whose work he was unaware, Davy began by looking for the attraction exerted by an electrified wire on iron filings, as well as for a magnetization effect on steel needles (Davy 1821d, 9–11). He studied whether such magnetic activity varied with the position of the battery relative to terrestrial magnetism (with a negative result [10]) and whether iron filings formed any particular patterns around the wire. Among other things, he observed that filings scattered on a glass plate placed above a horizontal wire would arrange themselves in straight lines perpendicular to the wire (11). Taking the quantity of filings attracted as a measure of the strength of the magnetic action, he sought after the factors on which this strength depended. It was independent of the material composition of the wire but directly dependent on the number and submersion depth of the battery plates. The magnetic action of the wire thus depended on the same parameters as the wire's heating and was undiminished by any increase in temperature (10–12). In this respect, it was markedly different from the action of conventional magnets, which always lose their magnetism when heated past a certain point.

In a further series of experiments, Davy attempted to produce magnetic effects by means of conventional electricity, that is, by discharging Leyden jars. As expected, given the rapidity of the discharge, he observed no deviation of magnetic needles, though he did observe residual magnetization (Davy 1821d, 13). He also explored arrangements of multiple parallel wires, determining that iron filings placed between wires in which the current ran in the same direction would attract each other, while those between wires with opposing currents repelled (16). Like many others, Ampère included, Davy attempted to produce an effect mechanically reciprocal to Ørsted's by observing a magnet displacing a movable electrified wire. His experimental setup consisted in two upright, horizontally positioned knife blades, each in contact with one of the battery poles, on top of which a wire segment could roll freely. At the approach of a magnet, the wire would indeed move (17). This assembly affords a remarkable solution to the oft-mentioned problem of simultaneously maximizing electrical connectivity and easy mobility.

Davy's (1821d) article also offered theoretical speculations. He hypothesized that terrestrial magnetism might be caused by electrical currents within the earth, where fluctuations in terrestrial magnetism could in turn be traced to chemical and thermal changes in the earth's interior (17). Ampère had thought much the same. Indeed, by the time of his November 16 lecture, Davy had an outline of Ampère's reflections in hand, the "Conclusions." There, Ampère had spoken of terrestrial currents moving "in the direction opposite to the apparent motion of the sun" (fifth claim of the "Conclusions"; see chap. 3). So when Davy asserts "that if strong electrical currents be supposed to follow the apparent course of the sun, the magnetism of the Earth ought to be such as it is found to be" (17), it strikes me as unquestionable that he was mirroring Ampère's conception and word choice, though he reversed the direction of the terrestrial currents. His failure to mention Ampère's name was quite typical—one need only recall his behavior in the electrolysis debate of 1806 (see chap. 1). Still, another of his speculations was truly original. Davy wondered whether the northern lights, which had increasingly begun to attract scientific interest, might be caused by electrical effects (17). This idea had not previously been formulated by anyone.

In a remarkable series of experiments undertaken in November, with some participation from Wollaston, he sought to determine a "law of the production of magnetism" (Davy 1821d, 14) by establishing the spatial distribution of electromagnetic action. The method familiar from conventional magnetism, that of visualizing the distribution of magnetic action by means of the "curves" formed by iron filings, was quickly ruled out for the magnetism of the current-bearing wire. In this case, the iron filings did not arrange themselves in recognizable patterns, instead simply forming a cylindrical clump around the wire. Davy thus developed a highly original arrangement for exploring Wollaston's suggestion that a kind of "revolution of magnetism" surrounded the wire. At various points on a paper disk he attached tangentially oriented steel needles. A wire ran through the center of the disk, perpendicular to it, and a Leyden jar discharged through the wire. Following the discharge, the steel needles had become magnetized. Furthermore, the direction of magnetization was independent of the terrestrial compass points, instead pointing toward a uniform direction of circulation. It proved difficult, however, to formulate these results coherently, let alone specify a general rule. Instead, Davy asked his readers to envision a complicated procedure (Davy 1821d, 15). First, they must imagine the paper disk lying in a horizontal plane, with the wire running vertically. The polarity of the needles following successful magnetization was then expressed by the direction of the (fictitious) motion of some arbitrary body from the north to the south pole of any one of the needles. When the upper end of the wire was connected to the positive terminal of the battery or electrostatic generator, this motion would run in the direction opposite that of the apparent daily motion of the sun. Inverting the polarity of the wire would reverse the direction of motion. This schema could be generalized: if the wire instead ran horizontally from east to west, with its eastern end connected to the positive pole and the paper disk positioned vertically, the motion would run from north to south in the lower half of the disk, and from south to north in the upper half.[27] Davy related experiments and results in verbal form only, providing no visual representations.

Once again, such cumbersome constructions are very typical of the epistemic situation of early research on electromagnetism, bringing to mind Ørsted's likewise cumbersome spatial descriptions of the magnetic action of the wire, Ampère's definition of current direction by reference to an imagined observer submerged in the current, or Biot's long-winded accounts. All of these cases point to the lack of a conceptual system or frame of reference adequate to the expression of the complex effects being observed. Geometry had no concepts that could be used to express the relation of three spatial directions, as typically was required in electromagnetic cases.[28] In physics, moreover, traditional concepts of attraction and repulsion proved inapplicable in the new domain, and no other, more flexible conception was yet in sight. Significantly, Davy, like Ørsted, explicitly formulated his "law" only for the two special cases in which the experimental apparatus was aligned with the compass points. While asserting that "the same law prevailed" (Davy 1821d, 15) in other cases, he had neither the concepts nor the language necessary to express it in full generality. At the conclusion of his struggle to articulate the law, Davy mentioned

Ørsted's idea of action propagating out from the wire in a spiral (15) but did not develop it further.

Davy's broad experimental studies bore all the hallmarks of "exploratory" experimentation. The systematic variation of numerous parameters, the search for regularities, and the development of appropriate concepts and frames of reference are highly typical for that process, as I discuss in more detail in chapter 7. As should be clear by now, Davy was in good company; many other researchers in Europe also took an exploratory approach, Ampère among them. Others also investigated electromagnetic reciprocity and the effects of conventional electricity. Davy knew none of this when he began his work. Unlike many others, he quickly began looking for a way to capture the new phenomena in quantitative terms. However, he placed no importance on sophisticated mathematical formalization.

The quantifying tendency emerged even more clearly in the second phase of Davy's work, in April and May 1821. The broad-based exploration that predominated in his earlier work was now only occasional. With Faraday's help and in the presence of other researchers at the London Institution, Davy (1821a, 427) undertook his experiments on the magnetic deviation of an electric arc, detecting attraction or repulsion "with a rotatory motion."[29] This description remained vague, however. He did not pursue the search for the "law" governing the direction of electromagnetic action any further. Instead, he took up the detailed investigation of the dependence of the effect on the strength, material composition, and temperature of the connecting wire. This effort shifted his focus toward the fundamental question of the relationship between temperature and the conductivity of metals—a question that could now, by means of electromagnetic effects, be treated in totally new ways. In these studies, electromagnetism was not so much the object of research itself but rather served as a tool for addressing other questions—a significant move that recalls Davy's approach to the voltaic pile two decades earlier, discussed in chapter 1. In the article that emerged out of this phase of work, pride of place was given to investigations of the relationships among "quantity," "intensity," and conductivity.[30] With respect to electromagnetism proper, it hardly went beyond the previous essay.

Both phases of Davy's research on electromagnetism were brief. Many of the topics he entertained and questions raised were neither pursued further nor satisfactorily resolved by him. The presidency of the Royal Society may well have placed such demands on Davy that he had relatively little time left for his own research. His position in the society presumably also explains his decision to publish his results in the *Philosophical Transactions*, though this effort could be expected to take some time, during which the field would progress further. While choosing another journal would have allowed for quicker publication, doing so was evidently not an option for Davy, despite his having observed in February, presumably with Ampère's and Arago's reports in mind, that his efforts had begun to lag behind: "I find that they have lost the merit of novelty, for most of my observations have occurred to other philosophers."[31] This was all the more true by June 1821, when his results finally appeared in print. From the outset, other authors had placed far more value than Davy on not only conducting original and important research but also publishing it as quickly and widely as possible.

Faraday's Survey of Electromagnetism

In July 1821, at the request of his friend Richard Phillips, copublisher of the *Annals of Philosophy,* Faraday began work on a survey article on electromagnetism.[32] The fruit of his efforts, the "Historical Sketch of Electro-Magnetism," was the most comprehensive report on electromagnetism to date, and it would have a profound impact on subsequent British research. Faraday's own research, begun immediately after completion of the "Historical Sketch," was also grounded in that report. Thus, in the following sections, I analyze that work in detail. It provides us with detailed information on the starting point for Faraday's further investigations.

Faraday pursued his survey with extraordinary diligence.[33] He not only systematically sought out the largest possible number of publications in the field but also replicated most of the experiments he found described in them—a point to whose significance I return later. In his literature search he completely scoured eight journals for contributions on electromagnetism. In addition to the most important British journals (*Philosophical Transactions, Annals of Philosophy, Philosophical Magazine, Edinburgh Philosophical Journal,* and *Quarterly Journal of Science*), these included the Paris-based *Annales de chimie et de physique* and *Journal de physique,* and the Geneva-based *Bibliothèque universelle.* His thoroughness reminds us of the years he had spent autodidactically mastering the whole field of chemistry. Taking into account that his assimilation of the new field took only three months, and not six years, Williams's description of a "Herculean task" seems even more applicable.

From the beginning, however, Faraday's survey was faced with one limitation. Contributions from the German-speaking world were considered only when cited in French- or English-language journals. He did not cover German-language journals, even those collected in the libraries of the Royal Institution or Royal Society.[34] First of all, though Faraday read French, he did not read German. But a further factor was that, while London enjoyed attentive and vigorous scientific exchanges with Paris, there was no comparable exchange with the German-speaking world, which lacked a comparable center of research.[35] Consequently, it had become generally accepted, even when it was not always true, that a survey of work performed in the great centers would suffice to capture the most interesting developments. Significantly, Faraday took that approach for granted to such a degree as not to mention the limitations of his literature search explicitly, thereby unwittingly contributing to the persistence of that picture.

Faraday's engagement with Ampère's work played a particular role in his literature search. He went through an offprint of Ampère's (1820e) first *ACP* article very thoroughly, making notations in the margins.[36] He not only regularly took note of experiments mentioned in Ampère's text but also always tried to describe them, for example, as "pile acting on a needle" (at Ampère 1820e, 11).[37] The resulting division and classification of the empirical findings was far more explicit than Ampère's own:

> attraction between two conductors (13–19)
> two effects of a helix (25–29)
> action of the earth
> action of the earth on one or more conductors (41–44, 45–47)
> conductor and magnet (48–54)
> magnetism of the earth (55–59)
> action of two conductors (60)
> action of conductor and magnet (60, 61, 66)
> action of two magnets (61)

These categories would continue to be very important in Faraday's own empirical studies.

We find fewer marginalia concerning Ampère's many theoretical speculations and surmises. Faraday's chief goal was to get a grip on Ampère's various efforts, for, as we recall, the structure of this article was far from transparent. His annotations included "nature of electric current" (at Ampère 1820e, 78), "a magnet what" and "magnet nature of" (31, 33, 59–62, 65–66), and "law of attraction" (29–34, 36–40). Faraday took equal interest in Ampère's general circuit hypothesis (21, 23, 31, 33), in his conjecture regarding terrestrial currents (41, 58), and his speculations on the nature of the stars (59). He took very careful note of Ampère's emphasis on the "instrumental" character of the concept of current (50). But to judge by the many notes, exclamation marks, and the like, he was most intensely interested in Ampère's project of reduction to two "general facts"—the very feature of his program whose importance Ampère had so consistently understated and which would recede so completely into the background over the coming years. For Faraday, in his own empirical research that immediately followed the survey, this aspect was of paramount importance. The same holds for Ampère's attempts to provide a detailed explanation of Ørsted's experiments with the vertical wire; here, too, Faraday's marginalia indicate his utmost interest (53). All in all, the particular focal points in Faraday's reading of Ampère convey a very clear impression of his interests. A French commentator would presumably have found greater significance in an entirely different set of passages.

The "Historical Sketch of Electro-Magnetism"

Faraday composed the "Historical Sketch of Electro-Magnetism" in three parts, and insisted, contrary to Phillips's intentions, on publishing it anonymously.[38] The first part was devoted to Ørsted's findings, the second to experiments subsequently performed by others, and the third to various proposed theories. The first two parts had been completed by the end of August, appearing in the September and October issues.[39] The third part, for whose overdue manuscript Phillips issued a renewed plea in a November letter to Faraday, did not appear until the February issue of the following year, along with an appendix in which Phillips himself reported on the discovery of electromagnetic rotation, which Faraday had made in the interim. Phillips had felt obligated to add this codicil when Faraday neglected to do so himself, as requested.

Even in the first part of the "Historical Sketch," Faraday went well beyond a simple précis of Ørsted's experiments. He commented critically on Ørsted's attempt to provide rules for the behavior of the needle in proximity to the wire, asserting that a "much simpler idea of the relative movement of the wire and the needle" could be gained by eliminating or compensating for the effects of terrestrial magnetism ([Faraday] 1821–22, 198). Following through on this proposal, he offered his *own* "much simpler" statement of the rule governing the "simple but important point of position" between needle and wire (199). This endeavor alone took up three of the five printed pages of the first part of his survey, turning it into anything other than a simple report of others' findings. It was extremely typical of Faraday's usual operating procedure and deserves careful scrutiny.

Without naming Ampère, he began with the experimental arrangement Ampère had described as his first "general fact": when the middle of the needle was directly opposite the wire, the needle always assumed a right angle to the wire. Unlike Ampère, he offered graphic illustrations of this rule, providing three different images. In his Figure 1, he presented the right-angle position for a special case: PN is a wire running from the positive to the negative battery poles, and SN is the magnet, with its south and north poles. His Figure 2 depicted the superposition of several different cases in a side view. The arrows indicate the directions in which the needle aligned itself on various sides of the wire. This, too, must have struck him as insufficiently clear, so he attempted a three-dimensional model in his Figure 3. On the back of a rectangular glass prism he symbolized the connection wire with a line running its length, negative pole on top.[40] A line parallel to the width of the front side of the prism signified the magnetic needle, whose north pole, indicated by an arrow, was to the right. The whole prism could be placed in any arbitrary spatial orientation. Because the relationship between the two directions remained unchanged regardless of the prism's orientation, it would always indicate the electromagnetic action of a correspondingly oriented wire. This model allowed him to free himself from all reference to the compass points, removing what had proved an obstacle to both Ørsted and Davy. Faraday's model, which he certainly also built in glass, was in some sense an embodiment of the rule expressed in Ampère's swimmer analogy. In place of a complex linguistic construction he opted for a visually intuitive or even a three-dimensional physical mode of representation. This was emblematic of his approach.

Ampère's results may have gotten him this far, but in his subsequent explication of the behavior of the needle when it was placed in an eccentric position relative to the wire, Faraday was forced to rely entirely on his own experiments. Ørsted and Ampère had offered only a few special cases, and Davy undertook no such experiments.[41] Here Faraday had made unusually extensive annotations to Ampère's expository sketch. Now he set about expressing a general rule. Suspending a magnetic needle horizontally by a thread attached to its middle, he studied its behavior in proximity to a vertical current-bearing wire. When one end of the needle

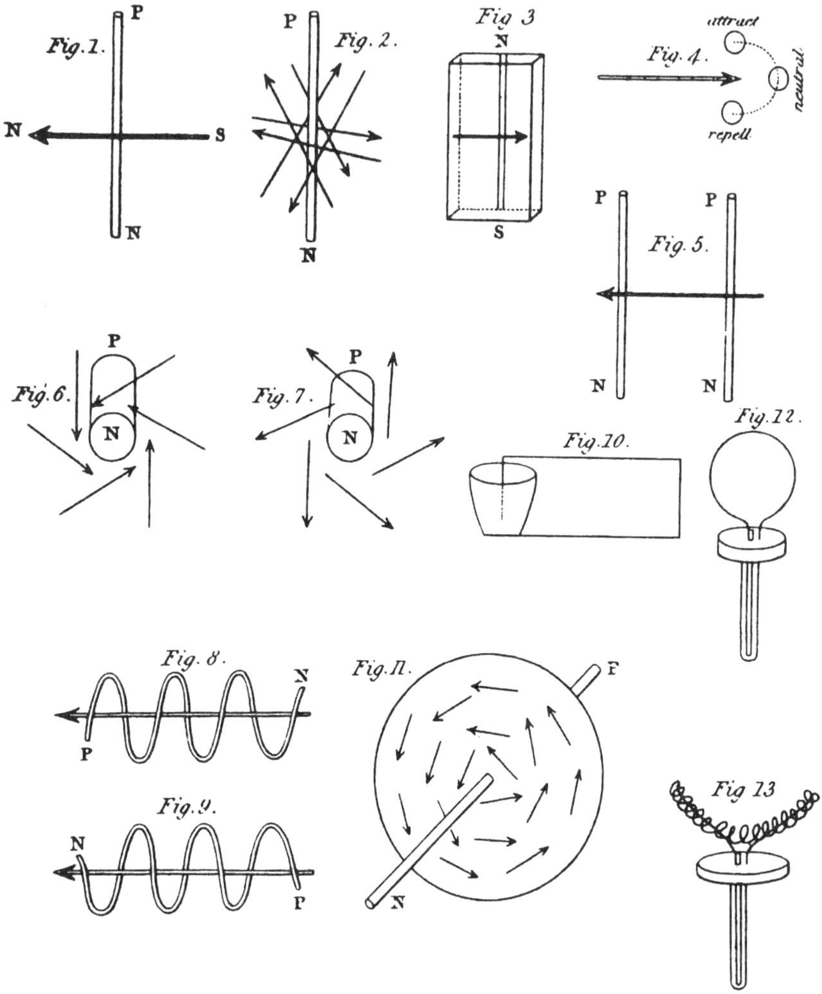

Fig. 5.1. Illustrations from Faraday's "Historical Sketch" (Faraday 1821–22, plate IX, 198 verso).

was close to the wire, there would clearly be various positions in which it was attracted or repelled by it. Faraday distinguished four different kinds of position, illustrating them graphically. In his Figures 6 and 7 (fig. 5.1), PN is the connecting wire, with connections to the positive and negative battery poles, and the various arrows indicate the magnetic needle in different possible positions, always with the arrow point at the needle's north pole. His Figure 6 shows all of the positions in which the north pole of the needle was attracted, and his Figure 7 shows the same for the south pole. In his Figure 4 (see fig 5.1) he abstracted still further from actual experimental conditions. Departing from the real experiments, in which the needle could be expected to move, he depicted it at rest, sketching around the needle three possible relative positions for the wire, assumed to run perpendicular to the plane of the illustration, with one position each for attraction, repulsion, and no interaction. This image had the advantage of simultaneously capturing an even larger range of experimental findings.

The role played by images here is noteworthy. Faraday was well aware that he lacked appropriate concepts for expressing the directions of electromagnetic action. In addition to being aware of Ørsted's, Ampère's, and Biot's attempts, he also knew well the limits on which Davy's effort to formulate a rule had foundered. He addressed this difficulty by means of graphic illustrations and three-dimensional physical mnemonics. He had no appropriate prototype to work from, but it is clear at the same time that for him these images were more than simply representations of experimental conditions. On them, and with them, he developed his generalizations and abstractions. By compiling as many individual observations as possible in a single picture, he crafted images that would play a central role in his search for regularities, binding all of the individual cases together. The image functioned as the instrument by which he developed his attempts at generalization. This point will soon show up again, when we turn to the novel research he would soon conduct.

In the second and longest part of the "Historical Sketch," Faraday gave an overview of developments since Ørsted's original discovery. He confined himself to a "historical sketch of the facts" as these were "of utmost importance" ([Faraday] 1821–22, 274). The separation of "facts" from "theories" reflected a general sentiment in England at the time.[42] The overview proceeded in chronological order, its sequence based on the dates of journal issues, dates given in Ampère's mémoire, and various accounts of sessions of the Académie des Sciences and Royal Society. The chronological sequence itself must have taken some effort. His overview was complete through April 1821, and at some points it went further.[43] For example, he mentioned Davy's lecture of July 5, 1821, but gave no reference, as the text was not yet available. A large number of studies and their authors were mentioned, including Ampère, Arago, Boisgiraud, Biot, Savart, Fresnel, von Buch (in Frankfurt), Lehot, Davy, Yelin, von Buch (in Utrecht), Gazzeri, Ridolfi, Antinori, La Borne, Berzelius, Berthollet, Schweigger, de la Rive, Moll, Offerhaus, Hare, Pictet, and Poggendorf. Not a single important name was missing, though he focused most keenly on the work of Ampère and Davy.

But again, Faraday went well beyond pure survey, inserting his own incisive judgments at numerous junctures. For example, he qualified Ampère and Arago's findings on magnetization as "natural consequences" of Ørsted's ([Faraday] 1821–22, 278), an assessment to which others, in addition to the two named authors, would have objected. His treatment of the work of Frankfurt professor von Buch was significantly more pointed, beginning with the claim that "M. Buch . . . in repeating Oersted's experiments, added nothing new to them" (281). To his account of the experiments of Gazzeri, Ridolfi, and Antinori, he appended the remark, "There is, probably, some mistake in this," but without offering any further explanation (286). Schweigger's experiments with the "multiplier" came off poorly, there being "nothing new in them after what has been said" (286).[44] Berzelius's experiments were similarly assessed (287). Faraday characterized Moll's interpretation of his experiments as "probably premature," going into further detail on one of them: "There can . . . be no doubt . . . that such a spiral is not necessary" (289).

Notwithstanding his predominantly critical tone, we also find some positive judgments. A small device by Charles-Gaspard de la Rive was called "very simple, easily made, and effectual" ([Faraday] 1821–22, 288). Faraday had received one—a floating wire loop highly responsive to a magnet's proximity—directly from de la Rive and had experimented with it.[45] This sampling of his judgments should suffice to show how firmly and, for a newcomer to the field, how surprisingly decisively Faraday had formed his own opinions on the various contributions.

As he had already done in the first part of his sketch, besides sorting and assessing the various results Faraday set about presenting them more clearly than their authors had. He paid particular attention of this sort to Davy's experiments on the "circular" nature of the action surrounding the wire. Once again, Faraday took recourse to visual representations. In his perspective drawing (labeled "Fig. 11" in our fig. 5.1), NP once again refers to the wire. Needles were placed tangentially on a paper disk at a right angle to the wire. Arrows indicate the polarity in which the needles were found to have been magnetized after current had flowed through the wire. This representation allowed Faraday not only to avoid Davy's cumbersome exposition but also, and more importantly, to eliminate all reference to the compass points. The sketch conveys the "rotatory" character of electromagnetic action far more clearly than Davy's verbal description.

Though clearly composed later, the third part of the "Historical Sketch" belongs to the same integrated whole as the first two. Here Faraday (1821–22, 107) concerned himself with "attempts at a theoretical explanation of the phenomena," continuing his sorting and evaluation. For example, he expressed a clear view, if not one generally shared, of the relationship between Ørsted's findings and his theory: "It cannot be doubted for a moment . . . that his theory rather led to the experiments, than the experiments to his theory" (107). But he also wrote, "I must confess I do not quite understand it," hoping for some further explication from Ørsted himself (107–8). Schweigger's theoretical reflections received a similar comment: "It is difficult to understand how the above theory is to explain the phenomena described by Oersted." On Ridolfi's theory, he remarked, "No experimental proofs of the correctness of the opinion were obtained" (111). Faraday devoted some space to Berzelius's reflections, concluding, "It is, however, probable that M. Berzelius has been himself too hasty in his conclusions. The state of the wire [described] above, is utterly incompatible with the experiments of Oersted and others. . . . There can, however, be but little doubt that M. Berzelius will correct his opinion, and contribute to the advancement of this branch of science, by something worthy of his great name" (109). A sentence like this last one, when expressed by a complete novice with respect to one of the most recognized chemists in Europe, is evidence of considerable self-confidence. I return to this matter later.

Faraday's (1821–22, 110) evaluations of London researchers were noticeably more reticent. He abstained from any assessment of Wollaston's views, instead simply quoting in full, with expressions of deference to Wollaston's authority, the short, cryptic report Brande (1821a) had given of his views in the *Quarterly Journal of Science*.[46] Wollaston was both esteemed and influential in London. Furthermore, at the time, Faraday was involved in a charge of having plagiarized Wollaston (see below). He devoted particular attention to Ampère's theory, with six of the ten pages of this third part of his "Historical Sketch" devoted to discussing it (while the other five researchers took up four pages altogether). Faraday justified this disproportion, claiming that Ampère's theoretical views were the "most extensive and precise" of all of them, "and have been tested by the application of facts and calculation very far beyond of the rest. Indeed, it is these alone among all those which have been given to the public, which deserve, if any do, the title of *A Theory*" ([Faraday] 1821–22, 111, original emphasis).

Despite this laudatory assessment, Faraday (1821–22, 112) maintained his critical approach, in particular toward what he saw as the inadequate articulation of the concept of current. His description of Ampère's assertion of the existence of currents for which there was no experimental evidence as a "bold thought" blended admiration for his audacity with concern for his carelessness (114). Faraday's analysis of Ampère's theory carefully distinguished two separate claims: the existence of two electrical fluids and the identity of electricity and magnetism. His concluding assessment asserted that "the first part of the theory [is] by no means sufficiently developed" (117).

The Significance of Faraday's Experiments

Faraday's decisive and occasionally harsh judgments were unusual, even daring for someone not only new to electromagnetism but also having no established place on the scientific landscape. This is presumably one of the reasons for which Faraday chose to publish the "Historical Sketch" anonymously. It's not that he was unwilling to take responsibility for his critical judgments, for in a different context he would shortly do precisely that, and quite decisively. He was more likely worried about being accused of immodesty in his handling of the results and reflections of those who, unlike him, had already engaged in productive, innovative research. It was important to him to have contributed something constructive to the field before voicing criticism of others in his own name.

Still, the certainty and definiteness of his judgments remains astonishing. As has thus far been overlooked, this confidence had much to do with his experimental work. His independent sorting and evaluation of results reported by others would not have been possible had Faraday not performed his own experiments. The reports and articles on which he drew were addressed to an audience with experimental experience in the field. Their descriptions of experiments in the various texts were often much too vague to convey, by themselves, enough detail for the reader to compare them with other arrangements described with equal vagueness. Had Faraday not seen the experimental constellations with his own eyes, he could never have attained such confidence and incisiveness. His intensive experimental engagement had a further consequence: he gained significant practical experience with experimental procedures and techniques. Besides the few studies he had performed with Davy, he had not previously done any laboratory work on electromagnetism. His preparations for the "Historical Sketch" left him with a clear sense of the technical state of the art.

The huge investment of experimental labor Faraday devoted to this first contribution to the field, the "Historical Sketch," prepared him to do his own research and to attempt his own conceptual restructuring.

Given all that, the manner in which Faraday presented his own experimental work—or rather hid it—is noteworthy. Nowhere was it explicitly mentioned in the "Historical Sketch." That text thus stands in marked contrast to Faraday's later contributions, in which his experimental work would take center stage. Only the most attentive and informed reader could be expected to discern the extent to which this article went beyond a mere survey, even with regard to its empirical, experimental background. An only slightly less informed reader might well come away with the sense that it really was just a summary of prior efforts and not original research. This was the impression Faraday sought to convey, if an erroneous one. This view continues to permeate the history of science literature, leading to the serious underestimation of the significance of the "Historical Sketch." L. Pearce Williams (1985a, 89), one of the few scholars to engage with that text to a significant degree, reaches the standard conclusion that "by and large, the first two parts of the Sketch merely summarized what had been done." Williams draws attention to the importance of only a few theoretical or conceptual points. But, like others, he does not pay attention to the autonomy of Faraday's judgment, his development of new means of representation, and the central role of Faraday's own experiments (Williams 1965, 153–56; 1985a, 88–91).[47] These were precisely the aspects of the "Historical Sketch" most important to Faraday. By such means could he catch up in an already maturing field, to the point where he could immediately set to work on his own novel research. The rapid progress of his ensuing studies cannot be understood apart from this point.

A Researcher Positions Himself in a New Field

Faraday began his research under very special circumstances. Launching such a labor-intensive enterprise in such a short time despite his duties at the Royal Institution would have been inconceivable with anything less than optimal resources at his disposal. He had access to the well-appointed library of the institution, with all of the pertinent journals in its collection, and above all to its laboratory. Accessing either resource required only that he descend a few floors. As superintendent he presumably enjoyed all-hours access. The laboratory was not only one of the best equipped in Europe, it was also a space with which Faraday was intimately acquainted after years of working there. He knew where to find everything and had extensive experience with all of the available materials and apparatus. By such means, and such means alone, could he begin to undertake meaningful experiments so quickly, even after his official duties were over.[48]

With the "Historical Sketch," Faraday positioned himself, albeit anonymously, within the existing landscape of research. He showed himself to be self-confident relative to most other authors and had no hesitation in voicing trenchant criticisms of even the "great" scientists. With regard to two prominent and, for Faraday, especially important actors, Davy and Ampère, matters were more complicated. He always spoke very highly of Davy's contributions, emphasizing the "deduction of fact from fact" ([Faraday] 1821–22, 283) by which Davy had arrived at his numerous results. Davy's modus operandi had been a striking model. But Faraday's admiration also created some bias. Davy had shown that iron filings clinging to two different conductors would attract when the current flowed in the same direction through both, and they would repel when the directions were opposed. In Faraday's account, the outcome of these experiments was more or less "similar" to Ampère's discovery of the mutual attraction and repulsion of electrified conductors: "these experiments . . . prove the same thing" (284). This is a gross overstatement. The connection between Davy's results and Ampère's was far from obvious, and Davy himself had never made any such claim. Faraday's assessment reveals his special relationship with Davy, one that did not yet permit much in the way of sober, critical distance. Given Faraday's career trajectory, this is hardly surprising.

Faraday's "Historical Sketch" engages with no other author as thoroughly as with Ampère. Despite all of his criticisms, Faraday's text reads like an expression of profound admiration. Next to Davy, he considered Ampère to be, without question, the leading authority in the field. Nonetheless, he found Ampère's work less accessible than that of other authors. For one thing, important parts of Ampère's reasoning—all those that involved mathematical considerations—remained beyond his reach. In addition, in working through Ampère's electrodynamic experiments, Faraday reached the limits of his own experimental capacity. The construction and operation of Ampère's apparatus were extremely laborious tasks, and Faraday appears to have been unable to replicate them as precisely as he wished. In a striking departure from his treatment of other authors, he offered no detailed commentary or critique of Ampère's experiments in the "Historical Sketch." However, where Ampère's enterprise ventured closest to his own interests—in the search for "general facts" or regularities—he was powerfully engaged. On this point he would shortly take the inadequacies of Ampère's account as the point of departure for his own research efforts.

As the "Historical Sketch" makes clear, by the time Faraday began his research in the immediate aftermath of this work, electromagnetism was in a completely different state than it had been a year earlier, when Ampère had made his own beginning. Whereas in the fall of 1820 there were very few reference points that might have given direction to experimental and conceptual work, by the fall of 1821 there were numerous more or less well established experimental techniques, concepts, and theoretical proposals by which to orient oneself, for or against. In breadth and depth, Faraday knew them like no one else. Consequently, he clearly saw where the need for further research was most urgent. In particular, he had a good notion of where, with his own specific research methods, he might best make his own contributions. Furthermore, unlike Ampère, who undertook his first ever experimental work in the fall of 1820, Faraday began his work on electromagnetism with a highly developed and consciously pursued research style. In addition, the research climate and competitive environment had changed. The feverish haste and pioneering spirit with which many had taken the field in 1820 had given way, even in Paris, to a more considered pace, one in which more evaluation and sorting of what had already been achieved were not only possible but necessary. The competitive environment had cooled down to its usual level, and the numbers of active researchers become manageable. Furthermore, as Faraday could clearly see, he faced no strong competition in England—with the exception of the most important actor of all, the highly esteemed Humphry Davy.

CHAPTER 6

Faraday's First Studies

Electromagnetic Rotation

IMMEDIATELY AFTER SUBMITTING THE FIRST TWO parts of his "Historical Sketch," Faraday began his own electromagnetic research.[1] Conditions at the Royal Institution allowed him to pursue his own research interests, at least for short periods. Over the course of an extremely intensive week he not only made a spectacular discovery but also completed a very wide-ranging article that, in print, would comprise twenty-two pages. The stunning novelty of electromagnetic rotation would immediately turn the European spotlight, and especially the Parisian spotlight, squarely on him. His finding, and to a greater extent, the way he presented it, posed a truly serious challenge to Ampère's theory. At the same time, Faraday's first contribution to electromagnetism already displayed basic features of his approach, one that would continue to shape his entire decades-long engagement with this field. Taken together, this weeklong episode, along with the briefer follow-up episodes through December 1821, are as central to understanding the development of electromagnetism in general as to the investigation of Faraday's science in particular. The core of the present chapter will thus be a comprehensive reconstruction of his work during this period.

I proceed in three sections, each corresponding to one of three clearly distinguished phases in Faraday's work. In the first section my focus is on the investigations that led to the discovery of electromagnetic rotation. The most important of these took place in one day, September 3, 1821. Though these studies have repeatedly been the object of historical analysis, previous efforts have neglected certain points whose proper appreciation casts new light on the entire episode. In the second section I carefully consider Faraday's activities in the days *following* his discovery. The fact that he did not immediately publicize his discovery, instead continuing to experiment for nearly a whole week, remains astonishing. But his goal was to situate the new effect within a broader context. His efforts in this direction have not yet been studied in any detail, though, as I show, they provide a rare view of the typical features of his modus operandi. Following his first publications Faraday was engaged in various attempts to stabilize the discovery and secure its general recognition, thus initiating a new round of research in December. The three-month period from October through December 1821 has never yet been sufficiently scrutinized. This is the subject of the third major section of this chapter. Taken together, the three sections provide the first comprehensive study of Faraday's entrance on the stage of electromagnetism. In a final section, I compare this initial period to that of Ampère's first inquiries, a year earlier.

The Reconstruction of an Initial Phase

A few remarks on methodology and the state of source materials are called for. By contrast with the case of Ampère, for Faraday the state of available sources is nearly optimal. The most important of these sources is his laboratory diary, which he began to keep quite early in his career and which, over the course of his life, grew to several volumes. He recorded every experiment with a date, the apparatus employed, its precise setup, and the experimental results. Very often, he added drawings and schematic sketches of arrangements at the margins of the page. In a few instances we also find notes on his reflections over the course of his work. Occasionally we also find expressions of excitement, satisfaction, or discontent. Through the end of Faraday's life, the diary would encompass five bound manuscript volumes. It was transcribed in its entirety and published in a reliable edition (Martin 1932) whose only imperfection is that Faraday's few deletions and corrections are not indicated as such.[2] In addition, Faraday had the habit of flagging those *Diary* passages that he had used for publication with a vertical line down the middle. Such marks, very useful for purposes of reconstruction, have also been omitted in the published edition. For this reason, an inspection of the original, preserved in the archive of the Royal Institution in London, was called for in several instances.

In the original manuscript, kept in the Royal Institution, Faraday's handwriting is very clear. Given the almost complete absence of deletions, corrections, and insertions, it reads like a clean copy. Even Faraday's sketches are executed clearly and carefully. The manuscript pages bear no trace of the daily laboratory routine, such as chemical stains, bleaching, or the like. In addition, some daily entries are preceded by a kind of précis, for example, "To-day went still more generall to work . . . and obtained I think very satisfactory and reconciling results" (D209). Such evidence suggests that he did not make entries on the fly, in the laboratory itself, but at the end of the day, in a different place. I assume that Faraday made provisional notes in the laboratory (though these have not been preserved), transcribing them each evening into his bound diary. The *Diary* thus represents a revised text, though the order of entries appears not to have been altered. There are sudden and unsystematic changes of subject of the sort that might be expected in the course of a day's work in the laboratory. Faraday also reports missteps and experiments that he later recognized as defective, sometimes on the same day. The order of entries in the *Diary* gives every indication of following the sequence of laboratory events, as Faraday occasionally stresses explicitly.[3] In any case, I am proceeding from this assumption in my reconstruction, like others before me. Nowhere have I found discrepancies or contradictions with his assertions elsewhere, in published materials or in letters.

Faraday's complete and masterfully edited correspondence (James 1991) has also been a central source for my reconstruction.[4] Finally, Faraday's published articles constitute my third main source. Because the manuscripts of these articles have typically not been preserved, however, we are in no position to retrace the precise steps of their composition.[5] Nonetheless, the availability of sources necessary for the reconstruction of Faraday's practices throughout this period is superb. This is particularly clear when we compare it with the case of Ampère, as described in chapter 3 and, in greater detail, in appendix A.

From Regularities to Electromagnetic Rotation, September 3, 1821

The Search for a Regularity

Faraday began his research on Monday, September 3, 1821, and soon attained spectacular results. Figure 6.1 reproduces his laboratory notes and sketches for the day, in their transcribed form. Faraday begins with a refer-

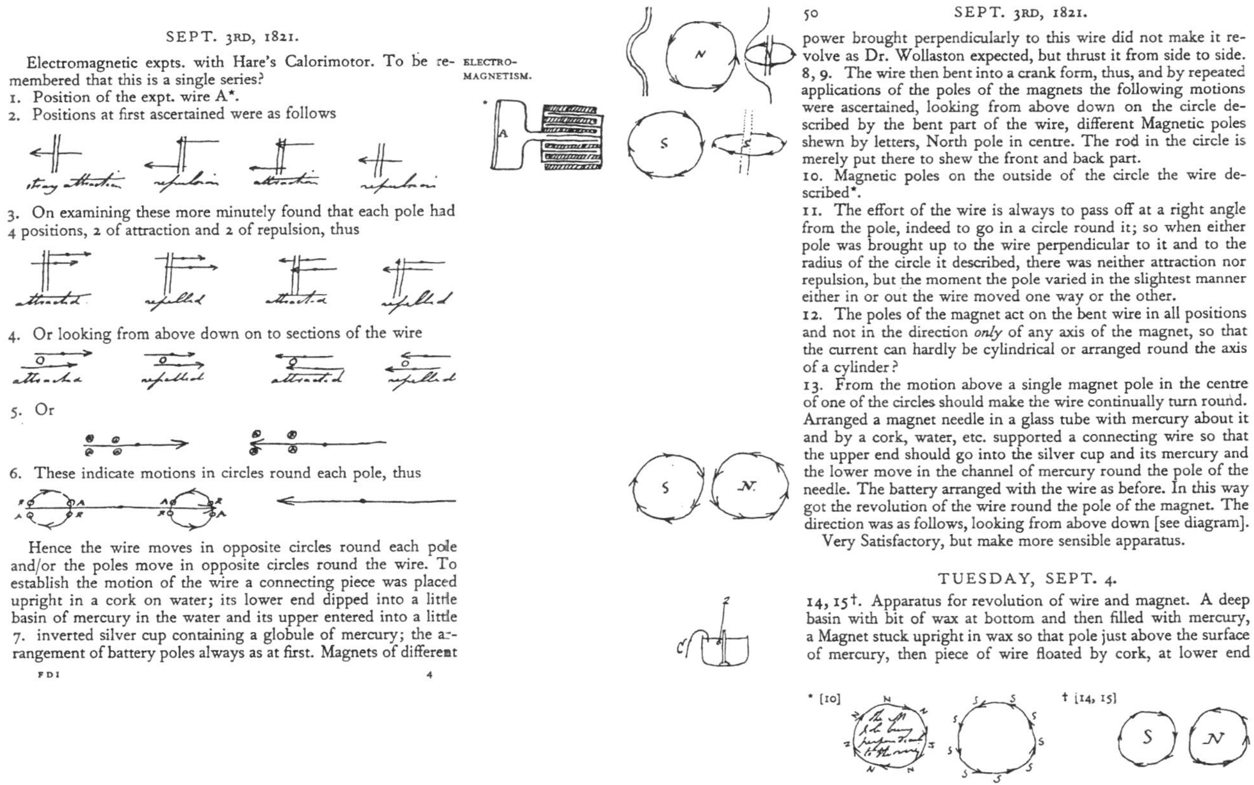

SEPT. 3RD, 1821.

ELECTRO-MAGNETISM.

Electromagnetic expts. with Hare's Calorimotor. To be remembered that this is a single series?
1. Position of the expt. wire A*.
2. Positions at first ascertained were as follows

3. On examining these more minutely found that each pole had 4 positions, 2 of attraction and 2 of repulsion, thus

4. Or looking from above down on to sections of the wire

5. Or

6. These indicate motions in circles round each pole, thus

Hence the wire moves in opposite circles round each pole and/or the poles move in opposite circles round the wire. To establish the motion of the wire a connecting piece was placed upright in a cork on water; its lower end dipped into a little basin of mercury in the water and its upper entered into a little
7. inverted silver cup containing a globule of mercury; the arrangement of battery poles always as at first. Magnets of different

FDI 4

50 SEPT. 3RD, 1821.

power brought perpendicularly to this wire did not make it revolve as Dr. Wollaston expected, but thrust it from side to side.
8, 9. The wire then bent into a crank form, thus, and by repeated applications of the poles of the magnets the following motions were ascertained, looking from above down on the circle described by the bent part of the wire, different Magnetic poles shewn by letters, North pole in centre. The rod in the circle is merely put there to shew the front and back part.
10. Magnetic poles on the outside of the circle the wire described*.
11. The effort of the wire is always to pass off at a right angle from the pole, indeed to go in a circle round it; so when either pole was brought up to the wire perpendicular to it and to the radius of the circle it described, there was neither attraction nor repulsion, but the moment the pole varied in the slightest manner either in or out the wire moved one way or the other.
12. The poles of the magnet act on the bent wire in all positions and not in the direction *only* of any axis of the magnet, so that the current can hardly be cylindrical or arranged round the axis of a cylinder?
13. From the motion above a single magnet pole in the centre of one of the circles should make the wire continually turn round. Arranged a magnet needle in a glass tube with mercury about it and by a cork, water, etc. supported a connecting wire so that the upper end should go into the silver cup and its mercury and the lower move in the channel of mercury round the pole of the needle. The battery arranged with the wire as before. In this way got the revolution of the wire round the pole of the magnet. The direction was as follows, looking from above down [see diagram].
Very Satisfactory, but make more sensible apparatus.

TUESDAY, SEPT. 4.

14, 15†. Apparatus for revolution of wire and magnet. A deep basin with bit of wax at bottom and then filled with mercury, a Magnet stuck upright in wax so that pole just above the surface of mercury, then piece of wire floated by cork, at lower end

* [10] † [14, 15]

FIG. 6.1. Faraday's *Diary* entries for September 3, 1821, in their published form (Martin 1932, 1:49–50).

ence to the galvanic device he was using, which he called "Hare's Calorimotor" (D1). This device, introduced by Robert Hare of Philadelphia in 1818, produced powerful current effects (including the thermal effects for which it was named). In modern terminology we would describe it as connecting its plates in parallel, so that the entire arrangement can be understood as a single pair of plates with a very large surface area.[6] For its time, it could be considered heavy equipment, as Hare's illustration makes clear (fig. 6.2). The use of such powerful devices was not the usual practice in electromagnetic research, and in his earlier experiments Faraday had always used a conventional trough device. His transition to the calorimotor was likely motivated by its exceptional strength.

Thus equipped, Faraday resumed studying the behavior of the needle in the vicinity of a vertically suspended wire. His experimental arrangement, which hearkened back to Ørsted, was the same as that employed in his "Historical Sketch" (see fig. 2.2), only now he sought to investigate the positions of attraction and repulsion "more minutely" than before (fig. 6.1, D3). He offered no explanation for his dissatisfaction with the earlier results, a matter to which I return later. In the first step of his

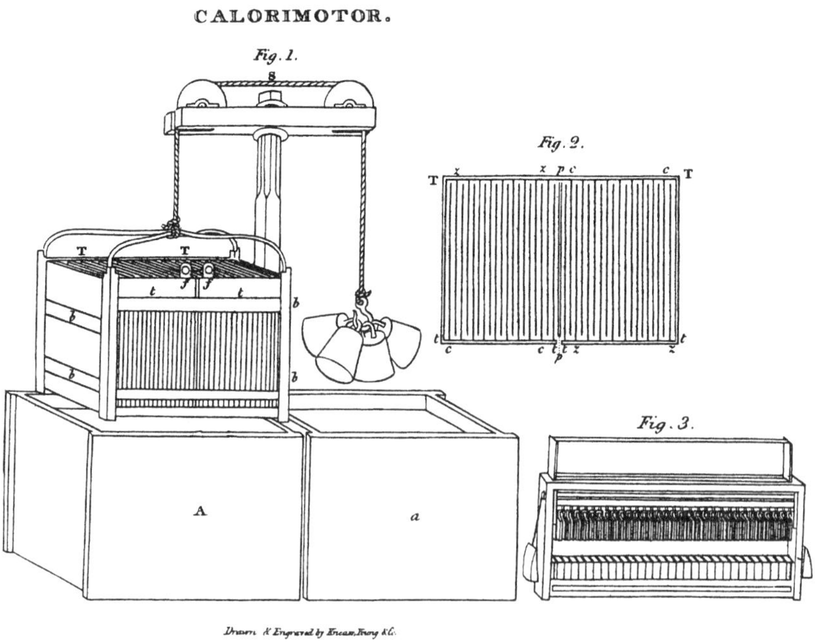

FIG. 6.2. Hare's "Calorimotor," used in Faraday's experiments (Hare 1819).

investigation (D2), he reproduced the main results of the "Historical Sketch": the attraction/repulsion effect in the central position, as well as the four positions of attraction/repulsion at the needle ends, as shown in his illustration. With regard to the second point, however, his sketches in the *Diary* coincided only partly with those in the "Historical Sketch." There, his Figures 6 and 7 (presented in fig. 5.1) had jointly treated the positions in which north and south poles of the needle were attracted to the wire. Now, in the *Diary*, in addition to the figure titled "attraction," he added another for "repulsion." While all of this remained, to some extent, a recapitulation of the achievements of the "Historical Sketch," he also undertook an experiment he had presumably performed only in passing for the earlier work. His article describes the procedure in detail: "allowing the needle its natural position across the wire [i.e., the position in which the center of the needle lay directly opposite the wire] . . . and then drawing the support away from the wire slowly, so as to bring the north pole, for instance, nearer to it, there is attraction, as is to be expected; but on continuing to make the end of the needle come nearer to the wire, repulsion takes place, though the wire still be on the same side of the needle" (Faraday 1821d, 75; reprint, 1839–55, 2:128).[7] In other words, he slowly moved the needle and its support along the wire, out of the central position (in which its center was closest to the wire), so that one of the poles approached it. In the process, he observed first attraction between needle and wire, but the attraction turned, at a greater distance, into repulsion. This much had been described, in outline, by Ørsted.[8] It is in his comprehensive pictorial illustration of the result that Faraday clearly moved beyond Ørsted. Indeed, in working with these illustrations he developed a remarkable series of generalizations that culminated in a fundamental proposal. He began with the statement "that each pole had 4 positions, 2 of attraction and 2 of repulsion" (D3). The sketch in paragraph D3 is similar to the earlier illustrations in D2 in that it represents, in a single sketch, two positions of attraction and repulsion, respectively. The difference lies in the fact that at the needle's end he now distinguished not one but numerous different positions of attraction or repulsion. His standard side-view illustration having thus become more difficult to interpret, Faraday drafted a second sketch that took a different perspective on the same constellation, which he described as "looking from above" on the experimental setup (D4). In this view, the wire has to be imagined as perpendicular to the paper and hence appears only in cross section, as a small circle. In a further step, Faraday attempted to condense all four sketches, performing the same kind of abstraction he had used in the "Historical Sketch": for purposes of illustration, the movable needle was held fixed, while the fixed wire was drawn at various positions around the needle. The resulting sketch (D5) represents the experimental findings in a highly compressed format. Because this sketch came about through much the same process as Figure 4 of the "Historical Sketch" (in fig. 5.1), the contrast between the two expresses the difference between the corresponding experimental results in a particularly vivid way.

The sequence of images through which Faraday "thinks his way" is highly remarkable and illustrates the peculiar and central use he made of graphic representations, a matter to which I return later. In the *Diary* (fig. 6.1, D3) he speaks of "4 positions, 2 of attraction and 2 of repulsion," and uses this same terminology in his published article: "instead of finding [the attractive and repulsive positions of the wire] to be four, one attractive and one repulsive for each pole, I found them to be eight, two attractive and two repulsive for each pole" (Faraday 1821d, 75; reprint, 1839–55, 2:128). Without explicitly referring to it, Faraday here invokes the rule formulated in the "Historical Sketch," for no one besides Faraday had ever spoken of four positions.[9] Though this result is more complex than the preceding one, it has been formulated in a similar way. In both cases regularities in the needle's behavior are captured in terms of concepts of attraction and repulsion, in complete accord with the prevailing conceptual framework of the time. The fact that Faraday's sketches at D5 present new and surprising experimental results without going beyond the "Historical Sketch" terminology of attraction and repulsion or altering the mode of representation has gone unnoticed in earlier studies. But this point is decisive if we want to understand how and why the innovations of September 3 took place.

Before proceeding, it is instructive to consider the reason why Faraday had now obtained a decidedly different result from that of a few weeks prior, when he had performed experiments of the same kind in preparing the "Historical Sketch." His experimental procedure had changed only slightly, and there are no grounds for thinking that Faraday was less attentive in performing the earlier series. The essential change in the experimental arrangement consisted in the use of a much more powerful galvanic device. As a re-creation of these experiments demonstrates, the observable attraction-and-repulsion effects are very feeble.[10] The needle can easily begin to oscillate or to respond to ambient air currents, and it is far from easy to distinguish such motion from actual electromagnetic effects. Faraday was well aware of the resulting uncertainty. Though he had not drawn attention to it in the "Historical Sketch," he had noted that in his approach to the problem of the "simple but important point of position," he had strived to convey a clear sense of what "I am at present able to do" (Faraday 1821–22, 199). This may have been intended as a disclaimer regarding the provisional character of the results. If his new series of experiments was geared toward investigating precisely these results "more minutely," this may have meant distinguishing the electromagnetic motions of the needle from other perturbations with greater certainty. Toward that end, the first course of action would be to strengthen the electromagnetic effects by employing a more powerful galvanic device. This path led to success. He did indeed obtain clearer, less uncertain, and more reliably reproducible experimental results. He could not have expected that they would differ so dramatically from his earlier results, but neither could he have ruled it out.

Circular Motion as a New Mode of Representation

Faraday now observed the behavior of the needle to be much more complicated than before. With his highly compressed representation of this behavior in hand, he considered the question of whether the observable attractive and repulsive motions might not be understood as effects of a more general action of potentially very different character. He formulated a proposal that went beyond preexisting concepts and made possible a new perspective on the experimental results: "These [positions] indicate motions in circles round each pole" (fig. 6.1, D6). His visual representation in the *Diary* makes matters clearer. Perhaps the eight positions described in the preceding paragraph might—so his proposal suggested—be interpreted as nothing but special points of a more general tendency toward circular motion. Attractive and repulsive motions would then become merely arc segments of this larger, circular motion. In this view, the centers of magnetic action would no longer be found in the needle ends themselves but rather in two points some distance nearer the center. Both the denial of the fundamental character of attraction and repulsion, as well as the displacement of the poles of a bar magnet away from its ends, broke with traditional conceptions. It is important to highlight that these ideas were formed in the context not only of intense experimentation but also of successive attempts to find the most general graphical presentation of the experimental results—attempts that involved a highly versatile use of various visual perspectives on one and the same experimental subject.

There is also a more general background to Faraday's ideas. His considerations were provoked by his recognition, as definitively demonstrated by his latest experiments, of the greater complexity of the needle's behavior. This complexity made its clear description in terms of attraction and repulsion particularly cumbersome. From the very beginning, electromagnetism had had to contend with the peculiar necessity of attributing both attractive and repulsive properties to one and the same point of the needle, depending on from which side one was viewing it. Toward that end, Biot had developed a specific if rather implausible construct, and Ampère's floating-man analogy had been an elegant approach to the same problem in a special case. But now Faraday's new findings alerted him to the fact that this already peculiar property of the needle end completely reversed itself within a short distance in the region surrounding the tip of the needle. His graphical compilation and compression of the results brought the problem into such stark relief that he felt prompted to ask whether the concepts of attraction and repulsion themselves were adequate to the task of capturing this behavior in a rule. This question was absolutely fundamental in that he, like everyone else at that time, had until then understood attraction and repulsion as basic modes of action. To be sure, he was less committed to this conception than those for whom it was bound up with a mathematical tool kit. Thus, he was more open to asking whether other conceptual structures might be found within which the findings would appear less problematic and their regularity expressed more clearly.

However, Faraday's greater willingness to embark on the quest for alternative means of expression does not yet explain his coming up with the specific idea of circular motion. There were many factors involved here. First, talk of circular action was not new to electromagnetism. Ørsted had spoken of the circular or spiral structure of the electrical conflict surrounding the wire.[11] For his part, Biot had repeatedly stressed this point, though without further specification or explanation. Wollaston had understood his "electro-magnetic current" as circulating around the long axis of the wire, and Brande had illustrated it graphically (see fig. 6.10 and Brande 1821a). Finally, Davy, as was well known to Faraday, more or less explicitly suggested a kind of circular action in his interpretation of his magnetization results, as so clearly depicted by Faraday in his illustration of Davy's experiments in the "Historical Sketch" (Figure 11, in our fig. 5.1). Via Davy, Faraday also knew of Wollaston's idea of producing a rotation of the wire around its own axis. In short, when it came to looking for possible alternatives to the conceptual framework of attraction and repulsion as basic modes of action, the idea of circular action in general was certainly in the air. However—and this is often overlooked—this idea is extraordinarily vague. Its diverse implementations, if intelligible at all, were very different and sometimes incompatible. Earlier such proposals had hardly anything in common with the kind of circular motion around a magnetic pole that Faraday now envisioned, except that all of these cases involved some sort of circular action. He can thus hardly be said to have simply adopted prior ideas.[12]

Furthermore—this is my second point—the illustration in Faraday's entry D5 (see fig. 6.1) itself evokes thoughts of circular motion. The immediate proximity of attractive and repulsive positions on opposite sides of the wire, as indicated by A and R, might easily suggest the prospect of a continuously propagating tendency toward motion. This suggestion is further reinforced by the fact that the same positions are reversed only a little way down the needle. All that remains is to deprive the needle end of the specific role previously ascribed to it and still taken for granted in Faraday's "Historical Sketch."

Given available information, when Faraday set down entry D6, the idea of a circular motion around the needle was still speculative, its validity yet to be demonstrated. Still, it must have been quite fascinating. It would offer not only a promising approach toward a comprehensive representation of the complex experimental givens but perhaps even the opportunity for a spectacular new discovery. Not surprisingly, determining the validity of this still speculative idea now became Faraday's priority. Toward that end, he turned next to an attempt to find the expected "neutral" position.[13] In the sketch in entry D6, A and R refer once again to points of mutual attraction and repulsion, respectively, between wire and needle. Assuming circular motion, there ought to be a neutral position between the two, one where neither attraction nor repulsion takes place. When Faraday's experiments confirmed this prediction, his confidence in the new idea increased. He immediately set about subjecting it to a still stronger examination.

From Representation to Physical Effect

For the remainder of the September 3 workday, Faraday concentrated on producing a direct demonstration of the circular motion. This effort signaled an important change in direction. The goal with which he had begun, that of establishing regularities in the motion of the needle, receded into the background for the time being. Instead, he set about establishing the real existence of a hypothesized motion. This second goal is related to the first but is significantly different.

He had a clear notion of the expected effect: "Hence the wire moves in opposite circles around each pole and/or the poles move in opposite circles round the wire" (fig. 6.1, D6). Simultaneously with the expected rotation of the wire around the magnet pole, Faraday also articulated a "reciprocal" effect in which a mobile magnet pole ought to move in circles about a fixed wire. This is a "mechanical" reversal of the sort that had already been repeatedly observed in electromagnetism. The clause "the wire moves" should presumably be read as shorthand for "the wire should move"; at this point, the effect still remained purely speculative.

To position a vertically oriented wire segment in a mobile configuration, Faraday ran it through a cork floating on water, with the upper and lower ends of the wire submerged in drops of mercury (D6, D7). This left the wire with lateral freedom of motion, within relatively narrow limits. This complex and unwieldy experimental setup constituted Faraday's solution to the problem that had plagued electromagnetic research since its inception, that of positioning a wire in such a way as to optimize both mobility and electrical connectivity with the battery at the same time.[14] The idea of submerging the wire ends in mercury could be found in the work of Ampère, while the technique of balancing the wire's weight with a floating cork was Faraday's. In the experiment, he brought a horizontally oriented magnet near the wire and observed the resulting motion.[15] It always moved sideways away from the magnet ("from side to side"), by which he presumably meant that the direction of motion was perpendicular to the axis of the bar magnet. His remark in the *Diary* that Wollaston's predicted rotation of the wire about its own axis had not been observed may be taken to imply that Faraday had still been prepared to admit that possibility. However, his own prediction, that of the wire's rotation *around the magnet pole,* could not be tested using this arrangement. The range of motion permitted by the small volumes of mercury was much too small.

This led Faraday to a second construction, one in which the wire segment, which still ran vertically and had both ends submerged in mercury drops, was bent into a "crank form" (D8) as I try illustrate in a sketch (fig. 6.3). The resulting wire "crank" could rotate without its ends moving transversely. When Faraday brought a horizontally held bar magnet toward the wire at the height of the crank, the expected rotation did indeed take place. (The weight-compensating cork float is not shown.) However, the rotation could not be continuous. The magnet stood in the way, and

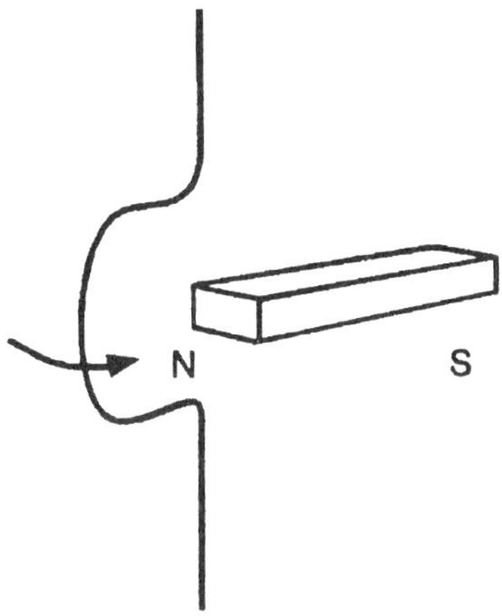

FIG. 6.3. Schematic drawing of Faraday's rotation experiments with wire bent into a "crank form."

the crank collided with it before completing a circle. Faraday then tried withdrawing the magnet shortly before the wire crank hit it, allowing the crank's inertia to carry it forward slightly, then returning the magnet to its initial position. His *Diary* speaks of "repeated applications of the poles of the magnets." In the text of the article he gives more detail: "a magnet pole brought to the centre of motion of the crank, the wire immediately made an effort to revolve until it struck the magnet, and that being rapidly brought round to the other side, the wire again made a revolution, giving evidence that it would have gone round continually but for the extension of the magnet on the outside" (Faraday 1821d, 77, ¶8; 1839–55, 2:129–30).

This procedure appears to have led to such reliable results that Faraday immediately began investigating the factors on which the observed motion depended. He switched the magnet's polarity, which reversed the direction of rotation (D9). The rotation was likewise reversed when the magnet approached the crank's circle of rotation "from the outside" (D10). His surmise that the wire would always move at a right angle to the direction of the bar magnet having been strengthened by these results, he finally tested it by holding the magnet at a tangent to the arc, or "perpendicular to [the wire crank] and to the radius of the circle it described" (D11). As expected, he observed no motion, regardless of which magnet pole faced the wire.

In entry D12, finally, Faraday investigated the influence of the direction of the magnet axis relative to the wire. The outcome of the experiment—that this parameter made no difference—prompted a theoretical reflection:

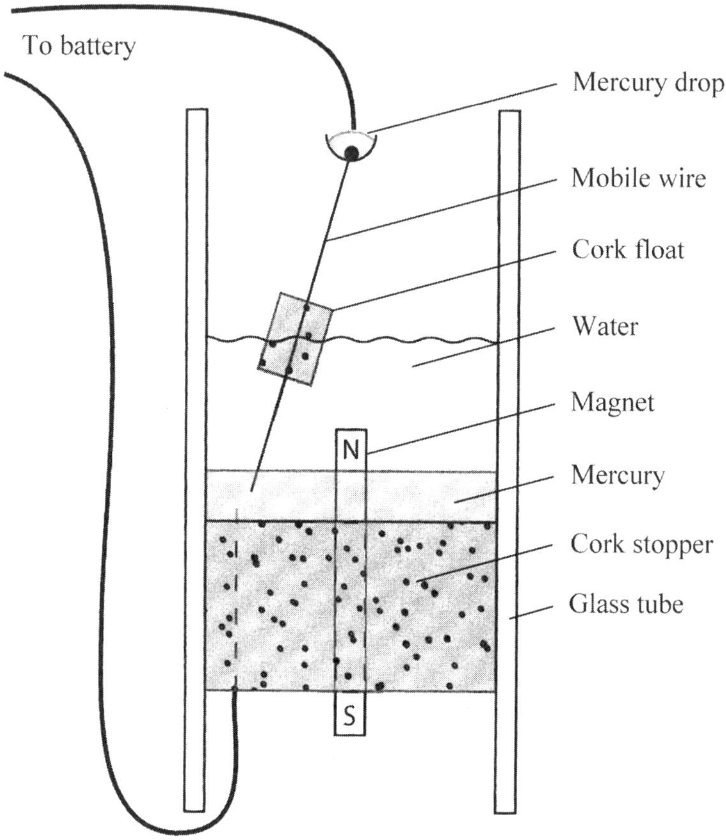

Fig. 6.4. Schematic drawing of Faraday's first rotation apparatus.

"So that the current can hardly be cylindrical or arranged round the axis of a cylinder?" (D12). It is not clear what Faraday meant by "current." It is certainly not the electrical current in the wire connected to the battery. I guess he was rather referring either to the circuits of current Ampère had hypothesized within magnets or to Wollaston's "electro-magnetic current" surrounding the electrified wire. In both cases, the fact that the interaction between magnet and wire still occurs when they no longer form a right angle, but are placed obliquely, could be taken as an argument against precise alignment. The question cannot be decided on the basis of the brief remark in the *Diary*, nor does Faraday's article discuss any further argument.

The experimental discovery that the direction of the magnet axis relative to the wire plays little role in the action on the wire also spurred Faraday toward completely different reflections. In the previous experiments the magnet always obstructed the continuous rotation of the wire crank. But if the orientation of the axis of magnetism is unimportant, the magnet could be placed vertically and thus perhaps bring about the wire's continuous rotation: "From the motion above, a single magnet pole in the centre of one of the circles should make the wire continually turn round" (D13). Faraday redesigned his apparatus accordingly (fig. 6.4). The assembly to compensate for the weight of the wire segment was the same: the wire was passed through a piece of cork floating in water. The floor of the water basin was covered with a layer of mercury (not a mere drop, as before) in which the end of the wire was submerged and, through it, connected to one pole of the battery. The upper end of the wire remained mobile, lodged in its own drop of mercury and, through it, connected to the other battery pole. In the middle of the basin, a magnetic needle or bar magnet could be inserted through the cork from below. On connecting the wire, Faraday did indeed observe continuous rotation of the wire around the upright magnet, in a direction dependent on the magnet's polarity. This apparatus was rather complicated and difficult to place in service. It required the simultaneous optimal configuration of many elements, including the water level, the height of the cork, the lodging of the upper wire end in mercury, and the connection to the battery. In his final entry of the day, Faraday's awareness of this problem blended with the great satisfaction he took in the success of his experiment: "Very satisfactory, but make more sensible apparatus" (D13).

In this development, Faraday's engagement with graphical representations is again highly remarkable. His laboratory record contains no drawings of the experimental setups themselves, only the occasional sketch of the shape of the wire segment. Of much greater importance are his sketches of the experimental *results*. As before, these alternate easily between side views and views from above. The side views are less abstract. But even in these drawings Faraday had to add an imaginary post in the center of each described rotation, so as to distinguish front from back and thereby specify the direction of rotation. Again, his sketches served as working media in which he developed stepwise abstractions. They played a constitutive role in the evolution of his view.

Faraday's first efforts on the following day, September 4, were devoted to the improvement of his rotation apparatus. He devised significant simplifications (see fig. 6.1, sketch for entry D14): instead of compensating for the weight of the wire segment by means of the cork float on water, he filled the entire basin with mercury, its greater density being sufficient to buoy the wire. The magnet could simply be placed in the basin; since the electrical contact was made entirely on the upper surface of the mercury, the lower pole of the magnet was far enough away so as not to interfere. This setup was not only more straightforward, it was also easier to handle and evidently quite reliable. In Faraday's subsequent experiments, the assembly of the rotation apparatus remained essentially unchanged. With the completion of the improved design on September 4 (D14, D15), Faraday had gone beyond simply demonstrating the rotation effect by making it stable and reliable enough for further investigation. His work had reached the point at which he was forced to consider the direction of his future research.

The Rotation Effect in Faraday's Published Article

Before I turn to Faraday's next steps, it is instructive to look ahead to the way he presented his own process of discovery in the article he would publish soon thereafter. The essay begins with a short historical account, beginning with his experiments on the attraction between needle and wire, as recorded in *Diary* entries D1 through D5. At the critical juncture following D5, he had immediately noted, "These [the experiments he had conducted so far] indicate motions in circles." In the article, by contrast, he devotes a great deal of space to laying out the reasoning leading to this surmise (Faraday 1821d, 76, ¶5; 1839–55, 2:128), distinguishing three separate conjectures. First, the "true pole" of a magnetic needle is not at its end but rather a little closer to the center. Second, a wire in proximity to a pole should be compelled to rotate. Third, the effect exerted by a needle pole on the wire should be independent of the position of the other pole of the needle. These are a lot of arguments, especially considering that Faraday formulates them before even introducing his readers to the rotation effect. His argument for the three claims rests mostly on graphical representations of the relative position of needle and wire, as sketched in the *Diary*. The published article reproduces his sketch with only a few added details (see his Figure 3 in our fig 6.5; compare the sketch at D6 in the *Diary*), with Faraday ascribing great persuasive power to it: "By attending to fig. 3, which represents sections of the wire in its different positions to the needle, all this will be plain" (Faraday 1821d, 76, ¶5; 1839–55, 2:128). Having thus motivated the *expectation* of rotational motion on the basis of his experiments, he does not immediately proceed to its experimental *realization*

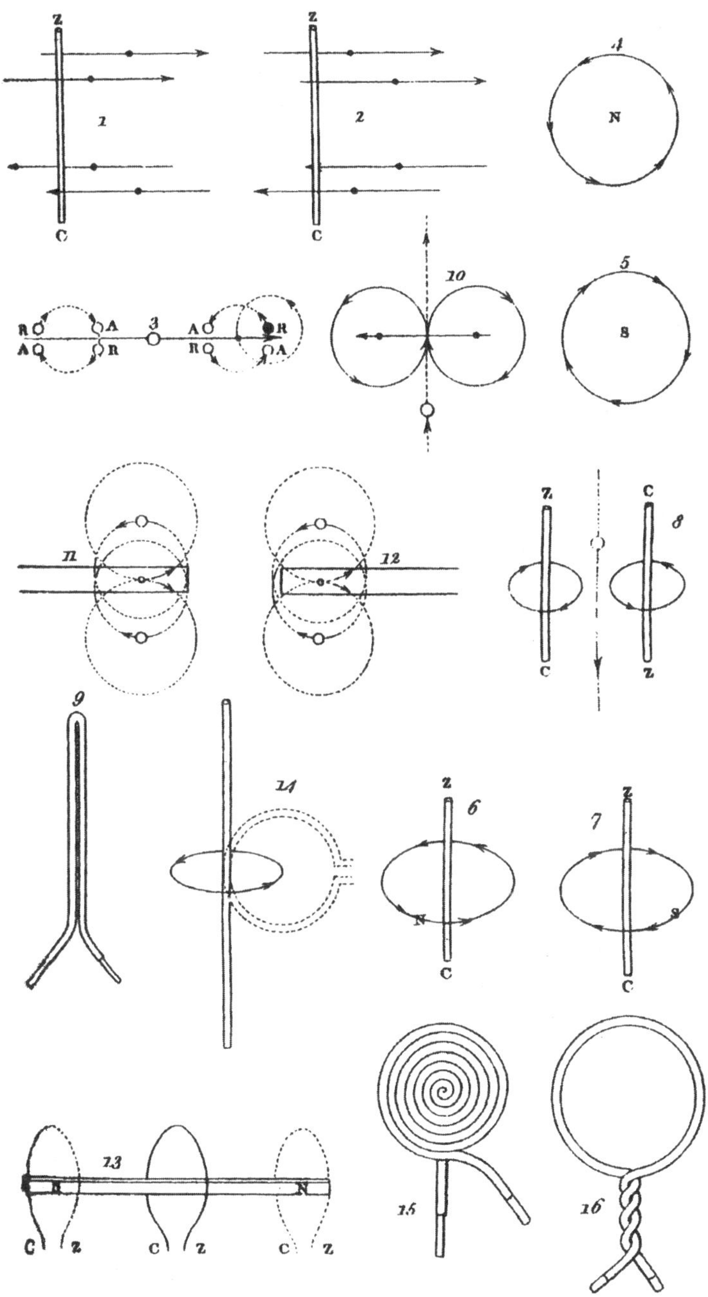

Fig. 6.5. Faraday's images in his first paper on electromagnetic rotation (Faraday 1821d, plate 3).

(as he had in the laboratory), instead injecting a general outlook. He formulates the four "conclusions" (1821d, 76, ¶7; 1839–55, 2:129) quoted below that strike him as following from these "facts" and that together proclaim the full scope of his new finding:

> There is no attraction between the wire and either pole of a magnet.
> Instead, there is a tendency toward rotational motion arising between the two.
> Both attraction and repulsion of connecting wires, and probably magnets, are compound actions.
> True magnetic poles are centres of action induced by the whole bar.

The first two theses are a sharp departure from all conventional ways of speaking and thinking about electromagnetism, that is, from the same conventions Faraday himself had invoked a few weeks earlier, in his "Historical Sketch." The third thesis poses a direct challenge to Ampère's theory, for in it Faraday describes the very phenomena Ampère had treated as elementary "compound actions," derived from other elementary constituents. The fourth thesis is quite vague. As Faraday discusses it no further, its meaning remains largely obscure. What he presumably means is that, though each pole of the magnet exists only in relation to the opposing pole, its effects are independent of the other. This is also an unconventional notion. In the received view, supported by the theory of magnetic fluids, magnetic poles always operate only in pairs.

Faraday is well aware of the provocation inherent in these theses and chooses their placement within the article carefully: he introduces them not at the beginning but at a place where the preceding account has informed and sensitized readers enough to prevent them from simply dismissing the theses as nonsense. At the same time, the boldness of these claims enhances the reader's curiosity about what is to come. Only after this rhetorically motivated preview does Faraday return to his laboratory findings. He describes the diverse variants of the assemblies for implementing rotation in great detail. These passages constitute a quasi-experimental argument for the first and second theses: Faraday (1821d, 76–78, ¶8–14; 1839–55, 2:129–31) explicitly stresses that "there was neither attraction nor repulsion" in these experiments, "but the wire endeavoured to pass off in a circle."

A comparison between Faraday's laboratory journal and his published article shows that the latter relates all of the steps Faraday saw as important to his discovery, in essentially chronological order. The account is abridged and didactic.[16] However, it still hews relatively close to the course of his research. Yet, Faraday does interrupt the historical narrative with a few far-reaching theses, giving the reader advance warning of the preeminent significance Faraday attributes to his effect. It is noteworthy to find Faraday presenting his graphical representations, which played such an essential role in his evolving reflections, without any modification. Compare, for instance, the illustrations in the published article (his Figures 1, 2, 3 in our fig. 6.5) with the corresponding sketches in the *Diary*. By placing them in the article he not only makes them public but also ascribes to them an essential role in his argument. His remark that "all this will be plain" given due attention to the illustrations reveals both their importance to him and his certain conviction that they will also be understood by others.

David Gooding, Faraday's Rotations, and New Aspects of a Discovery

The steps rehearsed in the preceding section have been variously retraced by historians before (Williams 1965, 156–58; Bradley 1989; see also Bradley 1991), but such efforts have either missed or otherwise failed to devote sufficient attention to important aspects. The most important studies are those of the late historian and philosopher of science David Gooding, one of the leading proponents of a new philosophy of scientific knowledge in general and experimentation in particular, grounded in careful historical scholarship. In 1985, he presented the first detailed study of Faraday's discovery of electromagnetic rotation. His 1989 follow-up study relates his replication of Faraday's experiments. His 1990 book also devotes a great deal of space to this episode, undertaking a schematic reconstruction by methods of artificial intelligence research, which is geared toward the computer simulation of discovery. In a shorter précis published that same year he considered the relationship between theory and experiment. Finally, in a 1998 article on the role of graphic representations in experimental research, he treated Faraday's "rotation period" as an important case study.[17]

I discuss Gooding's extraordinarily stimulating views in greater detail in chapter 7. For now, I shall discuss some points on which I find his historical investigation of Faraday's discovery of electromagnetic rotation to be imprecise and, on one critical point, erroneous and misleading. As noted above, the sketch in entry D6 of Faraday's *Diary* constituted a decisive step, one in which, by means of a condensed representation of the experimental results recorded in D5, he arrived at the idea of treating attraction and repulsion as consequences of a (still hypothetical) circular motion. The principle of the D6 sketch is the same as that in D5, in that the experimental setup with vertical wire and horizontal needle was drawn as if from above—as Faraday had explicitly asserted in D4. The small round circles, labeled "A" and "R" in D5, are thus intended as cross sections of the wire. All that is new in D6 is the idea of a circular motion of the wire around a point somewhat removed from the end of the needle. Gooding, however, sees this differently, repeatedly claiming that, in this sketch, wire and needle are in *the same plane.*[18] With regard to the orientation of this plane, Gooding's interpretation appears to change. In 1985 (118), he seems to treat it as horizontal, while in 1989 (77) he explicitly claims that it is vertical. In either interpretation, the meaning of Faraday's sketch remains murky. Gooding (1990a, 125) concludes that Faraday's talk of putative circular motion must be "literally false."[19] And indeed, as Gooding understands the arrangement, with needle and wire in the same plane there should be neither attraction nor repulsion. Faraday's experiment, however, involved a different spatial configuration, as shown above.

Gooding's misinterpretation may concern a tiny historical detail, but it is a detail of crucial importance in the evolution of Faraday's research. Gooding also sees its importance, and he draws far-reaching inferences from his (mis)interpretation. He speaks of the "mental transformation" necessary for the conception of a picture like that in D6 and finds, on further scrutiny, a transition "from the temporal sequence to spatial structure" (Gooding 1985, 118). He asserts that "the diagramme to para. 6 moves from real space and time into mental space" (Gooding 1990a, 126). This interpretation, in turn, serves as an important argument for his general thesis, in which scientific creativity is typically associated with transformations in the dimensionality of representations, such as the transition from two-dimensional sketches to four-dimensional spatiotemporal representations and back (Gooding 1998). Without wishing to discuss the general thesis here, I would suggest that the particular argument at issue seems to be unfounded, resting on the aforementioned false interpretation of Faraday's *Diary* entries.[20] The transition from entry D5 to D6 simply involves no change in mode of representation and no dimensional transformation.

The episode points to a more general and persistent problem: an insufficient recognition of the continuity between the work on electromagnetic rotation and the "Historical Sketch." Faraday did not suddenly invent fundamentally new kinds of compressed visual representation on that fateful September 3. He was recalling procedures he had already used with great success and, in the "Historical Sketch," had presented in print. What did take place on September 3, and especially in the transition from D5 to D6, was the insight that such representational methods, together with fine-tuned experimental results, led to pictures that cast doubt on the fundamental role of attraction and repulsion and thus opened the horizon to new modes of thought. This means that Faraday's innovative achievements, that appear to be concentrated in a single day, when September 3, 1821, is studied in isolation, were in fact prepared and incubated over a much longer period of time that began months earlier with Faraday's experimental work for the "Historical Sketch" and his concurrent attempts to present its results in a compelling, concise form.[21] Seen in this light, September 3 appears less as the miraculous day of new ideas and methods and more as the culmination of a long-term struggle toward adequate concepts and modes of representation.

That result also extends to Faraday's *intentions* far more than Gooding acknowledges. September 3 found Faraday doggedly continuing to pursue a question he had already addressed in the "Historical Sketch" but had not solved to his satisfaction. The central problem, both in the "Historical Sketch" and on September 3, was the articulation of a regularity for the behavior of the needle and the creation of concepts and visual aids adequate to that task. The idea of circular motion arose in this context. The fact that it quickly took on a dynamic all its own and steered Faraday's research interests in a new direction is not unusual in an experimental research practice. Given the underlying continuity in Faraday's modes of representation and research goals, it is hardly surprising to find continuity at the level of experimental technique. Faraday's ability to construct the experiment and run it so successfully, all on a single day, would have been inconceivable without his prior familiarity with the kinds of experiments with which he began that day. These were exactly the same sorts of experiments he had undertaken, in some depth, for the "Historical Sketch" a few weeks earlier. Faraday's work on September 3 emerges far more clearly than previously supposed as the direct continuation of these investigations. We ought thus to take Faraday's later assertion quite literally: "It was in consequence of repeating and examining a particular experiment, that I was led into the investigation given in my first paper" (Faraday 1823, 291–92).

Situating the Rotation Effect, September 4–10, 1821

On September 4, Faraday found himself in a very special situation. He had discovered a new effect and stabilized the experiment to the point where he could repeat it at will. Furthermore, this discovery promised to open up new perspectives on electromagnetic phenomena in general. Attraction and repulsion might lose their status as primitive modes of action, with circular motion taking their place. Faraday had already undergone this shift in perspective, at least tentatively. But he knew perfectly well that this outlook contradicted a generally accepted view and would place him in opposition to established authorities in the field of electromagnetism, Ampère chief among them.

At the same time, Faraday was aware that the rotation effect as such would attract widespread attention and bring him great recognition. Given his personal situation, currently ruled by a kind of pioneering spirit, this acclaim could only be welcome. The *Quarterly Journal of Science* (*QJS*) presented itself as a natural publication venue. It was effectively the in-house journal of the Royal Institution.[22] By that point it had already published numerous smaller contributions by Faraday, and more importantly, *QJS* offered a chance at very swift publication. The next issue would appear at the beginning of October, only a few weeks off. Faraday knew this and was doubtless already busy editing contributions to his own "Miscellanea" section.[23] Given his good relationship with Brande, the publisher, Faraday could also count on securing space for his essay, sure to prove sensational, among the feature articles. The issue after next of *QJS* would not appear until three months later, in January 1822. Given that the first part of the "Historical Sketch" had either just gone to press or was about to, this was a long time to wait.[24] In that text Faraday had laid bare the problem of the regularity of the needle deviation for all to see, and he had every reason to expect that, in the wake of its dissemination, others would soon address it. With Davy's example before him, he must surely have recalled how quickly original results in this field could be superseded and overshadowed.

Taking this sudden time pressure into account helps make sense of many of Faraday's ensuing activities. He immediately began work on his article, completing the text within a week. The fact that a few days later he would write a long, enthusiastic, and relieved letter to Charles-Gaspard de la Rive, in which among other things he reported on his new results in electromagnetism, makes it clear that he had brought his work to its first conclusion.[25] In other cases, too, Faraday undertook such communication only after attaining satisfactory results. As expected, the article was published (Faraday 1821c) a few weeks later, on October 1, 1821, in the *QJS*.[26]

How firmly Faraday was determined to have his text printed with all due haste, eschewing any further delay, becomes clear by a sequence of events that would have long-term consequences. Faraday's article repeatedly mentions the concept of "electro-magnetic current" and the "powers" attributed to it, terms that had been used by Wollaston. It also alludes to possible experiments on the rotation of a wire or magnet about its own axis, also proposed by Wollaston. Though Wollaston had not published anything on the subject himself, Faraday was aware of his thoughts from a brief reference to them in Brande's report, another in Davy's article, and perhaps also from Davy's and Wollaston's conversation in the Royal Institution.[27] So as to avoid any misunderstanding with an established authority like Wollaston, Faraday, as he would later report, would have preferred to show the manuscript to Wollaston before sending it for typesetting. According to this later account, he would also have liked to include some explicit reference to Wollaston's views and experiments. But in the absence of any publications, that would have required personal communication. Wollaston, however, would not be in town for some time to come, and even Davy, with whom Faraday would otherwise readily have conversed, was traveling. In this situation, urgency superseded caution, and Faraday resolved to publish his article regardless—a decision he would come to regret.[28]

There is another surprising aspect. Faraday did not spend all of his time over the following week composing his most persuasive text, as one might have expected. Instead, he also continued to work very intensely in the laboratory. And at the end of this phase, the text was not only rather long—comprising some twenty-two pages of printed text—but the lion's share of it also treated experiments he had performed only *after* discovering the rotation effect and constructing his first apparatus. Faraday must have worked day and night, which is all the more remarkable as Sarah, so soon after getting married, would probably have liked to occasionally find her Michael spending his evenings not only in the laboratory and at his desk.

To sum up, in addition to having every incentive to publish quickly and secure his priority claim, Faraday also had reason not to be satisfied with a published account of the effect alone and thus reason enough to warrant taking on so frantic a period of work. Previous work by historians of science has not taken these circumstances into account: while we find various studies of the discovery on September 3 and 4, there is almost nothing on his research over the following week.[29] No one has seriously asked the question as to what might have prompted him to take on such a massive strain. But as should become clear in what follows, it is in pursuit of this question that we see highly significant aspects of Faraday's modus operandi.

Four Problem Domains

A survey of Faraday's *Diary* entries for the full week provides a sense of the progress of Faraday's work. Table 6.1 encompasses the whole period, including September 3 and 4, already discussed. After devoting September 3 and 4 to the search for regularity and the idea and demonstration of rotational motion (D1–D15), Faraday treated other questions in the following days. On a first, rather rough pass, we may distinguish four problem domains, which Faraday pursued in sequence.

First, he turned to the production and investigation of further rotation effects. He reversed the rotation of the wire about a magnet pole, successfully causing a pole to rotate about the wire (D16). Such experimental demonstration of mechanical reciprocity would prove an important motif in Faraday's subsequent studies, in remarkable analogy to corresponding efforts by Ampère, of which Faraday had learned via a reference by Biot (1821a, 287). Unlike Ampère, however, Faraday set about multiplying the effects. He turned from the interaction of *one* wire with *one* magnet pole to that between *several* wires and *several* magnetic poles (D16–D37, D43–D48), developing a whole series of new arrangements. Variation and extension of the experiments went hand in hand with his struggle to systematize the multiplicity of electromagnetic motions appearing at each step. All motion, whether circular or rectilinear, was to be understood in terms of—or, in Faraday's language, "reduced" to—the "simple case" of circular motion.

Faraday next turned to attaining the new alignment and rotation effects with terrestrial magnetism. Such efforts were closely related to the previous problem domain, but, given the peculiarities of terrestrial magnetism, they required their own reflections and apparatus. The experiments recorded in D36–D37 and D49–D50 were only partly successful, and he shelved their further improvement for a later day.

Toward the end of his fourth day of experimenting, he recorded in the *Diary* his intimation of a fourth problem domain. Having already investigated the motion of a floating needle provoked by a half-submerged coil, Faraday concluded with the observation, "See if natural cylindrical magnet will do this" (D37). Thus announced, the program of comparing the action of particular configurations of currents with that of magnets began to occupy his time on the following day (D39–D42) and nearly all of it on

September 7. The background to these experiments was Ampère's theory that magnetism was caused by hypothetical circuits of electrical current; in one *Diary* entry Faraday referred to this, in one of two possible senses, as the "current view" (D58). Initially, he compared a coil with a corresponding bar magnet (D39–D42, D51–D54, D58, D65, D67–D68). Later he also compared a flat spiral with a magnetized steel disk (D59, D63–D64, D66, D69). In addition to studying the motion evoked by the proximity of a magnet pole, he scrutinized the patterns in which iron filings arranged themselves around these objects of study.

In the columns on the right of table 6.1 I have assigned each of Faraday's various activities to one of these problem domains. It is clear from this overview that the first strand, after being treated exhaustively on the first two days, never reappeared again. The second demanded much more time: "reduction" to the "simple case" occupied the second, third, and fourth days of his laboratory work. The comparison between variously configured currents and magnets also took a lot of time, especially on the fifth through seventh days. These last two, we recall, had also shaped much of Ampère's work. In sharp contrast to this intensity, the third topic—the investigation of terrestrial magnetism—appeared only in isolated episodes.

In what follows I will go into each of the four problem domains in greater detail. Because the *Diary* describes experimental assemblies and results but offers only sparing hints as to Faraday's thought process, this will require consulting additional sources. As neither the manuscript of Faraday's article nor any other notes have been preserved, the most important source to take into account is the published version of the article. Thus, I begin with some thoughts on the relationship between this text and Faraday's laboratory work.

Faraday's Published Article and the Diary

The title of Faraday's (1821d) article, "On Some New Electro-Magnetical Motions, and on the Theory of Magnetism," already names two distinct topics, each of which clearly corresponds to one of the two problem domains so prominent in the *Diary.* However, the article lacks any explicit division into sections and subsections beyond a rough division in two major parts, and the text itself is hardly structured transparently. For ease of exposition and reference I have thus structured it into fifteen thematically defined sections (numbered I through XV) and numbered the paragraphs continuously (1 through 59). Table 6.2 provides an overview.[30]

Table 6.1. Faraday's entries on electromagnetic rotation in his laboratory notebook, September 3–4, 1821

Date	Diary entries	Contents of diary entries	Problem domain			
			Rot	Red	TM	Amp
Monday, Sept. 3	D1–13	Search for regularity; discovery of the rotation of a wire about a magnet pole; first rotation device	•			
Tuesday, Sept. 4	D14–15	Improved rotation device; effect becomes stably repeatable	•			
	D15	Reciprocity of the effect—magnet pole rotates about wire		•		
	D17–18	Investigation of the interactions between multiple wires and magnet poles		•		
	D19	Failed attempt to rotate wire about its own axis				
	D20–25	Investigation of the interactions between multiple wires and magnet poles; continuation of experiments in D17–18		•		
Wednesday, Sept. 5	D26–34	Investigation of the interactions between a mobile coil ("de la Rive's curve") and one or more magnet poles		•		
	D35–37	Investigation of the action of a coil on a floating magnet needle; reciprocity of experiments in D26–34		•		
Thursday, Sept. 6	D36^{I}–37^{I}	Investigation of the alignment of a coil by terrestrial magnetism			•	
	D38	Investigation of the action of a wire loop on a floating magnet needle; results as in D35–37		•		
	D39–42	Comparison of the magnetic effects of coil and bar magnet				•

Table 6.1 (*continued*)

Date	Diary entries	Contents of diary entries	Problem domain			
			Rot	Red	TM	Amp
	D43–48	Investigation of the action of multiple wires and pieces of copper on a magnet pole; continuation of experiments in D20–25 from September 4		•		
Friday, Sept. 7	D49–50	Investigation of terrestrial effects on a straight wire: alignment and elevation			•	
	D51–54	Comparison of the magnetic effects of coil and bar magnet; continuation of experiments in D39–42 from September 6				•
Saturday, Sept. 8	D55–57[1]	Investigation of location of the pole in variously shaped magnets				•
	D58–59	Reflections and further experiments on the comparison between coil and bar magnet				•
Monday, Sept. 10	D60–69	Comparison (motion, patterning of iron filings) between the effects of permanent magnets and variously constructed electromagnets				•
Tuesday, Sept. 11		Article completed and presumably submitted for publication				

Source: Faraday's diary entries as edited and published (Martin 1932, 1:49–50).

Notes: Column headings: **Rot**, rotation effect, demonstration and stabilization of; **Red**, reduction of effects to circular motion and variation of the experimental setup; **TM**, terrestrial magnetism, investigation of the effects of; **Amp**, Ampère's theory relative to comparison between variously configured currents and magnets.

1. Faraday's numbering of Diary entries, begun on September 3, does not continue through September 8 and 10. For ease of exposition, I have extended the numbering to include those dates.

TABLE 6.2. The structure of Faraday's first article on electromagnetic rotation, Sept. 1821

Contents	(¶); page/ reprint page	Diary entries
I. Description of the rotation device, the experiments, and the path leading to the discovery	(1)–(6); pp. 74–76/127–31	D1–D6
II. Formulation of four inferences from this discovery	(7); p. 76/129	D20
III. Description of the physical realization of the rotation effect and its reciprocity: devices and experimental results	(8)–(14); pp. 76–78/129–31	D8–D16
IV. Description of the experimental attempt to rotate a magnet or wire about its own axis; inference from the negative result: motion belongs to the current, not the wire	(15); p. 79/131	D7, D9
V. Argument for the explicability of Ampère's attractions by means of Wollaston's notion of "electromagnetic current"; introduction of the rotation of wire and magnet pole as the "simple case" of magnetic motion	(16)–(18); pp. 79–80/131–33	D48
VI. Description of experiments with multiple wires and/or multiple magnet poles as "easily reducible" to the motion of wire and magnet pole, illustrating the aforementioned explication	(19)–(25); pp. 80–83/133–35	D16–D18, D21–D22, D32, D43–D45
VII. Description of the experiments with de la Rive's apparatus, all "reducible" to the rotation of a wire about a magnet pole	(26)–(28); pp. 83–84/135–36	D26–D34
VIII. Mention of the experiment to lift a magnet by means of a coil and its partial success	(29), p. 84/136	D33, D35
IX. Presentation of a classification of all electromagnetic motions by reference to simplicity and the number of participating "powers"	(30)–(35); pp. 84–85/137	

Table 6.2 (*continued*)

Contents	(¶); page/ reprint page	Diary entries
X. Announcement: detailed comparison to follow between the actions of magnet and coil, with the goal of evaluating Ampère's theory of circular currents	(36); pp. 85–86/137–38	
XI. Elaboration of the "reduction" of the effects of a coil to the simple rotation between wire and magnet pole	(37)–(40); pp. 86–88/138–40	D17–D18, D36, D44–D45, D58, D63
XII. Announcement: the similarity between coil and magnet supports Ampère's theory; description of experiments to demonstrate this similarity, which make the inference to similar causes plausible	(41)–(51); pp. 88–92/140–43	D35–D41, D46, D51, D58, D63–D65
XIII. Description of experiments demonstrating the differences between effects of coil and magnet	(52)–(55); pp. 92–94/143–45	D39–D42, D53–D62, D67–D69
XIV. Remark: Ampère is undecided on the exact nature of currents in magnets; thus, for lack of sufficient evidence, no theory is preferred	(56)–(57); pp. 94–95/145–46	
XV. Description of several partly successful experiments on the effects of terrestrial magnetism, inspired by Ampère	(58)–(59); pp. 95–96/146–47	D36[1]–D38, D49–D50

The sections of the text can be correlated with the four thematic fields:

Theme	Section
Demonstration and stabilization of the rotation effect	I, III
Variation of experimental setup and "reduction" to circular motion	II, V–IX, XI
Investigation of effects of terrestrial magnetism	XV
Comparison of variously configured currents and magnets, with a view toward Ampère's theory	X, XII–XIV

Note: The second column gives the paragraphs of the text (in brackets, with my numbering), the pages in the original *QJS* version, and the pages in the reprint version (Faraday 1839–55) in *Experimental Researches in Electricity*, volume 2. The third column gives the *Diary* entries in which the topics were treated.

Following his "historical" introduction (I) on the discovery of the rotation effect, Faraday introduces the four aforementioned theses (II), of which the far-reaching third thesis is directly targeted at Ampère. The promised experimental proof of these theses is more implied than explicit. Next Faraday presents his physical realization of the rotation effect, along with the relevant apparatus and technical details (III). The negative result of the following experiment (IV), on the rotation of a wire about its own axis, has nothing to do with Faraday's theses, serving instead to distinguish his position from Wollaston's. Nonetheless, in his argument in support of his third thesis (V), so challenging to Ampère, he uses without comment Wollaston's notion of an "electro-magnetic current" surrounding the wire. The argument remains vague, since Faraday makes no effort to clarify the notion, which he had also cited without comment in his "Historical Sketch." Contrary to Faraday's promise, the thorough systematic exposition of further rotation effects (VI and VII) that follows no longer refers to the third thesis, nor does his discussion (VIII) of an isolated experimental test of the expected levitation of a magnet by a coil. With the help of Wollaston's notion of "powers," Faraday then undertakes a classification of all electromagnetic motions (IX), including the attraction between two wires described by Ampère. This constitutes the final section of the first main part.

The second part of the article, devoted to the theory of magnetism, is structured no more clearly. After offering a sketch of Ampère's theory (X), Faraday announces his intention to compare the magnetic action of a coil with that of a conventional magnet, thus furnishing "much information respecting the competency of this theory." First, however, he offers a detailed explanation (XI) of how all of the effects of the coil might be traced back to "simple rotation" between magnet pole and wire. Only then does he systematically present the similarities in the behavior of coil and magnet (XII), which strike him as supporting Ampère's thesis, before turning to the differences (XIII). He withholds any final judgment on Ampère's theory, instead turning (XIV) to something Ampère had left open: the question of the precise nature of currents (concentric or molecular). The article concludes with a brief report (XV) on his only partly successful experimental attempts to align wire loops to terrestrial magnetism.

The four problem domains I have identified in analyzing the laboratory *Diary* are recognizable in the article, at least in broad outlines:

- Demonstration and stabilization of the rotation effect: sections I and III
- Variation of the experimental setup and "reduction" of effects to circular motion: II, V–IX, XI
- Investigation of the effects of terrestrial magnetism: XV
- Comparison of variously configured currents and magnets, with a view toward Ampère's theory: X, XII–XIV

The order in which these themes were explored in the *Diary* has been changed in the published article. They are not presented sequentially but instead in partly interlocking form, with the connections among them not always made plain.

In addition to the obvious haste of its composition, the lurching structure of the article points toward a tight connection with the *Diary,* which I have explicated in detail for the first part. This connection is further reinforced by the fact that, in the manuscript diary, each of these entries has been marked with the aforementioned vertical line, with which Faraday signaled passages he had flagged for publication. As the many terminological similarities demonstrate, Faraday drew directly on the diary in composing his article.

A further aspect of the article's structure also points in this direction. At two junctures Faraday mentions experiments not involved in any argument or attempt at systematic categorization, discussed in isolation. One of these is his failed attempt to produce the rotation of a wire about its own axis, as predicted by Wollaston (D19; ¶15 in the article). Faraday ventures no firm opinion on this prediction; his newly discovered rotation effects have a different character. He presumably undertook the experiment because he was working on rotation effects at the time, and Wollaston's proposal was in the air.[31] The other experiment involves an attempt to levitate a magnetic needle within a coil (D35; ¶29). Faraday's previous work would have led him to expect such an effect in principle, though he was well aware that it would involve stronger forces than any he had thus far observed. The optimization of his apparatus with a view toward producing the strongest possible mechanical forces had not previously been especially important. The success of this experiment would surely have been a nice result, but without far-reaching implications. Even its failure would have no systematic consequences. Accordingly, Faraday's remarks on its partial success are rather brief. Neither experiment is of any particular importance to his article. Neither in the *Diary* nor in the article does Faraday situate them in relation to his other experiments. At the same time, he doesn't want to sweep them under the rug. I mention them here only because Faraday presents them at the very junctures in his article text that correspond to the place of the experiments in the order of his laboratory research.

It is difficult to venture anything specific about the temporal sequence of the article's composition.[32] We can, however, rule out any suggestion that Faraday wrote the text in a single block on September 10 or 11. It is simply too long and too complicated to have been put together over so short a time. Furthermore, in the first part of the article, through section IX, there is no mention of any experiments performed after September 6. Faraday most likely completed this part on or shortly after that date. To some extent Faraday regarded section IX, which presents a general classification of all electromagnetic motion, as the conclusion or high point of the first phase of his work and of the first part of his text. The next part is devoted to the theory of magnetism, and it clearly corresponds to a change of emphasis in the *Diary*. The text of sections XI through XV refers to experiments not performed until Monday, September 10. This, in turn, suggests that Faraday composed the entire second part as a unit only after concluding all of his experiments.

To sum up, it would appear that the substantive reflections Faraday presents in his article are not fundamentally different in kind from those that had guided his work in the laboratory. In the sections that follow, with this working hypothesis in the background, I treat his published expositions and laboratory sketches as intimately connected.

Varying the Phenomena and a New Perspective on Rotation and Attraction

After stabilizing the rotation effect, Faraday's first priority was to determine the factors on which it depended by varying many experimental parameters one by one. The polarity of the battery and the choice between north and south magnet poles are obvious candidate parameters.[33] His previous experience with electromagnetism had led him to expect such. Faraday provides only a graphical sketch of this result (fig. 6.6, sketch D15), always supplementing a side view of his experiment with a more abstract diagram.[34] Next he attempts to demonstrate the reciprocal effect, with the magnet pole rotating around the wire ("Now magnet round wire," *Diary*, D16). With a small modification to his apparatus, he achieves the desired result: he makes the magnet mobile by disengaging its lower end from its anchor while weighting it with enough platinum to keep most of the magnet submerged in the mercury, with only a small segment of the upper pole above the surface. If the wire is now held fast in the middle of the basin and connected to the battery, the upper pole of the magnet does indeed rotate about the wire. Once again Faraday reverses the polarity of battery and magnet and represents the observed dependencies graphically (see fig. 6.6, sketch D16).

This setup struck him as suited to more far-reaching variations on the experiment, to which he then turns. His next steps are typical of his procedure in general. He progresses from the rotation of the magnet pole about *one* wire to a more complex arrangement, investigating its motion around *two* wires. There are naturally two variants of this case, one in which the two wires are connected to the battery in the same polarity and the other, in opposing polarity—or in which current flows through the two in the same or opposite direction. Faraday studies both. In the case of opposing polarity, the magnet pole moves in a straight line between the wires, with its direction once again dependent on the polarity of the wires and on which magnet pole is used (D17).[35] His sketch makes this case clear (see fig. 6.6, sketch D17). By contrast, when the current runs in the same direction through both wires, a magnet pole placed between them is unaffected (D18).

These observations prompted Faraday to an ambitious conjecture, which, in an unusual step, he set down in detail: "Everything tends to prove that there is no attraction between the poles of the magnet and the wire, but only motion in a circular direction, and all the motions of the magnet or its poles around the wires may be deduced from this. When the single pole was floating upon the mercury it shewed it both by revolving round single wires and passing through double ones" (D20). This idea was not entirely new to him. His experiments on September 3 and the hypothesis of circular motion had already cast doubt on the fundamental status of attraction and repulsion. But his latest experiments lent this conjecture support of a different and more powerful kind. In experiment D17, Faraday can literally *see* how attraction and repulsion might result from a combination of circular motions. In the second sentence of the above quotation from D20, Faraday rests his argument on precisely this point. The associated sketch is telling, displaying not one but three different tracks for the magnet pole. He thus brings three different experiments together: the two mentioned in D16, in which the south pole of a magnet rotates about a wire, and the new one, in which both wires act on the pole simultaneously,

D 15

D 16

D 17

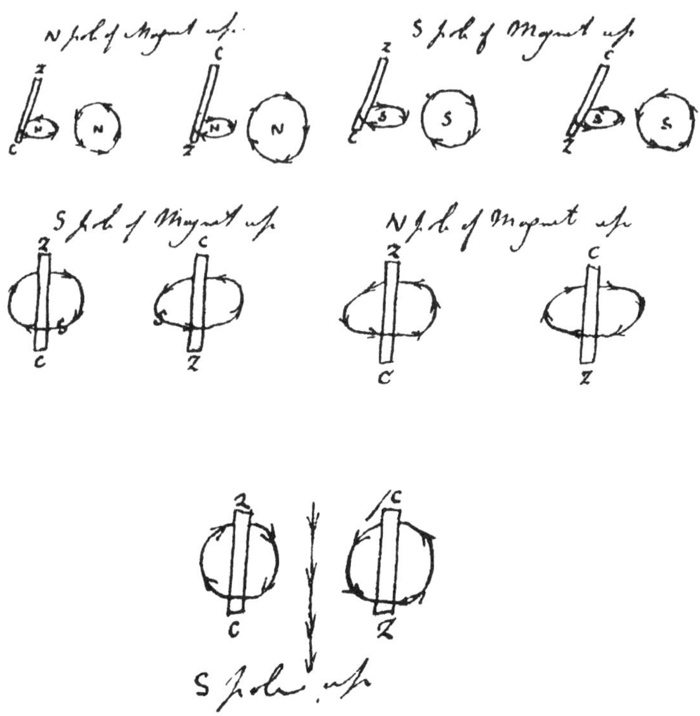

Fɪɢ. 6.6. Faraday's sketches of the rotation of a wire about a magnet pole (D15), of a magnet pole about a wire (D16), and a magnet pole about two wires (D17) (*Diary*, September 4, 1821).

in effect combining the two previous cases. As the sketch so compellingly suggests, the new result can itself be understood as a combination of these two, for the symmetrical superposition of the two circular motions would result in linear motion.

Faraday's sketch goes well beyond the abstract representation of an experimental result; in a sense, it encompasses an entire argument. What's more, this argument concerns a fundamental point. With this visual aid before his eyes, Faraday can see how rectilinear motions of attraction and repulsion can be generated by composing circular motions, that is, how to "deduce" the phenomena of electromagnetic attraction and repulsion from the superposition of more basic rotational effects. It is thus not surprising that even in his article, in which he necessarily had to be much more sparing in his use of illustrations, Faraday reproduces precisely this sketch (his Figure 8 in our fig. 6.5). He reproduces it again, immediately after the article's completion, in a letter he wrote to de la Rive on September 12, where, given the extreme brevity of his summary, this fact alone is significant. He also adds a thorough statement of his conclusion: "I find that all the usual attractions and repulsions of the Magnetic needle by the conjunctive wire are deceptions the motions being not attractions &c or repulsions nor the result of any attractive or repulsive forces but the results of a force in which the wire which instead of bringing the pole of the needle nearer to or farther from the wire endeavours to make it move around in a never ending circle and motion whilst the battery remains in action."[36] By referring to the attractive and repulsive effects as "deceptions," he clearly signals how fundamentally his view of the phenomena has shifted. Only a few weeks earlier he had attempted, in print, to capture the needle's behavior in terms of attraction and repulsion. Now his language shows how dramatic he sensed this transition to be. This episode also provides us with a paradigm case for what Faraday understood by the "deduction" or, more often, the "reduction" of phenomena to simple or elementary cases.

In the laboratory he proceeded directly to further variations of the experiment. Instead of a single magnet pole he studied the behavior of two, that is, the behavior of an entire magnetic needle, supported by a cork, floating on the surface of water (D21–D22). The motions performed by this needle under the influence of one or two wires are, of course, rather complicated. But in all cases Faraday convinces himself that the needle is behaving precisely as could be predicted from the superposition of the motions of individual magnet poles of opposite polarity, with which he was already familiar: "the apparent attraction and repulsion of the poles were all reduced into two circular motions about the wires" (D21). Faraday pays particular attention to the fact that the floating needle is attracted or repelled when it is placed in a central position against a wire. Once again referring to a sketch (fig. 6.7), he explains, "It is shewn that the attraction and repulsion, as they seem to be, are only the combined action of the two circles in which the poles endeavor to move round the wire" (D22).

The particular significance of the case is that this is precisely the effect that Ampère had so strongly emphasized and labeled as his second "general fact" (see above, chap. 3). His ability to show that this effect is not elementary but derived certainly increases Faraday's confidence in his new account. Unsurprisingly, he also discusses this "most instructive instance of attraction and repulsion" (D22) in his article, grounding his argument in another sketch (his Figure 10 in our fig. 6.5) with full confidence in its persuasive power: "This effect will be readily understood from fig. 10" (Faraday 1821d, 82, ¶22; 1839–55, 2:134).

In a sense, the experiments of the following day represent the reversal of those just discussed. In place of the movable magnet pole (or poles) and fixed wire (or wires), we now find movable configurations of wires in proximity to magnet poles or whole magnets. Faraday's instrument is a small device he had recently received from Charles-Gaspard de la Rive and briefly mentioned in the "Historical Sketch."[37] It is a single loop of copper wire (see his Figure 12 in our fig. 5.1), both of whose ends pass through a cork stopper. Underneath the stopper, each wire end is connected to one of two plates—one copper, one zinc. When the cork floats in a basin filled with a weak acidic solution, the two plates form a galvanic element, causing a constant current to flow through the wire loop. The whole device is thus a highly movable loop of current. In his experiments of September 5, Faraday carefully explores the reaction of this loop in a wide range of positions when a magnet pole is brought in proximity to it (D26–D34).

In an effort to interpret the various resulting movements, Faraday simplifies the action of the wire loop by imagining it as the superposition of the actions of two straight vertical wires with current flowing in opposite directions, connected on top by a straight horizontal wire.

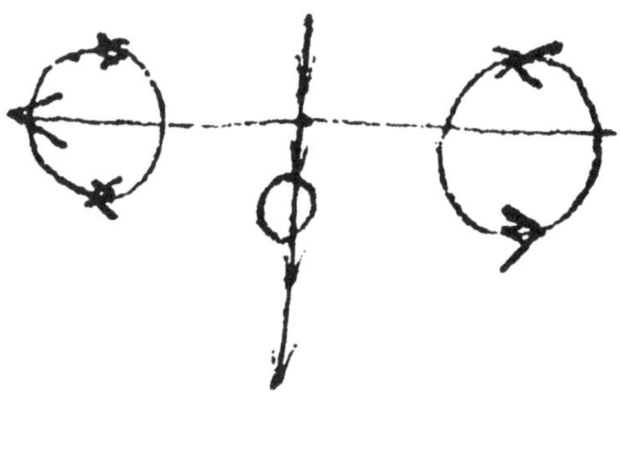

D 22

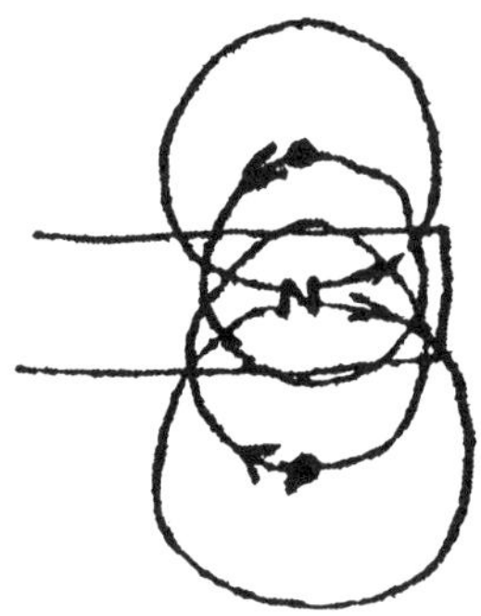

D 23

FIGS. 6.7 AND 6.8. Faraday's sketches of the experiments with the floating needle and a single wire (*Diary*, D22) and with a magnet pole and two wires (D23).

This simplification does, indeed, make sense of the complex motions of the loop. Faraday again makes the key point clear by means of a sketch (fig. 6.8). The rectangle signifies the magnet, with north pole "N," the points above and below it cross sections of the vertical wire segments running perpendicular to the plane of the paper. The circles surrounding each of the wire sections indicate the paths the wires would take around the pole were each of them independently movable. However, the distance between the two segments is fixed, and they can only move together. Regardless of whether we imagine the magnet as movable and the wire fixed (as in the experiments in D21–D22 discussed above) or the other way around (as in the experiments with de la Rive's device), the superposition of the two circles yields the resulting attraction or repulsion. Faraday reproduces the sketch in his article (his Figure 11 in our fig 6.5) and notes with satisfaction that though it initially seems "anomalous," on careful consideration the motion of de la Rive's device proves "easily reducible to the circular movement of the wire about the pole" (Faraday 1821d, 84, ¶27; 1839–55, 2:136). Faraday finally approaches an entire series of further experiments with this goal of "reduction," in turn examining the effects of coils (D39–D42), multiple parallel wires (D43–D44), and variously shaped pieces of sheet metal (D45–D47) on a movable magnet pole.

The Action of Coils, Reduced to the "Simple Case"

Departing from strict chronological sequence, I outline here another case of successful "reduction" to which Faraday returned several times during this period. It serves as an apt illustration of his modus operandi. This is the magnetic action of cylindrical coils. Faraday's special interest in coils may be traced to their prominent role in the context of Ampère's theory, in which they serve as the electrical analog of conventional bar magnets. Consequently, Faraday stresses that understanding the actions of coils will serve as a useful means for evaluating this theory. He thus sets out "to trace the action of an electromagnetic pole, either in attracting or repelling, to the circulating motion before described" (Faraday 1821d, 84, ¶36; 1839–55, 2:138).

In the article, he describes his procedure in four steps (Faraday 1821d, 86–88, ¶¶37–40; 1839–55, 2:138–40). First he extrapolates from the rotation of a magnet pole about a straight wire to the action of a wire loop, which

Faraday calls a "ring." Referring to a drawing (his Figure 14 in our fig. 6.5), he explains how the superposition of the action of its parts produces stronger action at the center of the loop, along with the appearance of an attractive or repulsive force. In his next step, he introduces the experimental finding that the action of several wires through which current flows in the same direction differs from the action of one only in its strength, and not in the trajectory of the resulting motion. This in turn makes sense of the fact that, while the action of a spiral is stronger than that of a single loop, it is qualitatively the same. A spiral (his Figure 15 in our fig. 6.5) can be understood as simply an assembly of concentric loops. As his third step, Faraday considers another way of combining loops—by winding them side to side in the form of a hollow cylinder. Once again, experiment confirms his expectation: the motion produced by this arrangement is amplified but not otherwise different. Finally, Faraday's deduction culminates by imagining the two sorts of amplification combined, inviting the reader to "consider the internal space of the cylinder filled up by rings or spirals, all having currents in the same direction." In the resulting coil, "the direction and kind of force would be the same, but very much strengthened." Its action would be strongest at either end, and a magnet pole brought in proximity to one of them ought to be attracted or repelled along a curved path, unless in line with the axis of the cylinder. In the latter case, it would simply be attracted or repelled in a straight line. Since all of this coincides with experimental results, Faraday can proceed: "Thus the phenomena of a helix, or a solid cylinder of spiral silked wire, are reduced to the simple revolution of the magnetic pole round the connecting wire of the battery" (all from Faraday 1821d, 88, ¶41; 1839–55, 2:140). I have summarized the course of Faraday's presentation and argument by means of a series of schematic drawings (fig. 6.9) closely modeled on Faraday's own. In most of them the wires are assumed to run perpendicular to the plane of the drawing. The lines surrounding them represent the paths taken by magnet poles in their proximity. Although Faraday himself did not present such a series of drawings in his article, his argument is certainly couched in pictorial language. Some of the sources for these illustrations may be found among the sketches in the *Diary*. Presenting them in series makes the visual aspect of his argument especially clear, while at the same time conveying in compressed format the procedure by which Faraday "reduced" compound actions to the "simple rotation" of a magnet pole about a wire.

Unlike the previous examples, the experiments to which Faraday here refers do not form a linear sequence in the *Diary*. They were carried out on

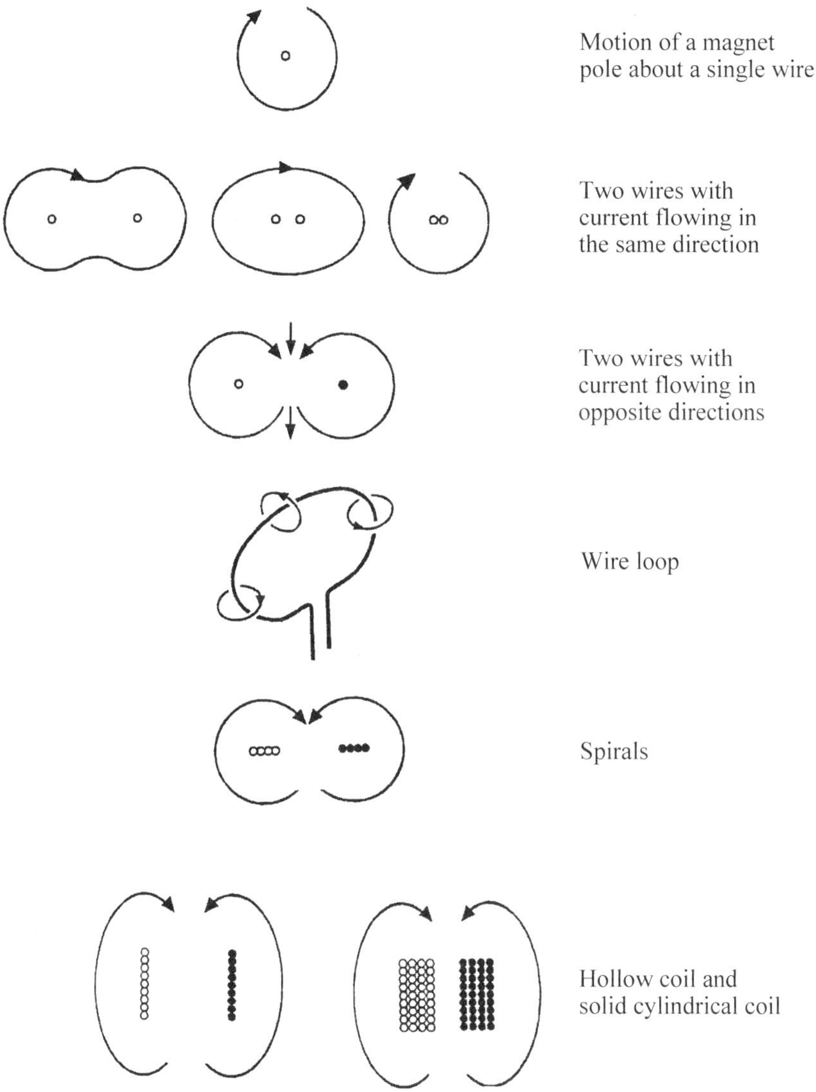

Fig. 6.9. Schematic drawings illustrating Faraday's "reduction" of the action of a coil to "simple" rotation.

different days and brought together for the article.[38] The idea of reducing the action of the coil occurred to him gradually, coming to a head when he decided to seriously engage with Ampère's theory. It was then that he realized that a detailed account of coils would prove very important and that toward that end he could draw on a wide range of results first obtained in other contexts.

The Electromagnetic Action of Terrestrial Magnetism

Faraday undertakes two series of experiments on terrestrial magnetism during this period. In the first, he investigates its alignment of a wire loop.[39] Such alignment was described by Ampère (see chap. 4) and achieved by de la Rive using the far simpler apparatus of a floating wire loop. Inspired by de la Rive, Faraday builds a small floating battery, short-circuited by such a loop. After demonstrating its alignment at a right angle to the magnetic meridian, he achieves the same result with de la Rive's device. In a further variation, he replaces the wire loop with a small coil, which, as expected, behaves exactly the same. The second series of experiments is a continuation of the first.[40] In line with his reductive aims, the natural next step would be the transition from experiments with loops to the "simpler" case of straight wires. Faraday thus attempts to demonstrate the action of terrestrial magnetism on a straight wire, either by alignment to the meridian or in the form of some vertical motion, in consequence of the large angle of magnetic inclination. He conceives of new experimental arrangements for both possibilities, but the experiments are unsuccessful. Faraday ascribes their failure to contaminated mercury with high surface tension (as we would call it nowadays): "mercury foul and surface skinned so as to prevent motion" (D49).

This problem could have been addressed easily enough, but Faraday sets it aside for the moment, instead turning to the thorough comparison of coils and magnets suggested by Ampère's theory, discussed above. On his fifth day in the laboratory, September 7, Faraday decides to devote the little time remaining to him to this challenge. Consequently, he offers only a brief mention of the partially successful experiments on terrestrial magnetism in his article (Faraday 1821d, 95–96, ¶¶58–59; 1839–55, 2:146–47). At the next opportunity (which would not come for several months, however), he would redouble his laboratory efforts on this issue.

Reduction to "Simple Cases" and "General Facts"

The period we have discussed so far illustrates the specific procedure that Faraday himself called "reduction to the simple case." It bears strong similarities to what Ampère called the reduction to "general facts," which I sketched in an interlude in the narrative in chapter 3. In both cases, experimental research is significantly shaped by the systematic variation of parameters and components. The goal consists in finding regularities and developing the visual and conceptual means necessary to represent them. Whereas Ampère followed this procedure for only a relatively short, if decisive period, Faraday persisted far longer, and with greater consistency. And where Ampère made a promise (which would remain unfulfilled) to reduce all electromagnetic phenomena to his two "general facts," Faraday sought much the same for the "simple rotation of magnet pole and wire."

In the process, he ascribes a very particular role to the rotation effect within the broader field of phenomena. In a sense, he situates it. His article presents the effect amid a whole field of related effects, all established by systematic experimentation, and explains the structure he has discerned among them. The phenomena are not created equal, for some of them are revealed as "simple" and thus fundamental, while others are "compound" and thus explicable in terms of the simple phenomena. The fact that the very effect that sparked the entire project would turn out to be the "simple" one is more coincidence than necessity. Matters stand differently in Ampère's case, in which his starting point, the effects discovered by Ørsted, would soon appear to him to be complex. The "general facts" from which they may be deduced are entirely different phenomena. Such cases, in which the point of departure for a long investigation turns out to be complex, can also be found in Faraday's later work. His study of electromagnetic induction is particularly illustrative. The famous induction ring with which he first demonstrated the phenomenon in August 1831 would turn out to be both complex and marginal within the broader scheme that eventually took hold (e.g., Steinle 1994, 1996). As "exploratory" research practice is a special feature of neither this period in the history of science nor of Ampère and Faraday in particular, in chapter 7 I consider this kind of research from a more general perspective. My goal is to shed light both on the role of experiments in such research and on the kind of explanatory structure to which it gives rise, which I call a "systematic phenomenology."

Previous studies of Faraday's discovery of the rotation effect have paid no attention to the procedure of "reduction to the simple case." Even Gooding's epistemologically oriented analysis, as discussed above, makes no effort to explicate it. In more sociologically inclined treatments, this neglect has led to serious misrepresentations.[41] But this procedure provides the key to understanding Faraday's historical agency. His work in the days following his initial discovery of electromagnetic rotation was so fundamentally shaped by it as to suggest that the epistemic goal of reduction to simple facts was his primary motivation for continuing his experimental work with such intensity, rather than focusing on finishing his article. Despite tremendous time pressure, he saw it as essential to assign his discovery its proper place within the appropriate domain of phenomena, rather than simply presenting it in isolation—though that would certainly have been spectacular by itself. He sought to grasp and convey the interrelations among effects not by theoretical means but by the phenomena themselves. His remark to Ampère the following year—"I am obliged to feel my way by *facts closely placed together*"—appears particularly pointed in the context of this process.[42]

Electromagnetic Rotation, "Ampère's Attractions," and the System of "Powers"

The reduction to simple cases allowed Faraday to make sense of many familiar and newly discovered electromagnetic motions. But the attractive and repulsive motions observed between electrified wires—phenomena Ampère had discovered and taken as the foundation for his theory—were not among them. Nor was it clear how they might be traced back to rotational motion. Faraday understood rotation as essentially involving a magnet pole, whereas the interactions between electrified wires did not involve magnets at all. Nevertheless, as Faraday announced in his theses, he intended to address them. In section V of his article (Faraday 1821d, ¶¶16–18) he briefly discusses this project, and in section IV (¶¶30–35) he describes in detail how he carried out the work. His experiments have shown, or so he asserts, that attraction and repulsion are not simple, but compound effects, as he proclaims in the statement quoted below. In order to extend this claim to attractions and repulsion that do not involve magnets, he reaches for very different tools from those deployed thus far. His key concepts are "electro-magnetic current" and the "powers" associated with it.[43]

> From the attraction of the north pole of a needle on one side [of] the wire and of the south on the other, and the repulsion of the poles on the opposite sides, Dr. Wollaston called this magnetism vertiginous,

> and conceived that the phaenomena might be explained upon the supposition of an electro-magnetic current passing round the axis of the conjunctive wire, its direction depending upon that of the electric current, and exhibiting north and south powers on the opposite sides. It is, indeed, an ascertained fact, that the connecting wire has different powers at its opposite sides; or rather each power continues all round the wire, the direction being the same; and hence it is evident that the attractions and repulsions of M. Ampere's wires are not simple, but complicated results. (Faraday 1821d, 76, ¶7; 1839–55, 2:129)

Even leaving aside for a moment the conclusion drawn in the final clause, this passage is not especially helpful when it comes to understanding what is meant by "powers." In particular, the idea that two opposing "powers" might be ascribed to different sides of the wire cannot easily be reconciled with the claim that "each power continues all round the wire."

In his reference to Wollaston, Faraday is referring to the report of Wollaston's views Brande had published in the *QJS:* "The phenomena exhibited by the electro-magnetic or conjunctive wire may be explained upon the supposition of an electro-magnetic current passing round the axis of the conjunctive wire, its direction depending upon that of the electric current, or upon the poles of the battery with which it is connected.—Dr. Wollaston" (quoted in Brande 1821a). Between this passage and the one that followed, Brande had printed the upper two of the drawings shown here in figure 6.10. "In the above figures," he noted, "such a current is represented in two sections at right angles to the axis of the wire, when similarly electrified, from which it will be apparent that the north and south powers meeting will attract each other. In the following figures, the sections of the wire are shown [the lower two drawings shown in fig. 6.10 were printed here] dissimilarly electrified, by which similar magnetic powers meet, and consequently occasion a repulsion" (Brande 1821a).[44]

The first passage speaks in rather general terms of an electromagnetic current flowing around the wire. By contrast, the second passage voices a concrete explanatory proposal for the attraction and repulsion between wires, articulated in terms of the "north and south powers" attributed to the sides of the two wires adjacent to each other when they are "similarly electrified." This explanation is illustrated by the accompanying figures, in which the "powers" are represented with arrows. Assuming that arrows pointing in opposite directions attract (like parallel magnets arranged with opposing polarity) and that the arrows *between* the wires exert the only (or at least the predominant) action, the images would indeed make the occurrence of attraction or repulsion plausible. But since such assumptions are nowhere explicitly articulated, the explanatory proposal remains exceptionally vague. In the preceding chapter I suggested that Faraday incorporated Brande's report unaltered into his "Historical Sketch" because he did not consider himself competent to make matters any clearer. Now, he again refers back to the report, importing whole sentences word for word. The concepts of electromagnetic current and the various "powers" are left every bit as vague

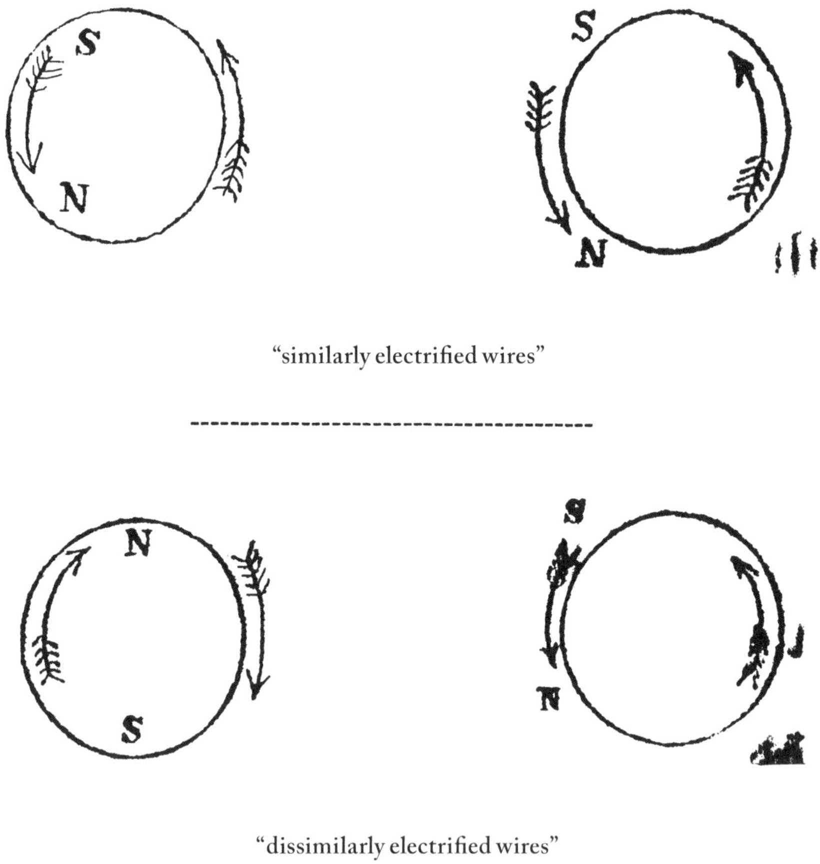

FIG. 6.10. Sketches accompanying Brande's report on Wollaston's views on electromagnetism (Brande 1821a, reproduced in Faraday's "Historical Sketch" [1821–22, 110]).

as before. But now Faraday draws far-reaching conclusions about Ampère's attractions and repulsions, which he represents without any justification as deducible from other premises. This suggests—and I return to this point later—that in this inference Faraday is drawing heavily on the authority of the report, which means on the authority of Wollaston himself.

Faraday's appropriation of Wollaston's concepts, despite their indeterminacy, is astonishing. I find two factors that may help shed light on it. First, Wollaston's approach is the only attempt, besides Ampère's theory, to relate the attraction and repulsion between wires to electromagnetic phenomena, thus seeking a unified explanation for both. This perspective is promising for Faraday and of sufficient interest to motivate him to buy into its basic concepts, despite their indeterminacy. This has to do with the second factor: the drawings provided with Brande's report on Wollaston immediately recall Faraday's own *Diary* sketches, reproduced in the article, of the motions described by a magnet pole under the influence of two wires (*Diary,* D16, D17 in our fig. 6.1; his Figures 6, 7, 8 in our fig. 6.5). The analogy between the two representations might be taken as support for Wollaston's notion of electromagnetic current and even suggest that such currents are, in a sense, made visible by the trajectory of a magnet pole moving under their influence.

At any rate, such an analogy would appear to be in Faraday's interests. In a *Diary* passage, and especially in the article, he argues that winding a wire into a coil pulls apart and spatially separates the two "powers," leaving one of them strongly concentrated at one end of the coil and the other widely removed from it.[45] This fits the experimental finding that each end of the coil exhibits the properties of one magnet pole; its "powers" acquire the characteristics of north and south poles. Faraday uses the analogy in both directions. On the one hand, as in ¶17, he wants to explain the attributes of a coil end by reference to the "powers." On the other, in the following paragraph, he *infers* from the analogy "that the pole of a magnetic needle presents us with the properties of one side of the wire" (¶18). It is not at all clear which side takes explanatory priority: the ascription of north and south pole attributes to the "powers" by virtue of a successful explanation of the properties of the coil ends, or the equivalence of coil end and magnet pole by virtue of the properties of the "powers." By contrast with other passages in the article, the goal of Faraday's argument in this section remains hazy and opaque.

Even so, he takes it a step further. Extending the analogy between coil ends and "powers," he turns to the "simple case" of rotation itself, attempting to explain it in terms of "powers" and thus to gain "a much more intimate knowledge of the state of the powers active in magnets." On his conception, a magnet pole in proximity to an electrified wire ought to always be attracted by one of the two "powers" and repelled by the other. Since the two "powers" are equally strong, the net result is neither attraction nor repulsion. But "the powers begin from opposite sides of the wire, the pole in its double effort to recede from one side and approach the other revolves in a circle, that circle being evidently decided by the particular pole and state of the wire" (Faraday 1821d, 80, ¶18; 1839–55, 2:132–33). Faraday offers no further elucidation of this challenging point.

In paragraphs 30–35, Faraday once again returns to the "powers" (Faraday 1821d, 84–85; 1839–55, 2:136–37). This passage appears at the end of the first part of the article, in which the focus is on electromagnetic motion. Faraday understands it as a conclusion of sorts, presenting a general system or classification of all "magnetic action." This comes as a bit of a surprise, as his earlier efforts toward reduction to simple cases had aimed at something similar. The new proposal is an alternative classification, encompassing not only electromagnetic motion but also the interaction between wires. The notion of "powers" plays a central role in its construction. I give an overview of the classification in table 6.3.

Faraday sorts the phenomena by their degree of "simplicity" but with a specific criterion of simplicity in mind. Simplicity is determined by the value of an integer designating the number of involved "powers," thereby imposing a hierarchical ordering of electromagnetic effects.[46]

Two Systems of Classification

Faraday's organization of magnetic effects by reference to "powers" provides a second system of classification for the field of electromagnetism, his reduction to simple cases having yielded the first. The structure of the article keeps the two systems clearly separate. Organization by "powers" occurs only in sections V and IX, which occupy relatively isolated positions within the text as a whole. By contrast, reduction to simple cases appears in sections VI, VII, XI, and XV, sometimes to be carried out in detail, sometimes only mentioned in passing. Faraday says nothing about the relationship between the two projects. Considering it carefully will prove very instructive.

Table 6.3. Faraday's classification by "powers"

Interaction	Effect	Number of powers
Between two coil ends ("poles"), ¶30	Attraction/repulsion	2
1 wire, 1 magnet pole, ¶31	Rotation	3
2 wires (Ampère's effects), ¶32	Attraction/repulsion	4
1 wire, 2 magnet poles, ¶33	Rotation and attraction/repulsion	4
2 wires, 1 magnet pole, ¶34	Rotation and attraction/repulsion	5
2 wires, 2 magnet poles, ¶35	Rotation and attraction/repulsion	6

Note: Numbers in the first column refer to the paragraphs in Faraday's (1821d, 1839–55, vol. 2) text.

Both approaches provide an attempt at classification, imposing some structure on the as yet incompletely demarcated field of electromagnetism with its confusing profusion of phenomena. Both also rest essentially on a criterion of *simplicity,* though not the same criterion. In the reduction project, the "simple cases" are specified by reference to a minimal set of experimental conditions. Complex cases must be "reduced" to simple cases by way of a chain of intermediate phenomena. Significantly, both complex and simple cases are *phenomena.* By contrast, in the second classification, phenomenological reduction is replaced by an abstract criterion: the number of implicated "powers." This puts the spotlight on a concept removed from the phenomenological level, for "powers" cannot be demonstrated directly. Instead, they are understood as entities standing *behind* the phenomena, not directly observable but physically efficacious. In the language of the philosophy of science, they are "theoretical quantities." With so abstract a concept at its core, the second system of classification stands in sharp and principled contrast to the reduction-based system. The fact that the experimentally demonstrable rotation effect, the irreducible "simple case" of the reductive project, is itself further explicated within the system of "powers," albeit only by recourse to phenomenologically inaccessible entities, is indicative of this contrast.

Turning to "powers" thus results in a simplicity criterion that is not only much harder to grasp but much further removed from the level of phenomena. As table 6.3 shows, effects placed close together within this system can present vastly different phenomena. Here, the proximity of two effects can no longer be ascertained by a glimpse at the corresponding experimental setups, as it could in the reduction-based system. For example, the interaction between a wire and two magnet poles lies at the same level of simplicity as Ampère's attraction and repulsion between wires. The two systems differ in one further, fundamental respect. As the structure of Faraday's presentation emphasizes, the "powers"-based criterion, which assigns an integer to every effect, imposes a linear, one-dimensional ordering. The reduction-based system, however, is multidimensional from the beginning. One may set out from the simple case in any number of directions, toward a spiral, hollow coil, sheet metal in any number of shapes, and so on. The simplicity criterion based on reduction to the simple case offers no way of distinguishing among such directions. It is thus less directly applicable and certainly not numerical.

But the most important difference between the two systems is that the classification by recourse to "powers," by virtue of its very abstractness, captures a greatly extended range of phenomena. At first blush, the reductive approach, with its requirement of phenomenological kinship, constrains this domain to strictly electromagnetic effects. By contrast, a system based on the more abstract "powers"-based criterion can even encompass effects involving no magnet pole at all, as well as, above all, Ampère's attraction and repulsion between two wires. With all their differences in degree of abstraction, proximity to the level of phenomena, and range of application, the two systems have little to do with each other. They coexist without mutual engagement.

In the laboratory, they manifest themselves in completely different ways. Especially in the first days of Faraday's laboratory work, nearly all experiments were explicitly geared toward reduction to the simple cases. By contrast, there is hardly any trace of the "powers"-based system in the *Diary*. We find a single reflection (D48) at the end of the entries for September 6, by which time Faraday had already taken his reduction quite far and was ready to transition to his second goal: the investigation of Ampère's theory.[47] Faraday's thoughts on a "powers"-based classificatory system did not evidently spur him toward any concrete experiments. The abstractness of this systematic project entails greater distance from the experimental goings-on of the laboratory.

The two systems' different relationship to laboratory work points toward a difference in the way Faraday argued for them. In the article, his discussion of reduction to simple cases makes frequent reference to empirical findings, and his claims are advanced with great confidence. For example, he asserts that the motions of the floating wire loop "are by a little attention easily reducible to the circular movement of the wire about the pole. I do not think it is necessary to state them particularly." With regard to the action of a coil, he concludes contentedly, "Thus the phenomena of a helix, or a solid cylinder of spiral silked wire, are reduced to the simple revolution of the magnetic pole round the connecting wire of the battery" (Faraday 1821d, 84, 88; ¶¶27, 41; 1839–55, 2:136, 140). Both of these remarks immediately follow the presentation of a long series of experiments.

In presenting the "powers"-based system, however, he takes a different tack, even in his language use. The text contains hardly any reference to experiments that might support the system. In their place we find, with suspicious frequency, a different kind of plea. For example, in introducing the "powers," Faraday states, "It is, indeed, an ascertained fact, that the connecting wire has different powers at its opposite sides." Or again, "It has been allowed, I believe, by all who have experimented on these phenomena, that the similar powers repel and the dissimilar powers attract each other" (Faraday 1821d, 79, 84, ¶¶16, 30; 1839–55, 2:131–32, 136). Nowhere else in Faraday's article do we find such appeals to generally accepted belief. Those arguments bespeak far less solid ground than that of the reduction project, signaling not only a trace of uncertainty but also Faraday's much more limited commitment to these beliefs. Toward the end of his article, Faraday is once again quite clear: "With regard to the opposite sides of the connecting wire, and the powers emanating from them, I have merely spoken of them as two, to distinguish the one set of effects from the other. The high authority of Dr. Wollaston is attached to the opinion that a single electro-magnetic current passing round the axis of the wire in a direction determined by the position of the voltaic poles, is sufficient to explain all the phenomena" (Faraday 1821d, 94–95, ¶57; 1839–55, 2:146).

Faraday could hardly distance himself any further from a commitment associated with his talk of "powers." In the context of this article, his reference to Wollaston's "authority" is both singular and highly suggestive. It stands in stark contrast to Faraday's claims on behalf of the primacy of the empirical, as articulated, for example, in the "Historical Sketch," and to his usual, far more assertive engagement with established authorities on matters scientific. A further indication of Faraday's uncertainty with regard to the "powers"-based system is the fact that there is not a word of it in the summary he sent to de la Rive.[48]

As this comparison makes clear, with regard to the two systems, Faraday is really only committed to the one based on reduction to simple cases. For its part, the "powers"-based classification is a welcome but rather half-hearted effort, one that allows him to also address Ampère's attractions. Furthermore—and this point is decisive—he can justify the thesis of the reducibility of Ampère's attractions to rotational motion, stated at the beginning of his essay, only by recourse to the "powers"-based system. This provocative thesis makes it worth venturing into territory whose contours he knows to be uncertain. But this excursion would not last long. He would never again return to this project, whereas reduction to simple cases would remain one of his most important approaches for the rest of his life. As early as his work on terrestrial magnetism in December 1821, discussed below, all talk of "powers" has disappeared, leaving reduction to the simple case as the overarching goal.[49] Faraday's contemporaries also took note of the different levels of certainty and gravity associated with the two systems. Ampère, for example, would react very determinedly to most of Faraday's experimental points but never once mention his "powers"-based explanatory efforts (discussed below).

Faraday's Discussion of Ampère's Theory

The second part of Faraday's work week was devoted to the investigation of Ampère's theory, which he understood as the thesis that all magnetic phenomena derive from circular electrical currents. He thus set about gathering the empirical material necessary for the evaluation of this thesis and subjecting it to a critical test. Faraday presented his results in the second part of his article.

His experiments proceeded from the reflection that, according to Ampère's theory, it should be possible to demonstrate complete equivalence between the magnetic behavior of permanent magnets and appropriate arrangements of electrified wires. This reflection may be found in Ampère's own work, though, as we recall, he had since pulled back from the idea of direct "imitation." Faraday's experiments place particular weight on the investigation of the coil, as it comes closest to the arrangement of circuits Ampère hypothesized for bar magnets, especially when configured as a "solid cylinder" rather than a hollow cylinder. Should he succeed in experimentally demonstrating the equivalence of the motions evoked by a coil with those evoked by a bar magnet, this would suffice, for Faraday, "to prove the identity of the two" (Faraday 1821d, 90, ¶44; 1839–55, 2:141). The corresponding experimental comparison between a spiral and a magnetized iron disk fails on the grounds that the disk cannot be magnetized with any discernible structure (Faraday 1821d, 94, ¶55; 1839–55, 2:145).[50]

Faraday's laboratory work on this point took on a much more focused character than earlier efforts, not least because it was constrained by a far more specific question. From September 7 on, nearly all of his experiments—eighteen *Diary* entries made over three days—were directly geared toward this question.[51] I do not consider each experiment individually but instead summarize the results as he would later present them in the article. Faraday enumerates the similarities between magnet and coil as follows:

- Like magnets, coils have one north and south pole each.
- They lift iron filings.
- They evoke rotational motion in electrified wires.
- They are attracted and repelled by electrified wires at four positions.
- "Filings of iron sprinkled on paper held over this cylinder, arranged in curved lines passing from one end to the other, shewing the path the pole would follow, and so they do over a magnet."[52] (Faraday 1821d, 88–92, ¶¶41–51; 1839–55, 2:140–43)

Given this list, Faraday stresses that the similarities are "so great, as at first to leave little doubt, that whatever it is that causes the properties of the one, also causes the properties of the other" (Faraday 1821d, 92, ¶51; 1839–55, 2:143).

The qualification, "at first," is deliberate, however, for immediately following this passage Faraday lists the differences between the behavior of the coil ("helix") and that of the magnet:

- "When the helix is held along-side the needle nearly parallel to it, and with opposite poles together, so that attraction should take place, and then the helix be moved on so that the pole of the needle gradually comes nearer to the middle of the helix, repulsion generally takes place before the pole gets to the middle of the helix, and in a situation where with the magnet it would be attracted."

- The poles of a coil are "at the extremity of its axis . . . whilst in the most regular magnets [they are] almost always . . . at some distance in from the end."
- "Similar poles of magnets, though they repel at most distances, yet when brought very near together, attract each other [weakly]." Thus positioned, they retain the ability to lift iron filings. No such effect is observed with coils.
- A magnetized cylinder of rolled steel behaves differently from a coil. All efforts to magnetize steel disks so as to compare their effects with those of spirals met without success.[53] (Faraday 1821d, 92–94; ¶52–54; 1839–55, 2:143–45).

Though these points pose clear problems for Ampère's theory, in his article Faraday eschews both the conclusive adjudication of the two lists and any overall judgment of the theory. In his private letter to de la Rive, however, he expresses an attitude of skepticism, or at any rate reticence: "therefore until the presence of Electrical currents be proved in the magnet by other than magnetical effects I shall remain in doubt about Ampère's theory."[54] That Faraday refrains from any such remarks in his article is surely largely a matter of diplomacy. After all, what he *had* done was daring enough, for no one had previously undertaken an experimental study of such breadth of the consequences of Ampère's theory. As subsequent developments would show, Ampère, for his part, immediately recognized the challenge. Whether one focuses on the positive or negative side of Faraday's result is a matter of inclination. Faraday would prove correct in his prediction, voiced immediately following his abovementioned skeptical remarks to de la Rive, that his experimental result "will be immediately applied by M. Ampère in support of his theory much more decidedly than it is by myself."

Faraday's challenge has a particular tone. He not only presents his experimental results as a critical test of consequences of Ampère's theory but also endeavors to derive these same results from completely different assumptions. As discussed above, he offers a detailed and persuasive "reduction" of the magnetic action of a coil to the "simple rotation" of wire and magnet pole (see fig. 6.9). This occurs at a crucial juncture in the article, immediately after the beginning of the second part, following Faraday's announcement of his intention to assemble the material necessary for an evaluation of Ampère's theory (Faraday 1821d, 85–86, ¶36;

1839–55, 2:138).[55] This placement is somewhat pointed, for the reduction to rotational motion has nothing to do with this evaluation. Instead, it paves the way for an alternative explanation of one of the phenomena central to Ampère's theory. Without having to stress the point, Faraday conveys the message that the behavior of the coil may be explained on the basis of different premises—premises contrary to Ampère's basic assumption that all electromagnetic action is ultimately attraction and repulsion. It is difficult to imagine that Faraday would have been unaware of this message when composing his article. Making it explicit, however, might have been a bit too daring, so he says nothing in the article about how his "reduction" relates to Ampère's theory. In private, in his brief summary for de la Rive, he is clearer: "Now I have been able experimentally to trace this motion into its various forms as exhibited by Amperes helix &c &c in all cases to shew that the attractions & repulsions are only appearances due to this circulation of the pole."[56] This is not meant as a conclusive judgment of Ampère's theory, for his remark on "appearances" refers only to the case of the coil, or "helix." Faraday is well aware that his more ambitious attempt to explain attraction between straight wires as epiphenomenal stands on much shakier ground. Be that as it may, his reductive enterprise does offer the prospect of an alternative explanation. Even if the explicit goal is not the refutation of Ampère's theory, Faraday's effort might nonetheless eventually supersede and replace that theory. For someone who has been active in the field for only a few weeks, this is no small claim. Faraday was certainly aware of this, and his reticence about pronouncing judgments on the public stage makes perfect sense in this context.

The Establishment of a New Result

The completion of the article manuscript on September 11 marked a shift in Faraday's activities. His duties at the Royal Institution presumably precluded continuing his research with the previous level of intensity, and we do not know whether that was indeed his desire. A newcomer to the field, he had just submitted an important article whose reception would have to wait. But his engagement with electromagnetism did not come to a halt; it simply took on new forms, progressing through several episodes scattered across the next four months. This period has garnered even less historical attention than the second phase, the intensive week of work just discussed. But as closer scrutiny quickly reveals, Faraday's activities during this period belong in important respects to his initial phase. They also yield significant insights into his modus operandi. This period is the subject of the third part of this chapter.

Demonstration Apparatus, Plagiarism Charge, and the Communication of Experiments

The first to learn of Faraday's discovery (besides Brande, to whom Faraday must have presented the text of his article as quickly as possible) was Charles-Gaspard de la Rive in Geneva, who had been in correspondence with Faraday for many years. I have already extensively cited the letter Faraday addressed to him the day following completion of his article, touching on the high points of his new results. The timing of the letter is typical. Throughout his life—and in this respect he differed from Ampère, for example—Faraday reported on new results only after bringing his thoughts to a conclusion of which he felt reasonably certain. The motto attributed to him, "work, finish, publish," presumably also meant, "work, finish, communicate."[57] Beyond the article and letter, he took no further steps in September toward communicating his results, for he knew the article could soon be in print. Instead, in the time available to him, he concentrated on a different matter: the development of the means for ensuring the general recognition for the rotation effect. His previous laboratory apparatus was suited only to his own observations and dependent on his highly specialized skill with materials and equipment.

He turned first to the development of a demonstration apparatus with which he could reliably display the effect on a larger scale. Toward that end he collaborated with an instrument builder named Newman, a frequent contractor for the Royal Institution with whom Faraday had often interacted. Newman's finished device (his Figure 1 in our fig. 6.11) could simultaneously demonstrate the rotation of a wire about a magnet pole, as well as the rotation of a magnet pole about a wire.[58] It was finished before the end of September, in time for Faraday to publish a short description in the very issue of the *QJS* containing his article. His brief description (Faraday 1821c) appeared in the "Miscellanea" section, for which Faraday was responsible and to which he could add short articles at a very late date, on his own authority. The date was, however, too late to allow for an illustration to be included. Faraday also published another short notice, in the same section, on techniques for the maintenance of very reliable elec-

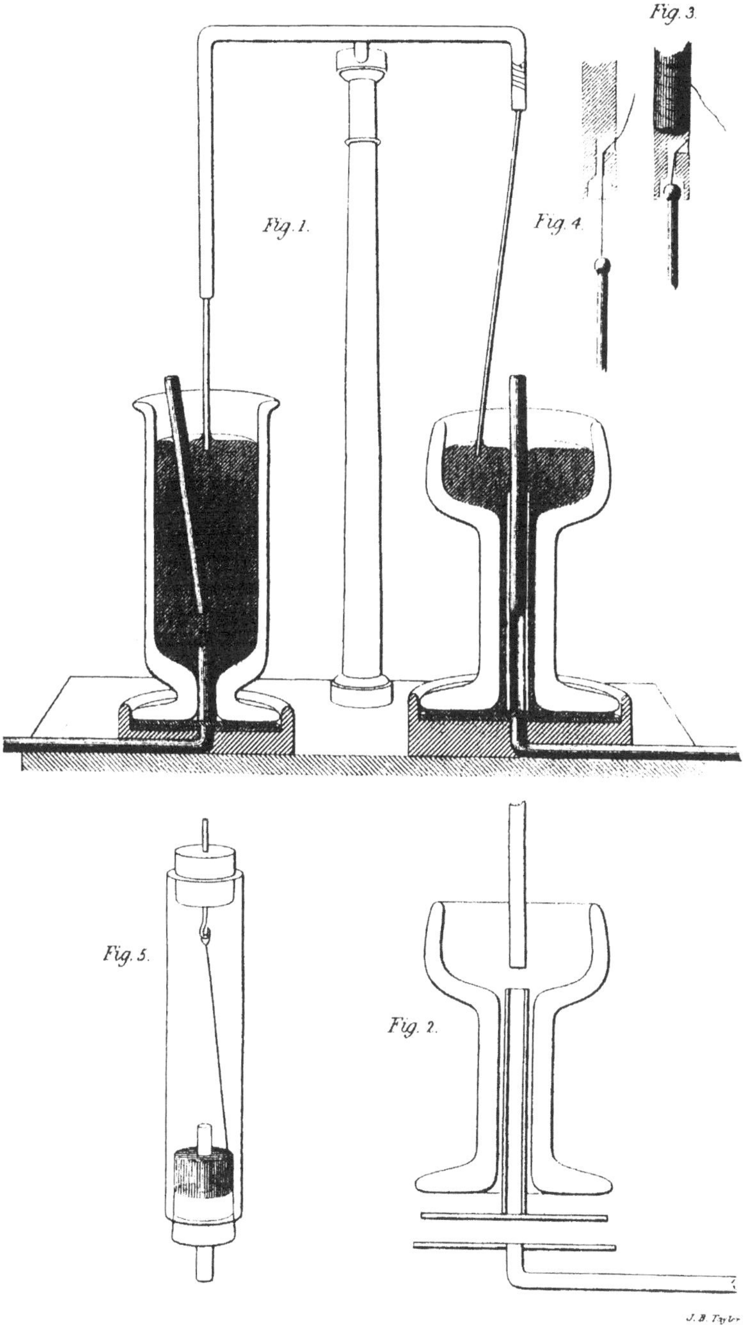

FIG. 6.11. Faraday's demonstration apparatus for electromagnetic rotation, along with the "pocket rotator" (Faraday 1822a).

trical connectivity in such experiments, a point whose importance would be difficult to overstate (Faraday 1821a).[59] At the end of his description of the demonstration apparatus, Faraday also alluded to a second track in his pursuit of technical improvements.

"This apparatus may be reduced in size," he wrote, "and made very much more delicate and sensible" (Faraday 1821c, 187). Such talk of reduced size coupled with greater sensitivity makes it clear that Faraday was working toward laboratory equipment suitable for use by other scientists in replicating his results and conducting their own further research. But before he could perfect the device, he faced other troubles.

The delivery of the September issue of *QJS* at the beginning of October placed Faraday's research before the academic public. Initial reactions, however, came from an entirely different direction than expected. They concerned neither his reduction project nor his critique of Ampère's theory. Instead, shortly after the article's publication, he became aware of a rumor that he had gotten the idea of rotation from Wollaston but had not credited him. Lending the rumor credence was the thought that no one who had entered the field as recently as Faraday could possibly have attained unaided such far-reaching results as those reported in the essay, and with such speed. But the charges remained unspecific; Faraday speaks of the "uncertain and indefinite form of the rumours which have come about me."[60] Most of those reacting to the rumor latched on to the fact that, in discussing the experiment on the rotation of a wire about its own axis, Faraday had not mentioned Wollaston's name, and it was evidently well known that such speculations originated with him. At the same time, Faraday's use of Wollaston's concepts of electromagnetic current and the "powers" met with no criticism, for in that instance he had credited his source explicitly.

Faraday was very sensitive to such accusations. As he wrote to his friend James Stodart, "You know perfectly well what distress the very unexpected reception of my paper on Magnetism in public caused me."[61] He viewed the profession of scientist as demanding high standards of moral character, and he applied these standards to himself. His remark in the same letter that, were there any truth to the accusations, he would feel obliged to abandon science strikes me as entirely serious. Still, he was aware of no misstep on his part and thus held the rumors to be groundless. So as to publicly clear the air, he attempted to arrange a meeting with Wollaston. When Stodart was unable to mediate one, Faraday wrote Wollaston himself about three weeks later, requesting an interview. Wollaston responded evasively and rather coolly.[62] Still, he agreed to a meeting, which presumably took place. This conciliatory effort appeared to clear matters up for the time being; at any rate, Faraday said nothing about it for a long while. But it left a lasting impression on him, and for the rest of his life he remained very careful and sensitive on such issues.[63]

If, however, he hoped to have put the charges to rest for good, he was in for a bitter disappointment. When he came up for election as a Fellow of the Royal Society a mere two years later, the accusation resurfaced as grounds for questioning his moral character. Only by means of a repeated public account of events, authorized by Wollaston, and by revealing his authorship of the "Historical Sketch" was he able to defuse the suspicion.[64] When the charge of plagiarizing the rotation effect reemerged in 1836, his reaction was much calmer. By then he was well known, however, and the accusations did not gain any broader traction.[65] But in October 1821, as Cantor (1991, 231) so aptly observes, "the bitterness [the affair] created affected Faraday deeply and robbed him of the joy of his first significant scientific discovery."[66]

In the first weeks of October, Faraday resumed work on the aforementioned second track of technical improvements to his apparatus. His published notice on the demonstration apparatus was not too helpful for experimental replication, as it did not include an illustration. Besides achieving greater sensitivity, Faraday's goal in reducing the dimensions of his device was to make it more portable, or even small enough to ship. These efforts met with success (see his Figure 5 in our fig. 6.11), and toward mid-October Faraday began sending small rotation devices to researchers across Europe. This was not unusual for the period; recall that he had received a small device from de la Rive a few weeks earlier.[67] In an October 18 letter to Ludwig Gilbert in Leipzig, Faraday described his new reduced device in some detail, noting that its sensitivity had been sufficiently improved for it to be driven by a single voltaic plate pair. In addition to being accompanied by an offprint of his article, this letter was presumably sent with an actual exemplar of the device.[68] That same day he also sent one to Hachette in Paris, another one two weeks later to Wollaston, and in two more weeks a third to de la Rive.[69] Faraday built these "pocket rotators" in various sizes. By its reported dimensions, the device he described to Gilbert, and presumably also sent him, was significantly smaller (by about a factor of two) than that received by Hachette.[70] We might suppose that Faraday progressed toward ever-smaller devices, also mailing out those in the intermediate stages.[71]

Such activities demonstrate that Faraday was well aware of the difficulties confronting any attempt at replicating his experiment. The device was extremely sensitive, and its successful operation was critically dependent on measures and size ratios. The suspension of the movable wire was particularly delicate, as it only exacerbated the problem others had encountered in maximizing electrical connectivity and mobility simultaneously. Faraday alluded to this difficulty in his first article (Faraday 1821d, 76, ¶8; 1839–55, 2:129), and, in his short notice "Contact in Voltaic Electricity" (Faraday 1821a), he described an amalgamation procedure for constructing highly conductive joints. The challenge of addressing this problem even with proper instructions can be assessed by attempting to reconstruct and operate such a device today.[72] Even making the small device, which Faraday considered especially sensitive, run successfully, requires extensive experience and much practice, and its operation is by no means straightforward. Faraday's contemporaries doubtless felt the same. By sending out copies of his device, Faraday was attempting, in a sense, to embody all his hard-won experience with his equipment in the object itself and thus to transmit it.

The "pocket rotators" could reach only a small circle of researchers. Broader communication required printed media. For the next issue of the *QJS,* Faraday prepared an article on the device, providing both a detailed description and a plate (see fig. 6.11) with two technical illustrations (Faraday 1822a).[73] He discussed and illustrated not only the demonstration apparatus but also the smaller rotator designed for laboratory and research use, which he had sent to other scientists. This article was submitted on time before the editorial deadline for the issue, presumably no later than the beginning of December, and it appeared as planned in the January issue of *QJS.*

The Rotation Effect in Paris and Back in London

Thus facilitated, the replication of the rotation effect appears nowhere to have met with great difficulties. Unsurprisingly, Faraday's announcement met with a particularly rapid and sustained response in Paris. Hachette replied promptly following the receipt of Faraday's device, but he decided to wait to replicate the experiment until Ampère's return from a business trip, presumably on account of his own lack of experience.[74] Immediately after Ampère's arrival, the two of them, together with another researcher, named Savary, attempted the experiment, attaining the expected outcome without any major trouble; Hachette would later write that it had "succeeded perfectly."[75] Greatly excited by the finding, Ampère immediately set to work on his own research. The mere existence of the rotation effect already challenged his theory, with its assumption of the primacy of attractive and repulsive forces. Faraday had also raised serious problems for his account of magnetism. As James Hofmann (1995, 288) stresses, the spate of research to which Ampère now felt himself called was no less intense and frantic than that of a year earlier (see chap. 3), with which he first began his work on electromagnetism. The rotation effect having been demonstrated beyond any doubt and being impossible to ignore, Ampère decided that for his first move the best defense would be to do his own research on the new effect. With experimental techniques more familiar to him, he constructed his own rotation apparatus.

In addition, despite the problems the new effect had raised for his theory, Ampère was very active in publicizing it. A few days after his successful replication of Faraday's experiment he reported it to the Académie des Sciences: "M. Ampère relates to the Academy a new and very remarkable experiment of M. Faraday's, which he has repeated along with M. Hachette and M. Savary, in which by the mutual action of a magnet and a voltaic conductor the rotation of one of the two about the other is produced, continuing constantly in the same direction so long as the pile remains active."[76] He delivered a similar report to the Société Philomatique a few days later, on November 24.[77] Before that, he prepared an article for the November issue of the *Annales de chimie et de physique* (*ACP*), which was about to go to press, taking advantage of this opportunity to include in his appendix a preliminary report (with illustrations) on the device received from Faraday and his own rotation apparatus.[78] Still not satisfied, he instigated, via Gay-Lussac, the publication of a French translation of Faraday's article in the next issue of the *ACP*.[79] Thanks to Ampère's promotion, Faraday's discovery gained enormous prominence on the Parisian scene. Faraday could not possibly have hoped for better publicity than a report to France's central scientific institution, delivered by the most important expert in the field, and supported by publication in one of the leading French journals. His efforts to make the effect easily replicable by others contributed significantly to this success. Ampère might well have reacted differently had the effect itself remained in serious doubt.

Ampère attached detailed remarks on his own reaction to the discovery to the aforementioned translation of Faraday's text. On the one hand, he insists (though without detailed explanation) that the rotation effect and many of Faraday's other experimental results can also be understood within the framework of his own theory. On the other, he vigorously disputes the elementary character Faraday ascribed to these effects. As macroscopic phenomena, they cannot by definition be primitive, for the only truly primitive effects are actions attributable to infinitesimal or pointlike centers, the *faits primitifs* discussed in chapter 4. This objection nicely illustrates the different concepts of simplicity then in use. Finally, Ampère addresses the differences between coils and bar magnets, noting that, if we assume microscopic rather than coaxial electrical currents as the cause of magnetism, they are not difficult to explain (Ampère 1822a, 371, 372, 376–79). Nowhere does Ampère engage with Faraday's "powers"-based attempt to treat the interaction between current-bearing wires as complex effects. Perhaps the vagueness of this approach struck him as too obvious to merit further discussion.

Faraday learned of Ampère's activities by a letter Hachette wrote on December 12, 1821, before the November issue of *ACP* had been printed.[80] Among other things, Hachette gave a thorough account of Ampère's new rotation device, to which Faraday reacted by publishing a notice on it in the "Miscellanea" section of the January issue of *QJS* (Faraday 1822b).[81] Hachette's letter also served notice that Faraday was now in the international limelight and could expect to face fierce competition. In particular, Hachette alerted him to Ampère's intention to drive his own rotation device by terrestrial magnetism, rather than with a conventional magnet. Faraday had touched on this very prospect in his essay, but time pressure had prevented him from pursuing it. Hachette's letter thus motivated him to resume his efforts with all due haste and publish as quickly as possible—perhaps even before Ampère could present a successful demonstration of his own.

This plan would require both time for research and an opportunity for swift publication. The next issue of *QJS* was scheduled to appear in a few weeks, so a full-scale article was out of the question, but there was time to get a notice in Faraday's "Miscellanea" section. Still, even that much would require time spent on research and writing, time Faraday evidently did not have until Christmas. The penultimate week of the year, December 21–25, found him in the laboratory experimenting on rotation effects evoked by terrestrial magnetism. Once again, intensive experimentation and writing took place in tandem. Spurred by the earlier plagiarism charge, he invited Wollaston to participate in his experiments, which he did, "two or three times" (Faraday 1823, 290). Faraday's research focused both on the demonstration of terrestrial rotation effects and on the anticipated debate with Ampère. His attainment of the desired result constituted not only the natural continuation of the research direction set out in his first article but also a renewed, explicit, and self-confident challenge aimed squarely at Ampère. I consider this work carefully in the next section.

His text did indeed appear in the January issue of *QJS*. A mere five days separated the finalization of his article text, which cannot have occurred before December 25, and the delivery of the issue on the last day of 1821. Faraday was using his privileged access to the journal to the fullest (as to what the print shop might have made of this, we have no way of knowing).[82] Not only was he stretching deadlines, he also did not hesitate to include under "Miscellanea" a text that did not really belong in that section. Despite its title, "Note on New Electro-Magnetical Motions," there is no disguising the fact that it was not a notice at all but an original, self-contained research article. At more than five printed pages, it was much too long for the section, which typically featured notices taking up no more than a half page. Faraday was well aware of all this divergence from standard practice. In his private notes he always placed this text on a level with his first article, frequently referring to it as his "second paper."[83] The fact that in the rush to publish, if possible before Ampère, he had stretched his privileges almost to the point of abusing them is a dramatic sign of his awareness of the competitive climate and the importance of securing priority claims.

Faraday's contributions to the January issue of *QJS* marked a milestone in his engagement with electromagnetism. With his article on the demonstration and research apparatus and his report on the action of terrestrial magnetism, both his full suite of rotation equipment and his theoretical orientation had been made public. His discovery had raised a considerable stir, and Faraday had successfully drawn the international attention of the most important researchers. His self-confidence in his engagement with Ampère in the report on terrestrial magnetism can be traced not least to the fact that by this point he knew how seriously Ampère had taken his work. At the end of Faraday's December campaign, his work on electromagnetism began a lengthy pause. The spring of 1822 found him occupied with correspondence generated by his publications. Over the course of the following years he continued to follow all new developments in the field, replicating the relevant experiments, but his duties at the Royal Institution left him time for only the most sporadic original research. Only in 1831 would he return to focus on his own work, garnering international recognition with his discovery of electromagnetic induction. However, Faraday's January 1822 publications marked the conclusion of his first phase of engagement with electromagnetism, just as January 1821 had brought an end to Ampère's.

The Action of Terrestrial Magnetism

The research Faraday undertook toward the end of December 1821 was a direct continuation of his work in September.[84] In the early autumn he had been unable to bring his investigation of the electromagnetic motions provoked by terrestrial magnetism to a conclusion, presenting them in his first article as incomplete. In the paper emerging from his latest efforts, this point is acknowledged: "the object of the present note is so far to complete that paper" (Faraday 1822c, 416; 1839–55, 2:151).

Faraday's goal was twofold: to demonstrate the motion of current-bearing wires under the influence of terrestrial magnetism and to relate that motion to electromagnetic rotation. Toward that end he built various devices. The first of these (fig. 6.12, sketch for December 21) was a kind of swing, inspired by Ampère, with its supports submerged in mercury beakers (points *d* and *e*) and its mobility enhanced by a counterweight (labeled *c* in the sketch). In the experiment, when the axle of the swing is oriented east to west, connecting the current causes its immediate deviation toward the north-south axis, irrespective of the direction of current. Faraday's remarks on the success of the experiment demonstrate the extent to which he was striving to relate it to his earlier work on rotation: "These motions are evidently the result of a rotative force emanating from the pole of the earth and acting on the wire; they account also for direction of Ampère's curve."[85]

The experiment itself hardly makes it obvious that any "rotative force" is in play. That the wire does in fact execute a partial rotation may be attributed mostly to the fact that it has no other degree of freedom. With this device it is impossible to distinguish horizontal from vertical motions or to detect an unexpected but not inconceivable rotation in the horizontal plane (e.g., alignment toward a particular compass point).

Faraday was certainly aware of these limitations, and he constructed a new apparatus the next day. A wire segment was suspended horizontally by a thread attached to its middle (see fig. 6.12, first sketch for December 22).

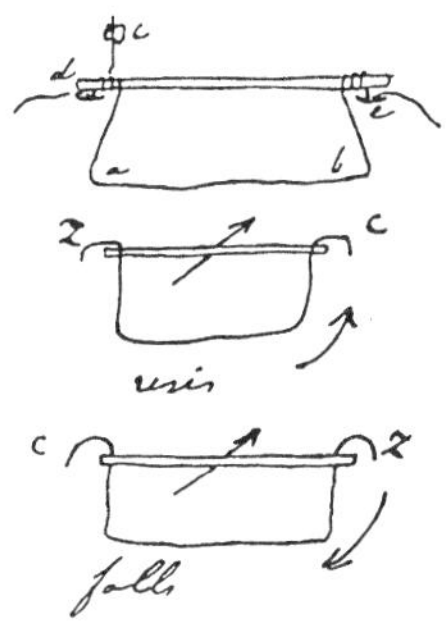

December 21

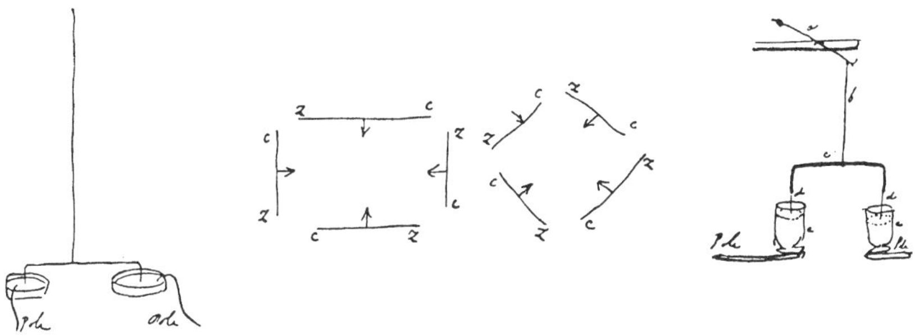

December 22

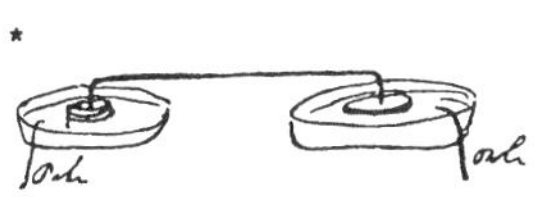

December 24
 December 25

Fɪɢ. 6.12. Faraday's sketches, accompanying *Diary* entries from December 21 through December 25, 1821 (*Diary*, 61–63).

Its two ends were bent downward at right angles and submerged in bowls of mercury. The surface of the mercury was improved by covering it with a thin layer of nitric acid. When the bowls were connected to the battery, electrifying the wire, the wire deviated in the horizontal plane, at a right angle to its direction. The effect was quite pronounced, with a deviation of up to three inches, and disappeared when the battery connection was

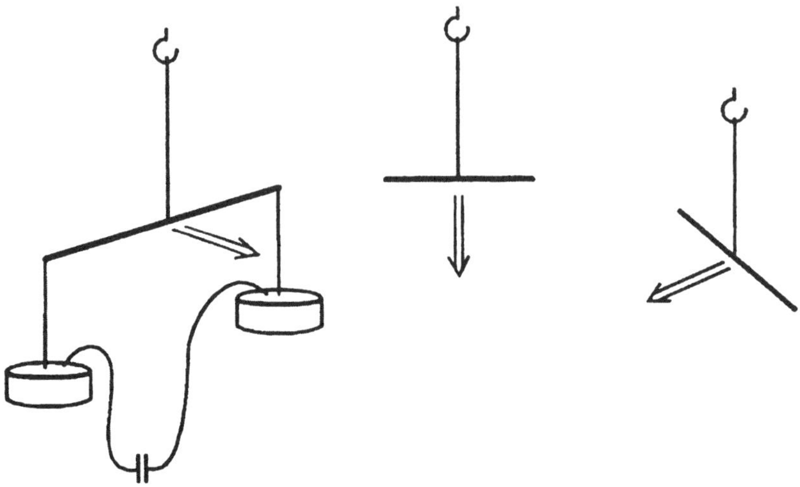

Fig. 6.13. Schematic illustration of Faraday's experiments on the motion of a suspended wire under the influence of terrestrial magnetism.

interrupted. Faraday studied the direction of motion for various initial positions of the wire (see my schematic illustration, fig. 6.13), compiling his results in an abstract sketch (see fig. 6.12, second sketch for December 22). Armed with this sketch, he carefully considered the expected direction of motion, concluding, "It is evidently the attempt of the wire to rotate round the pole of the earth."[86]

Given the minute proportions of the experiment, such a claim sounds peculiar, even far-fetched. Though the *Diary* does not record Faraday's reasoning leading to this conclusion, the published article does so in detail (Faraday 1822c, 416–17; 1839–55, 2:152). At its heart is an attempt to extrapolate from the familiar laboratory observation of the rotation of a wire about a magnet pole to an arrangement the size of the earth. Faraday begins by reviewing the most important characteristics of magnetic rotation. First of all, the rotation of a wire about a magnet pole is independent of the location of the second pole, provided only that it is far enough away to not interfere. Second, the wire does not move either toward the pole or away from it but always laterally, perpendicular to its own direction. The extrapolation to the terrestrial case assumes, first, that the inclination needle points toward the pole. The wire should thus be expected to move perpendicular to the direction of inclination, as well as perpendicular to the direction of the wire itself; taken together, the two conditions yield a fully determinate prediction of the wire's motion. Because in London the inclination needle has quite a large angle (Faraday measures it at 72° 30′), the plane perpendicular to the inclination needle is nearly horizontal. Given the vast distance to the pole, the radius of the predicted rotation is so long, and its curvature consequently so small, that any detected motion will appear flat. Faraday's experimental results coincide completely with this prediction.

Faraday was fully cognizant of the fact that this apparatus could display only the "horizontal" component of the expected motion since, given the small magnitude of the forces involved, the wire will move only in a horizontal plane. But as he explicitly notes, it follows from his reasoning that, when the wire deviates northward, it must also rise—or fall when deviating southward. So as to observe this motion, Faraday changed his experimental setup in such a way as to permit the detection of rising or falling caused by even the smallest vertical forces (see fig. 6.12, third sketch for December 22). This arrangement does indeed display clear vertical movement, but not the predicted movement: when the current is connected, the wire rises significantly, by a quarter inch. But this motion turns out to be independent of the direction of current and the alignment of the wire, so Faraday reasoned that it most likely had nothing to do with terrestrial magnetism. His search for alternative causes thus met with no immediate success: "could not make out the cause at all."[87]

Monday, December 24, found Faraday back in the laboratory after a Sunday off: "Got to work again." He constructed yet another device for the observation of motions provoked by terrestrial magnetism (see fig. 6.12, sketch for December 24). Both ends of the horizontal wire segment are bent downward at right angles and run through pieces of cork, which float on the surfaces of two bowls of mercury. Connecting the battery does produce an observable deviation, but it occurs very slowly, leading Faraday to immediately conclude there was "too much friction by the corks on the mercury." He sets the apparatus aside and returns to his search for the cause of the anomalous elevation of the wire on Saturday. Systematic variation of many parameters uncovers it: "found the surface of mercury altered whilst the current passed from the wire through it." This effect, which he describes in the article as "equivalent to diminution of the cohesive attraction of the mercury," lowers the mercury's adhesion to the submerged wire, causing it to rise. Further experiments allowed Faraday to determine the dependence of this effect on such conditions as battery strength.[88]

But he was still not completely satisfied with his experiments on terrestrial magnetism. He had verified the horizontal movement of the wire in the expected direction and also knew why the vertical component of the effect had resisted his detection. But the argument ascribing the observed horizontal motion to rotation about the magnetic north pole of the earth remained rather tenuous. He surmised that it would be far better to demonstrate actual rotation, as Ampère clearly intended to do with his own apparatus. Thus motivated, on December 25 Faraday constructed yet another device, a true rotation apparatus. He filled a shallow bowl with mercury, which he once again covered with a thin layer of nitric acid (see fig. 6.12, sketch for December 25). A very thin wire was submerged into the mercury from above, its weight buoyed by its being threaded through a floating piece of cork. In his *Diary*, Faraday describes the wire as suspended from above "as in the tube apparatus," thus drawing heavily on his experience with the earlier rotation device, in which the suspension of the wire was also the most critical element of the whole construction. Having mastered this difficulty, Faraday discovered that, when the wire forms a relatively acute angle with the mercury surface ("when the wire formed an angle greater than the dip of the needle"), it begins to rotate when electrified.[89] The direction of rotation depended on the direction of current and the speed of rotation, on the small angular distance (labeled "dip" in the sketch) between the wire and the angle of the inclination needle. Both dependencies precisely fulfilled Faraday's predictions of the rotation effect about the axis of magnetic inclination. With this fabulous success, he was finally satisfied. He ceased his laboratory work and set to writing with all due haste.

In the resulting article he mentions only two of the four devices with which he had successfully demonstrated rotation: the suspended wire of December 22 and the rotating wire of December 25. His omission of the swing was likely due to its greater complexity, as well as to the fact that the argument developed from its behavior was less than persuasive. The floating needle (December 24) he doubtless omitted on the very grounds given in the *Diary:* it was too unresponsive. Faraday's reasoning on the expected direction of motion and his argument that this motion is a genuine case of rotation are worked out in detail within the body of the article (Faraday 1822c, 416–18; 1839–55, 2:151–54). Remarks in the *Diary* make it clear that this same reasoning had guided his work in the laboratory, developed with the help of his *Diary* sketches. If he had had the opportunity to publish an illustration along with his text, he would presumably have used the second *Diary* sketch from December 22 (see fig. 6.12). His essay also devotes some space to the experiments on vertical motion and the unexpected lifting effect, along with his explanation of this effect as a consequence of a change in the surface properties of the mercury (Faraday 1822c, 419–21; 1839–55, 2:155–57).

Faraday's argument deserves some further attention. He was not only interested in demonstrating the existence of certain motions but also, and primarily, in showing them to be "related" to the magnetic rotations discussed in his first essay, indeed, that they *are* rotations about the magnetic north pole of the earth.[90] Such rotational motion cannot in principle be directly observed on a laboratory scale; direct observation would involve witnessing actual rotation about the North Pole. So Faraday had to argue indirectly. In his first step, he shows how the rotation he had demonstrated in the laboratory would appear if scaled up to the size of the earth, with only tiny arc segments available for observation. In the second step he then argues that the experimental results precisely coincide with these expectations, concluding, "The tendency, therefore, of the wire to revolve in a circle round the pole of the earth, is evident, and the direction of the motion is precisely the same as that pointed out in former experiments" (Faraday 1822c, 418; 1839–55, 2:155). That he takes the tendency as "evident," and not "proved" strikes me as an indication of his awareness that although the congruency of expected and observed effects is a necessary condition for his interpretation and increases its plausibility, one is not thereby compelled to accept it. The possibility of explaining the observed motions in other terms cannot be ruled out. And there certainly *are* other interpretations of motions provoked by terrestrial magnetism, with Ampère's being the most prominent among them.

This discussion may provide some background to Faraday's decision to take his essay even further and endeavor to show how alignment experiments long since reported by Ampère might be understood as involving rotation. The experiments in question involve a wire loop, which, if mounted on a vertical axle, aligns its plane to the east-west axis and, if mounted on a horizontal axle aligned to the east-west axis, moves its plane into a position perpendicular to that of the inclination needle.[91] To relate these effects to his now-established circular motion, Faraday argues, one must understand the ring-shaped wire loop as a polygon with an infinite number of sides. As suggested by the *Diary* sketch (see fig. 6.12, second sketch for December 22), the expected motion will be evoked for every one of these sides. If the loop is in a plane perpendicular to the magnetic inclination, the tendencies toward motion of all of the opposing pairs of wire segments compensate each other precisely. If the ring is not aligned, the net result is motion that eventually leads to its alignment (fig. 6.14). Faraday's summation of this result makes clear the extent to which his procedure of reduction to the simple case has guided his work here, too: "the traversing of the ring may be reduced to the simple rotation of the wire round a pole" (Faraday 1822c, 418; 1839–55, 2:153). The demonstration that Ampère's alignment effects could be traced back to Faraday's electromagnetic rotation constituted a further step toward establishing his conception as being on an equal footing with Ampère's. In the final paragraph of his article, he voices the hope that his law governing rotation about a magnetic pole of the earth or, "what is nearly the same thing," Ampère's law for the alignment of his wire loop, might be tested at various latitudes. The rhetorical equating of the two has been carefully considered. In a manner both delicate and, for anyone who knew the field, unmistakable, Faraday has laid claim to an alternative to Ampère's approach every bit as powerful.

A Comparison of Ampère's and Faraday's Initial Phases

Looking back, we see in detail the two periods in which Ampère and Faraday conducted their first work on electromagnetism. Both periods involved extraordinary innovation over a short period of time—innovation of the

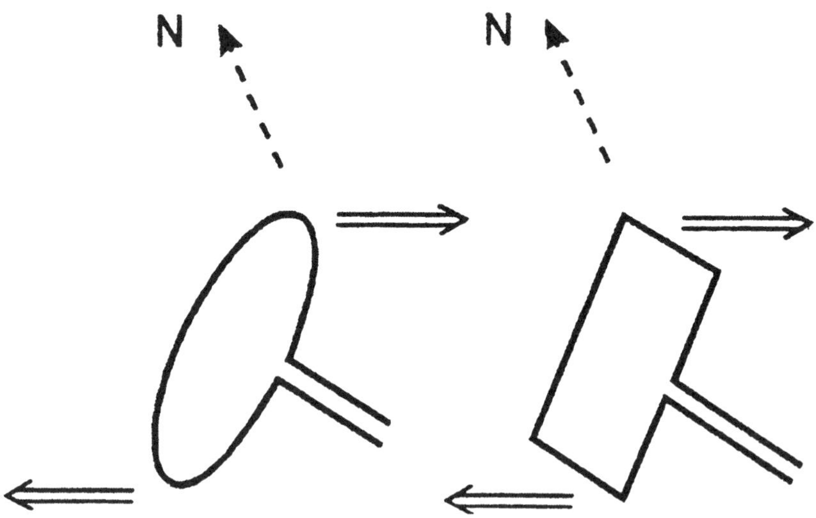

Fig. 6.14. Schematic illustration of Faraday's "reduction" of the alignment effect to "simple rotation."

sort that fundamentally transformed the perception of an entire field of phenomena. New concepts and means of representation were forged that reshaped all intellectual and experimental engagement with the field, molding it for a long time to come. To use a term I introduced in my introduction, it seems to me to be entirely appropriate to call such periods formative: in both cases, they were formative not only for the research field as such but also for the researchers themselves. The fact that their outcomes were so extraordinarily different makes their comparison all the more interesting.

Though separated by a year, the two periods are historically linked. Faraday, whose work began with a thorough survey of the field, engaged with Ampère's work more intensively than that of any other author. What's more, and has thus far been overlooked, Faraday began his investigation with the very same problem that Ampère had pursued in his own initial phase, though Ampère had set it aside before arriving at a solution: the problem of the motions of a needle in an eccentric position relative to a vertical electrified wire. For his part, Ampère reacted strongly to Faraday's results. Besides his own initial phase, he would never work so intensely on electrodynamics as in the fall of 1821, in response to Faraday's publications.

Both researchers began with clearly discernible phases of exploratory work. This had much to do with the novelty of the effect with which each began: in Ampère's case, Ørsted's findings; in Faraday's, the rotation effect. These exploratory phases were characterized by the expansion of the domain of phenomena by means of variation of experimental arrangements, as well as the construction of new means of representation that enabled the formulation and communication of regularities. The central challenge of the new field consisted in the development of a means of representing spatial relations adequate to grasping the complex dependency of needle behavior on position. Toward that end, Ampère proposed the concept of the circuit of electrical current, and with it he established "right" and "left" relative to the direction of current. Faraday contributed the idea of rotational motion as the basic effect of electromagnetism. In both developments, visual representations played a decisive part. In Faraday's case this fact is impossible to overlook, but even in Ampère's case it was the exploratory phase, and *only* that phase, that saw such abstract sketches as his *bonhomme* (see fig. 3.5) playing a constitutive role in his research. These visual representations gave rise, in turn, to hierarchies in which a multiplicity of effects could be traced back, or "reduced," to "general facts" or "simple cases." The exploratory phases were also characterized by great openness toward new and unexpected empirical findings and a willingness to revise and reconceive regularities and their representation. One might even speak of a reluctance to embrace any potentially premature stabilization of the resulting conceptions.

With regard to intensity and duration, however, the two exploratory phases were quite different. Ampère's saw the development of several important structural concepts and the outlines of a few successful cases of "reduction to general facts," quickly followed by the abandonment of unfinished exploratory work in favor of another promising project. In working out his theory of current-current interaction he soon achieved a stable conceptual framework, and thereafter his approach changed significantly. From then on he devised and evaluated experiments from within this framework. The potential significance of such experiments was determined in advance and required no further exploration of the terrain. By contrast, Faraday's work maintained a consistently exploratory approach throughout. He worked through many "reductions" in detail and with great experimental effort. Experimental breadth was not a transitory aspect but a goal in its own right. That he took up exactly the problem Ampère had abandoned was not a coincidence but a systematic consequence of his approach: the problem constituted a central, unresolved difficulty for any attempt at *completely* capturing the behavior of the magnetic needle on a broad experimental foundation.

We can find, in line with the differing intensity and duration of the two exploratory phases, stabilization along different timescales. The project of explanation in terms of current-current interactions, for which Ampère gave up his exploratory approach, was bound up with the swift and irreversible selection of a specific set of concepts and theories and a well-defined research program. Experimental breadth was no longer a goal; it became a distraction. Ampère's reaction to unexpected results was typical; the stabilization of his conceptual scheme and research program went along with an immunization against unexpected empirical findings. The dynamics of his research program no longer called for experimental innovations, in the sense of expanding toward qualitatively new phenomena. Instead, Ampère found himself called upon to react when such innovations confronted him from outside and proved impossible to dismiss, as was the case with Faraday's rotations. On the other hand, only this relatively swift stabilization allowed Ampère to make such rapid progress in the mathematical and experimental articulation of his theory.

In Faraday's case, we find no comparable stabilization process. All that remained stable throughout was his goal of constantly pushing back the boundaries of familiar phenomena, searching for the tools required to represent them, and, when necessary, revising those already established. In Faraday's work on electromagnetic rotation, this open-ended approach is clearly discernible in his investigation of the action of terrestrial magnetism, in which he constructed five different experimental arrangements. Given the brevity of his initial phase, we cannot trace that research practice over a long period of time. But we do find telling cases in his later laboratory work. For example, in his investigations of electromagnetic induction ten years later, he would be extremely cautious in proposing the concept of a "magnetic curve" and using it in his accounts. It would take two more decades of intensive work before he was prepared to take this conception as stable and to argue publicly for the existence of such curves. To a highly unusual degree, Faraday tended to resist stabilization, patiently enduring uncertainty at the level of basic concepts, all while continuing to research intensively. Elspeth Crawford (1985) has aptly named this tendency "negative capability." Maintaining such a stance comes with increased openness to unexpected experimental results or other empirical findings and stands in stark contrast to the immunization observed in Ampère's case. It also means that the eventual stabilization of concepts unfolds on a much longer timescale. Faraday's decades-long effort strikes me as significant in this regard.

Given the marked differences in intensity, duration, and relative significance of the two exploratory phases, the different outcomes of Ampère's and Faraday's initial periods of research are hardly surprising. Different ways of working might well generate different sorts of results. In Ampère's quest for a mathematical theory, experimental breadth was no longer a goal in its own right. The available mathematical tools would necessarily shape the resulting conception and representation of the phenomena in ways both fundamental and possibly unreflective. On the other hand, when the goal was the systematic development of new means of representation, spurred on by the constantly expanding breadth of phenomena, deviations from traditional conceptions and representational tools were much more readily to be expected. In such circumstances the conceptual tools themselves are subject to debate and revision. Faraday's transition from attraction and repulsion to rotational motion as a fundamental mode of action is only one telling example.

But this is only one aspect of the comparison between Ampère and

Faraday. Going one step further we might ask, What is the background to tendencies toward exploratory work or other approaches? As I show in the next chapter, regardless of all differences among local traditions, exploratory work *typically* takes place in epistemic situations in which uncertainty has arisen at the basic level of concepts, language, or means of representation. It is no coincidence that the initial response to Ørsted's findings throughout Europe was almost always exploratory. Almost no one had any sense of how to accommodate the effect within existing conceptual schemes. The most prominent exception, Biot in Paris, was made possible precisely by Biot's willingness to disregard all experimental aspects of the effect that might undermine confidence in the conceptual scheme he championed. However, the intensity and duration of exploratory work differed markedly among the various European agents. This had much to do with local scientific cultures, biographical particulars, and conditions within the scientific community.

This view holds for the comparison between Ampère and Faraday, too. Both were quite ambitious, and from the outset both saw their work on electromagnetism as bound up with the enhancement of their reputations. Both made targeted use of publication outlets, to which they had easy access and whose schedules frequently determined the rhythm of their research. Both were extremely interested in communicating their results as widely as possible, making their experiments easily replicable by others, and thereby gaining general recognition. Given his delicate apparatus, Ampère pursued this goal by means of long texts and detailed illustrations drawn precisely to scale. Faraday had these too, but he also developed versions of his device portable enough to be shipped. But the local cultures in which the two researchers worked and within which each sought to establish himself as a researcher to be taken seriously were quite different. In London, the report of an experiment might well have been sufficient to gain recognition, but not in Paris, at least not within the already mathematically oriented fields of electricity and magnetism. There, an approach aimed at mathematical formalization would be far more promising, whereas in London, even in 1820, such approaches remained of only secondary interest.

In addition, there were differences in biographical and institutional background. Researchers are likely to promote themselves with the kind of research in which they feel most competent and for which they possess the requisite means. This is all the more true when, as in both of the present cases and indeed in much of modern science, researchers have limited time at their disposal. Ampère was most comfortable in mathematics and far less so in experimental research. For exploratory experimentation, with all the improvisation, flexibility, and spontaneity that it required, he had only very limited means. With his modestly equipped laboratory, assembled piece by piece in his house, he could not go very far. When devices had to be commissioned elsewhere, this requirement placed sharp constraints on spontaneity and the variation of experimental parameters—a serious obstacle to exploratory work. In all of these respects, Faraday's situation was nearly the opposite. He was most comfortable in the realm of wide-ranging experimentation of the sort he had practiced for years, not only in chemistry but also in technology and materials science. He also had all the necessary resources at hand. Only a few floors down from his private rooms lay a laboratory stocked with equipment and materials sufficient for even the most spontaneous experimental notions. When necessary, he could spend hours there, building, rearranging, and otherwise varying one apparatus or turning to others. One could hardly wish for conditions more favorable to exploratory experimentation.

There is another biographical aspect. With Faraday, the resistance to conceptual stabilization was nearly programmatic. To persist in such an approach was made easier for him both by the fact that his position at the Royal Institution did not depend at all on his scientific success and that his personal equilibrium was vested in circles outside the practice of science, especially within the religious circles of the Sandemanian community.[92] This situation presents a sharp contrast to the highly competitive climate of the Paris academy, where an attitude that shied away from quick stabilization could put an end to professional advancement or even to an entire career. Given that background, Ampère's decision to bring his exploratory phase to a close after orienting himself and attaining a few new concepts seems perfectly reasonable. Given the expected competition with Biot that lay ahead, he was also well advised to turn to a project with some prospect for mathematical formalization on the horizon.

The two initial phases marked turning points in which the forging of new conceptual structures and experimental procedures would continue to shape the course of future research for the long term. In my introduction I claimed that the complex array of factors contributing to the development of a science becomes especially clear, even tangible, when we look to formative periods. The two initial phases discussed in this book serve as excellent illustrations. An integrated accounting of such heterogeneous factors as epistemic goals, theoretical proclivities, biographical peculiarities, experimentation talent, access to laboratories, local academic cultures, and contacts with craftsmen is required in order to study and understand such developments. Scientific change is not driven exclusively by epistemic or social, cultural, and material conditions but by a blending or amalgamation of all of them, however varied. Andrew Pickering's (1995b) talk of the "mangle of practice" seems especially apt, though I would still maintain that careful study allows for the particular strands of this process to be recognized and teased apart, and it should thus be undertaken. Conversely, this means that in the historiography of science, the quest for methodological purity is simply inappropriate. What is called for is instead an openness toward the most diverse factors, along with readiness to acknowledge the dynamics and limitations peculiar to each.

CHAPTER 7

Experiment and Concept Formation

The previous chapters have repeatedly raised questions regarding experimentation and its role in the generation of knowledge. In this final chapter I discuss these questions in more general terms, drawing on and extending aspects of the philosophy of science, epistemology, and historiography. In the first three sections I provide an overview of recent discussions of experimentation since the 1980s in the history, philosophy, and sociology of science. Against this backdrop, the following sections focus on specifically *exploratory* experimentation, a previously neglected form of experimentation whose characteristics emerge clearly out of my studies of research practice. I sketch the differences between exploratory and theory-oriented experimentation, a task that requires a nuanced differentiation of the concept of theory, as well as a discussion of epistemic goals and epistemic significance specific to exploratory experimentation. Finally, I survey the frequent occurrence of exploratory experimentation across the history of the sciences, asking why so significant a process could escape scholarly attention for so long.

Recent Debates on Experimentation

Though the history and philosophy of science paid little attention to the question of experiments through the early 1980s, the discussion has become quite lively in the thirty-plus years since then. One of its many starting points and foci was Bruno Latour and Steve Woolgar's (1979) study in which the two sociologists of science analyzed the work conducted in an experimental laboratory from an anthropological perspective, emphasizing the significance of communication and social interaction even in the "mere" establishment of experimental results. Ian Hacking approached experimentation from a different angle in his book *Representing and Intervening* (1983). He described a multiplicity of different epistemic functions for experimentation, gave experiment a central role in his argument for scientific realism, and captured his observations on the historical complexity of experimental research in his classic assertion that "experiment has a life of its own" (Hacking 1983, 150). The scholarship and debate that subsequently arose and still continue are directed at a variety of different topics. In lieu of a synthesis (which it is far from clear will ever be possible), I survey below several of these topics.

Gaining empirical results by means of experiment has turned out to be a much more complex process than envisaged in any of the empiricist proposals of the seventeenth through nineteenth centuries. Experiments aren't simply found; they are made. They are designed, constructed, conducted, and evaluated by human actors, always operating within a particular historical, social, conceptual, theoretical, and biographical situation. To experiment is to act, to intervene in the physical world by means of materials and objects, in order to obtain knowledge about it. This effort typically requires the aid of instruments, without which the desired intervention and observation would not be possible. Finally, there is a scientific community that must accept the outcome of an experiment as authoritative. With this focus on the materiality, agency, historicity, and communicability involved in experimentation, the historical and philosophical study of experiments faces new challenges.

Experiments require skills on the part of the experimenter. Though until a few decades ago most of these skills were manual in nature, they often now include such aptitudes as a particular knack for working with computer programs or other technology.[1] They are to be found only at certain times and places. There are thus experiments that simply cannot be performed at other times or in other contexts. For example, James Prescott Joule's marvelous determination of the mechanical equivalent of heat in the 1840s was made possible only by the experience with precision heat measurement he had gained working in his father's state-of-the-art brewery. What's more, it was in the basement of that brewery that he found the stable temperature conditions necessary for his work (Sibum 1995, 1998). The performance of this immensely influential experiment at precisely this moment, no earlier or later, had a great impact on subsequent conceptual and theoretical developments. The same holds for Coulomb's discovery of the inverse-square law for electrical attraction by means of his torsion balance, discussed in chapter 1. This would also have been inconceivable without his prior experience with torsion compasses (Heering 1998).

In both cases, the extent of the necessary skills, and thus the extent to which the experiments were specific to a given time and place, was made plain only by reconstructing the instruments and redoing the experiments themselves.[2] Beyond simply providing insight into experimental practices never recorded in writing, such replications open historical vistas that could never have been attained by means of written sources alone. It is only from such a perspective that questions regarding the accessibility of the materials and techniques, not to mention the historical and cultural conditions that gave rise to the specific design of an experiment and shaped its reception, truly become clear. Studying experiments by replicating them brings out the dynamics and consequences of experimental materiality, and thus the relationship between experimentation and material culture, in much sharper relief than otherwise attainable.[3]

Experimental results are obtained locally, at a specific time and place. They attain validity in science only through acceptance by a broader community of scientists. The touchstone of broad-based recognition has traditionally been replicability, ideally even actual replication by others. But replicability is not always a given, for there are cases in which most places simply lack the necessary experimental means. This was already true in the early modern era, with regard to experiments involving complicated new instruments such as the air pump or microscope—experiments that were the basis for far-reaching, revolutionary claims. Well aware of the lack of reproducibility, researchers used other means to gain credibility, such as invoking trustworthy witnesses or describing an experiment in such detail as to leave readers with no doubt about the trustworthiness of the report—a procedure called "virtual witnessing" by Shapin and Schaffer (1985) in their famous study of Boyle's air pump experiments.[4] Similar situations arise today, for example, in the field of high-energy physics, where present-day experiments, while replicable in principle, will still probably never be replicated because of their immense cost. The replication gap is only partly filled by sharing datasets and ensuring international representation on participating research groups. Just as for centuries the reliability of experimental results has been attained by appeal to high-ranking witnesses, so today the reputation of an experimenter (or research group) and its place within scientific networks play an important role. Recent scandals in the biosciences and physics, involving revelations of the forgery of experimental data that had gone undetected for years, dramatically underscore the continued importance of trust as a central category in contemporary scientific practice.

But even when material and monetary resources are no obstacle, replicability is hardly a trivial matter. The skills required for the construction of necessary equipment and the conduct of experiments often remain a critical point. Situations in which written or verbal reports have been insufficient to ensure replicability elsewhere are common and by no means confined to prior centuries. In those cases, instruments and materials (such as those of de la Rive and Faraday) have often been transported from place to place, and sometimes even to people possessing the requisite knowledge and skills. Of the many examples, the introduction of the cyclotron in the 1930s is particularly remarkable (Heilbron, Seidel, and Wheaton 1981).

Experiments play an important social role in the scientific community and in the cultural life of an era. Experiments also exercise a wide range of different, often very important rhetorical functions in the self-presentation of individual researchers, as in the promotion of research projects and disciplines and the account they give of themselves to society at large.[5] Some studies (e.g., Dörries 1998) have offered, moreover, a new view of the cultural significance of experiments, as well as their aesthetic aspects.[6] For all the importance of the cultural turn in science history, however, we should be careful not to take this approach as comprehensively capturing the experimental process. After all, researchers of every epoch have claimed to be learning something from their experiments—to have used them as a means for gaining knowledge. Even when such claims have been rhetorically exaggerated, we should not dismiss the epistemic aspect when studying experimentation.[7]

Experimental research harnesses objects and instruments. Various investigations have revealed the extent to which these items develop their own dynamics, thereby shaping the course of research. Hans-Jörg Rheinberger's (1992, 1994, 1997) work (including his collaboration [Rheinberger and Hagner 1993]) on the development of molecular biology constitutes perhaps the most thorough investigation of this process. Over the course of its evolution, every research project faces branching points at which the choice between alternative directions for future research must be made.

Which direction to pursue often depends on such "contingent" factors as the availability of materials, time, skills, and other resources. The selection of particular materials and objects can determine the direction of a research project for a long time to come. In the development of the study of both electricity and physiology, the frog may serve as an illustrative example (Holmes 1993b). The establishment of canonical experimental techniques allows some questions to be addressed, others generated, and still others excluded.

The intertwining of material resources, instrumentation, personal preferences, theoretical and conceptual structures, biographical particulars, and sheer historical chance is eminently typical of experimental practice and the progress of the experimental sciences. This insight has given rise to various historiographical approaches. Hans-Jörg Rheinberger has brought all of the complexities together under his concept of an experimental system, with its own internal dynamics and temporal evolution. He argues that the history of modern molecular biology can be written as the history of experimental systems, with their "objects, differences, and conjunctions."[8] Klaus Hentschel (2000) has attempted to further differentiate such experimental systems, not only drawing on Hacking's taxonomy of experiments but significantly expanding it. In a similar vein, but with a view toward large-scale, global developments, we find Andrew Pickering's (1995b) notion of the "mangle of practice."[9] It strikes me as providing scant structural guidance for the historiography of science. Peter Galison's (1995) concept of "constraint" refers to experimental, theoretical, or instrumental constraints that may be overcome only with great effort and may thus shape the course of developments. On very different grounds both Jed Buchwald (1995b) and Andrew Pickering (1995a) have criticized this approach as inadequate to other historical developments.

The Reproducibility of Experiments and Social Construction

Of all the theses that address the experimental generation of knowledge, the thesis of the "experimental regress" has perhaps been most widely discussed. In light of the insight that the reproducibility of experiments sometimes faces in-principle difficulties, as well as the consequent importance of sociological criteria such as trustworthiness or the institutional structure of science, some have questioned the independent reproducibility of experiments in general. The argument, roughly put, runs like this: because the deployment of instrumentation requires skills and qualifications that are only available locally, exact replication of experimental results at another locale is not possible. Therefore, replication is ruled out as a means of validating experimental results. Thus, the only way to determine whether an experiment leads to a valid outcome is to rely on theory: whether the result is "good" or not must be assessed relative to the theory against whose background the experiment was performed. But because this is the very theory the experiment was meant to test, the circle is complete: theory determines when a "good" experimental result has been obtained, but the tenability of the theory is established by "good" experimental results. This sort of "experimental regress" is, so the claim goes, inherent to experimental science and essentially unavoidable. The acceptance of an experimental result cannot rest on the (in principle unattainable) criterion of independent replicability but solely on a social process in which trustworthiness, status, and power relations are decisive. To put it provocatively, experimental results are not attained but negotiated. The most pointed presentation of this thesis is certainly given by Harry Collins, based on the example of Joe Weber's gravitational wave detectors, whose observational results have largely been discredited following a long debate that Collins has analyzed in detail.[10] The thesis is also supported by Latour and Woolgar's (1979) results: their quasi-ethnological study hinted strongly at the decisive importance of communication and social hierarchy in establishing an experimental result and forging consensus, even within a given laboratory.[11]

The purported universality of the experimental regress speaks to a central epistemic aspect of experimentation, seriously calling into question its status as an experiential tool. The experimental regress thesis, if taken as a universal claim, not only dispels any lingering optimism, left over from the seventeenth and eighteenth centuries, that empirical investigations might overcome human prejudice; in its place, it offers a diametrically opposed view in which the supposed empirical findings of science do not point to any extracultural nature but entirely and exclusively reflect a process of social construction, a process that would possibly yield completely different results under different cultural and social conditions. Reflections within the philosophy of language, setting out from a particular interpretation of Wittgenstein, have led to similar conclusions. In the so-called "strong program" of the Edinburgh school, they find expression in a comprehensive account of science.[12] Because empirical findings are about sociological conditions rather than any external world, the results of science must be understood as a purely social construct, with no particular correspondence to the nature science purports to study. Various social constructivist programs have been worked out since the 1970s. In all of them, we also find repeated emphasis on the constructive character of experiments.[13] It is noteworthy, however, that the far-reaching consequences of these approaches for our understanding of reality have been articulated in fundamentally different ways. In opposition to radical relativism (e.g., Bloor 1976), Latour (1994) has proposed a "symmetrical anthropology" in which the "nature pole" (*pôle nature*) is balanced by the knowing subject acting within a society, the "social pole" (*pôle société*).[14]

Given the far-reaching inferences drawn from the experimental regress thesis, it is hardly surprising that it has met with opposition. On the historical side, Allan Franklin (1994) has reanalyzed Collins's (1985) gravitational wave case, relying on different sources, and argued that the episode fails to justify Collins's radical conclusions in that scientists had good ("rational") reasons to dismiss Weber's results, even though replication was impossible.[15] In her philosophical critique, Mary Hesse (1986, e.g., 721, 725) has argued that while Collins's example might motivate the modification or fine-tuning of received accounts of experimentation, it by no means underwrites the inference to "radical conventionalism." Hans Radder (1992, 1995) comes to much the same assessment. It should be noted that Allan Franklin and Harry Collins (2016) have refined their opposing views in the meantime so as to agree on the sources of their disagreement and to find much more common ground than before in their analyses of research practice.

The historical cases I have presented in the preceding chapters suggest some further observations, which may help bring the problem into sharper focus. In those cases there is no trace of any experimental regress. Of particular interest in this regard is the reception history of Ørsted's discovery of electromagnetism, discussed in chapter 2. His results were accepted immediately and unreservedly throughout Europe, regardless of theoretical commitments or proclivities and despite the fact that Ørsted enjoyed neither particular prestige as an experimenter nor especially great influence. Indeed, all of these points raised serious doubts, especially in Paris. The fact that the result was nevertheless acknowledged so quickly, even in Paris, was entirely due to successful replication. But how could replication have succeeded so quickly, so universally, and so easily in a case in which experimenters had nothing at hand but a short text without any illustrations of the apparatus? It might be true to say that the experiment was simply "easy" to reproduce, but this claim does not really lead further.

Easiness is a relative notion, and the real question becomes, What is the background against which something appears easy? In the case of Ørsted's experiment, this may be stated quite precisely. The experiment involved only experimental procedures and techniques that had been well known for decades. The operation of voltaic devices, connecting wires, and magnetic needles was common knowledge and procedure in every laboratory. The same holds for the rest of the required equipment.[16] Avid communication about such matters, in journals, books, and letters, had spread across Europe, so that information was no longer bound to any particular localities. To be sure, Ørsted's experiments combined familiar elements in an entirely new way, but no novel laboratory procedures were required for their implementation. This was true throughout Europe, despite its many different scientific cultures: with regard to the experimental techniques needed to produce the effect, these cultures were not different at all. Only against this background could Ørsted's experiment be considered "easy," as well as amenable to such widespread and immediate replication.

In Collins's (1985) case study of gravitational wave detectors, conditions were entirely different. Weber's setups involved almost no standard equipment. Most instruments had to be developed specifically for a particular experiment, and procedures for their proper handling had to be learned and practiced for that specific equipment. This holds especially true for procedures to test the equipment's correct operation. In such a particular set of circumstances, the challenge of an experimental regress in which the validation of findings was intertwined with (equally uncertain) theory strikes me as perfectly plausible. But such circumstances are hardly typical of experimental research in general. The case indicates that the experimental regress thesis is in need of further qualification, with considerable attention to the extent to which a given experimental finding relies on generally established experimental techniques. Such qualification would, on the one hand, undermine its claims to global applicability but allow, on the other hand, for the thesis to be formulated much more precisely for those cases in which it actually holds.

Such analysis sheds light on the general question of how replication serves to secure the recognition of experimental results. New experiments arise within local contexts, with specific resources and persons having specific skills. Experimenters seeking recognition thus always face the task of liberating their experiments from the specificities of locality. At least two distinct strategies, often pursued in parallel, may be discerned in my historical studies. First, researchers often strive to make their experiments independent of strictly local skills, working to allow them to be performed with broadly distributed standard techniques. This was Ørsted's strategy, and he found it relatively easy to pursue, falling back extensively on familiar components. Circumstances are not always so favorable. When they aren't, a second strategy becomes visible: the attempt to communicate locally acquired know-how. Both Ampère, with his complicated devices, and Faraday, whose rotator was so difficult to operate, pursued this strategy. Ampère had scale-accurate drawings prepared, with some considerable effort, and complemented them with detailed descriptions. For his part, Faraday knew that the most delicate aspect of the rotation device was the maintenance of electrical connectivity in the suspension of the wire.[17] So as to make the techniques he had developed toward that end accessible to others, he first published a notice dedicated exclusively to this problem. Second, he sent rotation devices to several European researchers. Both Ampère and Faraday were well aware of the problem of replicability, and they addressed it head on. As their success demonstrates, successful replication by others is a significant, nontrivial, and sufficient condition for the general acknowledgment of experimental findings. In addition, as both cases make clear, theory need not necessarily play an important role in this process.[18]

Experiments and the Generation of Knowledge

Ian Hacking's reference to the "Baconian fluster" of different functions for experimentation points toward a vast field that has yet to be fully explored. The relationship between experiment and theory, or, more generally, the manner in which experimentation gives rise to knowledge, has turned out to be considerably more complex than the "standard view" might lead one to expect. Since the 1980s, a wide range of philosophical studies has led to the awareness of a "new experimentalism" in the philosophy of science (Ackermann 1989). Many of these studies—and this strikes me as a common thread among them—approach general questions on the basis of thorough historical investigations. Cartwright's, Galison's, and Franklin's attentions have been directed to various fields of modern physics.[19] Gooding's (1990a, 1992, 1998) attention has been fixed primarily on early nineteenth-century electromagnetism. More recently, the focus on physics has been broadened, beginning with the works of Rheinberger and Burian.[20] Latour, Mayo, Heidelberger, and Graßhoff draw on historical investigations of various disciplines in their treatment of various general philosophical problems.[21]

Diverse as these efforts are, their most important common feature is the tendency to take the complexity of experimentation more seriously than in previous work within the philosophy of science.

By way of illustration rather than in an attempt at systematic survey, I highlight here a few prevailing themes. Nancy Cartwright (1983) has critically investigated the relationship between very general physical laws and those that are less general but closer to the level of phenomena, and she has questioned the greater importance traditionally ascribed to the former. In his studies of high-energy physics, Peter Galison (1987, 1988) has raised the question of when experimenters take their experiments as having come to a result. Though trivial within the traditional conception of experimentation, the answer to this question turns out to depend on complex and varied criteria. The view of the relationship between theory, experiment, and instrument that emerges from Galison's work gives far greater independence to each aspect than has traditionally been assumed. Michael Heidelberger (1998) has proposed distinguishing the productive, representational, and constructive functions of instrumentation. Cutting across traditional philosophical categories, Hans-Jörg Rheinberger (1992, 69–72) has introduced the distinction between "epistemic things" and "technological objects" involved in the research process. The distinction is functional, not material: epistemic things are those toward which inquiries are aimed, or "what remains unknown" (be it material structures, laws, chemical reactions, formulas, or biological functions), while technological objects refer to those that are stable and well defined and without whose foundation no research could take place (stabilized concepts, theories, apparatus, procedures, etc.). It is the very functional character of this analytical distinction that makes it especially attractive.[22]

Taken together, recent work on experimentation has cast profound doubt on the primacy of theory over experiment so long taken for granted within twentieth-century philosophy of science. The evolution of science is not just the evolution of theory, with experimentation as a boundary condition. Instead, experiments serve an essential constructive function. They are a focal point at which theoretical reflections and targeted manipulation come irrevocably together with materials, instruments, specific historical situations, and the peculiarities of human actors. This insight does not, however, necessarily compel us to a constructivist conclusion like that of Peter McLaughlin (1993), in which experimentation can only tell us something about the world forged in the very process of experimentation itself.[23] It rather poses an important challenge to the philosophy of science—the challenge of doing justice to experimental practice in ways that lead us beyond debates over construction or representation.

One point strikes me as being of particular importance here: a better understanding of experiments cannot be attained without at the same time refining our considerations of theory, since the crude dichotomy between experiment and theory, familiar as it may be, is of no further help. There have been various attempts at a more detailed investigation of the epistemic domain lying between phenomena and theory. Hacking (1983, e.g., 159, 165) went back to the traditional distinction between "empirical laws" and systematic "theory," properly so called, a distinction erased by the indiscriminate appeal to the theory ladenness of all observation.[24] Cartwright, in turn, clearly demonstrated that empirical regularities can be far more important to scientific practice than higher-level theoretical statements. Later on, Hacking (1992, 44–45) proposed a more detailed differentiation. In place of the conventional notion of theory, he identified at least "three distinct kinds of knowledge": "background knowledge," or knowledge presupposed as stable and self-evident, and not questioned; "systematic theory," or high-level theory, far removed from the phenomenal level and without any direct empirical consequences; and, finally, as a third category, "topical hypotheses." It is these provisional hypotheses, still open to revision, that provide the only bridge between theory and empirical findings. These "hypotheses" point toward the *phenomenological* realm, though Hacking eschews this adjective because of its links to a specific philosophical tradition. This intermediate category between theory and individual empirical findings strikes me as particularly important, and it is no surprise to also find other sophisticated approaches targeting it. David Gooding (e.g., 1986), whose notion of "construals" bears strong resemblance to Hacking's "topical hypotheses," deserves particular mention and is discussed further below. Even authors who don't embrace new experimentalism felt the need to further differentiate their notion of theory. Martin Carrier (1998, esp. 182ff), for example, has revived the older category of "observation theories" mediating between "raw data" and theoretically significant magnitudes. It is significant that a more detailed differentiation of epistemic categories has proved particularly urgent in the study of experimental practice, even among those who wish to uphold the primacy of theory.

In relative isolation from the international debate, a philosophical approach initiated in the 1920s by Hugo Dingler (1928) has found new resonance in the German-speaking world. Starting from the insight that experimentation is a rule-governed activity, this approach sought a normative account of the basic concepts of both science and daily life in terms of elementary actions. Under the name "proto-physics," this project was carried through for elementary concepts of geometry, time, and mass, motivating the conclusion that *all* concept formation follows this same pattern.[25] More recently, a similar attempt has been undertaken for the basic concepts of chemistry and biology (e.g., Weingarten 1997). Such efforts bring to the fore the relationship between concept formation and experimentation, which has been of such interest to me here, and place experimental agency in the center of its analysis. However, there remains a strongly normative, universalistic tendency, one that overrides the demand of doing justice to historical cases.[26] The complexity of research activity and the epistemic process can scarcely be grasped in this way.

We are still a long way from a nuanced philosophy of experimentation capable of grasping the complexities of experimental knowledge generation and taking into account the new perspectives opened by the history of science.[27] Deborah Mayo (1994, 271, 272) has, in a survey article, called for further research on precisely this question. Drawing on Hacking, she has compiled a list of the various functions of experimentation beyond the testing, confirmation, and completion of theories. These functions are the validation of instruments, exclusion of interference factors, estimation of precision, distinction between facts and artifacts, and the evaluation of the influence of background factors. I find this list well conceived but in important respects still incomplete. To begin with, it cannot capture the experimental practice of the early working phases of either Ampère or Faraday.[28] That would require deeper analysis, both of the epistemic processes taking place between the establishment of an individual result and the formulation of a complete theory, as well as of the assumptions that went into the formulation of the individual result in the first place. In the sections that follow I concentrate on these points, drawing on my historical cases.

Exploratory Experimentation

Experiments certainly can be determined by a theory in the narrow sense, that is, they can be conceived, designed, carried out, and evaluated within the framework of a theory already defined and established. Such instances need not always involve a *test* of the theory in question. For example, the experimental determination of certain numerical parameters is often guided by theory. Among numerous examples within the history of science, I cite only two, taken from the preceding chapters. Ampère's experiment on the attraction between two spirals (shown in fig. 3.7) was designed specifically for the purpose of demonstrating the existence of an effect predicted by theory and thus supporting that theory. When he failed to achieve the expected result with this experiment, Ampère optimized it with a view toward the desired effect. Biot's experiments and measurements (see figs. 4.9 and 4.10) were also entirely framed by his theoretical reflections. In his quest for a force law operating between microscopic elements, the general form of any such law had already been determined by the Laplacean framework prior to any experiment. It had to be a simple power law, and the sole purpose of the experiment was to determine the numerical value of the exponent.

Such cases illustrate what is often meant by talk of experiments as guided by theory, as, for example, in the work of Bas van Fraassen (1981, 673), who views experimentation as the prosecution of theorizing by other means. However, the view he and others have defended—that *all* experimentation is of this sort—strikes me as much too narrow and one-sided. The history of science is replete with examples of experimental endeavors whose course was largely independent of such theories. Hacking (1983, 159) speaks of "pre-theoretical experiment" and Gooding (1990a, 88), of experimenting "while the conceptual necessities of theory are held at arm's length." Such experimentation is by no means necessarily a matter of mindless playing around with the apparatus or merely a trial-and-error process, as Martin Carrier (1988) has suggested. To be sure, as active intervention, experimentation is always guided by thought and epistemic goals. But not all thought is theory, or at any rate such an understanding does not seem to me to be particularly useful, as I discuss in greater detail in the following section. The thoughts and epistemic goals that guide an experiment need not be theoretical in the strict sense. Sometimes, as when no remotely applicable theory is available, they simply can't be theoretical. There is a broad range of potentially action-guiding thoughts and epistemic interests between mindless playing around with apparatus on the one hand and properly theoretical reflection on the other.

My studies of early electromagnetism provide a rich source from which to explore this range in more detail. A procedure I have called exploratory experimentation emerges with particular prominence. The cases of Ampère and Faraday enable an elucidation of its characteristic features. In the first phases of their experimentation, both researchers strove to discover empirical regularities and to form the conceptual tools required for their expression. The principles guiding their experimentation were general and methodological. Most important are these:

- the systematic variation of a large number of different experimental parameters one by one, leaving all others;
- the establishment of stable empirical rules, typically of the form, "If particular conditions obtain, then this or that effect may be observed";
- the determination of which of these experimental conditions is necessary for the occurrence of the effect in question and which only modify it;
- the formation of adequate representational schemes that allow empirical rules to be formulated as generally as possible; and
- the development of experimental arrangements incorporating only those conditions necessary for the effect to occur and that thus allow the general rule to be expressed with particular clarity. Such arrangements are characterized as "simple," "general," "elementary," or "pure" cases. They serve as the core to which a multiplicity of phenomena may be phenomenologically "reduced."

Exploratory research thus characterized differs in at least three important respects from research determined by theory in the strict sense. First, there is its *epistemic goal,* namely, the search for dependencies between experimental parameters and the establishment of laws. This goal is pursued by the systematic variation of parameters, under ceteris paribus conditions as broad as possible. Only if those attempts fall short may researchers consider the possibility that the failure might have something to do with deficiencies in the basic concepts of the field and thus feel encouraged to form and introduce new concepts that capture experimental findings under the smallest possible number of rules. Typical examples are Ampère's new concepts of "right and left of current" or his concept of a current circuit. In addition to establishing stable empirical rules (with or without new concepts), the exploratory enterprise often aims at a specific kind of "explanation" of a large domain of phenomena: explanation by "reduction" to such rules. In doing so, it engenders a hierarchy of phenomena in which "elementary" or "simple" phenomena are considered central, while others are understood as derived from them, as special cases or superpositions of elementary phenomena. I consider this form of explanation in greater detail below. In my many discussions of these issues with Lorenz Krüger, we dubbed such a hierarchy of phenomena a "systematic phenomenology."[29]

Exploratory experimentation—this is my second point—also has its characteristic features when it comes to *experimental practice* in the laboratory. The guidelines sketched above are rather general and nonspecific. They encompass a plethora of different experiments whose precise details are not determined by the guidelines, even if their direction is. Experimental work might begin even when no established conceptual hierarchy or basic framework for describing the phenomena is available. On the contrary, such hierarchies may themselves be established over the course of broad, systematic experimentation, and thus they typically appear only at the end of such work, not at the beginning. The situation is completely different in cases in which experimental work has been determined by theory. By the time Ampère came to consider possible consequences of his circuit hypothesis and had contrived experimental arrangements to test them, the design and especially the evaluation of these experiments required a well-formed conceptual scheme and a theoretical framework, if only a provisional one. When thus determined by theory, experiments are typically few in number and carefully targeted, remaining from conception through evaluation within the bounds set by well-formed concepts and theory. In such cases there is no trace of the wide-ranging collection and classification of findings found in exploratory episodes.

The third significant difference between exploratory and theory-guided experimentation concerns the character of the *instruments and apparatus* employed. As an illustration, consider Ampère's "astatic needle" (see fig. 3.1), with which he demonstrated the alignment effect. The instrument made possible a wide range of experimental variations, especially with regard to the orientation of the connecting wire. The constraints imposed by the apparatus itself were relatively loose, leaving room for a large range of possible experimental outcomes. Much the same holds for most of Faraday's experimental setups, but Ampère's apparatus for demonstrating the attraction between two spirals (see fig. 3.7) had a different character. This arrangement was so specifically geared toward the demonstration of a particular effect as to leave hardly any room for variations in the experimental conditions, except for optimizations. Consequently, while this apparatus worked extraordinarily well in the search for and demonstration of the expected attraction and repulsion between spirals, this was the *only* effect for which it was suited. A rotatory or lateral movement of the spirals, while not expected, could not be ruled out beforehand, could not have been detected with this equipment. The specificity of the experiment, as proposed by theory, was reflected in the specificity of the instrumentation, obtained at the cost of a significant loss in flexibility and openness to unexpected findings.

The term "exploratory experimentation" is not original to me. As Rose-Mary Sargent (1995, chap. 7) has noted, Robert Boyle used that term for all experiments that were designed to gain knowledge about the world, as contrasted with "probatory experiments" that sought a better understanding of the experimental apparatus and its functions. Gooding (1990a) spoke of exploratory experiments in a general sense to point to experiments that were not essentially framed by existing theory, but he left the concept very open. Given that background, it was indeed a much more specific concept of "exploratory experimentation" that Richard Burian (1997a) and I (Steinle 1997) proposed in 1997, independently of each other. Both proposals were based on in-depth analysis of historical experimental practice. Burian developed his concept over the course of a detailed historical study of an episode of histochemical research on protein synthesis, conducted by the Brussels researcher Jean Brachet and his group in the 1940s, an episode that yielded a first map outlining the basic elements of those cellular processes important for protein biosynthesis. While Burian's concept of exploratory experimentation is more open than mine, both emphasize the difference between exploratory and theory-driven experimentation and highlight the point that exploratory experimentation is not random trial and error but constitutes a systematic procedure.

Theory, Law, Concept, and the Language of Research

The distinction between exploratory and theory-driven experimentation may appear naïve. After all, it is widely taken for granted within the philosophy of science that *all* experimentation, indeed all observation, is theory laden, as was compellingly asserted by Karl Popper ([1934] 1969, §30), if not earlier. But this claim can be understood in a number of different ways, giving rise to frequent misunderstandings, and it is here that the aforementioned need to further differentiate the concept of theory becomes urgent. In one interpretation, the claim simply asserts that, in experimenting, one is always thinking something.[30] While hardly anyone would want to dispute this weak claim, it is neither especially instructive nor helpful for further analysis. Only when we begin to ask what sorts of thinking are involved does the thesis become more interesting. There is also a strong interpretation of the theory-ladenness thesis, in which experimentation is guided by sophisticated scientific theories, on the level of Hendrik A. Lorentz's electron theory or quantum mechanics. More precisely, in this interpretation experiments are always *test cases* for such theories, or at least for the expectations derived from them. While questions of whether experiments always work as tests, or whether such tests are even in principle possible, have been considered extensively, the central question from my point of view is what might reasonably be considered a theory.[31]

Ever since the seventeenth century, the concept of theory has been notoriously vague. One point that is perhaps common to the various understandings is the claim that theories should serve to provide explanations, which shifts the problem to that of articulating what exactly counts as acceptable explanation. A specific form of theory and explanation has been particularly prominent since the eighteenth century. It was widely employed in the nineteenth century and became of increasing importance in recent philosophy of science. Canonical examples include the various vortex, effluvia, or imponderable fluid theories of electricity, the corpuscular and wave theories of light, the caloric and kinetic theories of heat, and so on. Such theories typically strive to capture whole domains of phenomena by recourse to entities—be they light corpuscles or waves, electrical or magnetic fluids, central forces between pointlike particles, or the like—that lie behind the level of the phenomena themselves, no longer accessible to experience: theoretical entities in the traditional sense of the philosophy of science, or what Theo Arabatzis (e.g., 2006, 2011a) has called hidden entities.

Theories of this sort are not the only or even the most important object of scientific reflection. Of at least equal importance are the regularities traditionally called "laws" or "empirical laws" that specify the conditions under which a particular phenomenon occurs. Such rules typically take the form of "if-then" statements: if conditions A, B, C, and so on obtain, then effect D may also be observed. Most important for our purposes is the fact that A, B, C, and D are all observable conditions, at the level of phenomena. In physics, laws of this kind are usually expressed in mathematical equations. Though this adds the additional dimension of quantitative precision in the relationship among conditions, it does not change the underlying notion of law. These laws demand no recourse to theoretical or hidden entities. Science since the seventeenth century has offered up an abundance of such laws, often associated with the names of particular researchers, from Kepler, Snell, Galileo, Boyle, Hooke, and Newton to

Planck and Balmer or Paschen.[32] Many of these investigators either explicitly distanced themselves from the search for a deeper explanation by recourse to the micro level or at least deferred that search to a later stage. Through the end of the nineteenth century, the distinction between theory and law remained prevalent in the physical and chemical sciences and even in philosophy; Duhem retains it explicitly (Duhem [1908] 1978). After receding into the background in the twentieth century, it has been revived in recent debates in the philosophy of science, particularly in the context of attempts to gain a better understanding of experimentation. What I have called an empirical regularity falls under Hacking's category of "topical hypothesis" and Gooding's notion of "construal" but is significantly more specific than either.

However, the notion of a phenomenological regularity has its own problems, as the missteps of nineteenth-century empiricism taught us. It must not be taken in the naïve sense as referring to a kind of rule that becomes apparent simply by looking.[33] After all, the formulation of a law requires that we use language and presuppose concepts and categories capable of imposing a preliminary structure on the domain in question. The employment of concepts is by no means epistemically "innocent." Concepts and categories do not arise on their own; they are actively forged and shaped by the approaches of their time. The importance of systems of classification is well known from the history of natural history, as well as from several experimental sciences, such as chemistry. In the physical sciences, this point has not often been appreciated. But even such basic distinctions as William Gilbert's (1600, chap. 2) separation of electric and magnetic phenomena or the early eighteenth-century division of all materials into conductors and nonconductors (Heilbron 1979, chap. 8) were once *forged* by historical actors, thenceforth shaping all further research. Furthermore, as the examples make clear, the process of classification frequently goes hand in hand with intensive experimentation.[34] The formulation and stabilization of concepts in the course of experimental work is a remarkable process, one that deserves much more philosophical attention than it has received so far (Feest and Steinle 2016).

Categories, means of representation, concepts, conceptual schemes, empirical regularities, laws, theories, and theoretical entities: there are many diverse layers to scientific knowledge. "Epistemic things," as Rheinberger calls them, come about and can be created, shaped, stabilized, or even discarded, on wildly different levels. While there may be no way of drawing rigid boundaries between them, this does not mean that we ought to give such distinctions up entirely in favor of some all-encompassing notion of "theory." The resulting loss in our capacity to differentiate would make it difficult, if not impossible, to classify experimental efforts by reference to different possible epistemic goals or to appreciate the powerful extent to which such differences can work their way into the laboratory.

In particular, the distinction between theories, laws, and concepts strikes me as highly significant. The ways in which theories and concepts are formed, stabilized, and used differ markedly. Moreover, concepts, classifications, and laws have immense, and often greatly underestimated, epistemic significance. Once formulated and stabilized, they shape all subsequent research at a fundamental level. In their use, they regulate various manipulations and constitute the language in which research questions are formulated. The critical revision of this language, unlike that of theories properly so called, is no longer part of the everyday research enterprise. The formation, establishment, and stabilization of concepts and conceptual schemes establish the framework within which the direction of future research may range.

Exploratory experimentation typically aims at the level of laws, and sometimes that of concepts, but not of theories. My emphasis on this point has given rise to misunderstandings that illustrate the difficulties with the concept of a theory. For example, my view has been portrayed as asserting the existence of "theory-free" experiments (Radder 2003, 9, 161). But this description is both imprecise and misleading, as if saying that an experiment is not theory driven were the same as saying it is theory free—a leap that would immediately identify my view as a naïve, empiricist position. In the absence of a careful differentiation of the levels of knowledge, discussions of theory-ladenness of experiment will continue to produce misunderstandings of this sort and continue to be unproductive.

Finally, it should not be overlooked that such differentiations are themselves inherently historical in character. Even in the nineteenth century, the period in which even philosophers may have placed greatest emphasis on the distinction between empirical rules or laws and theories properly so called, some protagonists (if not all) were well aware that the distinction was not fixed but historically variable. One period's theory may well become a later period's unquestioned fact, empirical rule, or law. William Whewell (1840) clearly described and discussed such a process, one that in the later parlance of Husserl might be called sedimentation. Whereas Whewell optimistically viewed this process as proof of the progress of the sciences, in more recent historical investigations instances of such stabilization and sedimentation serve as instructive cases in our effort to gain insight into the factors at work in the evolution of science (Steinle 2010). To give one small example, the case of Ampère's introduction of the terminology of "galvano-magnetic" effects and his eventual transition to "electro-magnetic" effects illustrates how, over countless tiny steps, this process manifests itself in language use.

Language and the Constructive Role of Experimentation in the Works of Mill and Gooding

The constructive role of experimentation in the emergence of concepts and theories was quite prominent in nineteenth-century empiricism. In some respects, my account of exploratory experimentation may recall John Stuart Mill's detailed reflections on experiments. Indeed, revisiting them will help to sharpen my own views. The systematic variation of factors plays a central role in exploratory experimentation, especially in the establishment of correlations and laws. This same process is central to Mill's account. The four experimental methods he works out in detail all involve the targeted variation of experimental arrangements, geared toward the analysis of experimental conditions, with the goal of determining which contribute to the occurrence of an effect, either as necessary or as modifying conditions. Mill (1843, Book III, chaps 7 and 8) speaks of establishing "causal relations."[35] This point plays an equally important role in my account of exploratory experimentation.

There is, however, one significant aspect that cannot be found in Mill. He treats the language and concepts with which empirical findings and regularities are expressed as stable and given. He is well aware of the necessity and significance of having a conceptual scheme with which to structure a given domain of phenomena, describing the process by which it comes about as a kind of contemplation, which must always be undertaken before experimentation begins. He does not consider the possibility that the concepts used in conceiving experiments and formulating their results might themselves be subject to revision during experimentation. This prospect, however, has emerged as a significant feature of the experimental work undertaken in the formative periods I have analyzed here. The study of experimentation must also take situations into account in which hierarchies and classifications remain unstable, remaining open to question in the course of experimental work. This is the decisive point on which my investigations go beyond older empiricist approaches, opening up fundamentally new perspectives. In this respect my view of exploratory experimentation also goes beyond that of Gerd Graßhoff, who has otherwise offered a superb analysis of the role of experimentation in establishing causal relations (Graßhoff, Casties, and Nickelsen 2000 (Graßhoff 2011).

Twentieth-century philosophy has been much concerned with the constitutive role of language and concepts in our comprehension of the world, and with good reason. The unquestioned category of the empirical finding, not to mention that of the empirical rule, so firm and certain within the empiricist tradition, began to waver, revealing itself as far more complex and problematic than the logical empiricists had imagined. Now, thanks to studies of experimentation, an additional factor, generally neglected, has emerged with particular clarity. Not only can an empirical finding as such not arise without our concepts and language, but, in the course of experimentation, it is *called forth* in the first place by our actions, our active engagement with the material world. And the experimental intervention through which this comes about is guided by thought, which involves concepts and is formulated linguistically. But when these concepts are not stable and language itself is open to revision, the epistemological situation is precarious. Action-guiding concepts and action itself are intimately connected and must, thus tightly entangled, mutually stabilize or destabilize each other. These are precisely the processes at work in exploratory experimentation, which is one important reason for the preeminent importance of this practice. The resulting view no longer bears much resemblance to the traditional empiricist picture.

We are far from a detailed understanding of the process by which concepts are stabilized in the course of experimental activity. Though the New Experimentalism has acknowledged the problem, it continues to receive little attention. The most thorough efforts to date have doubtless been those of the late David Gooding, the title of whose book, *Experiment and the Making of Meaning,* describes a research program. Gooding (1986, 219n36) stresses that the early stages of empirical investigations in the natural sciences are far less "a matter of trial and error and intuition" than they are often presented to be. This point, which he so aptly expresses, was also the point of departure for the work presented in this book. As an instrument for the study of the early phases of research, Gooding proposes the concept of "construals": these are early expositions of and "interpretive possibilities" for new phenomena, provisional in nature and wide open to revision. Of particular importance is his insistence that this first approach take place without attention to theory: construals "may be compatible with several theories or with none." One of their functions is to permit a firm grasp on experimental results, but they have a heuristic function of at least equal importance, that of giving rise to "ways of modifying and exploring the phenomenal domain" (Gooding 1986, 219).[36] Applied to the early phases of research, such descriptions are extremely apt.

However, Gooding's (1986) analysis does not go far enough in my view. He describes the function of construals as being "to enable an ascent from the immediate and concrete world of the laboratory to a world of representations (words, images, symbols)" (215). While this indicates a clear direction, it remains unspecific. Gooding views construals as the first step in the "cognitive processing" that leads, in four stages, to the formulation of theories: construals, interpretation, definitive interpretation, and exemplary or paradigmatic interpretation (219). But he presents the first step as essentially a transitional phase, paying too little attention to its significance.[37] He thus misses essential points, such as the fact that the first step is about the discovery of regularities and that, toward that end, preliminary concepts and means of representation are formed. Gooding hence leaves the domain of exploratory experimentation largely unanalyzed. As a result, and though Gooding's work rests significantly on his study of Faraday, there remain serious gaps—he nowhere considers, for example, Faraday's central project of reduction to simple cases. At the same time, his general reflections miss out on an opportunity to work on the gap he himself so sharply diagnoses, by studying the very first phases of experimental research in much greater detail.

Reduction to Simple Cases and the Power of a Systematic Phenomenology

As the cases of Ampère and Faraday make clear, there are enormous differences in intensity and sustainability between episodes of exploratory experimentation. While Ampère acknowledged the epistemic goal of reduction to "general facts," he made only sketchy gestures toward pursuing it. By contrast, in the case of Faraday, for whom exploratory experimentation became a central, long-term endeavor, the pursuit and attainment of this goal may be traced in considerable detail, through successive concrete examples. That is my task in this section (see also Steinle 1995a). In Faraday's own writings we find only scattered remarks on methodology. Like many other scientists, he practiced specific procedures and had particular explanatory goals, while only rarely articulating them in explicit terms. In what follows I offer a reconstruction of his explanatory approach derived from my analysis of his research practice.

The terms "reduction" and "reductionist" typically refer to modes of explanation in which particular phenomena are supposed to be explained by recourse to other phenomena of a different kind, or to hypothetical ("theoretical") entities not themselves on the phenomenal plane. For example, we speak of optical phenomena as reducible to electromagnetic waves, of heat, to the movement of tiny particles, or of chemical phenomena, to physical effects. The notion of reduction being at work (and explicitly used) in Faraday (and Ampère) is completely different: these reductions take place entirely within one and the same domain of phenomena. Both *explanans* and *explanandum* are phenomena of the same kind. Electromagnetic motions are reduced to other electromagnetic motions and not to anything qualitatively different.

This explanatory approach involves a hierarchy of phenomena *within* the field in question (electromagnetic motion): while some phenomena demand explanation, others serve as the ground of such explanation without themselves being further reduced. Terminologically, this hierarchy is marked by the distinction between "simple," "elementary" (Faraday), or "general" (Ampère) phenomena versus those deemed "complicated" or "compound." The latter are to be "reduced" to or "deduced" from the former.

The criterion for distinguishing elementary from derived phenomena refers also to the phenomenological-experimental domain. By itself, Faraday's talk of "simplicity" offers little, remaining open to very different interpretations. But analyzing the way he actually uses the category makes it clear that, in the context of his reductive enterprise, "simple" phenomena are distinguished by the fact that they involve only the minimal set of experimental conditions that are necessary to produce the defining effects of the relevant domain—electromagnetic motions. Faraday regarded the mutual rotation of wire and magnet pole as the "simple case" of electromagnetic motion because this effect involved the smallest number of experimental factors: only *one* wire and *one* magnet pole. Neither factor could be dispensed with without eliminating electromagnetic rotation altogether. Other kinds of electromagnetic motion always involved more complex experimental arrangements, demanding more experimental factors—more wires or more magnet poles. Such cases were considered complex and thus deducible from the "simple case."

This minimalist condition for "simplicity" is nontrivial and not always easy to grasp. In the case of electromagnetic rotation, the purely numerical restriction to only one wire and one magnet pole is insufficient. That condition is met not only by the setup in which a pole rotates about a straight wire but also by one in which a wire loop is used instead. The two differ geometrically, and Faraday does not consider them comparable. There must thus be further criteria, never stated explicitly, by which various geometries may be arranged in order of simplicity. One of those criteria is obviously the general principle, demonstrated by experiment, that distance plays a decisive role: what is closest exerts the strongest action. Of even greater importance is another insight well established for electromagnetism, that the orientation of the wire is crucial. In the case of a straight wire, only short segments, all oriented in the same direction, prove efficacious. In a loop, however, all segments are equally efficacious, but all have different orientations. And indeed, Faraday concludes, the loop setup can be "reduced" to the case of the straight wire, thus explaining even such aspects as the greater intensity of motion at the loop's center. This reduction does not work the other way round, and thus the arrangement with the straight wire is the simple one, not the one with the loop.

Given this criterion, the "simple case" displays the conditions under which electromagnetic motion occurs with particular clarity and serves as a kind of direct manifestation of the law underlying such motion. Indeed, Faraday (1821d, 77–78, ¶¶9, 10; 1839–55, 2:130) describes the simple case of the rotation of wire around a magnet pole immediately prior to stating the law governing that motion: the *speed* of rotation decreases as its radius increases (where radius is in turn determined by such parameters as the length of the wire), while its *direction* is given by magnet and battery polarity. Faraday's visual illustrations of these dependencies (see his Figures 4 and 5 in our fig. 6.5) are abstracted from the experimental setup of the "simple case." The precise description of the simple case and the dependencies obtaining in it is equivalent to a statement of the general law. In place of reduction to the "simple case," Faraday sometimes also speaks of a reduction to the "simple law" of electromagnetic motion: "The law of revolution and to which all the other motions of the needle and wire are reducible is simple and beautiful."[38]

A phenomenon is "reduced" to the simple case when it is understood as either a modification of that case or as a superposition of such cases. The rules whereby the simple case is modified by the addition of additional conditions, or by which its instances are superposed, must be established experimentally. So Faraday considers, for example, what happens to the motion of a magnet pole when it is acted on, not by *one* wire but by several wires, either with the same or opposing direction of current. He first determines that its motion is amplified under the influence of several wires through which current flows in the same direction but is not qualitatively changed. Under the influence of several wires with opposite directions of current, however, it exhibits a qualitative change that can be grasped in a rule. Taken together, these two results yield a rule applicable to a wide range of multiwire experimental arrangements, tracing them all back to the simple case. Between the simple case and any given phenomenon, he can stipulate a chain of intermediate effects, demonstrating the deducibility in incremental steps. Ideally—and this reveals its phenomenological character—reduction would proceed by actually producing every intermediate effect in the chain.

A claim regarding the simplicity or complexity of a given effect can be made only in the context of a larger field of phenomena. So long as the effect remains isolated, there is nothing to learn. When Faraday produced the rotation effect for the first time, he had no way of knowing that this would be the effect that proved central, or simple, for the entire field. One important incentive for the vast experimental effort he made immediately after his discovery of the rotation effect, with a view toward scouting out its "surroundings," was his need to learn more about its status and significance. Only with those results in hand was he in a position to assign the effect a particular place within the larger field. Over the course of the efforts that made this possible, he had structured and organized that field. As a consequence, his article presented a multiplicity of different phenomena together with the hierarchical structure he had developed for articulating the relationships among them. The notion of a "systematic phenomenology" refers to precisely such a network of phenomena in which multiple regularities are knit together.

This kind of structure points to a specific kind of explanation, one strikingly different from explanation by recourse to the invisible level as described above. In the mode now outlined, the question as to why a given effect occurs in a particular way will be answered by pointing toward a particular modification of the simple case, or to the superposition of several instances thereof. Ampère's explanation of the behavior of the needle under the influence of a vertical wire as the superposition of his two "general facts" is one example, as is Faraday's deduction of the behavior of the coil or the alignment of a wire loop from the "simple case" of the rotation of a wire about a magnet pole. In such cases, to explain an effect is to situate it within a structured domain of phenomena.[39]

A systematic phenomenology also has predictive power. With the background of an established network of empirical rules, it is possible to envision what will happen in a particular experimental arrangement. Ampère explicitly laid claim to such predictions, and Faraday's reflections on the rotation evoked by terrestrial magnetism also serve as a telling example. He went beyond regularities observed in the laboratory, extrapolating from them to an as yet unknown case of an entirely different scale and spatial configuration. Such predictive power is also important in a common sort of technical application, one in which the goal is to produce an effect desired but as yet unobserved. The network of rules might yield an instruction to improve one set of components, remove another, and so on. This kind of reflection was of great significance in the work of the Paris instrument builder Antoine Hippolyte Pixii, son of Nicolas Constant Pixii, when in 1832, a few months after Faraday's announcement of the discovery of electromagnetic induction, he built the first electromagnetic generator. At the time, nothing remotely resembling a "theory" of this effect was on the horizon. What Faraday had provided was a set of rules and laws describing the experimental conditions under which the induction effect took place and in which it was amplified or weakened. For purposes of optimization toward a particular goal, that boundary-setting structure was both necessary and sufficient.

In addition to enabling reliable handling of the phenomena and their technical applications, the importance of systematic phenomenology to the pursuit of research resides in its establishment of a stable and empirically grounded conceptual foundation. It provides the language on which subsequent research will rely—a most fundamental epistemic achievement. Moreover, and in many cases, theories have been devised only when experimental and conceptual conduct in the field has been stabilized in this way. The fact that concepts are stabilized in the context of experimental action has a further significant side effect: the conceptual groundwork thus established is useful for both theory building and technical application at the same time. The example of Pixii's electromagnetic generator is particularly telling.

Exploratory Experimentation in the Sciences

Exploratory experimentation is much more common in the history of science than has previously been acknowledged. A few examples from the history of electricity may serve to illustrate this point. When Charles Dufay turned to the study of electricity in 1730s Paris, his declared goal was to establish laws for electrical attraction and repulsion, thus bringing order and clarity to a domain of phenomena that had previously seemed irregular. This endeavor was similar to his earlier work on the phosphorescence and luminescence of Bolognese stones (Daston 1997). His experimental approach was shaped by those efforts and involved the systematic, broad-based variation of multiple factors, especially materials. In the end, he succeeded in formulating a law only by introducing a new conceptual framework. He proposed replacing the *single* electricity of previous accounts with *two* electricities, which he called "resinous" and "vitreous," with wax and glass representing two classes of materials that exhibited the two electricities. By using terminology so immediately bound to experimental manipulations, he underscored his avowed intention to avoid committing his distinction to any micro-scale theory of electricity. His new framework allowed him to organize a vast number of experimental effects in a perspicuous way, as well as to formulate regularities (Heilbron 1979, chap. 9; Steinle 2006). This virtue explains its relatively swift universal adoption.

As a further example, consider the work carried out across Europe in reaction to Galvani's spectacular if irritating discovery of "animal electricity" in the 1790s. Maria Trumpler (1992, chap. 2; 1997, 1999) has systematically investigated the experimental efforts of Pfaff, Humboldt, and Ritter, characterizing them as verification, simplification, optimization, and variation. All of these processes may be understood as instances of a characteristic approach to exploratory experimentation, as described above. Significantly, all of the researchers discussed by Trumpler were engaged in the development of adequate means of representation, with Ritter attaining particularly spectacular innovations. The same applies to the numerous experimental endeavors throughout Europe provoked by Ørsted's discovery of electromagnetism (see chap. 2); most had an entirely exploratory character.

In both intensity and duration, Faraday's exploratory efforts were highly unusual. In addition to the case of electromagnetic rotation discussed earlier, his 1831 investigations of electromagnetic induction loom large. The newly discovered induction effect presented a range of extremely confusing features, and Faraday spent months struggling to formulate a rule capable of capturing a large number of experimental findings. His eventual success turned on his introduction of a new, unconventional means of representation: the "magnetic curve." This step, which he took almost reluctantly, would prove in retrospect to have been one of the most important milestones along the path to field theory (e.g., Steinle 1996; Romo and García Doncel 1994). Similarly fundamental shifts may also be found in his work on electrostatics, as in his concept of electrical charge (Gooding 1985).[40]

Julius Plücker's studies of electrical discharges in rarefied gases are yet another example of exploratory experimentation. With absolutely no theory, he formulated many phenomenological regularities, in the process developing the concepts that would shape all future work in the field (Hiebert 1995). Recent studies of Heinrich Hertz's work on high-frequency electrical oscillations in the 1880s have shown that these efforts also had a fundamentally exploratory character, long before they culminated in the discovery of electromagnetic waves, and that in this context Hertz repeatedly felt called upon to revise even long-accepted concepts (Buchwald 1994). Even the history of a thoroughly mathematized field like optics provides noteworthy examples. Étienne-Louis Malus's (1808) discovery of optical polarization initiated a wide range of experimental studies involving polarized light, whose reflection and refraction displayed baffling features that eluded the grasp of received concepts. In David Brewster's systematic and highly successful pursuit of regularities, the fact that he was a committed adherent of the corpuscular theory of light played no role. As Hacking (1983, 157) pointedly puts it, Brewster studied "how light behaves." In the process, he deferred all questions concerning the relationship between the established regularities and his theory to a later date. In the end they would prove to be the very laws that revealed the inadequacy of the corpuscular theory, allowing the wave theory to prevail.

Turning to still another field, in the early twentieth century, as Klaus Hentschel (2000, 339) has shown, spectroscopists concentrated on "phenomenological classifications," while setting aside "theoretical models" when confronted with inexplicable wavelength shifts in gas spectra. A bit removed from the scientific mainstream, the characteristic features of an exploratory approach may be found in Goethe's work on color. His reflections on the role of the "primordial phenomenon" (*Urphänomen*) expressed these features fairly explicitly.[41] He was, after all, concerned with establishing a new or alternative *conceptual* foundation for the field. In Goethe's case, as in the others I have mentioned, the idea of a few exemplary phenomena played an important role. Ampère called them *faits généraux* and Faraday, "simple" or "elementary" cases. Philip Lenard (1906), who was notorious for later activity in what was called "German" or "Aryan physics," discussed his own concept of "pure" or "simple" cases quite thoroughly.

From chemistry and microbiology, I shall mention three especially noteworthy cases. When the new field of organic chemistry branched off in the 1830s, it struck many as a "jungle" or "boundless tangle."[42] Once again, the very first forays into this unknown field were generally exploratory in nature, focusing on the systematic variation of individual parameters and the search for dependencies, even without theoretical reflection. New means of representation and concepts had to be crafted. The work performed during this period by the likes of Justus von Liebig and André Dumas provides a compelling illustration of the way in which means of representation developed during experimentation, or what Ursula Klein (1998) has called "paper tools," imperceptibly evolved their own internal dynamics. For the twentieth century, we need only recall the work of Hans Krebs, which led to the discovery of the urea cycle in 1935. Recent studies of his laboratory practice have shown that at several decisive points in his research there were periods in which he systematically varied one particular parameter, such as a substance, with no other goal but the discovery of potential correlations or dependencies (Holmes 1993a; Graßhoff, Casties, and Nickelsen 2000). Finally, as mentioned above, the molecular biologist Jean Brachet intensively engaged with questions concerning protein biosynthesis. Analyzing that episode, Richard Burian has proposed to describe the special role of experimentation in this work by calling it "exploratory experimentation," with many characteristics similar to those I have stated above. Burian's (1997b, 27, original emphasis) insistence that "the style of exploratory experimentation . . . *should* be of great historical and philosophical interest" captures perfectly well my motivations in writing this book.

As such cases illustrate, exploratory experimentation is hardly restricted to less-developed sciences but may well occur even in more recent periods. My final case highlights this point. In 1986, the physicists J. G. Bednorz and K. A. Müller, at the privileged IBM laboratory at Zurich, announced their discovery of a superconducting material with a critical temperature of 35 K, a temperature higher than anything allowed by theory. Quickly, the new research field of high-temperature superconductivity (HTSC) became established and developed with dramatic speed, not the least because of the enormous technical applications it brought into perspective. At the same time, theory could not help much in directing experiments, and as a matter of fact much of the immense experimental effort had (and continues to have) an exploratory character, with attempts to find rules and laws, along with concepts with which to formulate them (Holton, Chang, and Jurkowitz 1996; Di Bucchianico 2014). The fact that, despite the still murky theoretical situation, the first large-scale commercial application of HTSC was recently launched in the German city of Essen strikingly illustrates my claim that such experimentation provides foundations for both science and technology.

Exploratory experimentation is a modus operandi to be found throughout the natural sciences. It is practiced in various periods and in various fields, from early modernity to the present, and from physiology to chemistry and physics. Exploratory periods are typically devoted to the study of effects or problems for which there is no well-formed theory available, and even the traditional conceptions of the field may be called into question. Linguistic and conceptual foundations formerly thought secure may be questioned and made subject to revision. In such periods, exploratory experimentation strives toward a preliminary orientation and organization, removed from any theory. It is often in the context of just such exploratory efforts that fundamental shifts in perspectives take place and new concepts and means of representation are developed. Exploratory phases thus play a decisive role in the evolution of any field. When Michael Heidelberger (1998) stresses that experiments serve a kind of "generative function" in the formative phase of concepts and theories, I believe exploratory experimentation will prove one of the most important processes in the performance of that function.

Differentiating Exploratory Experimentation

Since its introduction, the idea of exploratory experimentation has been taken up by others and developed further, based on historical case studies, mainly in the biological sciences. Beyond noting the richness of studies performed, I shall highlight three significant general issues.

The first concerns the rejection of a theory-centered vision of scientific experiments. One question is whether exploratory experiments should be understood in the mode of a radical Baconianism, which constructs all of science—so to speak—from the bottom up. Countering this reading, Laura Franklin (2005) distinguishes between a theoretical background—which can direct or constrain an experiment—and a local hypothesis under test. Drawing on case studies from molecular biology, she argues that scientists will refer to experiments as exploratory when there is no theory in the latter sense but that this does not rule out the existence of theories in the former sense. Much in the same vein, Koray Karaca (2013), analyzing modern particle physics, differentiates between "theory-driven experiments" and "theory-laden experiments." While exploratory experimentation is always, and inevitably, a theory-laden effort, it is so only in what he calls the weak sense of theory-ladenness, while theory-driven experimentation is theory laden in a strong sense.

Second, further historical studies (Burian 2007; Waters 2007; Peschard 2012) have led to the articulation of a variety of forms of exploratory experimentation. In order to deal with that variety, Maureen O'Malley (2007) points to the choice between introducing further types of experimentation and widening the notion of exploratory experimentation. Kevin Elliott (2007) goes for the second option, answering the ensuing question of what keeps the family of exploratory experimentation together by providing two criteria: the epistemological criterion of not being theory driven and the methodological characteristic of having variation of parameters as central procedure. With those in mind, Elliott provides a full scheme of characteristics for exploratory experimentation. To his two criteria, I might add two further points: the flexibility and openness of the conceptual framework dealt with, as well as the characteristic that, while individual experiments carry little weight in exploratory experimentation, it is chains, series, or networks of experiments that lead to conclusions.

Third, the question of concept formation has repeatedly been addressed, and indeed I have characterized this point as one of utmost interest, not the least since it is here that the concept of exploratory experimentation definitively exceeds the bounds of all traditional accounts. It includes at least two challenges, however. The epistemological challenge lies in the observation that there is no experiment possible without concepts; concepts guide experimental activity at every step. But how can the very concepts that framed the work from the outset be revised in the course of experimentation? Metaphorically, how can one navigate in unknown territory after realizing that available navigational instruments do not work properly within that territory? This epistemologically critical situation poses a challenge that has yet to be adequately addressed. One possible angle would be to pursue Gooding's (1990a) approach of finding new ways to account for the role of agency in experiment.[43] Some recent efforts (e.g., Feest and Steinle 2012) have also taken up the issue. The other challenge is related but concerned with the dynamics of scientific research. As all studies to date have shown, revising existing concepts and creating new ones is a risky process: it carries the potential and the risk of calling into question much prior knowledge. There have been only a few analyses of the typical epistemic challenges that drive a researcher to take that risky path. When researchers move into a new domain with the goal of discovering its structures, they necessarily begin with the existing conceptual framework and hope that a structuring of the new field can be attained within that framework. Only if this proves intractable (a point that is difficult to decide), some (certainly not all, since there are other options, such as leaving the question open or waiting for better options) muster the courage to open the conceptual toolbox for revision. This uncomfortable situation is usually not sought out in the first place: in historical case studies performed to date, we do not see researchers starting out with the goal of creating new concepts. The historical observation that conceptual revisions often occur in the course of exploratory experimentation should not mislead one to infer that this was the initial goal (Steinle 2009, 2013). Concept revision is rather undertaken as a reaction to an epistemic challenge, and we still need to better understand the dynamics of that process.

A major recent challenge concerning exploratory experimentation is the question of Big Data: How can exploratory work take place in fields that come with large amounts of data, what forms can it take, and what new possibilities and restrictions arise? These topics have been partly addressed by L. Franklin (2005) for genomics and by Karaca (2013) for high-energy physics, but further analysis is called for, especially since the whole question of the characteristics of data-driven science has so recently come into focus (Leonelli 2014).

The Epistemic Significance of Exploratory Experimentation

Advocates of the received view on the role of experimentation may well agree that experiments serve a generative function. But this view could still contravene the epistemic significance of exploratory experimentation, operating as it does in the *generation* of concepts and theories. And according to the well-established distinction, articulated by Reichenbach, between "context of discovery" and "context of justification," from the epistemic point of view, that which contributes to the generation of theories is neither analyzable nor relevant.[44] In this view, the stable progress of science takes place only in the context of rigorous justification, in which competing theories confront each other and their contest is decided largely by targeted experiments. Although this crude picture of the evolution of science now has hardly any serious defenders, the context dichotomy continues to shape many debates, especially in the philosophy of science.[45]

One central point that has been overlooked in the context distinction is the fact that no theory can be formulated without a language, or more specifically, without a rigorous and, in most cases, highly specialized conceptual framework, one that is held to be stable throughout discussion and no longer subject to revision or debate. But this kind of conceptual scheme is not just there as a sign. It had to be proposed, shaped, and stabilized to the point where it could be used without itself remaining an object of critical scrutiny in its own right. Edmund Husserl's notion of "sedimentation," which he introduced in the context of his reflections on the role of language in science, strikes me as particularly apt.[46] A conception thus established can continue to shape the research on a particular domain for a very long time. For some examples, we need only recall Newton's concept of absolute space, or of a uniformly refrangible ray of light, Dufay's two electricities, or the concept of a chemical reaction. The processes by which concepts and basic classifications form and are stabilized are anything but epistemically "innocent" in the sense that they would set no bounds on future research. On the contrary, the process of stabilization imposes far-reaching constraints. On the one hand, the establishment of a firm conceptual scheme makes the choice of a particular research direction possible, while on the other, it rules out other directions. Research questions can be posed only with particular concepts. In the context of another conceptual scheme they may well fail to make sense, in which case they elude attention. It is important to note, however, that such frameworks are not fixed once and for all but may change, even drastically, though on long time frames, as illustrated by the cases of magnetic poles (Steinle 2012) and absolute space.

For another example, consider the contrast between Biot and Ampère in their first engagement with electromagnetism. Biot remained squarely within the conceptual foundation of the Laplacean tradition, which he believed secure. For him, the concept of a purely distance-dependent central force acting between pointlike centers was not up for discussion. It served as the basis for his theory of electromagnetic interaction, on which the qualitative multiplicity of electromagnetic effects was simply of no interest. By contrast, Ampère faced this multiplicity, took a step back, and remained open, at least in his first, exploratory phase, to the idea of other modes of action. The concept of a single circuit of current, forged during this phase, quickly became public property. Later, when Ampère reached back to the traditional conceptual scheme, it was only by remaining open to revising basic concepts that he was able to consider the prospect of central forces acting between *straight segments.* Highly unusual though this prospect was in 1820, a mere ten years later it was entrenched to the point of being self-evident.

I have called periods of fundamental conceptual formation or reorientation formative phases. The choice of research direction to which such a phase gives rise is not always a conscious, carefully considered decision. It may involve a diverse range of important contributing factors, including peculiarities of biographical background, theoretical tradition, and scientific culture. The processes involved are also fundamentally social, as Ludwik Fleck ([1935] 1980) stressed in introducing the terms "style of thought" (*Denkstil*) and "thought collective" (*Denkkollektiv*). Fleck's notions (which served as important inspirations for Thomas Kuhn's concept of paradigm) refer to the linguistic and conceptual tools of a research field, the very tools taken for granted by the actors themselves and thus no longer considered open for revision. Disregarding these processes of formation, stabilization, and sedimentation, as has been traditional, means overlooking important aspects. The context of justification, as traditionally understood, is confined to the study of arguments and debates that themselves can take place only on already solidly structured conceptual terrain that—to remain in the metaphor—no longer permits arbitrary free motion but, by its already prepared structure, preferentially fosters motion in certain directions, while making other directions more challenging or impossible to pursue. Even if, contrary to indications garnered from scientific practice, there has somewhere and at some time been a pure context of justification, it would have rested on important epistemic decisions made in the past. From an epistemic perspective, the process of the formation and stabilization of concepts, in all its historical and social complexity, is of utmost interest and significance. For the study of this process, a distinction between discovery and justification is of little value (Steinle 2006).

If, as is often but not always the case, such processes take place in the context of experimentation, this experimentation typically has an exploratory character. Concepts and means of representation are devised and stabilized, and sometimes destabilized, modified, or discarded, in the constant interplay of conceptualization and experimental action. Concepts become especially interesting when they enable stable engagement with a field of phenomena by, for example, allowing the formulation of a regularity that captures many individual cases. Experimental action and conceptual development are intimately intertwined, and exploratory experimentation plays a key role in the process of forming and stabilizing concepts and conceptual schemes in establishing the language of research.

How Exploratory Phases Disappear

Given that exploratory work is both highly significant and very widespread, we must ask why it has heretofore received so little attention. One significant reason was doubtless the separation of context of discovery from context of justification, which excluded the realm of theory and concept generation from systematic analysis and thus stimulated a bias against it. The philosophical authors of the distinction were certainly concerned with science and, most of all, with the particular science that struck them as most advanced—physics.[47] The fact that they nonetheless disregarded such important aspects has much to do with the self-image that the sciences have generated. The discrepancy between this self-image and actual research practice, which I have discussed in my introduction, strikes me as playing a decisive role.

The self-presentation of science has typically given very little space to exploratory experimentation. Most of the aforementioned examples are known in detail only from historical studies of laboratory research practice, not from the published works of the authors themselves. This holds even for very productive exploratory periods. We need only recall Ampère's important concepts of "galvano-magnetic" effects or the circuit of electrical current: they were based on exploratory experiments, but in Ampère's publications there was no remaining trace of that background. And this is merely a particularly compelling example of a widespread phenomenon. Quite often only the *product* of an exploratory phase comes to the surface in the form of new or modified concepts, which are simply used without comment. On other occasions, we find authors relaying an anecdote of the particular moment at which a concept simply occurred to them. Famous examples include Newton under the apple tree, F. A. Kekulé's perceiving the benzene formula, or Carl Correns's (re)discovery of the Mendelian laws during an early morning awakening. True as these narratives might be, it must be recognized that they shift focus away from the fact that such insights take place only after careful preparation. A lengthy process in which various possibilities are tried out and discarded or stabilized and in which thorough, systematic experimentation takes place typically precedes them (Holmes 1981). The fact that the personal insight into the adequacy and solidity of particular conceptions and the inadequacy of others rests on the foundation of such longer-term processes is all too easily obscured. There are some exceptions, of course. In his *Astronomia Nova*, for example, Johannes Kepler led his reader through all the reflections and wanderings in his quest for regularities in the orbit of Mars—an exploratory quest in the theoretical realm.[48] But this kind of self-presentation is certainly unusual. Even in the absence of a comprehensive historical study of the conventions of self-presentation, it strikes me as plain that, in most cases, exploratory experimentation simply vanishes.

As to the background of that disappearance, we may discern three interconnected factors, all of which lead to the point that, in exploratory phases, deep-seated epistemic elements—classifications and concepts—are objects of study and potential revision. First, it is impossible to give an argument, properly so called, in support of the adequacy of a concept or language. Concepts are rather a precondition and important component of arguments. A particular vocabulary can be neither true nor false but only adequate or inadequate, useful or useless. And the criteria for adequacy and solidity are highly variable, depending on the specific goals of a given research activity. In many cases, the stabilization of action and manipulation plays a decisive role. If the research is directed toward the search for laws, the solidity of a conception is demonstrated by showing how it allows for a broad range of phenomena to be captured. In that case, researchers typically present the results of their exploratory work by couching their laws in the terms of the language they have just developed, with its new concepts, and hence implicitly arguing for the validity of those concepts. But the exploratory work itself and the process of concept formation are no longer given in the presentation.

Second, for the researchers themselves, the stabilization of classifications and concepts occurs in a particular way. Personal insight into the adequacy of some concepts and the inadequacy of others typically develops slowly, against the background of a large range of empirical findings. The most important question is whether a specific concept is helpful in the comprehension and classification of a large number of such findings. In the course of this process, the researcher constantly explores and tests, progresses, discards, and begins anew. When stabilization finally occurs, it is not always clear what has brought it about. Sometimes all it takes is a small step, a slight change, to bridge a gap in a much larger domain of previously considered effects. The formation or revision of a concept has consequences not at one place but simultaneously at many places within the field. The reflections leading up to it constitute a web, not a linear sequence. Such reflections naturally resist expression.

Third, exploratory work goes hand in hand with uncertainty on the fundamental level of concepts. Addressing that level in the first place already indicates doubts about the suitability of existing concepts, and that kind of worry is not easily allayed: will the newly formed or revised concepts really fare better, or might they in the end obstruct a clearer view? Such uncertainty on the level of language is difficult to address. Laboratory logs reveal researchers sometimes inventing their own languages or systems of signs, designed only for their own provisional use and difficult to transmit beyond the walls of the laboratory. It is particularly challenging to report on such a period of uncertainty or even speechlessness retrospectively, when the perspective may have shifted altogether, when a new view of the field has been attained, and when uncertainty and speechlessness have been overcome. Ludwik Fleck ([1935] 1979, 86) describes the problem of hindsight as the product of a situation in which "we can no longer express the previously incomplete thoughts with these now finished concepts." He once again draws our attention to the level of expression and concepts. Unless they happen to be deeply interested in epistemology or epistemic psychology, researchers will seldom take the trouble to find the words with which to describe their former situation of speechlessness.

Exploratory periods thus disappear from accounts of the research process and often even from the general picture of science itself. Insofar as historical and epistemological analysis fails to look past self-presentations, it retains a significant blind spot in which the formation of concepts is overlooked and, with it, the roles experimentation can play in that process.

Scientific Methodology and Historical Constellations

My reflections on the role of exploratory experimentation should not be understood as implying that a researcher in this sort of epistemic situation has no choice but to adopt an exploratory approach. Again, the case of early electromagnetism is instructive. The research undertaken throughout Europe in response to Ørsted's report was mostly but by no means entirely exploratory. Biot clung unshakably to the Laplacean framework, even when doing so required disregarding the complexity of the new effects. By contrast, Ampère, like many others in Europe, took the qualitative multiplicity of the phenomena seriously and adopted an exploratory strategy. When shortly thereafter the prospect of a mathematically formalizable theory of circuits arose, his priorities shifted in the direction of this option, which appeared much more attractive to him. With this switch, the attempt to grasp the broad diversity, and with it the exploratory strategy, receded into the background.

The episode illustrates the variety of criteria on which the stability of a conceptual scheme, and its suitability as a foundation for future research, can be assessed by researchers. Comprehending the broadest possible range of diverse phenomena is only one of them. All decisions to open concepts for revision and all processes of forming new concepts are situated within a specific historical situation. Biographical particulars, the general and local state of research, methodological standards, intellectual traditions and proclivities, community expectations, career interests, and the like, all play their roles in a researcher's decision on which concepts to call into question and which to take as an appropriate foundation for subsequent research. They are thus equally important in the selection of a particular research direction and in the dismissal of alternatives. Historical exigencies and contingencies are an ineliminable part of the scientific process, of which my studies of early electromagnetism provide telling examples. If this last chapter has strongly emphasized the epistemic dimension, it should not be misunderstood as claiming that this is the only dimension of interest (Steinle 2002b; Steinle and Burian 2002). But keeping it in view and taking it seriously strike me as essential to our historical understanding of science.

APPENDIX A

Source Materials for the Early Phases of Ampère's Research in Electrodynamics

A1. The Académie des Sciences Collection of Ampère's Materials

The collection of Ampère's materials (the Dossier Ampère) held by the Académie des Sciences in Paris consists of thirty-nine boxes. During a 1936–37 survey, its contents were classified thematically and sorted into more than four hundred folders (*chemises*). The materials within each folder are not numbered or otherwise labeled. The classification is mostly but not always reliable, and boxes 27 through 37, marked as containing "various drafts" (*brouillons divers*), are simply segregated by topic, without folders.

Folders 156 through 208bis (fifty-eight folders in all), filling boxes 8 through 11, are devoted to "electricity and magnetism." Each folder contains, on average, fifteen to twenty individual documents (single sheets, sketches, bound sheets, or even entire thread-bound booklets), yielding a total of approximately a thousand documents on the topic in question, not even counting those consigned to the "various drafts." This rough estimate may convey a sense of scale.

The archivists labeled each folder, but there are some misleading groupings and a great deal of overlap. This has less to do with any assumed carelessness on the part of the archivists than with the peculiarities of the materials in question:

- With a very few exceptions, Ampère did not date these documents. Other direct means of determining dates are also extremely sparse.
- The two sides of a given sheet are often devoted to entirely different topics. Ampère would frequently reuse the blank sides of pages whose fronts he had already inscribed, evidently in an effort to save paper.
- Individual parts have often been removed from the documents to which they originally belonged, either to allow their paper to be reused elsewhere or to insert the text into another context.

Such issues, combined with the sheer scale of the collection, make the precise attribution of particular documents to specific periods or thematic contexts extremely difficult and, at a minimum, very labor intensive.

So as to facilitate clear and unambiguous reference to particular items within this vast array of Ampère materials held by the Académie des Sciences (AS), I have assigned additional letters to the individual documents of particular interest within a given folder. For example, by "AS 205(b)" I mean the particular document I have designated as "(b)" within folder 205; correspondingly, "AS 208bis(a)" refers to document (a) in folder 208bis, and so on.

In investigating the early period of Ampère's work, I have drawn on a very diverse range of material. The manuscript of his first *mémoire* in the *Annales de chimie et de physique* (*ACP*) survives in three large segments, AS 208bis(f), 205bis(b), and 159(a), and two fragments, AS 208bis(g) and 208bis(j). These documents also constitute part of the manuscript of the first two academy lectures, the rest of which is scattered throughout other folders and boxes: AS 208bis(g) and 205bis(j). Other pertinent working notes may be found in AS 206bis(m) and 205(b). The archive preserves his published "Conclusions" and "Notes," both in manuscript and in offprint. In sections A6 to A11 below, I present and discuss in detail some pieces of particular importance:

AS 208bis(f) (the *ACP* manuscript)
AS 208bis(g) (fragment of the *ACP* manuscript)
AS 205bis(j) (fragment of the *ACP* manuscript)
AS 206bis(m)
AS 205(b)

In addition, I present a source from another archive. In October 1820, Ampère sent a letter to William Hyde Wollaston in London. The original has been lost, but an English translation, presumably commissioned immediately upon receipt, has been preserved in the archives of the Royal Society.

A website (http://www.ampere.cnrs.fr/) constructed by CNRS researchers in France presents immensely rich material on Ampère and, in particular, all his manuscripts in facsimile and partly in transcription. Among others, all the manuscripts mentioned in this book can be accessed online at that website; table A1 gives an overview of their specific locations.

TABLE A1. Locations of Ampère manuscripts on the CNRS website

		Image number		
Manuscript number per Steinle system	Webpage heading	Start of manuscript	Start of relevant part	End of relevant part
159(a)	Chemise 159: "Mémoire lu le 30 octobre 1820 (images 2 à 13)"	2	2	24
205(b)	Chemise 205: "Notes d'Ampère sur les instruments qu'il a imaginé (images 7 à 11)"	7	7	11
205bis(b) fragment of *ACP* manuscript, 1st part	Chemise 205: "Note d'Ampère (action de la pile) (images 218 à 237)"	218	218	237
205bis(j) fragment of *ACP* manuscript	Chemise 205: "Plusieurs notes et fragments sur l'attraction et la répulsion, discussion de la théorie de Biot, des hypothèses (?) de Demontferrand, etc. (images 238 à 261)"	238	238	239
206bis(m)	Chemise 206: "Notes et brouillons se rapportant à la détermination de la formule de l'action mutuelle de deux éléments de courant (images 286 à 295)"	286	286	286
208bis(f) *ACP* manuscript	Chemise 208: "Suite du mémoire sur l'action mutuelle de deux courants électriques, d'un courant électrique sur un aimant et de deux aimants l'un sur l'autre (images 28 à 105)"	28	70	103
208bis(g) fragment of *ACP* manuscript	Chemise 208: "Fragment d'un mémoire d'Ampère sur l'identité de l'électricité et du magnétisme (images 106 à 129)"	106	122	129

Note: The CNRS website, Ampère et l'histoire de l'électricité, is at http://www.ampere.cnrs.fr. All manuscripts on electricity and magnetism of that website are grouped at http://www.ampere.cnrs.fr/manuscrits/parcourir/chemises-theme/index.php?&page=5&total=54&theme=12.

A2. Dating the Manuscripts

In general it is not possible to date a given manuscript to a particular day. Ampère almost never dated his writings, with "20.7.^{bre}," the title of manuscript 206bis(m), constituting a rare exception. Nonetheless, it remains possible to sort the materials into distinct phases of one or more weeks in duration. For a first approximation, I distinguish four phases, chiefly with a view toward the composition of texts that contributed to Ampère's first mémoire:

- I. Prior to September 18, the first academy lecture
- II. Between September 18 and the time of the second academy lecture, on September 25
- III. From September 25 to the composition of the first part of the *ACP* mémoire, presumably around the middle of October 1820
- IV. From this point to the composition of the second part of the *ACP* mémoire, roughly the second week of November. Period IV thus comprises the four weeks from the middle of October through the middle of November.

Each of the three dates separating one phase from the next provided Ampère with an incentive to return to the manuscript of the lectures of September 18 and 25 and possibly to revise it. In some cases, more detailed sequencing within particular phases is also possible.

A3. Ampère's Use of *Galvanique* and *Électrique*

Ampère's terminology, which changed significantly over the course of his work, provides a significant point of chronological reference. In particular, Ampère decided at some juncture, its precise date still unknown, to consistently use the adjective *électrique* in place of his previously habitual *galvanique*. His initial reference to *courant galvanique* and *conducteur galvanique* (what I call the "G terminology") soon gave way to *courant électrique* and *conducteur électrique* (which I call the "E terminology"). He made this choice deliberately, drawing explicit attention to it in his first mémoire by referring to "*le courant galvanique*, dénomination que j'ai cru devoir changer en celle de courant électrique."[1] Ampère was well aware that he had not merely picked another word but taken an important theoretical step. His new terminology implied an assertion whose truth had by no means been settled and at which he himself had arrived only in the course of his work: the assertion that the effects of the closed pile, or currents, were effects of electricity in the prior sense of the word. The very fact that Ampère initially used *galvanique*, a more neutral term or at any rate a term with less far-reaching implications, and was thus being decidedly more cautious than Ørsted, shows the extent to which he was cognizant of the full significance of such terminological choices.

Having once taken this step, Ampère consistently followed through. All notes taken after this moment contain the E terminology exclusively. A few weeks later, when he returned to the drafts of his earliest academy lectures in preparing the text of the second part of his first mémoire, he changed the terminology in every passage selected for the published text. Consequently, excluding the passage cited above, in which he makes his decision explicit, the word *galvanique* appears almost nowhere in the mémoire. The sole exception, a single reference to *conducteur galvanique* (p. 201 in the print version), can presumably be blamed on a proofreading oversight. By contrast, in those passages deleted during his preparation of the final text, the older G terminology remains (see, e.g., *m43** = *r32*). In those instances where he neglected to make this change in the manuscript, it was nonetheless carried out later, presumably while correcting the galleys, and is reflected in the printed text (as, e.g., in *m35* = *r42* as compared to p. 205 in the print version).

The precise time at which Ampère decided to change his terminology is constrained at both ends of an interval. He was still using the G terminology throughout the September 25 lecture. This is evident from the manuscript reconstruction and from manuscript AS 205(b), closely associated with the lectures. In this regard, passage *r48* = *m41*, in which Ampère reported on his successful experimental demonstration of attraction between two spirals, is of particular note, for this experiment did not succeed until September 25. In his description of it, Ampère continued to use the G terminology, and he did likewise in a letter to his son written on the evening of that same day.[2] The same also holds for the "Conclusions" with which he closed his September 25 lecture. But at the same time, the "Conclusions" help constrain the date of his decision at the other end: by October 11, he had received the special printing of the "Conclusions" and was sending it to other researchers. This pamphlet was Ampère's very first printed text on electromagnetism, and it used the E terminology throughout, as did its manuscript (AS 158[c]), presumably composed before his academy lecture of October 9. The status of October 9 as the *terminus ante quem* of the terminological shift is reinforced by the fact that the surviving pieces of the text prepared for the October 9 lecture, AS 208bis(f), pages 1–34 and 71–78, consistently use the E terminology. The date of the terminological shift may thus be narrowed down to the two-week interval between September 25 and October 9.

The fact that Ampère composed the first part of his first *ACP* mémoire (Ampère 1820d) during this same interval yields further evidence. It was there that he first set out his unified theory of electricity and galvanism, in which the assertion of the "electrical" nature of galvanic phenomena played a central role and found a theoretical explanation. The surviving manuscript fragment for this text, AS 205bis(b), uses the E terminology from the outset, and it seems highly probable that the decision to undertake the terminological shift took place in the course of the very reflections set down in the text. The shift of terminology is of great significance in dating Ampère's early manuscripts in that it provides a provisional but precise way of distinguishing the very earliest among them from those that followed.

A4. Ampère's First Mémoire and the Corresponding Manuscripts

Ampère's first published mémoire on electromagnetism was printed in two parts, appearing in successive issues of *ACP*:

> Part 1, September 1820 issue, pp. 59–76 (Ampère 1820d)
> Part 2, October 1820 issue, pp. 170–218 (Ampère 1820j)

The two parts are quite different in character. In the first, Ampère provides a comprehensive synthesis and develops, at considerable remove from any immediate accounts of experiments, an overarching theoretical framework: a theory of electric current encompassing the phenomena of both conventional and galvanic electricity. Only a few experimental particulars are mentioned at all, and the only instrument presented in any detail is the later so-called "current balance" for the demonstration of attraction and repulsion between rectilinear conductors. By contrast, as Ampère himself states explicitly, the second part is devoted to material from his academy lectures of September 18 and 25, October 9 and 30, and November 6. Consequently, many passages refer closely to the course of his research, and we find numerous accounts of experiments and instruments. At the same time, it is hard to find a common thread running through them all. On occasion, a passage will refer to matters not properly introduced until later passages. There are also some repetitions. As Ampère was well aware, this part of the mémoire had been cobbled together in great haste from texts already at hand, unlike the first part, which had been composed as a whole.

As to the manuscript of the mémoire, most parts have been found, albeit in scattered fragments, including some of part 1 and all of part 2 (see table A2). The character of the manuscripts of the two parts is as significantly different as the character of the texts themselves. The surviving manuscript portion of part 1 contains the printed text from page 64 on, uninterrupted and containing comparatively few corrections. In its structure, the final published text largely corresponds with the manuscript; differences are usually merely stylistic. It appears to have been written from the outset with a view toward publication, with no discernible recourse to any previously drafted texts. Matters are different with respect to part 2, in which the fragmentary quality of the printed text mirrors the fragmentation of the manuscript segments. The final manuscript surviving today is composed of pieces taken from the texts prepared for Ampère's various academy lectures. But in stitching them together, Ampère made various significant changes, introducing new passages while striking, reworking, and updating others and, in two instances, removing long blocks of text. I was able to locate the deleted segments elsewhere in the archive: AS 208bis(g) and AS 205bis(j) (see below, sections A8 and A9). Table A2 provides an overview of source materials for the first mémoire. The fourth column of the table lists the available dating specifics. It should also be noted that the two pertinent issues of *ACP* were recorded as received at the academy approximately five weeks apart:

September issue (including part 1 of the mémoire): November 13, 1820
October issue (including part 2 of the mémoire): December 18, 1820

These dates, together with a glance at the contents of the two parts, make it clear that the second was composed significantly later than the first. The second part discusses the lecture of November 6, by which time the issue containing the first part was about to be delivered.

For a more precise dating of the composition of the *first part* of the mémoire, we have Ampère's own assertion, in his October 11 letter to Wollaston, that a mémoire was in press (for a transcription of this letter, see below, section A11). This claim can refer only to the first part of the mémoire. Though in Ampère's case such claims must be treated with

Table A2. Extant manuscripts for Ampère's first mémoire in *Annales de chimie et de physique*

Segment of published text	Content	Extant manuscripts	Dating
First part	Comprehensive theory of electrical currents and their magnetic effects		Printed prior to Nov. 13, 1820
"§ Ier: De l'action de deux courans électriques," (beginning), pp. 59–64		—	
Ibid., pp. 64–76 (end)		205bis(b)	Composed prior to Oct. 11
Second part			Full text edited mid-Nov., printed before Dec. 18
"Continuation du § Ier," pp. 170–88	Draws on the lectures of Oct. 9 and Nov. 6	208bis(f), pp. 1–34	Composed before Oct. 9 and Nov. 6, respectively
"§ II. Direction des courans électriques par l'action du globe terrestre," pp. 188–96	Draws on the lecture of Oct. 30	159(a) (one page of the manuscript is missing)	Composed shortly before Oct. 30
"§ III. De l'action mutuelle entre un conducteur électrique et un aimant," pp. 196–209	Draws on the lectures of Sept. 18 and 25	208bis(f), pp. 35–65	Composed during the first three weeks of research
Ibid., pp. 209–12	"Conclusions" of the September 25 lecture	208bis, pp. 67–70 (special offprint)	Offprint delivered by Oct. 11
Ibid., pp. 212–18	Draws on the lecture of October 9	208bis(f), pp. 71–86	Composed shortly before Oct. 9
Deleted from published version			
		208bis(g)	Composed before Sept. 25
		205bis(j)	Composed before Sept. 25

caution and checked against an analysis of content, in this instance there are several reasons for taking the claim seriously. The letter to Wollaston lists, in highly condensed form, some of the same results set down in the first part of the mémoire. In both mémoire and letter, Ampère draws particular attention to his experiment on attraction and repulsion between currents. A rather problematic date is claimed in the title of the published text: "Mémoire, Présenté à l'Académie royale des Sciences, le 2 octobre 1820." But there is no confirmation for any such dating to be found in the proceedings of the academy—no mention either of an oral presentation or of the submission of a written text. Accordingly, in this instance I have discounted Ampère's claim. To sum up, we may infer that Ampère composed the first part no later than October 11 and submitted it very soon thereafter to Arago, publisher of the *ACP*. This yields an interval of somewhat more than four weeks between the submission of the manuscript and the delivery of the printed text.

Since the *second part* refers to the lecture of November 6, Ampère's final assembly or revision of that part has to be dated later. At the same time, as it contains no reference to the lectures of November 13 and thereafter, I infer that by then the assembly was at least in progress, if not actually completed. Another circumstance pointing to the second week of November is the fact that Ampère was also engaged in composing his first account of the progress of his investigations (the "Notes" [Ampère 1820h]), for which he in all probability returned to the originals of the texts he had prepared for his lectures. It follows that during the week in question, he not only had these texts at hand but was also in the midst of an attempt, parallel to that of the mémoire, to situate them within a reasonably coherent narrative. Finally, the completion of the manuscript no later than the middle of November would also have left enough time for the printed journal issue to be delivered on December 18; a later date might have made this difficult.

For purposes of an investigation of the earliest period of Ampère's work, the section of part 2 of his first mémoire designated "§III" is of particular importance, as is the corresponding part of manuscript AS 208bis(f), along with manuscripts 208bis(g) and 205bis(j). With considerable effort, these sources collectively make it possible to reconstruct a manuscript for the academy lectures of September 18 and 25. Appendix B provides a detailed discussion of the reconstruction procedure. The remainder of appendix A is devoted to discussion and transcription of the manuscript sources themselves.

A5. Conventions Used in the Transcription of Manuscript Texts

Except where explicitly noted, all of my transcriptions of archival sources render their text in the form in which it survives today. In presenting them, I rely on the following conventions:

> For each document, the beginning and end of transcription is indicated by a horizontal rule, ————————————.
>
> With respect to capitalization, accents, punctuation, and even obvious errors in grammar and spelling, each text has been transcribed precisely as written. In a very few instances of striking divergence from conventional usage or grammar, I have inserted a "[*sic*]" in square brackets.
>
> Passages struck through by the author are ~~struck through~~ in transcription; those underlined by the author are underlined in transcription. Later insertions, whether written above the line or inscribed over the prior text, are enclosed in wavy braces, { }.
>
> Illegible passages are rendered as "xxx," or, if they are struck through, as "~~xxx~~." Where I have a plausible but not definitive reading of such passages, I enclose the reading in square brackets, followed by a question mark.
>
> Page breaks in the original are indicated by //.
>
> All paragraph breaks are present in the originals, but not line breaks.

So as to facilitate reference to my transcription of Ampère's manuscripts AS 208bis(f), 208bis(g), and 205bis(j), I have numbered their paragraphs in accordance with the following conventions:

> Paragraph numbers beginning with "m" refer to manuscript 208bis(f), pages 35–66, in the state in which it survives today, including paragraphs later struck through or pasted over by Ampère.
>
> Paragraph numbers followed by an asterisk (*) indicate paragraphs Ampère later struck through in their entirety.

Paragraph numbers followed by a "+" indicate paragraphs Ampère later pasted over. Both sorts of deletion are also marked by a vertical line in the left margin.

Supplemental numbers beginning with "r" refer to the paragraph numbering of my reconstruction, reproduced in its entirety in appendix B, section B2.

Finally, Ampère's manuscripts are not paginated. I have introduced page numbering and indicate the beginning of each new page with a "//" followed by the number of that page.

A6. Manuscript AS 208bis(f), the "*ACP* Manuscript"

This manuscript is a diverse collection of pieces bound together with needle and thread. It consists of fifty pages, constituting the textual basis for §I and §III of the second part of Ampère's first mémoire.[3] The individual pieces from which this whole was assembled are usually easy to identify as such. Some are also bound together as units or written on paper sheets that clearly belong together. In some cases such indicators as the size or type of paper, the ink used, and so on also suffice to make it clear where Ampère inserted new pages over the course of his revision. For ease of exposition, given the frequency with which I refer to this source, I often call it simply the "*ACP* manuscript."

For the purpose of my investigation, the portion of the manuscript belonging to §III and drawing on the academy lectures of September 18 and 25 is of particular importance. It constitutes pages 35–66 in my pagination. I have transcribed this portion only. For more detailed discussion of the text, see appendix B.

---

m1 (p. 35)[4]
§. III.
~~je commencerai ce paragraphe par le récit~~
de l'action mutuelle entre un conducteur électrique et un aimant.
c'est cette action découverte par Mr Oersted, qui m'a conduit à reconnaitre celle de deux courans électriques l'un sur l'autre, celle du glôbe terrestre sur un courant, et la manière dont l'électricité produisait tous

les phénomènes que presentent les aimans, par une distribution semblable à celle qui a lieu dans le conducteur d'un courant électrique, suivant des courbes fermées perpendiculaires à l'axe de chaque aimant. ces vues, dont ~~une partie ne~~ la plus grande partie n'a été que plus tard confirmée par l'expérience, furent communiquées à l'académie royale des sciences //(36) lu dans sa séance du 18 septembre 1820, et que je vais transcrire sans autres changemens que la suppression des passages ~~relatifs à la déscription des~~ {qui ne seraient qu'une répétition de ce qu'on vient de lire, et en particulier ceux où je decrivais les} appareils que je me proposais de faire construire, ~~qui~~ {ils} l'ont été depuis et ~~se trouvent~~ {la plûpart sont} décrits dans les paragraphes précédens. {on pourra par ce moyen se faire une idée plus juste de la marche que j'ai suivie dans mes recherches sur le sujet dont nous nous occupons.}

m2

les expériences que ~~je presente~~ {j'ai faites} sur l'action mutuelle des conducteurs ~~etant~~ qui mettent en communication les extrémités d'une pile voltaïque, m'ont montré que tous les faits relatifs à cette action peuvent être ramenés à deux résultats généraux, qu'on doit considérer d'abord comme ~~deux lois empiriques xxx~~ uniquement donnés par l'obsérvation, en attendant qu'on puisse les ramener à un principe unique comme j'essayerai bientôt de le faire. je commencerai par ~~les exprimer~~ ~~les donner~~ les énoncer sous la forme qui me parait la plus simple et la plus générale.

m3 (= r1)*

//(37) lu le 18.7.bre

m4 (= r2)*

dans le travail {que j'offre aujourd'hui à l'académie} sur l'action réciproque du galvanisme et de l'aimant ~~que j'offre aujourd'hui à l'académie royale de sciences, je me suis proposé~~: {découverte par M. Oersted, je me suis proposé:}

m5 (= r3)*

1.° de distinguer les deux résultats généraux de cette action, auxquels peuvent être ramenés tous les faits observés par l'auteur de {cette} ~~l'~~importante découverte ~~dont M.r Oersted xxx d'enricher la physique~~ {l'une de plus remarquables de la physique moderne,} et de ramener l'énoncé de ces deux résultats à l'expression la plus simple et la plus générale.

m6 (= r4)*

2.° de décrire les expériences que j'ai faites et les instrumens que j'ai imaginés, pour les présenter isolés l'un de l'autre et de l'action du glôbe terrestre sur l'aiguille aimantée, qui se combine avec eux dans les expériences de M.r Oersted, et en altère les effets {ou en} ~~xxx~~ dissimule les lois.

m7 (= r5)*[5]

{3.° ~~de ramener~~ [?] {d'expliquer} ces deux résultats généraux en les ramenant à une même cause, qui rend raison //(38) en même tems d'un assez grand nombre de phénomènes connu depuis longtems ou plus recemment constatés:}

m8 (= r6)*

§. 1.er

~~xxx~~ exposition des deux résultats généraux au moyen desquels on peut rendre raison de diverses expériences de M.r Oersted et prévoir ce que l'action réciproque ~~du galvan~~ d'un aimant et d'un conducteur galvanique, dans les cas qu'on n'a pas encore soumis à l'obsérvation.

m9 (= r7)

ces résultats consistent, d'une part dans ~~xxx~~ l'action directrice d'un de ces corps sur l'autre, de l'autre part, dans l'action attractive ou répulsive qui s'établit entre eux, suivant les circonstances.

m10 (= r8)

<u>action directrice:</u> lorsqu'un aimant et un conducteur ~~galvanique~~ ~~{électrique}~~ agissent l'un sur l'autre, et que l'un d'eux, étant fixe, l'autre ne peut que tourner dans un plan perpendiculaire à la plus courte distance du conducteur//(39) et de l'axe de l'aimant, celui qui est mobile tend à ~~xxx~~ {se mouvoir} de manière que les directions du conducteur et de l'axe de l'aimant forment un angle droit, et que le pole de ~~l'aiguille~~ l'aimant qui regarde habituellement le nord soit à gauche de ce qu'on ~~a nommé~~ {appèle ordinairement} le courant galvanique {, dénomination que j'ai cru devoir changer en celle de courant électrique}, et le pole sud à sa droite. {Bien entendu que la ligne qui mesure la plus courte distance du conducteur et l'axe de l'aimant, rencontre la direction de cet axe entre les deux pôles.}

m11 (= r9)

pour conserver à cet énoncé toute la généralité dont il est susceptible, il faut distinguer deux sortes des conducteurs ~~galvaniques~~, 1°. la pile même, dans laquelle ce qu'on nomme {ordinairement} le courant ~~galvanique~~ {électrique} se porte du pôle où il se produit de l'hydrogène dans la décomposition de l'eau à celui d'où l'oxigène se degage. 2°. le fil métallique qui unit les deux extrémités de la pile, et où l'on doit alors considérer le même courant comme se portant au contraire du pôle qui donne de l'oxigène à celui {qui} developpe //(40) de l'hydrogène.

m12^{+} *(= r10; later pasted over with m14)*

je n'emploie au reste ici le mot de courant galvanique que comme une expression convenue pour définir les directions dont il s'agit. cette ~~xxx~~ expression pourrait être employée ~~xxx~~ comme ~~xxx~~ d'un usage commode lors même qu'on n'admettrait pas l'éxistence réelle d'un pareil courant.

m13^{*+} *(= r11 [beginning])*

pour préciser ~~le sens dans lequel {que} je donne ici~~ {le signification que j'attribue} à ces mots la droite, la gauche d'un courant galvanique, il faut faire une supposition analogue à celle à l'aide de laquelle on définit la droite et la gauche d'un fleuve, en concevant un homme qui le descend la face tournée vers son embouchure. on supposera de même un homme placé dans le sens du courant galvanique de manière que ce courant aille de ses pieds

m14 (= r10a; replacement for m12+)

on peut comprendre ces deux cas dans une même définition en disant qu'on entend par courant ~~galvanique~~ {électrique} la direction suivant laquelle l'hydrogène et les bases des sels sont transportés par l'action de toute la pile, en ~~la~~ concevant {celle-ci} comme formant avec le conducteur un seul circuit, ~~si~~ [?] ~~à un point quelconque de~~ {lorsqu'on interrompt} ce circuit ~~on l'interrompait~~ pour y placer soit de l'eau soit une dissolution saline ~~qui seraient~~ {que cette action} décompos~~és~~{e}. au reste tout ce que je vais dire ~~dans ce paragraphe et dans les deux suivans~~ {sur ce sujet} ne suppose aucunement qu'il y ait réellement un courant dans cette direction, et on peut ne considérer que comme une manière commode et usitée de {la} désigner ~~cette direction~~, l'emploi que je fais ici de cette dénomination: {de} courant ~~galvanique~~ {électrique}.

m15 (= replacement for m13^{*+}*)*

dans les expériences de Mr Oersted cette action directrice ~~xxx~~ se combine toujours avec celle ~~du~~ {que le} glôbe terrestre exerce sur l'aiguille aimantée qu'il emploie, et quelques fois en outre avec l'action que je décrivai tout à l'heure sous la dénomination d'action attractive ou répulsive, ce qui conduit à des résultats compliqués dont il est difficiles [*sic*] de d'analyser les circonstances et de reconnaitre les lois.

m16 (= replaces m13^{*+})[6]
pour pouvoir observer les effets de l'action directrice d'un ~~conducteur~~ {courant} électrique sur un aimant, sans qu'ils fussent altèrés par ces diverses causes, j'ai fait construire un instrument que j'ai nommé aiguille aimantée astatique. cet instrument représenté pl.4. fig.8. consiste dans une aiguille aimantée tirée perpendiculairement à un axe qu'on peut au moyen d'un mouvement semblable à celui du pied d'un télescope et de deux vis de rappel placer dans

m16a (= r20 [end])[7]
//(41) {la} direction que l'on veut, en sorte que l'aiguille aimantée ne peut se mouvoir qu'en tournant dans un plan perpendiculaire à cet axe. il faut que son centre de gravité soit placé exactement dans l'axe, ~~xxx xxx~~ {en sorte} qu'avant qu'elle soit aimantée ~~xxx~~ on puisse s'assurer que la pesanteur n'a aucune action pour mouvoir l'aiguille. on l'aimante alors, et cet instrument peut servir à vérifier que tant que le plan où se meut l'aiguille n'est pas perpendiculaire à ~~l'aiguille~~ la direction de l'aiguille d'inclinaison, le magnetisme terrestre tend à faire prendre à l'aiguille aimantée la direction de celle des lignes traceés sur le plan qui est le plus rapprochée possible de la direction de l'aiguille d'inclinaison, c'est-à dire la projection de cette direction sur le même plan. on place ensuite l'axe parallelement ~~xxx~~ à la direction de l'aiguille d'inclinaison, le plan où se meut l'aiguille aimantée lui étant alors perpendiculaire le magnetisme terrestre n'a plus //(42) aucune action pour diriger l'aiguille aimantée, qui devient ainsi complètement astatique. le ~~xxx~~ {meme appareil} porte dans le plan de cette aiguille un circle divisé en degrés, sur lequel sont fixé deux petits barreaux de verre pour attacher les conducteurs galvaniques, dont l'action directrice agit alors seule, et sans complication avec la pesanteur et le magnétisme terrestre.

m17 (= r21)
la principale expérience à faire avec cet appareil est de montrer que l'angle entre ~~l'aiguille~~ les directions de l'aiguille et du conducteur est toujours droit quand l'action directrice est la seule qui ait lieu.

m18 (= r22 [beginning])*[8]

Le second instrument que j'ai nommé: appareil pour les attractions et répulsions magnético-galvaniques à aimant mobile, consiste dans un pied auquel on peu [sic] suspendre ~~xxx~~ d'aucune situation verticale, et à

m19 (= r11 [end])*

//(43) à sa tête, et qu'il ait la face tournée vers l'aiguille, alors sa main droite sera du coté que j'appèle la droite du courant, et sa main gauche marquera la gauche du courant.

m20 (= r12)

action attractive où répulsive. ce second résultat général consiste: 1.° en ce qu'un ~~courant galvanique~~ {conducteur joignant les deux extremités d'une pile voltaïque} et un aimant, dont l'axe fait un angle droit avec la direction ~~de ce~~ {du} courant qu'on imagine dans ce conducteur d'après le définitions précédentes, s'attirent quand le pôle nord est à gauche du courant, c'est-à-dire quand la position est celle que le conducteur et l'aimant tendent à prendre en vertu de leur action mutuelle, et se repoussent quand le pole nord de l'aimant est a la droite du courant, c'est-à-dire quand le conducteur et l'aimant sont {maintenus} dans la position opposée à celle qu'ils tendent à //(44) se donner mutuellement.

m21 (= r13)

on voit par l'énoncé même de ce deux résultats que l'action entre le conducteur et l'amant est toujours réciproque. c'est cette réciprocité que je me suis d'abord attaché à vérifier, quoiqu'elle me parut assez évidente par elle-même, quelques unes des expériences que je vais décrire ont ~~xxx~~ pour objet de la constater.

m22 (= r14)

les deux modes d'action entre un amant et un conducteur galvanique que je viens d'exposer, en les considérant comme de simples résultats de l'expérience suffisent pour rendre raison ~~de tous les faits xxx que nous a fait connaitre~~ {des faits observés par} Mr Oersted, et pour prévoir ce qui doit arriver dans les cas analogues à l'égard de{squels} on n'a point encore fait d'obsérvation. ils indiquent par exemple d'avance tout ce qui doit arriver quand un ~~conducteur~~ {courant électrique} //(45) ~~galvanique~~ agit sur l'aiguille d'inclinaison. je n'entrerai dans aucun détail à cet égard, puisque tout ce que je pourrais dire sur ce sujet, découle immédiament des enoncés précédens. je dirai seulement qu'après avoir déduit seulement le premier résultat ~~des expériences de M~~. général de la note de Mr Oersted, j'en déduisis l'éxplication ~~dont~~ {des phénomènes magnétiques dont} j'exposerai les bases dans ~~le 3.me paragraphe~~ {la suite} de ce mémoire, que cette explication me conduisit au second résultat général, et me suggéra, pour le constater, une expérience qui réussit complètement. lorsque je la communiquai à Mr Arago, il me fit remarquer avec raison, que cette attraction ~~et cette répulsion~~ entre un aimant et un conducteur ~~galvanique~~ {électrique} placés à angles droits dans la direction où ils tendent à se mettre mutuellement et {cette répulsion} dans la direction opposée, pouvait [*sic*] seule rendre raison des résultats //(46) publiés par l'auteur de la découverte, dans le cas où l'aiguille aimantée étant horizontale, on en approche un conducteur galvanique dans une situation verticale, et qu'on pouvait même déduire aisement cette loi, de l'une des expériences de Mr Oersted, celle qu'il enonce ainsi: posito autem filo ~~(accipiente~~ (cuius extremitas superior electricitatem à termino négativo apparatûs galvanici accipit) è regione puncto inter polum et medium acùs sito, occidentem versus agitur.

m23 (= r15)

car ce mouvement ~~xxx~~ {de} l'aiguille aimantée, indiqué comme ayant lieu, soit que le conducteur soit à l'occident ou à l'orient de l'aiguille, est dans le premier une attraction parce que le pôle nord se trouve à la gauche du courant, et dans le second une répulsion parce //(47) qu'il se trouve à droite.

m24 (= r16)

mais en convenant de la justesse de cette obsérvation, il me semble que la distinction que j'ai faite des deux résultats {généraux} de l'action mutuelle d'un aimant et d'un conducteur galvanique, n'en devient que plus importante pour expliquer en montrant que dans ce cas c'est tantôt une attraction et tantôt une répulsion, toujours conformement à la loi ~~que je viens d'exposer~~ {du second résultat général} que je viens d'exposer, tandis que dans l'expérience que Mr Oersted énonce immédiament avant en ces termes: quando filum conjungens perpendiculare ponitur è regione polo acùs magneticae, et extremitas superior fili electricitatem à termino negativo apparatûs galvanici accipit, polus orientem versus movetur, ce mouvement n'a lieu que pour que //(48) l'aiguille aimantée prenne à l'ègard du conducteur la direction déterminée par le premier résultat général, avec toutes les circonstances que j'ai comprises dans son énoncé, et en particulier la remarque qui le termine{.} ~~, celle remarque le second.~~

m25 (= r17)*[9]

la condition exprimée dans cette remarque ne semble pas liée nécessairement à cette détermination de la direction de l'aiguille, mais elle resulte immédiament ~~du principe que j'ai donné~~ {des} considérations par lesquelles j'ai ramené à une cause unique, les deux résultats généraux que je viens d'exposer, et qui seront développées dans le 3.me paragraphe de ce mémoire. {je reviendrai alors sur ~~cette expérience~~ {l'explication}, ainsi que ~~xxx~~ {sur} les expériences de M.r Oersted avec des conducteurs recourbés en fer à cheval de manière que les deux branches soient paralleles entre elles.}

m26[10]

//(49) il me reste à décrire l'instrument avec lequel j'ai constaté l'existence de cette action entre un courant électrique, que j'ai désignée dans ce qui précède sous le nom d'action attractive ~~et~~ {ou} répulsive, et ~~que~~ j'en ai observé les effets sans que l'action directrice vint les altérer en se combinant avec elle. cet instrument est représenté pl.4 fig.9.

m27 (= r34)*[11]

//(51) Lu le 2x 7.bre

§.III.

explication des phénomènes précedens reduction des deux résultats généraux qu'elle ramène à un principe unique à l'aide duquelle on rend raison ~~des plusieurs autres~~ non seulement de ces phénomènes mais des ~~plusiers~~ autres faits dépendant de la théorie de l'aimant.

m28 (= r35)

(1)[12] la première réflexion ~~qui m'a conduit à cette explication est~~ {que je fis lorsque je voulus chercher les causes des nouveaux phénomènes découverts par M.r Oersted, est} que l'ordre dans lequel on a découvert deux faits, ne faisant rien aux conséquences des analogies qu'il [*sic*] présentent, nous pouvions supposer qu'avant de savoir que l'aiguille aimantée prend une direction constante du sud au nord, on avait d'abord connu la propriété qu'elle a d'être amenée par un courant ~~galvanique~~ {électrique} dans une situation perpendiculaire ~~à xxx~~ {à ce courant}, de manière que le pôle nord de l'aiguille fût porté à gauche ~~de l'aig~~ du courant, ~~xxx xxx xxx~~ {et qu'on découvrit ensuite la} propriété qu'elle a de tourner constamment au nord celle de //(52) ses extrémités qui se portait à gauche du courant, l'idée la plus simple et celle qui se présenterait immédiament à celui qui voudrait expliquer la direction constante du nord au sud, ne serait-elle {pas} d'admettre dans la terre ce qu'on appèle un courant galvanique dans une direction telle que le nord se trouvât à gauche d'un homme qui couché sur sa surface pour avoir la face tournée du coté de l'aiguille, recevrait ce courant dans la direction de ces pieds à sa tête, en sorte que ce courant se porterait de l'est à l'ouest ~~en xxx~~ dans une direction perpéndiculaire au méridien magnétique.

m29 (= r36)

*[13] cette hypothèse ~~xxx~~ {devient d'autant} plus probable ~~en faisant~~ {qu'on fait plus} attention à l'ensemble des ~~phénomènes qu'xxx à l'obsérvateur~~ faits connus, ce courant s'il existe doit être comparé à //(53) celui que j'ai montré dans la pile ~~xxx~~ agir sur l'aiguille aimantée comme ~~xxx~~ {se} dirigeant ~~l'aiguille aima~~ du pôle de l'hydrogène à celui de l'oxigène, {quand on établissait un coducteur [*sic*] entre eux et } ~~et~~ qui aurait lieu ~~xxx~~ de même si la pile formant une courbe fermée il venaient se réunir en un couple semblable aux autres, car il n'y a ~~rien dans~~ probablement rien dans notre glôbe qui ressemble à un conducteur continu et homogène, mais les matières diverses dont il est composé sont précisement dans le cas d'une pile voltaïque ~~composé~~ formée d'élémens disposés au hazard, et qui revenant sur elle-même formerait comme une cinture continue tout autour de la terre. des élémens ainsi disposés donnent moins d'énergie électrique sans doute que s'ils l'étaient dans un ordre périodiquement régulier, mais il faudrait qu'ils fussent arrangés à dessein pour ~~xxx~~ que dans une série de substances différentes, formant une courbe fermée autour de la terre, il n'y //(54) eut pas courant dans un sens ou dans l'autre. il se trouve que d'après l'arrangement des substances de la terre ce courant a lieu de l'est à l'ouest, et qu'il dirige partout l'aiguille aimantée perpendiculairement à sa propre direction ~~qui trace~~ {. cette direction trace} ainsi sur la terre un parallèle magnétique, et de manière que le pôle ~~xxx~~ de l'aiguille ~~xxx~~ {qui doit être} à ~~xxx~~ gauche {~~dirige xxx xxx~~}, ~~xxx xxx~~ {se trouve par là constamment porté} vers le nord {, dans celle du méridien magnétique.}

m30 (= r37 [beginning])

je ferai remarquer à ce sujet que les effets produits par les piles de la construction anglaise où l'on brule un fil ~~xxx~~ {fin de} métal ~~xxx~~ même avec une seule ~~xxx~~ {paire} dont le zinc et le cuivre plongent dans un acide, prouve suffisament que c'est une supposition trop restrainte de n'admettre l'action électro-motrice qu'entre les métaux, et de ne regarder le liquide interposé que comme //(55a) conducteur. Il y a sans doute action entre deux métaux, Volta la démontré da la manière la plus complètc, mais est-ce une raison pour qu'il n'y en ait pas entre eux et d'autres corps? Ou entre ceux-ci seulement? Il y en a ~~sans doute~~ {probablement} entre tous les corps qui peuvent conduire plus ou moins

m31⁺ (= r38 [middle])

//(55c) tension double de celle du second sur le premier, et ainsi de suite. Pour que ces tensions s'établissent il faut que les deux extrémités de la pile ne communiquent pas, des qu'on les fait communiquer, il n'y a plus de tension, mais un courant dans

m32⁺ (= r39 [end])

//(55e) du courant galvanique, malgré l'action de cette plaque de zinc vers laquelle il se porte alors, et se ferait contre elle que se dégagerait l'hydrogène tant que le circuit serait fermé.

m33 (= r40)

//(55f) l'électricité galvanique. ~~xxx~~ cette action est plus sensible dans les piles composées des métaux et d'acides étendus, tant parce qu'il parait que ce sont les substances où elle se développe avec le plus d'énergie que parce que ce sont celles qui conduisent le mieux l'électricité.

m34 (= r41)

les ~~xxx xxx~~ {diverses arrangemens} que nous pouvons donner à des corps non métalliques ne sauraient produire une action électro-motrice comparable à celle d'une pile voltaïque à disques métalliques séparés alternativement par des liquides, à cause du peu de longeur qu'il nous est permis de donner à nos appareils, mais une pile qui fait le tour de la terre conserverait sans doute quelqu'intensité lors même qu'elle ne serait pas composée de métaux, et que les élemens en seraient //(56) arrangés au hazard, car sur une si grande longeur il faudrait que l'arrangement fût fait à dessein pour que les actions dans un sens fussent exactement détruites par les actions dans l'autre.

m35 (= r42)

je crois devoir faire observer à ce sujet que des courans galvaniques dans un même corps, ne peuvent être indépendans les uns des autres, à moins qu'ils ne fussent séparés par des substances qui les isoleraient ~~xxx~~ ~~{et encore d'aucun cas le même, il y}~~ ~~autrement c'est dans la direction où il y aurait le plus~~ {completement dans toute leur étendue}, et encore dans ce cas la même ils devraient influer les uns sur les autres, puisque leur action se transmet à travers tous les corps, à plus forte raison lorsqu'ils coexistent dans un ~~corps~~ {glôbe} dont toutes les parties sont continues, doivent-ils se diriger dans tous dans le même suivant la direction que tend à leur donner //(57) ~~xxx xxx~~ la ~~xxx~~ {réunion} de toutes les actions électromotrices de ce glôbe. je suis bien loin au reste de croire que ce soit seulement dans ces actions que réside la cause des courans galvaniques qui y sont indiqués par la direction de l'aiguille aimantée à chaque point de la surface de la terre, je crois au contraire que la cause principale en est toute différente comme je le dirai ~~tout à l'heure~~ dans la suite de ce mémoire, mais comme ~~xxx~~ cette cause, dépendante de la rotation de la terre, donnerait en chaque lieu une direction constante à l'aiguille, ce qui est contraire à l'obsérvation, je regarde l'action électromotrice des substances dont se compose la planète que nous habitons, comme se combinant avec cette action générale et en en expliquant les variations à mesure que l'oxidation fait des progrès dans l'un ou l'autre //(58) région continentale de la terre.

m36 (= r43)
quant aux variations {diurnes} elles s'expliquent aisement par le changement de températures alternatif de ces ~~xxx~~ {deux régions} pendant la durée d'une rotation de la terre, d'autant plus facilement qu'on connait depuis longtems ~~l'action~~ l'influence de la température sur l'action ~~galvanique~~ {électromotrice} sur laquelle Mr Dessaignes a fait des obsérvations très interessantes. au reste il faut compter parmi les actions ~~galvaniques dans~~ {électromotrices des} differentes parties de la terre celle ces minerais aimantés qu'elle contient et qui doivent comme nous allons voir être considérés comme autant de piles galvaniques.

m37 (= r44)[14]
l'élevation de température qui a lieu dans les conducteurs ~~xxx xxx~~ {des} courans ~~galvaniques~~ {électriques} doit avoir lieu aussi dans ceux du glôbe terrestre, ne serait-ce pas là la cause de cette chaleur interne constatée récemment par les expériences rapportées dans une //(59) des dernières séances le l'académie par un de ses membres ~~qui xxx xxx~~ dont les travaux sur la chaleur ~~ont portée la lumière crée une nouvelle branche~~ ont fait rentrer cette partie de la physique dans le domaine des mathématiques? et quand on fait attention que cette élevation de température ~~xxx~~ {produit}, quand le courant est assez énergique, une incandescence {permanente} accompagnée de la plus vive lumière, sans combustion où déperdition de substance, ne

pourrait-on pas en conclure que les ~~corps xxx~~ {glôbes opaques} ne le sont qu'a ~~xxx~~ {cause} du peu d'énergie des courans ~~galvaniques~~ {électriques} qui s'y établissent, et trouver dans des courans plus actifs la cause de la chaleur et de la lumière ~~du xxx lumineux~~ {des globes qui brillent par eux-mêmes}? ~~maintenant si c'est la cause le l'action directrice que la terre exerce sur l'aiguille aimantée, comment ne serait pas aussi cette~~

m38 (= r45)
on sait qu'on expliquait autre fois par des courans, qu'on nommait magnétiques, //(60) mais on les supposait parallèles à l'axe de l'aimant, pas de possibilité sans se croiser et se détruire, d'ailleurs pourquoi plus tôt du ~~nord~~ {sud} au nord que du nord au sud.

m39, first part (= r46)
maintenant si c'est là la cause directrice de la part de la terre ce le sera aussi de la part d'un aimant d'où il suit qu'un aimant est un ~~composé~~ [?] {assemblage} de courans ~~galvaniques~~ {électriques} qui ont lieu dans des plans perpendiculaires à son axe, dirigés de manière que le pôle austral de l'aimant, qui se porte du coté du nord se trouve à droite de ces courans puisqu'il est toujours à ~~droite~~ {gauche} d'un courant placé hors de l'aimant

m39, second part (= r46a)
{et qui lui fait face dans une direction parallèle, ou plutôt ces courans s'établissent d'abord dans l'aimant suivant les courbes fermées les plus courtes soit de gauche à droite, soit de droite à gauche, et alors la ligne perpendiculaire aux plans de ces courans devient //(61) l'axe de l'aimant, et ses extremités en font les deux poles.} Ainsi à ~~xxx~~ {chacun} des poles d'un aimant les courans galvaniques dont il se compose sont dirigés suivant des courbes fermées concentriques, j'ai imité cette disposition autant qu'il était possible avec un courant ~~galvanique~~ {électrique} en en pliant le fil ~~en spirale~~ conducteur en spirale. ~~xxx que~~ cette spirale {était} formée avec un fil de laiton ~~dont les extremites etaient renfermées xxx~~ {et terminèe par deux portions rectilignes de ce même fil, qui étaient renfermées dans} deux tubes de verre afin qu'elles ne communiquassent pas entre elles et pussent être attachées aux deux extrémités de la pile.

m40 (= r47)
suivant le sens dans lequel on fait passer le courant dans une telle spirale elle est en effet {fortement} attirée ou repoussée par le pole d'un aimant

qu'on lui présente de manière que la direction de son axe soit perpendiculaire au plan de la spirale. {selon que les courans électriques de la spirale et du pôle de l'aimant sont dans le même sens ou en sens contraire.}

m41 (= r48)[15]

//(62) en remplaçant l'aimant par une autre spirale, dont le courant soit dans le même sens que le sien on a les mêmes attractions et répulsions, c'est ainsi que j'ai découvert que deux courans ~~galvaniques~~ {électriques} s'attiraient quand ils avaient lieu dans le même sens, et se repoussaient dans le cas contraire. je me propose de le verifier sur des courans rectilignes.

m42 (= r49)[16]

en remplaçant au contraire dans l'expérience de l'action mutuelle d'un des pôles d'un aimant et d'un courant dans un fil métallique plié en spirale, cette spirale par un aimant on a encore les mêmes effets soit en attraction soit en répulsion, conformément à la loi des phénomènes connus de l'aimant, il est évident d'ailleurs que toutes les circonstances de ces phénomènes, sont une suite nécessaire de la disposition des courans ~~galvaniques~~ {électriques} dont ils se composent, d'après la manière dont ceux-ci s'attirent et se répoussent.

m43 (= r32 [end])*

//(63) proposer aux physiciens qui s'occupent d'expériences sur la pile galvanique, d'y adapter une aiguille aimantée pour voir à chaque instant par la position qu'elle prend si ~~xxx~~ le courant galvanique est établi, et quelle est son intensité, précisement comme on adapte un électroscope à une machine électrique. ce petit instrument que je crois par analogie devoir nommer galvanoscope, peut au reste ne différer d'une boussole ordinaire que par l'usage qu'on en fait. {on peut en rendant très mobile et très legère, rendre cet instrument propre à reconnaitre l'existence de courans galvaniques très faibles, et c'en est le seul moyen.}

m44 (= r33)

je fais dans ce moment ~~executer plusieurs autres appareils, deux entre autres, où le~~ {j'ai construit un autre appareil où le fil} conducteur {est} ~~galvanique étant~~ plié en hélice {~~à pas très fins~~} autour d'un tube de {verre,} ~~carton ou de verre doit~~ d'après la théorie que je me suis faite de ces sortes de phénomènes, {ce conducteur doit} presenter quand on y fera passer le courant ~~galvanique~~ {électrique}, une action ~~xxx~~ //(64) semblable à celle d'une aiguille ou d'un barreau aimanté, dans toutes les circonstances où ceux-ci agissent sur d'autres corps, ou sont mus par le magnétisme terrestre. (1)[17] j'ai {déja[18] observé} ~~donné à ces deux appareils~~ {une partie des effets que j'attendais de l'emploi d'un conducteur plié en helice} ~~les noms d'helice galvanique fixe et d'helice galvanique mobile. je vais donner quelques détails sur les expériences que j'ai faites avec ces instrumens, {effets que j'attendais de cet instrument;}~~[19] {et je ne doute pas que plus on ~~xxx conformement à~~ {variera les expériences fondées sur} l'analogie qu'etablit la théorie entre cet instrument et un ~~aimant~~ {barreau aimantée}, plus on obtiendra de preuves que l'existence des courans électriques dans les aimans ~~xxx~~ est la cause {unique} de tous les phénomènes magnetiques.}

m45
(1) quand j'écrivais cela je ne connaisais pas bien l'action ~~de l'hélice dans le sens longitudinal~~ exercée par les spires d'hélice {parallelement à son axe} et je croyais qu'on pouvait la négliger, ce qui n'est pas, mais tout ce que je dis ici sera vrai si on l'entend[20] d'une hélice ou l'on ai detruit cette action par un courant ~~longitudinal et~~ {rectiligne} opposé dans le tube de verre qu'elle entoure de ses spires en sorte qu'il ne reste plus que[21] //(71) l'action qu'exerce chaque sprire dans un plan perpendiculaire à l'axe de l'helice ainsi que je l'ai expliqué dans le premier paragraphe de ce mémoir

m46
//(65) je terminai la lecture que je fis à l'académie de ce que je viens de transcrire, par un resumé ~~xxx~~ ou j'en deduisais les conclusions suivantes: 1.° deux courans electriques s'attirent . . . (placez ici les conclusions que M.^{r} fergueray a imprimées).

A7. Manuscript AS 208bis(g) (*ACP* Manuscript Fragment)

Manuscript 208bis(g) consists of four loose leaves, inscribed on both sides. All of the leaves show clear tear marks along one side, showing that each was originally torn from a folded sheet. I initially identified the document as belonging to the *ACP* manuscript on the basis of an analysis of its contents. But an additional persuasive piece of evidence is the fact that complementary matches for all of the tears may be found in the *ACP* manuscript itself, in places where the content fits and where word fragments left isolated by the tearing may be pieced back together. Since the document so clearly belongs to this manuscript, I have continued the pagination used there.

---

r18 (p. 87)
§.2.
description des appareils et ~~des~~ {indication des} expériences qui me paraissent les plus propres à ~~démontrer la généralité des~~ établir les deux résultats généraux exposés dans le paragraphe précedent.

r19
{dans} les observations de Mr. Oersted la position ~~de~~ {que prend} l'aiguille aimantée soit qu'elle dépende de l'action directice, ou de l'action attractive ou répulsive, est influencée par ~~xxx~~ la force qui dirige cette aiguille dans l'état ordinaire, et ~~quelques fois par les poids~~ il {n'}est ~~difficile~~ [?] {pas toujours} de distinguer ~~dans~~ [?] ~~chaque expérience ce qui~~ celles des ses expériences dont les résultats sont dus à l'action directrice et celles qui dépendent de l'action attractive ou répulsive. il y a même un de faits qu'il rapporte où le poids de l'aiguille aimantée {n'}a ~~xxx~~ permis ~~empeché~~ {au} conducteur ~~galvanique~~ {du courant électrique} de l'abaisser ou de la soulever, conformement aux lois de son action sur elle, que quand il en était très rapproché, //(88) probablement à cause de la distance à laquelle le centre de gravité de l'aiguille se trouvait de son point de suspension. j'ai donc d'abord cherché à construire ~~des~~ {deux} appareils ~~xxx~~ {tels que} l'action ~~répulsive~~ directrice agit seule dans l'un, sans que la force qui dirige {constamment} l'aiguille aimantée, ni les poids de cette aiguille, ne pussent influer sur elle, et xxx {que} ~~xxx~~ l'autre montrât les effets de ~~xxx~~ l'action attractive ou répulsive en équilibre avec le poids de l'aiguille, sans que la direction de l'aiguille fût influencée par d'autres forces.

r20 (beginning)
le premier de ces appareils, que j'ai nommé aiguille aimantée astatique, consiste dans une aiguille aimantée fixée perpendiculairement à un axe ~~auquel~~ {qu'}on peut au moyen d'un mouvement semblable à celui du pied d'un télescope et de deux vis de rappel, ~~xxx donner~~ placer dans la

r22 (end)[22]
//(89) assez grande hauteur par un fil de soie, des aiguilles aimantées semblables à des aiguilles à coudre pour la forme et la grandeur, excepte qu'elles ont une chasse à chaque extremité pour être suspendues à volonté

par l'une ou l'autre extrémité. le même pied porte deux supports de verre sur lesquels on place le conducteur galvanique dans une position horizontale, comme il se trouve alors à angles droits avec l'aiguille, on vérifie dans toutes les situations qu'on peut donner aux extrémités tant du conducteur que de l'aiguille, qu'il y a constamment attraction quand le pôle nord est à gauche du courant et répulsion quand il est a droite. {cette expérience que j'ai faite chez moi a été répetée hier à l'obsérvatoire.}

r23
après ces deux appareils j'en ai imaginé deux autres, pour constater par l'expérience que les deux actions, ~~xxx~~ {savoir}: l'action directice et l'action attractive ou répulsive, sont réciproques //(90) entre l'aiguille aimantée et le conducteur.

r24
pour l'action directrice l'appareil se compose de deux canaux, presque semicirculaires ~~xxx~~ en bois, qui sont situés sur la circonférence d'un même cercle, mais séparées par un intervalle d'un demi-pouce. on remplit ces canaux de mercure et ~~xxx~~ {on les} fait communiquer ~~l'xxx xxx xxx~~ {avec les deux} extrémités de la pile. au centre du cercle est un pivot qui supporte une aiguille en cuivre dont chaque extrémité porte une petite roue en platine très mince qui plonge dans le mercure d'un quart de ligne.

r25
cette aiguille galvanique est influencé par un barreau aimanté, comme le serait une aiguille aimantée par un conducteur galvanique. j'ai nommé cet appareil: conducteur galvanique mobile, l'expérience de sa direction //(91) au moyen d'un barreau aimanté a aussi été faite hier à l'obsérvatoire.

r26
les morceaux de platine placés aux deux extrémités du conducteur mobile, n'étant pas disposées convenablement, l'appareil était très peu sensible relativement à ce qu'il le sera quand ~~xxx~~ on aura rémédié à cet inconvénient, j'y fais travailler dans ce moment, et dès que l'instrument sera achevé, j'espère dès demain j'examinerai si le magnétisme du glôbe terrestre suffit pour ~~xxx~~ diriger le conducteur, et par conséquent pour l'amener dans une position voisine de la perpendiculaire au plan du méridien magnétique.

r27[23]
pour montrer que l'action attractive ou répulsive, est également réciproque, j'emploie un instrument semblable à l'appareil pour les attractions et répulsions magnético-galvaniques, à aimant mobile, mais dont toutes les parties sont dans une //(92) disposition inverse. le pied de cuivre est remplacé par une baguette de verre, le fil de soie par un fil métallique, l'aiguille aimantée par une aiguille de cuivre de même grandeur et de même forme, également à deux chasses, et le conducteur galvanique par un barreau aimanté soutenu par deux supports en cuivre.

r28
c'est alors l'appareil pour les attractions et répulsions magnético-galvaniques, à conducteur mobile.

r29
il est inutile de dire que quand on adaptera aux deux chasses de l'aiguille de cuivre, les deux fils qui se rendent aux deux extrémités de la pile, on observera les mêmes attractions et répulsions que dans l'appareil où c'est l'aiguille aimantée qui est mobile.

r30
//(93) une autre partie du premier des deux résultats généraux exposés cidessus, me restait à vérifier, c'est celle qui attribue à la pile elle même la même action directrice qu'au conducteur, en faisant attention que dans la ~~pile le courant va du pôle négatif~~ pile la direction que j'ai indiquée par le nom de courant ~~galvanique~~ {électrique}, va ~~du pôle de l'hydrogène à celui de l'oxigène~~ {de l'extrémité cuivre à l'extrémité zinc}, et que c'est le contraire dans le conducteur. j'avais fait construire pour cela un petit instrument, et l'expérience a encore complètement réussi, en sorte que la pile ~~et le conducteur~~ étant, pour rendre l'effet plus sensible, placée dans la direction du méridien magnétique, en mettant une aiguille aimantée sur la pile, son pôle nord était porté à gauche du courant de la pile dès qu'on établissait //(94) la communication des deux pôles par un conducteur, et que l'aiguille revenait à sa première direction de que cette communication était interrompue.

r31
en dirigeant aussi parallelement au méridien magnétique, le milieu de ce conducteur, on constatait aisement l'opposition dans le sens des deux courans,

car en laissant toujours l'aiguille aimantée sur la pile, et {en} plaçant {une} ~~l'~~autre alternativement au dessus et audessous de conducteur, celle-ci était détournée {de sa position naturelle,} dans le même sens que la première quand elle était sous le conducteur, et du coté opposé quand elle était audessus.

r32 (beginning)
lorsque j'ai montré cette expérience à Mr Arago, il m'a dit qu'on en avait fait d'analogues à genève, et je n'en parle ici que pour

---

A8. Manuscript AS 205bis(j) (*ACP* Manuscript Fragment)

Manuscript 205bis(j) is part of a larger leaf. It has been cropped on top and bottom and inscribed on both sides. The cropped borders are an exact match to those of the *ACP* manuscript sheet I have numbered pages 55–56. The match is further confirmed by the complementarity of isolated word fragments on each. The text is discussed in appendix B.

Recto (= p. 55b)

r37 (end)
//(55b), bien l'électricité ~~voltaïque~~ {galvanique}. Seulement comme le courant ne peut exister sans une serie de ces corps formant une courbe fermée, il faut en revenir à un meme corps, et la difference des tensions qu'il éprouve est la somme des tensions résultant de chaque contact, prises avec les signes convenables, comptons donc ainsi, par exemple, cuivre, zinc, eau acidulée, cuivre, le 2.^{d} cuivre acquercera une tension électrique {positive} egale à celle du zinc sur le cuivre moins ~~cette~~ {la tension négative} de l'eau acidulée sur le zinc, plus celle du cuivre sur l'eau acidulée {qui est positive, car tout métal plongé dans un acide devient positif à l'égard de cet acide. c'est même ~~pour cela que l'eau le rend plus oxidable~~ pour cela que le contact de l'acide le rend plus oxidable qu'auparavant}

r38 (beginning)
si aprés le 2.^{d} cuivre, on place encore zinc, eau acidulée, cuivre, ce 3.me cuivre aura sur le premier une

Verso (= p. 55d)

r38 (end)

//(55d) le sens où serait porte ~~l'oxigène~~ {l'hydrogène} si de l'eau interposée dans le circuit voltaïque venait à être decomposée.

r39 (beginning)

la direction suivant laquelle l'hydrogène et les bases sont transportés dans ce cas, est la definition plus précise qu'on puisse donner du courant galvanique. ~~bien entendu qu'il faut supposer l'eau interposée dans un {des} intervalle{s} de la pile occupé par le liquide,~~ car ~~xxx xxx~~ {lors meme qu'on} séparerait deux des plaques, l'une de cuivre l'autre de zinc, qui se trouvent en contact dans la pile pour y interposer de l'eau, le reste de la pile la decomposerait {toujours} ~~en sens opposé~~, en portant ~~au contraire~~ l'hydrogène du ~~coté de la plaque de zinc~~ {meme coté et dans le sens}

A9. Manuscript AS 206bis(m)

This manuscript is one side of a folded (four-sided) sheet, of which only the first side has been inscribed, the rest remaining blank. The text breaks off in the middle of a sentence, and its continuation has not been found.

20.7.^{bre}

j'appèle circuit ~~voltaïque~~ [?] électromoteur un assemblage de corps conducteurs de l'électricité disposés en une courbe fermée et ou, par une des causes qui vont être indiquées, ou par toute autre capable de produire les mêmes effets, il se développe une action, que je nomme électromotrice, et qui ne nous est connu que par les effets qu'elle produit. On ne peut donc la définir que par ces effets dont voici les principaux.

1.° si l'on interrompt le circuit à un point quelconque, ~~une des~~ les deux extrémités du circuit interrompu, ~~devient vitreuse~~ quand elles sont isolées, deviennent électriques, l'une positivement l'autre negativement, et quand l'une des deux est en communication avec le reservoir commun, l'autre manifeste des signes d'une électricité de même nature que dans le cas précédent, mais plus intense.

2.° si l'on rétablit la continuité du circuit en plaçant dans l'interruption, de l'eau, un acide, un alcali ou une ~~sel~~ {dissolution salcine}, qui communiquent avec les deux extremités du circuit interrompus, ils seront décomposés,"

Discussion

In this text Ampère is attempting the systematic exposition of his reflections on a unified treatment of the effects of open and closed piles, or electrical and galvanic phenomena. The concept of a current circuit encompassing both connecting wire and battery is an essential presupposition of this approach. Ampère had only recently developed this notion, and in this document he describes the very experiments that lent it further plausibility. The key point is that the effects produced by opening or closing the circuit may be observed at *any* point in the circuit—"à un point quelconque"—be it in the connecting wire or within the battery. Ampère's electro-motive circuits (*circuit électromoteur*) and electro-motive action (*action électromotrice*) also constitute a terminological extension of the circuit concept.

This document is clearly closely related to paragraphs r37–r39 of the lecture manuscript. Those paragraphs had begun with general reflection on the addition of tensions in galvanic combinations and on the prospect of universal electro-motive action. But he had cut that discussion short, mentioning the whole topic only briefly in his lecture. I surmise that the text transcribed above constitutes his first attempt to work it out in its own right. I also infer that his removal of paragraphs r37–r39 from the manuscript of the coming lecture coincides very closely with the time of composition of the manuscript here transcribed. As this latter manuscript bears a date, it provides a basis for dating the composition of the lecture manuscript itself.

A10. Manuscript AS 205(b)

This manuscript is a once-folded, four-sided sheet of which every side has been inscribed in Ampère's hand. One side contains calculations for the integration of an irrational function, their orientation upside-down relative to the remainder of the text. These calculations appear to have no bearing on the rest. The main text consists of two numbered lists, which I have called list A and list B. Using braces in the left margin, Ampère has grouped several of the points in list B together, briefly describing each group. Since the manuscript provides one of the rare points of reference for dating, as I show, I discuss it in some detail. I have transcribed Ampère's descriptions of each group in italics above the items ascribed to each group. In order to convey a clearer sense of the structure of this manuscript, I have also provided a facsimile (figs. A1 and A2).

(List A)

---

1. aiguille aimantée astatique.
2. aiguille de boussole remplacée par un conducteur se dirigeant de l'est à l'ouest. {la diriger avec un fort aimant.}
3. attractions et répulsions par le conducteur de l'aiguille à coudre aimantée, {aiguille à coudre à 2 chasses.}
4. construire une aiguille de boussole en bois, papier, manganèse, {étain, ou cuivre et zinc.}
5. essayer à l'école de medecine {sur l'aiguille aimantée} le conducteur déchargeant une batterie.
6. aiguille aimantée sur la pile, en sens opposé. a faire demain. {opposition dans la pile à godet}
7. potasse dissoute entre cuivre et zinc. le contraire.
8.[24] mon explication de la chaleur, vibrations des molécules dans les particules.
9. un seul fluide électrique.
10. courans galvaniques terrestre de l'est à l'ouest, c.a.d. du nord est au sud ouest.
11. prendre chez pixii son appareil d'un couple, en lui y faisant ajouter une petite aiguille aimantée au dessus du conducteur.
12. essayer de diriger l'aiguille aimantée en déchargeant de fortes jarres à travers un demi conducteur, afin que le courant galvanique dure.

---

13. mettre une graduation à l'aiguille sur une vitre ronde pour en faire un galvanomètre
14. mettre des roues de platine à l'appareil 2. aiguille galvanique rectiligne.
15. aiguille galvanique en hélice.
16. helice galvanique pour aimanter une aiguille d'acier.

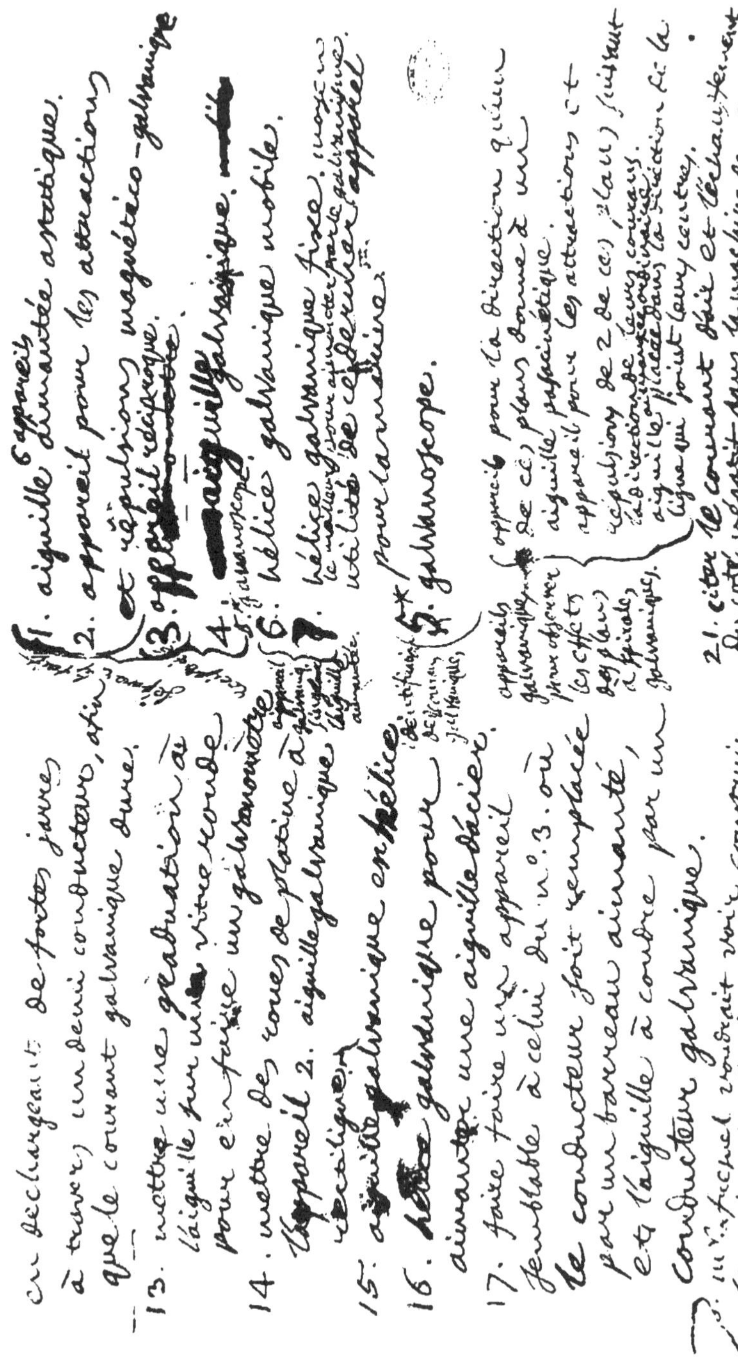

en déchargeant de fortes jarres
à travers un demi conducteur, afin
que le courant galvanique dure.

13. mettre une graduation à
l'aiguille sur une vitre ronde
pour en faire un galvanomètre

14. mettre des roues de platine à
l'appareil 2. aiguille galvanique
rectiligne.

15. aiguille galvanique en hélice

16. hélice galvanique pour
aimanter une aiguille d'acier.

17. faire faire un appareil
semblable à celui du n.° 3. où
le conducteur soit remplacé
par un barreau aimanté,
et l'aiguille à coudre par un
conducteur galvanique.

M. Fechner voudrait voir construire
la vraie aiguille à piles galvaniques
circulaires
horizontal à spirales.
20. expliquer ce mouvement

1. aiguille aimantée astatique. (6 appareils)
2. appareil pour les attractions
et répulsions magnético-galvanique
3. appareil réciproque.
4. aiguille galvanique.
5. galvanoscope.
6. hélice galvanique mobile.
7. hélice galvanique fixe. moyen
le meilleur pour aimanter par le galvanisme.
utilité de ce dernier appareil
pour la marine.

5. galvanoscope.

appareils galvaniques pour observer les effets des plans à spirales galvaniques.
appareil pour la direction qu'un
de ces plans donne à une
aiguille ...
appareil pour les attractions et
répulsions de 2 de ces plans suivant
la direction de leurs courans.
aiguille aimantée ordinaire placée dans la direction de la
ligne qui joint leurs centres.

21. citer le courant d'air et l'échauffement
du côté négatif dans la machine de verre.

22. mesurer la déviation de l'aiguille à
mesure qu'on rapproche les 2 fils
plongeant dans l'eau.

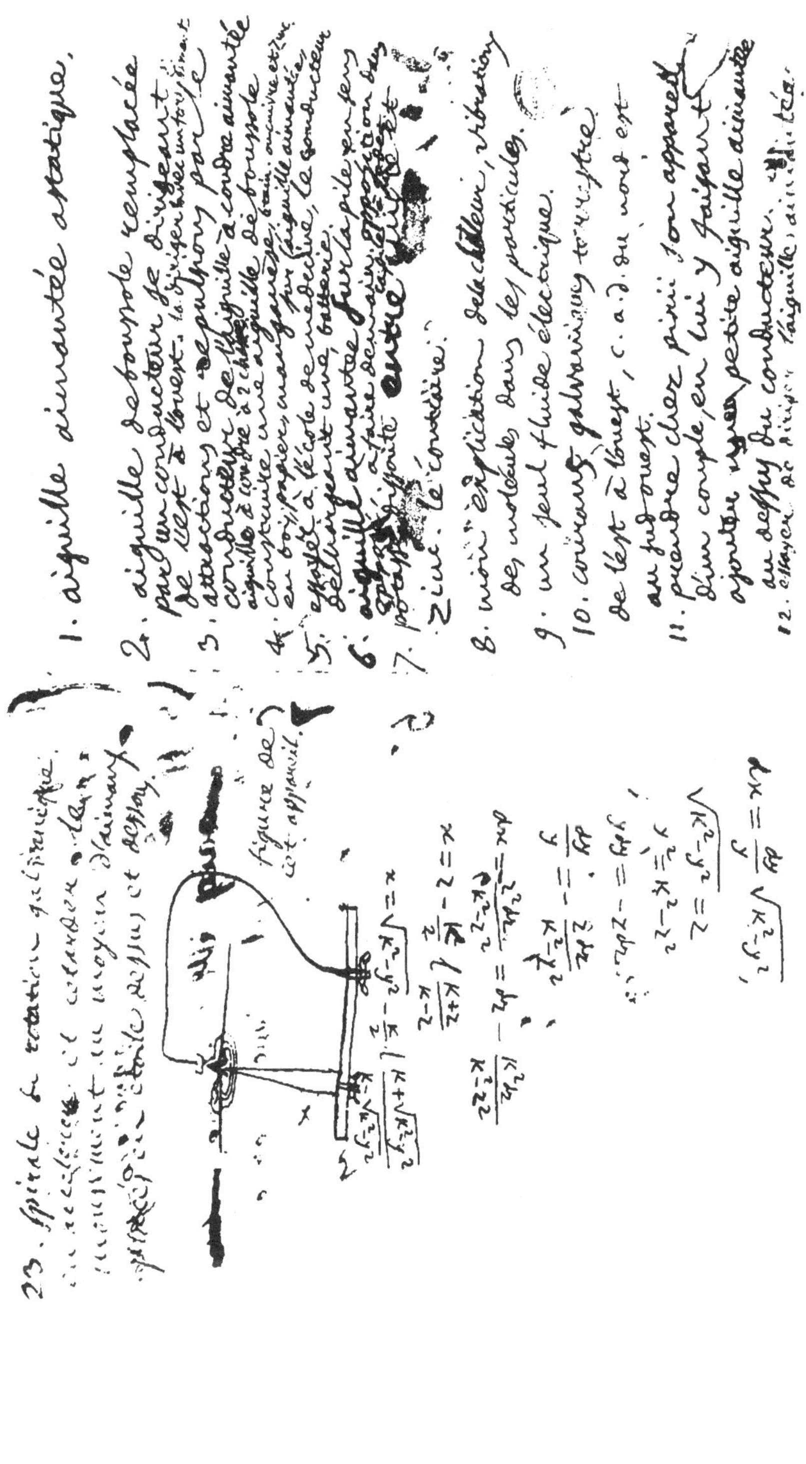

FIGS. A1 AND A2. Ampère's manuscript, front (*left*) and back (*right*) (AS 205[b]), ©Académie des Sciences–Institut de France).

17. faire faire un appareil semblable à celui du no. 3. où le conducteur soit remplacée [*sic*] par un barreau aimanté, et l'aiguille à coudre par un conducteur galvanique.
18.[25] Mr. Fresnel voudrait voir construire la vraie aiguille à piles galvaniques circulaires.
19. disque horizontale à spirales [Sketch inserted here]
20. expliquer ce mouvement
21. citer le courant d'air et l'echauffement du coté négatif dans la machine de verre.
22. mesurer la déviation de l'aiguille à mesure qu'on rapproche les 2 fils plongeant dans l'eau.
23. spirale de rotation galvanique en accélerer et retarder le mouvement au moyen d'aimans placés en etoile dessus et dessous.
[Sketch inserted here] figure de cet appareil

(List B)[26]

---

6 appareils

{séparer les faits}
1. aiguille aimantée astatique.
2. appareil pour les attractions et répulsions magnético-galvaniques.

{réciprocité}
~~2.~~ 3. ~~galvanomètre~~ appareil réciproque.
4. ~~??conducteur??~~ {aiguille} galvanique. ~~mobile.~~

{5.* galvanoscope.}

{appareil galvaniq. singeant l'aiguille aimantée}
~~5.~~ 6. hélice galvanique mobile.
~~6.~~ 7. hélice galvanique fixe. {moyen le meilleur pour aimanter par le galvanisme.} utilité de ce dernier appareil pour la marine.
{identification des courans galvaniques}
~~7.~~ 5.* galvanoscope

{appareils galvaniques pour observer les effets des plans à spirales galvaniques.} appareils pour la direction qu'un de ces plans donne à un aiguille pasacinétique.
appareil pour les attractions et répulsions de 2 de ces plans suivant la direction de leurs courans.
aiguille {aimantée ordinaire} placée dans la direction de la ligne qui joint leurs centres.

General Discussion

The two lists differ in both length and character. List A contains twenty-three items, of which some simply name instruments or experiments, while others describe tasks to be performed ("6. . . . a faire demain"; "11. prendre chez pixii son appareil . . ."; "17. faire faire un appareil . . ."). This is clearly a list with which Ampère actively worked, jotting down results and intentions using keywords. His language is often extremely terse, plainly intended only for personal use. I read this list as notes toward Ampère's private research agenda. It has, for the most part, been written as a continuous whole, containing few corrections or later modifications. Points 19 and 23 are also accompanied by schematic sketches. On purely formal grounds, the list may be divided into four parts. In items 1 through 7, Ampère's script is cramped, whereas in items 8 through 12 it takes on a very different, more open character. Following item 12, Ampère himself placed a horizontal stroke. Following item 17, the script is once again smaller and was apparently written with a different quill. The resulting division into four parts, 1–7, 8–12, 13–17, and 18–23, is also reflected in their contents (see below).

List B is much shorter, consisting of seven items and an unnumbered insertion. Braces in the left margin join items into groups, each of which is described with a few keywords. The items in this list are all instruments—there are no tasks of the sort found in list A. Unlike list A, list B shows clear signs of multiple revisions (strikethroughs, insertions, renumberings). Leaving aside the unnumbered insertion, the contents and ordering of the final version of the list correspond precisely to the detailed descriptions of instruments in §2 of Ampère's manuscript of the academy lecture of September 18. It is easy to surmise that he intended this list, from the outset, as a structured outline for this lecture (more precisely, for the part of the lecture devoted to his instruments), and as such it inevitably had to be changed several times. The last part of the list, however, was surely written later, most likely arising in connection with the lecture of September 25.

With regard to the dating of this manuscript, it should be noted that throughout both lists, Ampère uses the terminology of "galvanic currents," "magneto-galvanic interactions," and the like. The terminology of "electrical currents," which he introduces shortly after September 25 and thereafter consistently maintains, does not appear. Even without further detailed analysis, this consideration alone is enough to show that we are dealing with very early materials. In the next sections, I shall discuss the two lists in detail and start with list B, for it is closer to Ampère's working process.

Discussion of List B

The numerous discernible signs of correction (strikethroughs, later insertions, renumberings) allow us to reconstruct four successive versions of the list. Without going through the requisite analysis in any detail, I reproduce the product of my reconstructive efforts below:

First Version (up to Ampère's mark)

1. aiguille aimantée astatique.
appareil pour les attractions et répulsions magnético-galvaniques.
2. galvanomètre

Second Version (expanded, with a title added)

6 appareils
1. aiguille aimantée astatique.
2. appareil pour les attractions et répulsions magnético-galvaniques.
3. appareil réciproque.
4. aiguille galvanique [or originally: conducteur galvanique mobile].
5. hélice galvanique mobile.
6. hélice galvanique fixe. utilité de ce dernier appareil pour la marine.
[Some time later, one further item was added:]
7. galvanoscope

Third Version (further insertions, with a narrower quill; renumbering of items 5–7; division into groups)

6 appareils
séparer les faits.
1. aiguille aimantée astatique.
2. appareil pour les attractions et répulsions magnético-galvaniques.
réciprocité
3. appareil réciproque.
4. aiguille galvanique.
identification des courans galvaniques.
5. galvanoscope.
appareil galvanique singeant l'aiguille aimantée.
6. hélice galvanique mobile.
7. hélice galvanique fixe. moyen le meilleur pour aimanter par le galvanisme. utilité de ce dernier appareil pour la marine.

Fourth Version (items 1–7 unchanged; expansion by the insertion of unnumbered points)

appareils galvaniques pour observer les effets des plans à spirales galvaniques
appareils pour la direction qu'un de ces plans donne à un aiguille pasacinétique.
appareil pour les attractions et répulsions de 2 de ces plans suivant la direction de leurs courans.
aiguille aimantée ordinaire placée dans la direction de la ligne qui joint leurs centres.

The first version lists only the instruments for establishing the "general facts," and the instrument for establishing the action of the pile itself, which he already calls a "galvanometer." The second expands the list significantly, including the two reciprocal devices and the two implementations of the helix. Ampère also speaks of naval applications, by which he presumably envisions the magnetization of needles by coils. The third introduces no new items, instead presenting a rearrangement, systematization, and classification of the various instruments. It seems likely to me that Ampère undertook this rearrangement in the course of his immediate preparation for his academy lecture. At any rate, the third version of the instrument list already mirrors the lecture manuscript precisely.

Finally, the fourth version introduces three new instruments geared toward the action and interaction of galvanic spirals (as clearly distinguished from coils [hélices] wound into cylinders). The first of these is designed to investigate the action of such a spiral on a magnetic needle (Ampère's neologism, *pasacinétique*, presumably means something like "astatic"). The second targets attraction and repulsion between two spirals, while the third contemplates the action of two spirals on a magnetic needle placed between them, perhaps with a view toward the kind of magnetization he had already obtained with coils. Only for the second of these do we have any further information: this is the very instrument with which Ampère first demonstrated the interaction between currents to the academy, at his lecture of September 25. The other two are also geared toward electromagnetic interactions, but Ampère does not mention them anywhere else. These additions to the fourth version of list B were presumably made in the context of the September 25 lecture, that is, the day on which Ampère first observed the interaction between currents, a success that opened up entirely new vistas for research. Even if he had already built the two instruments for demonstrating the electromagnetic action of spirals (for which there is no evidence), in light of the spectacular discovery of the direct interaction of spirals they would have lost much of their interest.

Discussion of List A

My discussion is in four parts, following the division of the list described above. There are neither formal nor thematic indications for any more detailed division. I suspect Ampère wrote down each of the four parts as an integrated whole.

Part 1, Items 1–7

In part 1 Ampère takes note of several different instruments and experiments. Some of them (1, 2, 3) have clearly already been realized, while others (4, 5, 6) are explicitly marked as still pending. The status of item 7 can be determined only from context. The items already realized are (1) the astatic needle, (2) the first form of its reciprocal inversion, and (3) the experiment on the attraction or repulsion of the needle as a whole. In item 4, Ampère states his intention to construct needles out of other materials. The phrase *aiguille de boussole* (compass needle) makes it clear that he envisioned needles mounted in a mobile configuration, with a view toward establishing their capacity for alignment. The materials listed include some that are neither galvanic conductors nor magnetic, as well as nonmagnetic metals.

Item 5 is difficult to understand on account of Ampère's cryptic language. His unusually long formulation, *conducteur déchargéant une batterie* (conductor discharging a battery) suggests that he was not referring to a galvanic battery, which he always called a *pile* (this applied even to trough batteries, which he called *piles à auges*; see first mémoire, part 1, p. 68). Instead, he must have meant a "battery" in the traditional sense of an array of Leyden jars. His intention was to investigate the action of discharging such a battery on a magnetic needle, which explains why he wanted to make the attempt elsewhere, rather than at home, like his other experiments. His home laboratory included a voltaic pile but certainly no Leyden jar battery, which was practically a high-tech device. He would return to this task in item 12.

Items 6 and 7 concern magnetic effects exerted by the pile itself. Item 6 proposes to determine the direction of the deviation of the magnetic needle, which he expects to be the inverse of the direction of the action exerted by the wire in the same current circuit, an expectation he also plans to verify using a different kind of battery, the *pile à godet*, presumably a kind of trough device. In his first mémoire he would later stress the fact that piles of this sort, now called *piles à auges*, are especially suited to the observation of such magnetic action (first mémoire, part 1, p. 68).[27] In item 7 he seems most likely to have envisioned an investigation into whether and how the salts within "electrolysis" cells incorporated into the pile are decomposed. Given the expected "inversion" of the direction of current within the pile, this question is of particular importance. Such an experiment would permit the mutual confirmation of electromagnetic and electrochemical methods of establishing direction of current.

Part 2, Items 8–12

Items 8 and 9 contain several theoretical notes. It's not clear whether they represent future tasks or results already obtained. Item 8 mentions the account of heat he had worked out many years earlier. There he had attempted simultaneously to address heat, electricity, and magnetism by means of a single ether theory.[28] Item 9 is related, for in the manuscript in question Ampère had attempted to derive electricity, along with all the other interactions, from the effects of a *single* fluid, and had thus taken a stand in the old dispute between one- and two-fluid theories. But his theoretical inquiries quickly moved in a very different direction, and such reflections are absent from his later work.

Items 10 and 11 are concerned with the electrical circuit hypothesis, and especially with the representing of the direction of current within such circuits. In item 10, Ampère takes note of his reflections on the direction of currents in terrestrial circuits, then proposes a more precise experimental investigation of the direction of current within the pile. Fortunately, the galvanic apparatus mentioned in item 11, a device with only a single pair of plates, is ideally suited to this purpose, especially when a magnetic needle is brought into proximity with it, as Ampère plans. It also allows for the constant monitoring of the pile's activity, the very task for which he originally conceived the instrument he called a "galvanometer."

In item 12 Ampère returns to the task he had set himself in item 5, that of investigating possible magnetic action on the part of a wire discharging a Leyden jar battery. Here he adds only the further important stipulation that the battery be discharged through a *demi conducteur* (half conductor), so that "the galvanic current lasts." We may infer that he had actually performed the experiment described in item 5, without observing any effect, and then ascribed the failure to the fact that under normal conditions the discharge takes place too quickly to produce a motion on the part of the magnetic needle. The procedure sketched in item 12 would thus be an attempt to perform a modified version of the experiment, prolonging the duration of the discharge by means of a poor conductor.

The fact that item 11 is, in a sense, an extension of item 6, just as item 12 extends item 5, and that in both cases he has most probably conducted experiments in the interim, makes it clear that the first two parts of this list are intimately connected but were written down in two different phases. The two were separated by an experimental phase.

Significantly, neither of these first two parts as yet contains any mention of the phenomenon of magnetization by galvanic currents, as investigated by Arago. Nor is there any mention of experiments with coils of the sort that would soon prove so decisive. The idea of a circuit of current appears only with respect to terrestrial currents and not in its full generality. The first two parts of list A are thus a good fit with the first version of list B: they describe a state of research when the only experimentally confirmed results were the astatic needle, the attraction effect, and the magnetic action of the pile (which Ampère immediately thought to exploit in devising a measurement device). The corresponding instruments, and only those instruments, are precisely those listed in the first version of list B.

Part 3, Items 13–17

To an even greater extent than the first two, this part of the text constitutes a pure enumeration of tasks to be performed. Ampère has observed that the instrument proposed in item 11 does not yet fully deserve to be called a galvanometer, and in item 13 he considers ways of improving it, at least by providing it with a scale. Item 14 concerns the "reciprocal" instrument proposed in item 2, in which the mobile needle was to be replaced with a mobile conductor. Ampère now envisions mounting these conductors on platinum wheels. As background to this idea, one might imagine serious difficulties involved in producing the desired effect and in Ampère's ensuing search for ways of improving the mobility of his apparatus.

Instruments with coils appear for the first time in item 15. Ampère's proposal for a "galvanic needle in the form of a coil" shows that he was thinking of using a sufficiently mobile coil like a magnetic needle. This item is closely related to remarks in the manuscript of his September 18 lecture, in which he alludes to efforts to improve this instrument that had yet to yield any satisfactory results. In item 16 he considers another possible use for the coil: magnetizing steel needles. By this point he must already have been in close contact with Arago, who had performed precisely these experiments. Finally, in item 17 he proposes the construction of a reciprocal apparatus for investigating the attraction-repulsion effect described in item 3. It is possible that he expected it to be easier to demonstrate the reciprocity of electromagnetic action using this device than with the previously envisioned reciprocal inversion of the astatic needle, whose experimental implementation had proved challenging.

Part 3 of list A stands in close correspondence with the second version of list B, in which Ampère has added to his list of instruments the two reciprocal devices, the galvanic helix in both of its applications, and finally the galvanoscope. These are the very instruments with whose development and improvement part 3 is concerned. In addition, part 3 evidently reflects a close exchange with Arago. We know the two collaborated on experiments in the Obsérvatoire on September 17. That this list was composed around this time, or at any rate shortly before the lecture of September 18 was drafted, may also be inferred from its congruence with the instruments and experiments mentioned in the manuscript of this lecture (see app. B, §B2, r33).

Part 4, Items 18–23

This part is diverse and occasionally cryptic. Its first item (18) alludes to a communication with Fresnel. The *vraie aiguille* (true needle) is geared toward the analogy with which Ampère had pictured terrestrial circuits of current: a very long, circular pile doubling back upon itself. Several items are devoted to experiments with spirals. Unlike those described in the fourth version of list B, these have to do not with the interaction between spirals or their action on magnets but with a rotatory motion on the part of the spirals themselves. The experiment sketched in items 19 through 21 appears to have involved the rotation of a horizontally mounted galvanic spiral about a vertical axis. In item 21, Ampère appears to be considering other possible causes, besides electricity and magnetism, that might bring about such motion. Item 23 imagines a similar experiment, this time with the motion caused or influenced by nearby magnets arranged in a star pattern. Nowhere else does Ampère mention these experiments, nor are the theoretical considerations that might have led him to anticipate their results particularly transparent. It may be that they or at least the idea for them also arose in conversation with Fresnel, who was evidently much quicker to open speculation than Ampère himself.

It was also Fresnel who, in connection with the study of electromagnetic effects, performed experiments aimed at using magnetism to produce electricity (see my sketch in chap. 4). In particular, he looked for signs of water decomposition at the submerged ends of a wire wound around a magnet. It may be that Ampère was considering such experiments in item 22, which remains the most cryptic in this list. Here, too, we have an experimental arrangement with two submerged wire ends.

The arrangement of the text on the manuscript page makes it clear that part 4 was written last and, in particular, after completion of the fourth and final version of list B. This makes it likely that part 4 was added after September 25 and possibly considerably later.

Dating and Collating the Two Lists

As the discussion makes clear, the two lists were composed not consecutively but in parallel. They are closely related both to each other and to the manuscript of Ampère's academy lectures. Table A3 provides an overview of the results of their comparison.

TABLE A3. Lists A and B from Ampère's manuscript AS 205(a)

List A	List B	Dating
1–7		
8–12	First version (two instruments)	
13–17	Second version (expanded: galvanic coils)	Shortly before September 18
	Third version (rearrangement)	Around September 18 (revision of the lecture manuscript)
	Fourth version (expanded: instruments with spirals)	Around September 25
18–23		After September 25

Given that various elements of lists A and B came about before September 18, they offer a significant way to make out a chronological structure within the first two weeks of Ampère's investigation, a structure hardly discernible within the lecture manuscript—the only other direct source we have for this period.

A11. Ampère's October 11, 1820, Letter to William Hyde Wollaston

The original of this letter has not survived, nor is it listed in Ampere's correspondence (AC) collected for publication by Louis de Launay. It exists only in the form of a notice in the archive of the Royal Society of London (designated there as AP 9.2), which contains an English translation of the letter. It was presumably prepared in anticipation of the November 9 meeting of the Royal Society, at which Ampère's letter was read.

The text consists of three pages of clear script, without corrections. Its status as literal translation, rather than paraphrase, is made more probable by the fact that the scribe has included such details as the polite closing salutation.

The "copies of a printed note" to which Ampère refers in the letter are the offprint of his "Conclusions," his very first printed text on electromagnetism.

Withdrawn

Subject of a Letter from Monsr Ampère to Dr. Wollaston P.R.J.
dated Paris 11 Oct 1820

Sir

Enclosed I send you two copies of a printed note requesting your acceptance of one of them & begging that you will present the second in my name to the Royal Society of London.

The experiments of which it contains the principal results will be described in a Memoir now in the press, of which I will also forward two copies as soon as they are ready for delivery.

The principal facts which I have established, & which I think new, are that when a wire conveying a current of Electricity between the two extremities of a Voltaic pile is adjacent to another wire conveying a similar current parallel to the former, these conducting wires attract, or repel each other accordingly as the currents move in the same or in opposite directions.

The attractions & repulsions which form the subject of the present notice, differ essentially from those which are known to take place between bodies positively & negatively electrified in the ordinary way.

First, they do not take place between these conductors of the Voltaic pile till the circuit is complete between the two extremities of the pile; consequently not until the electric tension, on which common electric attraction & repulsion depend, is at an end or nearly so.

The second difference observable is that this species of attraction takes place between wires similarly affected, & repulsion where they are dissimilarly situated; but, in common Electricity, the reverse is the case: bodies similarly Electrified repel each other, while attraction occurs only between bodies oppositely electrified.

Thirdly, this attraction does not cease upon contact as in common cases of Electric attraction, but the wires which have come into contact remain firmly attached to each other.

The principal experiments on which these results are founded have been performed before the Institute at two of their sittings in Sept[r] & October, and they appear to me of the more importance, because they lead to the conclusion that the phenomena of the magnet are but effects of a particular case of Electric attraction.

I shall be extremely happy if you will do me the favor to accept a copy of a work in which I flatter myself that I have made some progress in the new path of science opened to us by the learned Danish Professor Oersted, & I shall be much honored if the Royal Society think my labour not beneath its notice.

I am Sir with all respect, &c &c

A. Ampère, Member of the Institute ~~xxx~~

APPENDIX B

Reconstruction of the Manuscript of Ampère's Lectures of September 18 and 25, 1820

B1. Sources, Methods, and Tools

Document AS 208bis(f), the manuscript of the second part of the *Annales de chimie et de physique* (*ACP*) *mémoire*, is the most important single source for this reconstruction. I have transcribed it in its entirety in appendix A, in the form it had taken on by November 1820 and in which it survives today. In this piece (in the section he labeled "§III"), Ampère referred explicitly to the lectures of September 18 and 25, and indeed he had used the text originally drafted for those lectures in preparing the manuscript for the printed text. In the course of that transformation, however, he modified its original structure in various ways. Many passages have been edited, deleted, and inserted. Large segments were rearranged or struck. Such segments have been located elsewhere among Ampère's papers, either in the immediate vicinity of the *ACP* manuscript or in other folders. AS 208bis(g) and 205bis(j), also transcribed in appendix A, are of particular significance.

Reconstructing the original manuscript out of the various extant pieces required a range of different procedures:

- recovering the original ordering of the text
- identifying and re-inserting deleted passages
- identifying and removing later insertions

- deciphering passages, including whole sections, that had been struck through
- recovering and deciphering pasted-over passages
- identifying and reversing Ampère's later edits

The goal of these procedures was to rearrange the text and reconstitute the original phrasing, as it was written down for the lectures of September 18 and 25, 1820. Diverse criteria factored into the rearrangement, ranging from associations of text and content to such material traces as co-occurrence on the same sheet of paper, tear and cut marks, kind of paper used, and so on. Typically, the supposed connection between two pieces can and must be established by reference to the agreement of multiple criteria. One very important indicator is Ampère's shift from the word *galvanique* to *électrique*, discussed earlier. As late as September 25 he continued to refer to *courant galvanique*, *action magneto-galvanique*, and the like, reserving the word *électrique* for conventional electricity. This practice changed soon thereafter, so that by the time the mémoire went to press, nearly all instances of *galvanique* had been replaced by *électrique*. This shift yields a first criterion for excluding passages from the reconstruction: all passages in which the term *électrique* is used from the outset must have been added after September 25 and could not have been part of the original manuscript.

Applying such methods to the *ACP* manuscript, our principal source, is no easy matter, however. The all-important §III comprises some thirty-five pages, some in consecutively bound sheets, others on loose leaves, and without pagination or paragraph numbering. To make unequivocal reference to all this material possible, I have introduced my own pagination. In addition, for §III, the portion of chief interest, I have numbered the paragraphs. I have likewise numbered the paragraphs of my final reconstruction of the original lecture manuscript. Finally, to permit cross-referencing with the *published* version of the *ACP* mémoire, I have also numbered the paragraphs of §III as they appear there. In what follows, I will thus be working with three different series of paragraph numbers, of which table B1 gives a concordance:

- Paragraph numbers a1, a2, a3, and so on refer to §III of the *published* version of the *ACP* mémoire, beginning on page 196.
- Paragraph numbers m1, m2, m3*, and so on refer to the corresponding portion of the *ACP manuscript* in its present form. This numbering includes paragraphs deleted in their entirety, with the numbers of such paragraphs followed by an asterisk (*). The numbers of paragraphs pasted over later are followed by a plus sign (+). Where a paragraph

begins a new manuscript page, I insert my own page number in parentheses, following the paragraph number.

- Paragraph numbers r1, r2, r3, and so on refer to the *reconstructed manuscript* of the lectures of September 18 and 25 (see the text of the reconstruction below). It is important to recall that this numbering, unlike the other two, refers not to any extant document but to the product of a reconstruction.

Subsequent to the rearrangement of extant manuscript segments, reconstruction demands the recovery of the original text. As already indicated, the manuscript text that survives today exhibits the traces of numerous corrections, changes, deletions, and insertions. But not all of these were

TABLE B1. Concordance of paragraph numberings in the printed version, in the existing manuscript, and the reconstructed version of Ampère's lecture manuscript

<table>
<tr><th>Text printed in ACP (Ampère 1820j): paragraph (page)</th><th>ACP manuscript, 208bis(f): paragraph (page)</th><th>Reconstructed version of lecture manuscript of Oct. 25, 1820: paragraph</th></tr>
<tr><td>a1 (196)</td><td>m1 (35–36)</td><td></td></tr>
<tr><td rowspan="7">a2 (196–97)</td><td>m2 (36)</td><td></td></tr>
<tr><td>m3* (37)</td><td>r1 (“Lu le 18.7bre”)</td></tr>
<tr><td>m4* (37)</td><td>r2</td></tr>
<tr><td>m5* (37)</td><td>r3</td></tr>
<tr><td>m6* (37)</td><td>r4</td></tr>
<tr><td>m7* (37–38)</td><td>r5</td></tr>
<tr><td>m8* (38)</td><td>r6 (“§1”)</td></tr>
<tr><td>a3 (197)</td><td>m9 (38)</td><td>r7</td></tr>
<tr><td>a4 (beginning) (197)</td><td>m10 (38–39)</td><td>r8</td></tr>
<tr><td rowspan="3">a4 (halfway) (197–98)</td><td>m11 (39–40)</td><td>r9</td></tr>
<tr><td>m12+ (40)</td><td>r10</td></tr>
<tr><td>m13*+ (40)</td><td>r11 (beginning)</td></tr>
<tr><td>a4 (end) (198)</td><td>m14 (40)</td><td>r10a</td></tr>
<tr><td>a5 (198)</td><td>m15 (40)</td><td></td></tr>
<tr><td>a6 (beginning) (198–99)</td><td>m16 (40)</td><td></td></tr>
<tr><td>a6 (end) (199)</td><td>m16a (41–42)</td><td>r20 (end)</td></tr>
<tr><td rowspan="3">a7 (199)</td><td>m17 (42)</td><td>r21</td></tr>
<tr><td>m18* (42)</td><td>r22</td></tr>
<tr><td>m19* (43)</td><td>r11 (end)</td></tr>
</table>

TABLE B1 (*continued*)

Text printed in *ACP* (Ampère 1820j): paragraph (page)	*ACP* manuscript, 208bis(f): paragraph (page)	Reconstructed version of lecture manuscript of Oct. 25, 1820: paragraph
a8 (beginning) (199–200)	m20 (43–44)	r12
a8 (end) (200)	m21 (44)	r13
a9 (200–201)	m22 (44–46)	r14
a10 (201)	m23 (46–47)	r15
a11 (201–2)	m24 (47–48)	r16
	m25* (48)	r17
a12 (202)	m26 (49)	
	(50) empty	
	m27* (51)	r34 ("§III")
a13 (202–3)	m28 (51–52)	r35
a14 (203–4)	m29 (52–54)	r36
a15 (beginning) (204)	m30 (54–55a)	r37
	m31+ (55c)	r38 (halfway)
	m32+ (55e)	r39 (end)
a15 (end) (204–5)	m33 (55f)	r40
a16 (205)	m34 (55f–56)	r41
a17 (205–6)	m35 (56–58)	r42
a18 (beginning) (206)	m36 (58)	r43
a18 (end) (206–7)	m37 (58–59)	r44
a19 (207)	m38 (59–60)	r45
a20 (207–8)	m39 (60–61)	r46, r46a
a21 (beginning) (208)	m40 (61)	r47
a21 (end) (208)	m41 (62)	r48
a22 (208)	m42 (62)	r49
	m43* (63)	r32 (end)
a23 (208–9)	m44 (63–64)	r33, r33a
a24 (footnote, 209)	m45 (64)	
a25 (209)	m46 (65)	

Note: Paragraph numbers followed by an asterisk (*) indicate paragraphs Ampère later struck through in their entirety; those followed by a plus sign (+) indicate paragraphs Ampère later pasted over.

made in the course of Ampère's November 1820 preparation of the *ACP* manuscript; some must have been made earlier. Some of the discernible changes had already been incorporated into the text prepared for the September 25 lecture. As this was the text I had set out to reconstruct, changes made prior to September 25 had to be distinguished from those introduced at a later date and the later changes, to some extent, reversed. In this process I was guided by the following considerations:

- Ampère had various occasions for reviewing and, where appropriate, revising his manuscript prior to September 25. After all, it consists of the drafts of *two* lectures, on each of which he continued to work until shortly before its presentation. But after September 25 he had no reason to return to this material for some time. Only when he began to assemble the *ACP* mémoire did he have need to carefully review it, and he then made substantive changes.
- Consequently, we may start from the assumption that changes within paragraphs later scored through in their entirety may have been made prior to September 25. Only in a few instances is this assumption plainly false, as when Ampère penned a new paragraph for publication, only to discard it while editing. This is the case on several pages of manuscript 208bis(g), for example, page 87, r19, or page 93.
- Once again, Ampère's terminological shift is of capital importance. His nearly complete replacement of *galvanique* by *électrique* can only have been carried out after September 25, most likely in the course of preparing the *ACP* mémoire for publication. Paragraphs not intended for inclusion in the printed mémoire still exhibit the original terminology. This is particularly clear in the main portion of 208bis(g), pages 81–92.
- Numerous smaller insertions (such as words inscribed between lines of text) can be categorized by reference to other changes made in the same paragraph or in its immediate vicinity, where peculiarities of script or quill point show that they were made at the same time. In such instances, the dates ascribed to one set of changes may be transferred to the other (as, e.g., in r8 and r10a).
- In the absence of any point of reference for dating a given edit, I have ascribed it to the earlier period—with the exception of purely stylistic edits, which, given the fact that Ampère was clearly far more concerned with (relatively) fluid style in preparing the text for publication, I have consigned to the later period.

On the basis of such considerations, I have been able to reconstruct the text Ampère had in hand on September 25 in its entirety, with very little residual uncertainty. This reconstruction has, for the first time, made possible a detailed analysis of the contents of Ampère's lectures. Its full text is reproduced in the following section. Table B2 provides an overview of the archival location of each paragraph.

TABLE B2. Location of the paragraphs of the reconstructed version in the *ACP* manuscript and other manuscripts

Reconstructed version of Ampère's lecture manuscript of Oct. 25, 1820: paragraph	*ACP* manuscript, 208bis(f): paragraph (page)	Other manuscripts
r1 ("Lu le 18.7bre")	m3* (37)	
r2	m4* (37)	
r3	m5* (37)	
r4	m6* (37)	
r5	m7* (37–38)	
r6 ("§1")	m8* (38)	
r7	m9 (38)	
r8	m10 (38–39)	
r9	m11 (39)	
r10	m12+ (40)	
r10a	m14 (40)	
r11 (beginning)	m13*+ (40)	
r11 (end)	m19*(43)	
r12	m20 (43–44)	
r13	m21 (44)	
r14	m22 (44–46)	
r15	m23 (46)	
r16	m24 (47–48)	
r17	m25* (48)	
r18 ("§2")		208bis(g), 87
r19		208bis(g), 87–88
r20 (beginning)		208bis(g), 88
r20 (end)	m16a (41–42)	
r21	m17 (42)	
r22 (beginning)	m18* (42)	

Table B2 (*continued*)

Reconstructed version of Ampère's lecture manuscript of Oct. 25, 1820: paragraph	*ACP* manuscript, 208bis(f): paragraph (page)	Other manuscripts
r22 (end)		208bis(g), 89
r23		208bis(g), 89–90
r24		208bis(g), 90
r25		208bis(g), 90–91
r26		208bis(g), 91
r27		208bis(g), 91–92
r28		208bis(g), 92
r29		208bis(g), 92
r30		208bis(g), 93–94
r31		208bis(g), 94
r32 (beginning)		208bis(g), 94
r32 (end)	m43* (63)	
r33	m44 (63–64)	
r34 ("§III")	m27* (51)	
r35	m28 (51–52)	
r36	m29 (52–54)	
r37 (beginning)	m30 (54–55a)	
r37 (end)		205bis(j), 55b
r38 (beginning)		205bis(j), 55b
r38 (halfway)	m31+ (55c)	
r38 (end)		205bis(j), 55d
r39 (beginning)		205bis(j), 55d
r39 (end)	m32+ (55e)	
r40	m33 (55f)	
r41	m34 (55f–56)	
r42	m35 (56–58)	
r43	m36 (58)	
r44	m37 (58–59)	
r45	m38 (59–60)	
r46	m39 (60–61)	
r47	m40 (61)	
r48	m41 (62)	
r49	m42 (62)	

B2. Complete Text of the Reconstructed Lecture Manuscript

What follows is the text Ampère very probably had on hand when he delivered his lecture at the Académie des Sciences on September 25, 1820. Changes and insertions made prior to this date have been marked as such, following the conventions set out in appendix A, section A5.

---

r1 (= m3, p. 37)*
lu le 18.7.^{bre}

r2 (= m4)*
dans le travail {que j'offre aujourd'hui à l'académie} sur l'action réciproque du galvanisme et de l'aimant ~~que j'offre aujourd'hui à l'académie royale de sciences, je me suis proposé:~~ {découverte par M. Oersted, je me suis proposé:}

r3 (= m5)*
1.º de distinguer les deux résultats généraux de cette action, auxquels peuvent être ramenés tous les faits observés par l'auteur de {cette} ~~l'~~importante découverte ~~dont M.r Oersted xxx d'enricher la physique~~ {l'une de plus remarquables de la physique moderne,} et de ramener l'énoncé de ces deux résultats à l'expression la plus simple et la plus générale.

r4 (= m6)*
2.º de décrire les expériences que j'ai faites et les instrumens que j'ai imaginés, pour les présenter isolés l'un de l'autre et de l'action du glôbe terrestre sur l'aiguille aimantée, qui se combine avec eux dans les expériences de Mr Oersted, et en altère les effets {ou en} xxx dissimule les lois.

r5^{1} (= m7)*
{3.^{o} ~~de ramener~~ [?] {d'expliquer} ces deux résultats généraux en les ramenant à une même cause, qui rend raison //(38) en même tems d'un assez grand nombre de phénomènes connu depuis longtems ou plus recemment constatés:}

r6 (= m8)*
§. 1.er
~~xxx~~ exposition des deux résultats généraux au moyen desquels on peut rendre raison de diverses expériences de M.r Oersted et prévoir ce que l'action

réciproque ~~du galvan~~ d'un aimant et d'un conducteur galvanique, dans les cas qu'on n'a pas encore soumis à l'obsérvation.

r7 (= m9)
ces résultats consistent, d'une part dans ~~xxx~~ l'action directrice d'un de ces corps sur l'autre, de l'autre part, dans l'action attractive ou répulsive qui s'établit entre eux, suivant les circonstances.

r8 (= m10)
action directrice: lorsqu'un aimant et un conducteur galvanique agissent l'un sur l'autre, et que l'un d'eux, étant fixe, l'autre ne peut que tourner dans un plan perpendiculaire à la plus courte distance du conducteur//(39) et de l'axe de l'aimant, celui qui est mobile tend à ~~xxx~~ {se mouvoir} de manière que les directions du conducteur et de l'axe de l'aimant forment un angle droit, et que le pole de ~~l'aiguille~~ l'aimant qui regarde habituellement le nord soit à gauche de ce qu'on a nommé le courant galvanique, et le pole sud à sa droite.

r9 (= m11)
pour conserver à cet énoncé toute la généralité dont il est susceptible, il faut distinguer deux sortes des conducteurs galvaniques, 1°. la pile même, dans laquelle ce qu'on nomme le courant galvanique se porte du pôle où il se produit de l'hydrogène dans la décomposition de l'eau à celui d'où l'oxigène se degage. 2°. le fil métallique qui unit les deux extrémités de la pile, et où l'on doit alors considérer le même courant comme se portant au contraire du pôle qui donne de l'oxigène à celui {qui} developpe //(40) de l'hydrogène.

r10 (= m12+; later pasted over with 10a)
je n'emploie au reste ici le mot de courant galvanique que comme une expression convenue pour définir les directions dont il s'agit. cette ~~xxx~~ expression pourrait être employée ~~xxx~~ comme ~~xxx~~ d'un usage commode lors même qu'on n'admettrait pas l'éxistence réelle d'un pareil courant.

r10a (= m14; replacement for r10)
on peut comprendre ces deux cas dans une même définition en disant qu'on entend par courant galvanique la direction suivant laquelle l'hydrogène et les bases des sels sont transportés par l'action de toute la pile, en la concevant celle-ci comme formant avec le conducteur un seul circuit,{.} ~~si~~ [?] ~~à un point quelconque de~~ {lorsqu'on interrompt} ce circuit ~~on l'interrompait~~ pour y placer soit de l'eau soit une dissolution saline qui seraient décom-

posés. au reste tout ce que je vais dire dans ce paragraphe et dans les deux suivans ne suppose aucunement qu'il y ait réellement un courant dans cette direction, et on peut ne considérer que comme une manière commode et usitée de désigner cette direction, l'emploi que je fais ici de cette dénomination: courant galvanique.

r11 (= m13*+ and m19*)
pour préciser ~~le sens dans lequel {que} je donne ici~~ {le signification que j'attribue} à ces mots la droite, la gauche d'un courant galvanique, il faut faire une supposition analogue à celle à l'aide de laquelle on définit la droite et la gauche d'un fleuve, en concevant un homme qui le descend la face tournée vers son embouchure. on supposera de même un homme placé dans le sens du courant galvanique de manière que ce courant aille de ses pieds //(43) à sa tête, et qu'il ait la face tournée vers l'aiguille, alors sa main droite sera du coté que j'appèle la droite du courant, et sa main gauche marquera la gauche du courant.

_r12 (= m20)_
action attractive où répulsive. ce second résultat général consiste: 1.° en ce qu'un courant galvanique et un aimant, dont l'axe fait un angle droit avec la direction de ce courant qu'on imagine dans ce conducteur d'après le définitions précédentes, s'attirent quand le pôle nord est à gauche du courant, c'est-à-dire quand la position est celle que le conducteur et l'aimant tendent à prendre en vertu de leur action mutuelle, et se repoussent quand le pole nord de l'aimant est a la droite du courant, c'est-à-dire quand le conducteur et l'aimant sont {maintenus} dans la position opposée à celle qu'ils tendent à //(44) se donner mutuellement.

_r13 (= m21)_
on voit par l'énoncé même de ce deux résultats que l'action entre le conducteur et l'amant [_sic_] est toujours réciproque. c'est cette réciprocité que je me suis d'abord attaché à vérifier, quoiqu'elle me parut assez évidente par elle-même, quelques unes des expériences que je vais décrire ont ~~xxx~~ pour objet de la constater.

_r14 (= m22)_
les deux modes d'action entre un amant et un conducteur galvanique que je viens d'exposer, en les considérant comme de simples résultats de l'expérience suffisent pour rendre raison ~~de tous les faits xxx que nous a fait~~

~~connaitre~~ {des faits observés par} M.r Oersted, et pour prévoir ce qui doit arriver dans les cas analogues à l'égard de{squels} on n'a point encore fait d'obsérvation. ils indiquent par exemple d'avance tout ce qui doit arriver quand un conducteur //(45) galvanique agit sur l'aiguille d'inclinaison. je n'entrerai dans aucun détail à cet égard, puisque tout ce que je pourrais dire sur ce sujet, découle immédiament des enoncés précédens. je dirai seulement qu'après avoir déduit seulement le premier résultat ~~des expériences de M~~. général de la note de M.r Oersted, j'en déduisis l'éxplication dont j'exposerai les bases dans le 3.me paragraphe de ce mémoire, que cette explication me conduisit au second résultat général, et me suggéra, pour le constater, une expérience qui réussit complètement. lorsque je la communiquai à M.r Arago, il me fit remarquer avec raison, que cette attraction et cette répulsion entre un aimant et un conducteur galvanique placés à angles droits dans la direction où ils tendent à se mettre mutuellement dans la direction opposée, pouvait [*sic*] seule rendre raison des résultats //(46) publiés par l'auteur de la découverte, dans le cas où l'aiguille aimantée étant horizontale, on en approche un conducteur galvanique dans une situation verticale, et qu'on pouvait même déduire aisement cette loi, de l'une des expériences de M.r Oersted, celle qu'il enonce ainsi: posito autem filo ~~(accipiente~~ (cuius extremitas superior electricitatem à termino négativo apparatûs galvanici accipit) è regione puncto inter polum et medium acùs sito, occidentem versus agitur.

r15 (= m23)

car ce mouvement ~~xxx~~ {de} l'aiguille aimantée, indiqué comme ayant lieu, soit que le conducteur soit à l'occident ou à l'orient de l'aiguille, est dans le premier une attraction parce que le pôle nord se trouve à la gauche du courant, et dans le second une répulsion parce //(47) qu'il se trouve à droite.

r16 (= m24)

mais en convenant de la justesse de cette obsérvation, il me semble que la distinction que j'ai faite des deux résultats {généraux} de l'action mutuelle d'un aimant et d'un conducteur galvanique, n'en devient que plus importante pour expliquer en montrant que dans ce cas c'est tantôt une attraction et tantôt une répulsion, toujours conformement à la loi ~~que je viens d'exposer~~ {du second résultat général} que je viens d'éxposer, tandis que dans l'expérience que M.r Oersted énonce immédiament avant en ces termes: quando filum conjungens perpendiculare ponitur è regione polo acùs magneticae, et extremitas superior fili electricitatem à termino negativo apparatûs galvanici accipit, polus orientem versus movetur, ce mouvement n'a lieu que pour que //(48) l'aiguille aimantée prenne à l'ègard du conducteur la direction déterminée par le premier résultat général, avec toutes les circonstances que j'ai comprises dans son énoncé, et en particulier la remarque qui le termine{.} ~~, celle remarque le second.~~

r17 (= m25)*

la condition exprimée dans cette remarque ne semble pas liée nécessairement à cette détermination de la direction de l'aiguille, mais elle resulte immédiament ~~du principe que j'ai donné~~ {des} considérations par lesquelles j'ai ramené à une cause unique, les deux résultats généraux que je viens d'exposer, et qui seront développées dans le 3.me paragraphe de ce mémoire. {je reviendrai alors sur ~~cette expérience~~ {l'explication}, ainsi que ~~xxx~~ {sur} les expériences de Mr Oersted avec des conducteurs recourbés en fer à cheval de manière que les deux branches soient paralleles entre elles.}[2]

r18 (p. 87)

§.2.

description des appareils et ~~des~~ {indication des} expériences qui me paraissent les plus propres à ~~démontrer la généralité des~~ établir les deux résultats généraux exposés dans le paragraphe précedent.

r19

{dans} les observations de Mr. Oersted la position ~~de~~ {que prend} l'aiguille aimantée soit qu'elle dépende de l'action directice, ou de l'action attractive ou répulsive, est influencée par ~~xxx~~ la force qui dirige cette aiguille dans l'état ordinaire, et ~~quelques fois par les poids~~ il {n'}est ~~difficile~~ [?] {pas toujours} de distinguer ~~dans~~ [?] ~~chaque expérience ce qui~~ celles des ses expériences dont les résultats sont dus à l'action directrice et celles qui dépendent de l'action attractive ou répulsive. il y a même un de faits qu'il rapporte où le poids de l'aiguille aimantée {n'}a ~~xxx~~ permis ~~empeché~~ {au} conducteur galvanique de l'abaisser ou de la soulever, conformement aux lois de son action sur elle, que quand il en était très rapproché, //(88) probablement à cause de la distance à laquelle le centre de gravité de l'aiguille se trouvait de son point de suspension. j'ai donc d'abord cherché à construire ~~des~~ {deux} appareils ~~xxx~~ {tels que} l'action ~~répulsive~~ directrice agit seule dans l'un, sans que la force qui dirige {constamment} l'aiguille aimantée, ni les poids de cette aiguille, ne pussent influer sur elle, et ~~xxx~~ {que} ~~xxx~~ l'autre montrât les effets de ~~xxx~~ l'action attractive ou répulsive en équilibre avec le poids de l'aiguille, sans que la direction de l'aiguille fût influencée par d'autres forces.

r20 (second part = m16a)
le premier de ces appareils, que j'ai nommé aiguille aimantée astatique, consiste dans une aiguille aimantée fixée perpendiculairement à un axe ~~auqel~~ {qu'}on peut au moyen d'un mouvement semblable à celui du pied d'un télescope et de deux vis de rappel, ~~donner~~ placer dans la //(41) direction que l'on veut, ~~xxx xxx~~ {en sorte} que l'aiguille aimantée ne peut se mouvoir qu'en tournant dans un plan perpendiculaire à cet axe. il faut que son centre de gravité soit placé exactement dans l'axe, en sorte qu'avant qu'elle soit aimantée ~~xxx~~ on puisse s'assurer que la pesanteur n'a aucune action pour mouvoir l'aiguille. on l'aimante alors, et cet instrument peut servir à vérifier que tant que le plan où se meut l'aiguille n'est pas perpendiculaire à ~~l'aiguille~~ la direction de l'aiguille d'inclinaison, le magnetisme terrestre tend à faire prendre à l'aiguille aimantée la direction de celle des lignes traceés sur le plan qui est le plus rapprochée possible de la direction de l'aiguille d'inclinaison, c'est-à-dire la projection de cette direction sur le même plan. on place ensuite l'axe parallelement ~~xxx~~ à la direction de l'aiguille d'inclinaison, le plan où se meut l'aiguille aimantée lui étant alors perpendiculaire le magnetisme terrestre n'a plus //(42) aucune action pour diriger l'aiguille aimantée, qui devient ainsi complètement astatique. le ~~xxx~~ {meme appareil} porte dans le plan de cette aiguille un circle divisé en degrés, sur lequel sont fixé deux petits barreaux de verre pour attacher les conducteurs galvaniques, dont l'action directrice agit alors seule, et sans complication avec la pesanteur et le magnétisme terrestre.

r21 (= m17)
la principale expérience à faire avec cet appareil est de montrer que l'angle entre ~~l'aiguille~~ les directions de l'aiguille et du conducteur est toujours droit quand *l'action directrice* est la seule qui ait lieu.

r22 (second part = m18)*
Le second instrument que j'ai nommé: appareil pour les attractions et répulsions magnético-galvaniques à aimant mobile, consiste dans un pied auquel on peu [*sic*] suspendre ~~xxx~~ d'aucune situation verticale, et à //(89) assez grande hauteur par un fil de soie, des aiguilles aimantées semblables à des aiguilles à coudre pour la forme et la grandeur, excepte qu'elles ont une chasse à chaque extremité pour être suspendues à volonté par l'une ou l'autre extrémité. le même pied porte deux supports de verre sur lesquels on place le conducteur galvanique dans une position horizontale, comme il se trouve alors à angles droits avec l'aiguille, on vérifie dans toutes les

situations qu'on peut donner aux extrémités tant du conducteur que de l'aiguille, qu'il y a constamment attraction quand le pôle nord est à gauche du courant et répulsion quand il est a droite. {cette expérience que j'ai faite chez moi a été répetée hier à l'obsérvatoire.}

r23
après ces deux appareils j'en ai imaginé deux autres, pour constater par l'expérience que les deux actions, ~~xxx~~ {savoir}: l'action directice et l'action attractive ou répulsive, sont réciproques //(90) entre l'aiguille aimantée et le conducteur.

r24
pour l'action directrice l'appareil se compose de deux canaux, presque semicirculaires ~~xxx~~ en bois, qui sont situés sur la circonférence d'un même cercle, mais séparées par un intervalle d'un demi-pouce. on remplit ces canaux de mercure et ~~xxx~~ {on les} fait communiquer ~~l'xxx xxx xxx~~ {avec les deux} extrémités de la pile. au centre du cercle est un pivot qui supporte une aiguille en cuivre dont chaque extrémité porte une petite roue en platine très mince qui plonge dans le mercure d'un quart de ligne.

r25
cette aiguille galvanique est influencé par un barreau aimanté, comme le serait une aiguille aimantée par un conducteur galvanique. j'ai nommé cet appareil: conducteur galvanique mobile, l'expérience de sa direction //(91) au moyen d'un barreau aimanté a aussi été faite hier à l'obsérvatoire.

r26
les morceaux de platine placés aux deux extrémités du conducteur mobile, n'étant pas disposées convenablement, l'appareil était très peu sensible relativement à ce qu'il le sera quand ~~xxx~~ on aura rémédié à cet inconvénient, j'y fais travailler dans ce moment, et dès que l'instrument sera achevé, j'espère dès demain j'examinerai si le magnétisme du glôbe terrestre suffit pour ~~xxx~~ diriger le conducteur, et par conséquent pour l'amener dans une position voisine de la perpendiculaire au plan du méridien magnétique.

r27[3]
pour montrer que l'action attractive ou répulsive, est également réciproque, j'emploie un instrument semblable à l'appareil pour les attractions et répulsions magnético-galvaniques, à aimant mobile, mais dont toutes les parties

sont dans une //(92) disposition inverse. le pied de cuivre est remplacé par une baguette de verre, le fil de soie par un fil métallique, l'aiguille aimantée par une aiguille de cuivre de même grandeur et de même forme, également à deux chasses, et le conducteur galvanique par un barreau aimanté soutenu par deux supports en cuivre.

r28
c'est alors l'appareil pour les attractions et répulsions magnético-galvaniques, à conducteur mobile.

r29
il est inutile de dire que quand on adaptera aux deux chasses de l'aiguille de cuivre, les deux fils qui se rendent aux deux extrémités de la pile, on observera les mêmes attractions et répulsions que dans l'appareil où c'est l'aiguille aimantée qui est mobile.

r30
//(93) une autre partie du premier des deux résultats généraux exposés ci-dessus, me restait à vérifier, c'est celle qui attribue à la pile elle même la même action directrice qu'au conducteur, en faisant attention que dans la ~~pile le courant va du pôle négatif~~ pile la direction que j'ai indiquée par le nom de courant galvanique, va du pôle de l'hydrogène à celui de l'oxigène, et que c'est le contraire dans le conducteur. j'avais fait construire pour cela un petit instrument, et l'expérience a encore complètement réussi, en sorte que la pile ~~et le conducteur~~ étant, pour rendre l'effet plus sensible, placée dans la direction du méridien magnétique, en mettant une aiguille aimantée sur la pile, son pôle nord était porté à gauche du courant de la pile dès qu'on établissait //(94) la communication des deux pôles par un conducteur, et que l'aiguille revenait à sa première direction de que cette communication était interrompue.

r31
en dirigeant aussi parallelement au méridien magnétique, le milieu de ce conducteur, on constatait aisement l'opposition dans le sens des deux courans, car en laissant toujours l'aiguille aimantée sur la pile, et {en} plaçant {une} ~~l~~'autre alternativement au dessus et audessous de conducteur, celle-ci était détournée {de sa position naturelle,} dans le même sens que la première quand elle était sous le conducteur, et du coté opposé quand elle était audessus.

r32 (last part = m43)*
lorsque j'ai montré cette expérience à Mr Arago, il m'a dit qu'on en avait fait d'analogues à genève, et je n'en parle ici que pour //(63) proposer aux physiciens qui s'occupent d'expériences sur la pile galvanique, d'y adapter une aiguille aimantée pour voir à chaque instant par la position qu'elle prend si ~~xxx~~ le courant galvanique est établi, et quelle est son intensité, précisément comme on adapte un électroscope à une machine électrique. ce petit instrument que je crois par analogie devoir nommer galvanoscope, peut au reste ne différer d'une boussole ordinaire que par l'usage qu'on en fait. {on peut en rendant très mobile et très legère, rendre cet instrument propre à reconnaitre l'existence de courans galvaniques très faibles, et c'en est le seul moyen.}

r33 (= m44)
~~je fais dans ce moment~~ ~~executer plusieurs autres appareils, deux entre autres, où le~~ {j'ai construit un autre appareil où le fil} conducteur {est} ~~galvanique étant~~ plié en hélice {~~à pas très fins~~} autour d'un tube de {verre,} ~~carton ou de verre doit~~ d'après la théorie que je me suis faite de ces sortes de phénomènes, {ce conducteur doit} presenter quand on y fera passer le courant galvanique, une action ~~xxx~~ //(64) semblable à celle d'une aiguille ou d'un barreau aimanté, dans toutes les circonstances où ceux-ci agissent sur d'autres corps, ou sont mus par le magnétisme terrestre. j'ai {déja observé une partie des} ~~donné à ces deux appareils les noms d'helice galvanique fixe et d'helice galvanique mobile. je vais donner quelques détails sur les expériences que j'ai faites avec ces instrumens,~~ {effets que j'attendais de cet instrument.}

r34 (= m27, p. 51)*
Lu le 25.7.^{bre}
§.III.
Explication des phénomènes précedens reduction des deux résultats généraux qu'elle ramène à un principe unique à l'aide duquelle on rend raison ~~des plusieurs autres~~ non seulement de ces phénomènes mais des ~~plusiers~~ autres faits dépendant de la théorie de l'aimant.

r35 (= m28)
la première réflexion ~~qui m'a conduit à cette explication est~~ {que je fis lorsque je voulus chercher les causes des nouveaux phénomènes découverts par Mr Oersted, est} que l'ordre dans lequel on a découvert deux faits,

ne faisant rien aux conséquences des analogies qu'il [*sic*] présentent, nous pouvions supposer qu'avant de savoir que l'aiguille aimantée prend une direction constante du sud au nord, on avait d'abord connu la propriété qu'elle a d'être amenée par un courant galvanique dans une situation perpendiculaire ~~xxx~~ {à ce courant}, de manière que le pôle nord de l'aiguille fût porté à gauche ~~de l'aig~~ du courant, ~~xxx xxx xxx~~ {et qu'on découvrit ensuite la} propriété qu'elle a de tourner constamment au nord celle de // (52) ses extrémités qui se portait à gauche du courant, l'idée la plus simple et celle qui se présenterait immédiament à celui qui voudrait expliquer la direction constante du nord au sud, ne serait-elle {pas} d'admettre dans la terre ce qu'on appèle un courant galvanique dans une direction telle que le nord se trouvât à gauche d'un homme qui couché sur sa surface pour avoir la face tournée du coté de l'aiguille, recevrait ce courant dans la direction de ces pieds à sa tête, en sorte que ce courant se porterait de l'est à l'ouest ~~en xxx~~ dans une direction perpéndiculaire au méridien magnétique.

r36 (= m29)
cette hypothèse ~~xxx~~ {devient d'autant} plus probable ~~en faisant~~ {qu'on fait plus} attention à l'ensemble des ~~phénomènes qu'xxx à l'obsérvateur~~ faits connus, ce courant s'il existe doit être comparé à //(53) celui que j'ai montré dans la pile ~~xxx~~ agir sur l'aiguille aimantée comme ~~xxx~~ {se} dirigeant ~~l'aiguille aima~~ du pôle de l'hydrogène à celui de l'oxigène, {quand on établissait un coducteur entre eux et } ~~et~~ qui aurait lieu ~~xxx~~ de même si la pile formant une courbe fermée il venaient se réunir en un couple semblable aux autres, car il n'y a ~~rien dans~~ probablement rien dans notre glôbe qui ressemble à un conducteur continu et homogène, mais les matières diverses dont il est composé sont précisement dans le cas d'une pile voltaïque ~~composé~~ formée d'élémens disposés au hazard, et qui revenant sur elle-même formerait comme une cinture continue tout autour de la terre. des élémens ainsi disposés donnent moins d'énergie électrique sans doute que s'ils l'étaient dans un ordre périodiquement régulier, mais il faudrait qu'ils fussent arrangés à dessein pour ~~xxx~~ que dans une série de substances différentes, formant une courbe fermée autour de la terre, il n'y //(54) eut pas courant dans un sens ou dans l'autre. il se trouve que d'après l'arrangement des substances de la terre ce courant a lieu de l'est à l'ouest, et qu'il dirige partout l'aiguille aimantée perpendiculairement à sa propre direction ~~qui trace~~ {. cette direction trace} ainsi sur la terre un parallèle magnétique, et de manière que le pôle ~~xxx~~ de l'aiguille ~~xxx~~ {qui doit être} à gauche ~~xxx~~ {~~dirige xxx xxx~~}, ~~xxx xxx~~ {se trouve par là constamment porté} vers le nord {, dans celle du méridien magnétique.}

r37 (first part = m30)

je ferai remarquer à ce sujet que les effets produits par les piles de la construction anglaise où l'on brule un fil ~~xxx~~ {fin de} métal ~~xxx~~ même avec une seule ~~xxx~~ {paire} dont le zinc et le cuivre plongent dans un acide, prouve suffisament que c'est une supposition trop restrainte de n'admettre l'action électro-motrice qu'entre les métaux, et de ne regarder le liquide interposé que comme //(55a) conducteur. Il y a sans doute action entre deux métaux, Volta la démontré da la manière la plus complète, mais est-ce une raison pour qu'il n'y en ait pas entre eux et d'autres corps? Ou entre ceux-ci seulement? Il y en a ~~sans doute~~ {probablement} entre tous les corps qui peuvent conduire plus ou moins

r37 (end)

//(55b), bien l'électricité ~~voltaïque~~ {galvanique}. Seulement comme le courant ne peut exister sans une serie de ces corps formant une courbe fermée, il faut en revenir à un meme corps, et la difference des tensions qu'il éprouve est la somme des tensions résultant de chaque contact, prises avec les signes convenables, comptons donc ainsi, par exemple, cuivre, zinc, eau acidulée, cuivre, le 2.^{d} cuivre acquercera une tension électrique {positive} egale à celle du zinc sur le cuivre moins ~~cette~~ {la tension négative} de l'eau acidulée sur le zinc, plus celle du cuivre sur l'eau acidulée {qui est positive, car tout métal plongé dans un acide devient positif à l'égard de cet acide. c'est même ~~pour cela que l'eau le rend plus oxidable~~ pour cela que le contact de l'acide le rend plus oxidable qu'auparavant}

r38 (middle = m31+)

si aprés le 2.^{d} cuivre, on place encore zinc, eau acidulée, cuivre, ce 3.me cuivre aura sur le premier une //(55c) tension double de celle du second sur le premier, et ainsi de suite. Pour que ces tensions s'établissent il faut que les deux extrémités de la pile ne communiquent pas, des qu'on les fait communiquer, il n'y a plus de tension, mais un courant dans //(55d) le sens où serait porte ~~l'oxigène~~ {l'hydrogène} si de l'eau interposée dans le circuit voltaïque venait à être decomposée.

r39 (last part = m32+)

la direction suivant laquelle l'hydrogène et les bases sont transportés dans ce cas, est la definition plus précise qu'on puisse donner du courant galvanique. ~~bien entendu qu'il faut supposer l'eau interposée dans un {des} intervalle{s} de la pile occupé par le liquide,~~ car ~~xxx xxx~~ {lors meme qu'on} séparerait deux des plaques, l'une de cuivre l'autre de zinc, qui se trouvent en contact dans la pile pour y interposer de l'eau, le reste de la pile la decomposerait {toujours} ~~en sens opposé~~, en portant ~~au contraire~~ l'hydrogène du ~~coté de la plaque de zinc~~ {meme coté et dans le sens} //(55e) du courant galvanique, malgré l'action de cette plaque de zinc vers laquelle il se porte alors, et se ferait contre elle que se dégagerait l'hydrogène tant que le circuit serait fermé.

r40 (= m33)
//(55f) l'électricité galvanique. ~~xxx~~ cette action est plus sensible dans les piles composées des métaux et d'acides étendus, tant parce qu'il parait que ce sont les substances où elle se développe avec le plus d'énergie que parce que ce sont celles qui conduisent le mieux l'électricité.

r41 (= m34)
les ~~xxx xxx~~ {diverses arrangemens} que nous pouvons donner à des corps non métalliques ne sauraient produire une action électro-motrice comparable à celle d'une pile voltaïque à disques métalliques séparés alternativement par des liquides, à cause du peu de longeur qu'il nous est permis de donner à nos appareils, mais une pile qui fait le tour de la terre conserverait sans doute quelqu'intensité lors même qu'elle ne serait pas composée de métaux, et que les élemens en seraient //(56) arrangés au hazard, car sur une si grande longeur il faudrait que l'arrangement fût fait à dessein pour que les actions dans un sens fussent exactement détruites par les actions dans l'autre.

r42 (= m35)
je crois devoir faire observer à ce sujet que des courans galvaniques dans un même corps, ne peuvent être indépendans les uns des autres, à moins qu'ils ne fussent séparés par des substances qui les isoleraient ~~xxx {et encore d'aucun cas le même, il y}~~ ~~autrement c'est dans la direction où il y aurait le plus~~ {completement dans toute leur étendue}, et encore dans ce cas la même ils devraient influer les uns sur les autres, puisque leur action se transmet à travers tous les corps, à plus forte raison lorsqu'ils coexistent dans un ~~corps~~ {glôbe} dont toutes les parties sont continues, doivent-ils se diriger dans tous dans le même suivant la direction que tend à leur donner //(57) ~~xxx xxx~~ la ~~xxx~~ {réunion} de toutes les actions électromotrices de ce glôbe. je suis bien loin au reste de croire que ce soit seulement dans ces actions que réside la cause des courans galvaniques qui y sont indiqués par la direction de l'aiguille aimantée à chaque point de la surface de la terre, je crois au contraire que la cause principale en est toute différente comme

je le dirai ~~tout à l'heure~~ dans la suite de ce mémoire, mais comme ~~xxx~~ cette cause, dépendante de la rotation de la terre, donnerait en chaque lieu une direction constante à l'aiguille, ce qui est contraire à l'obsérvation, je regarde l'action électromotrice des substances dont se compose la planète que nous habitons, comme se combinant avec cette action générale et en en expliquant les variations à mesure que l'oxidation fait des progrès dans l'un ou l'autre //(58) région continentale de la terre.

r43 (= m36)
quant aux variations {diurnes} elles s'expliquent aisement par le changement de températures alternatif de ces ~~xxx~~ {deux régions} pendant la durée d'une rotation de la terre, d'autant plus facilement qu'on connait depuis longtems ~~l'action~~ l'influence de la température sur l'action ~~galvanique~~ {électromotrice} sur laquelle Mr Dessaignes a fait des obsérvations très interessantes. au reste il faut compter parmi les actions ~~galvaniques dans~~ {électromotrices des} differentes parties de la terre celle ces minerais aimantés qu'elle contient et qui doivent comme nous allons voir être considérés comme autant de piles galvaniques.[4]

r44 (= m37)
l'élevation de température qui a lieu dans les conducteurs ~~xxx xxx~~ {des} courans doit avoir lieu aussi dans ceux du glôbe terrestre, ne serait-ce pas là la cause de cette chaleur interne constatée récemment par les expériences rapportées dans une //(59) des dernières séances le l'académie par un de ses membres ~~qui xxx xxx~~ dont les travaux sur la chaleur ont ~~portée la lumière crée une nouvelle branche~~ ont fait rentrer cette partie de la physique dans le domaine des mathématiques? et quand on fait attention que cette élevation de température ~~xxx~~ {produit}, quand le courant est assez énergique, une incandescence {permanente} accompagnée de la plus vive lumière, sans combustion où déperdition de substance, ne pourrait-on pas en conclure que les ~~corps xxx~~ {glôbes opaques} ne le sont qu'a ~~xxx~~ {cause} du peu d'énergie des courans galvaniques qui s'y établissent, et trouver dans des courans plus actifs la cause de la chaleur et de la lumière ~~du xxx lumineux~~ {des globes qui brillent par eux-mêmes}? ~~maintenant si c'est la cause le l'action directrice que la terre exerce sur l'aiguille aimantée, comment ne serait pas aussi cette~~

r45 (= m38)
on sait qu'on expliquait autre fois par des courans, qu'on nommait magnétiques, //(60) mais on les supposait parallèles à l'axe de l'aimant, pas de

possibilité sans se croiser et se détruire, d'ailleurs pourquoi plus tôt du ~~nord~~ {sud} au nord que du nord au sud.

r46 (= m39)

maintenant si c'est là la cause directrice de la part de la terre ce le sera aussi de la part d'un aimant d'où il suit qu'un aimant est un ~~composé~~ [?] assemblage de courans galvaniques qui ont lieu dans des plans perpendiculaires à son axe, dirigés de manière que le pôle austral de l'aimant, qui se porte du coté du nord se trouve à droite de ces courans puisqu'il est toujours à ~~droite~~ {gauche} d'un courant placé hors de l'aimant

r46a

{et qui lui fait face dans une direction parallèle, ou plutôt ces courans s'établissent d'abord dans l'aimant suivant les courbes fermées les plus courtes soit de gauche à droite, soit de droite à gauche, et alors la ligne perpendiculaire aux plans de ces courans devient //(61) l'axe de l'aimant, et ses extremités en font les deux poles.} Ainsi à ~~xxx~~ {chacun} des poles d'un aimant les courans galvaniques dont il se compose sont dirigés suivant des courbes fermées concentriques, j'ai imité cette disposition autant qu'il était possible avec un courant galvanique en en pliant le fil ~~en spirale~~ conducteur en spirale. ~~xxx que~~ cette spirale {était} formée avec un fil de laiton ~~dont les extremites etaient renfermées xxx~~ {et terminèe par deux portions rectilignes de ce même fil, qui étaient renfermées dans} deux tubes de verre afin qu'elles ne communiquassent pas entre elles et pussent être attachées aux deux extrémités de la pile.

r47 (= m40)

suivant le sens dans lequel on fait passer le courant dans une telle spirale elle est en effet {fortement} attirée ou repoussée par le pole d'un aimant qu'on lui présente de manière que la direction de son axe soit perpendiculaire au plan de la spirale.[5]

r48 (= m41)

//(62) en remplaçant l'aimant par une autre spirale, dont le courant soit dans le même sens que le sien on a les mêmes attractions et répulsions, c'est ainsi que j'ai découvert que deux courans galvaniques s'attiraient quand ils avaient lieu dans le même sens, et se repoussaient dans le cas contraire. je me propose de le verifier sur des courans rectilignes.[6]

r49 (= m42)

en remplaçant au contraire dans l'expérience de l'action mutuelle d'un des pôles d'un aimant et d'un courant dans un fil métallique plié en spirale, cette spirale par un aimant on a encore les mêmes effets soit en attraction soit en répulsion, conformément à la loi des phénomènes connus de l'aimant, il est évident d'ailleurs que toutes les circonstances de ces phénomènes, sont une suite nécessaire de la disposition des courans galvaniques dont ils se composent, d'après la manière dont ceux-ci s'attirent et se répoussent.

B3. Specific Reconstruction Notes

It is neither possible, nor would it make for palatable reading, to provide a thorough accounting of all of the minute considerations that went into the reconstruction reproduced above. Instead, I content myself with explaining several larger-scale rearrangements and discussing a few exemplary minutiae. This should suffice to convey a sense of the style of the reflections at work throughout.

r1–r6 These paragraphs were written continuously, without interruption, and later struck out in their entirety. The first (r1) appears at the beginning of a page, beneath the title, "Lu le 18 7bre" ("Read on September 18"), and in r2 Ampère refers to the research to be presented to the academy "today."

r7–r9 The text of the preceding paragraphs is continued, with the difference that, rather than striking them, Ampère retains these paragraphs in his later draft. The sentence in m10 = r8 regarding the terminological shift ("dénomination . . . électrique") is definitely a later insertion. Given the style of the script, the same may be inferred for the sentence beginning and ending, "Bien entendu . . . deux pôles."

r10–r11 The page on which these paragraphs appear (p. 40) was, on two separate occasions, pasted over with new text. In the upper half, the terminological evidence is enough to show that r10 was pasted over with r10a before September 25. Later changes to r10a presumably did not take place until the *ACP* manuscript was in preparation, for Ampère struck out his reference to "le deux [paragraphes] suivans" (the two following paragraphs), having adopted different paragraph breaks for the print version. Similarities of script allow this dating to be transferred to the other changes made to r10a as well. The lower half of the page must have been pasted over at the same time, as the terminology definitively shows. The page had originally ended with the first part of r11 (later struck out and pasted over), and its text continued on page 43 (one sentence spans the end of page 40 and the beginning of page 43). The loose leaf containing pages 41–42 was interposed later. Furthermore, pages 39–40 and 43–44 all belong to a single folded sheet. I will discuss the substance of these changes below, in section B6.

r12–r16 These paragraphs are continuous and uninterrupted. Ampère's few changes must have been made in editing the *ACP* manuscript; once again, he has struck a reference to the "3.me paragraphe" (3rd paragraph), having changed his paragraph breaks.

r17 This paragraph, which occupies page 48, was later struck in its entirety, without replacement. It concluded the first section of the tripartite division Ampère had announced at the beginning.

r18–r33, summary overview With a few exceptions (see below), these paragraphs did not make it into the later text, nor do the sheets on which they are written belong to the manuscript of the *ACP* mémoire. They have been archived separately, though in close proximity to it. Despite their status as separate archival documents, their intimate substantive connection to the manuscript is indisputable, and so I have continued my pagination, numbering them pages 87–94.

r18–r19 This passage begins with the heading, "§2: Description des appareils et indication des expériences qui me paraissent les plus propres à établir les deux résultats généraux exposés dans le paragraphe précedent" (Description of the apparatus and account of the experiments that seem to me most appropriate to establishing the two general results expounded in the previous paragraph). Ampère is here referring back to the tripartite division described in r3–r5.

r20–22 The first part of r20 is on page 88, the last sentence of which is divided by the page break. Though page 89 begins in the middle of a sentence, it is neither the same sentence begun on page 88, nor is it even on the same topic. It is clear from context that the continuation of page 88 may be found on page 41. The sheet containing pages 41–42 was relocated to its current position later. These pages begin with the end of r20 and continue with r21 and the beginning of r22. Once again, the page break at the end of page 42 interrupts a sentence, the continuation of which may be found on page 89. Though Ampère retained r20 and r21 in his later text, he struck out the entirety of r22, which was written on page 42, part of the manuscript delivered for publication.

r23–r31 These paragraphs form an uninterrupted sequence on pages 89–94. In preparing the manuscript for his *ACP* mémoire, Ampère originally intended to include r30 (his corrections update the terminology and include a redescription of the battery by reference to copper and zinc poles, instead of hydrogen and oxygen poles). But he decided not to.

r32–r33 Paragraph r32 begins on page 94, only to be interrupted in the middle of a sentence by the page break. The conclusion of the sentence, and of the remainder of r32, may be found on page 63, on which the entirety of the text belonging to the paragraph has been struck. Pages 63–64 are on a loose leaf, later relocated to its current position. The last sentence of m44 = r33 was added later (terminological evidence). Paragraph r33 concludes the second of the three parts Ampère had advertised.

r34 Page 51 begins with this paragraph, which has been struck out in its entirety. It contains the heading, "Lu le 2xx 7bre. § III. Explication des phénomènes précedens. Reduction des deux résultats généraux qu'elle ramène à un principe unique à l'aide duquelle on rend raison non seulement de ces phénomènes mais des autres faits dépendant de la théorie de l'aimant" (Read on September 2xx. §III. Explication of the preceding phenomena. Reduction of the two general results to which they point to a single principle, by means of which one may make sense not only of these phenomena, but of other facts depending on the theory of the magnet). Though the striking of this text has left the date illegible, there can be no doubt that the heading refers to the third of the three parts announced in r5, delivered in the lecture of September 25.

r35–r36 These paragraphs were written in continuous succession on pages 51–54 and display no major deletions or restructuring.

r42–r49 This text has been written in continuous succession, displaying no major restructuring. To judge by script alone, r44 represents the beginning of a new session. Paragraph r46 must presumably have originally ended on page 60, but later sentences were added to it in a clearly different script, continuing onto page 61. Paragraphs r47 and r48 were written without interruption on pages 61–62. In revising the *ACP* manuscript, Ampère added one additional sentence to r47 (terminological evidence). The third part of the lecture ends with r49.

On September 25, by his own account, Ampère followed §III with a reading of his "Conclusions," also reproduced in the published mémoire immediately following r48. But the "Conclusions" are not in the lecture manuscript. For the *ACP* manuscript, Ampère pasted in a copy of his special offprint.

B4. Control Criteria and Comparison with Published Sources

Ordering of the Manuscript Sheets

A further investigation provides additional support for this reconstruction. The ordering of sheets and pages in portions of the reconstruction diverges widely from that of the final manuscript. Ampère frequently accomplished the reordering by tearing the original sheets, which had usually been folded to form four sides. The complementary tear marks he made in the process provide a clear indication of which pages originally belonged together. On occasion I have needed to test a page ordering reconstructed on the basis of other considerations by recourse to such evidence. The fact that this test always proved successful greatly reinforces the validity of the reconstruction. It is completely consistent with my analysis of the physical evidence for the ordering of pages. This evidence shows that Ampère wrote the original lecture manuscript in two matching booklets, §1 in the first and §§2 and III in the second. Both booklets consisted of bound sheets folded into four sides, and, unlike the later manuscript, they were not supplemented by loose leaves.

Comparison of Ampère's Accounts in the AGSP Essay and the "Notes"

Ampère's later account in the *Annales générales des sciences physiques* (*AGSP*) essay, the first part of which reports on the lectures of September 18 and 25, provides important support.[7] It is highly probable that he consulted the original of his lecture manuscript in preparing this text. A comparison of the topics treated in this essay with those of the reconstructed lecture manuscript shows that their ordering in the published text largely coincides with the ordering in my reconstruction. The essay does not mention any topics not also covered in the reconstructed lecture, though it does contain some omissions. On the one hand, these may be explained by the demands of brevity to which a general overview must conform. But they also reveal characteristic features of Ampère's self-presentation.

Further confirmation may be gained by comparing the reconstructions with the short account of his lectures to date in the "Notes," which Ampère assembled at Hachette's request in November 1820 (Ampère 1820h). The "Notes" contain no topics that are not also covered in my reconstruction, though they also leave out a great deal. With respect to chronology, the "Notes" display a mild tendency to date the theory of circuits of electrical current earlier, while also downplaying the initial task of "reduction." Both features are entirely consistent with the tenor of Ampère's self-presentation in November 1820, some six to seven weeks after the events, as is his pervasive reference to *courans électriques* rather than *galvaniques*.

B5. More Detailed Dating of the Manuscript

The reconstructed text provided above reflects the state of the manuscript Ampère had on hand at the time of his academy lecture of September 25, that is, at the end of what I have called phase II. By this time, as the text shows, he had already reworked, modified, and enlarged it. For purposes of my investigation of the first three weeks of Ampère's research, it was important to date these changes more precisely, at least to the point of assigning each of them to either phase I or phase II.

As both manuscript and later account make clear, on September 18 Ampère read only §§1 and 2 (paragraphs r1–r33). On September 25 he read §III (paragraphs r34–r48). Ampère cited time limits as grounds for confining his first lecture to §§1 and 2. So it is not out of the question that at the time of this lecture, September 18, he may already have had a text that went beyond these first two sections. Indeed, he announced the tripartite division quite early.[8] Throughout the September 18 lecture he repeatedly looked ahead to the planned third section (e.g., in r14 and r17). A special inquiry into the real scope of the text Ampère had at hand on September 18 is thus called for. After a few general remarks, I discuss paragraphs of particular importance separately.

I take it as given that all of paragraphs r1–r33 belong to phase I, as do the changes made to these paragraphs. In phase II, between the first and second lectures, Ampère had no reason to expand or modify the text of the lecture he had already delivered. Of primary significance are the contents envisioned for each of the three parts announced at the beginning of the lecture. The first part was to discuss the two *faits généraux*, the second, their experimental basis, and the third, theory. The substance of this division is strictly observed in the first two parts but interrupted toward the end of the third. Here, contrary to Ampère's advertised intentions, we again find a discussion of instruments (r46a–r49). It is thus highly probable that these paragraphs came about only *after* the first lecture. On that occasion he had already read the section on instruments, so that the presentation of any new instrument devised thereafter would have to be deferred until the delivery of §III, contrary to his original division.

Table B3 provides an overview of the outcome of this investigation. Those edits made with a view toward improving style, but without introducing any change in content, I have marked with the sign *. Paragraphs r10–r11, r37–r39, and r46, however, display important substantive changes, which I thus consider separately.

Table B3. Paragraph chronology for Ampère's lecture manuscript

	Phase I		Phase II		
Paragraph	**Composed**	**Revised**	**Composed**	**Revised**	**Remarks**
r1 ("Lu le 18.7^{bre}")	·				
r2	·	*			
r3	·	*			
r4	·				
r5	·				
r6 ("§1")	·				

TABLE B3 (*continued*)

	Phase I		Phase II		
Paragraph	Composed	Revised	Composed	Revised	Remarks
r7	•				
r8	•	*			
r9	•				
r10	•	•			Later replaced by r10a
r10a	•	•			
r11	•	*			
r12	•				
r13	•				
r14	•	*			
r15	•	*			
r16	•	*			
r17	•	•			
r18 ("§2")	•	*			
r19	•	*			
r20	•	*			
r21	•				
r22	•	•			
r23	•	*			
r24	•	*			
r25	•				
r26	•				
r27	•				
r28	•				
r29	•				
r30	•				
r31	•				
r32	•				
r33	•	•			
r34 ("§ III")			•		
r35			•		
r36			•		

Table B3 (*continued*)

	Phase I		Phase II		
Paragraph	Composed	Revised	Composed	Revised	Remarks
r37 (beginning)			•		
r37 (end)			•	•	Later wholly removed
r38			•	•	Later wholly removed
r39			•	•	Later wholly removed
r40			•		
r41			•	*	
r42			•	*	
r43			•	•	
r44			•	•	Likely the beginning of a new session
r45			•		
r46			•	•	
r46a			•	•	
r47			•		
r48			•		
r49			•		Likely the beginning of a new session

B6. Paragraphs r8–r11

1. Reconstruction of the successive versions

In its present condition, page 40 was pasted over twice. It is possible to detach the pasted sections and, by means of such features as sentence continuity, strikethroughs, overlapping pastings, and so on, to reconstruct the earlier stages of the manuscript. This process shows that the current manuscript represents the last of four successive versions. The sequence of changes encompasses pages 40–43 in their entirety.

- The *first* version contained neither pastings nor longer deletions. After page 40, its text continues on page 43, as it survives today.

- For the *second* version, the upper half of page 40 was pasted over with new text (m14 = r10a), with the text of the lower half remaining as in the first version. This text also continues on page 43.
- In the *third* version, the entire text of r11, on the lower half of page 40 and the upper half of page 43, was scored out without replacement.
- Finally, in the *fourth* version, the lower half of page 40 was pasted over with new text (m15 and the beginning of m16) and continued on the freshly inserted pages 41–42. This leaf was not originally composed for this purpose but was removed from an earlier text, and its last part (p. 42, m18* = the beginning of r22) struck through on insertion in the present manuscript. In this condition, the text largely corresponds to the published version in *ACP*.

TABLE B4. Condition of Ampère's lecture manuscript for r8–r11

Page	Paragraphs	Version			
		(1)	(2)	(3)	(4) = present condition
39	m10 = r8	•	•	•	•
	m11 = r9	•	•	•	•
40	m12^{+} = r10	•	Pasted over with m14	Pasted over with m14	Pasted over with m14
	m13^{*+}= r11 (beginning)	•	•	Struck	Pasted over with m15, m16
	m14 = r10a		•	•	•
	m15				•
	m16 (beginning)				•
41	m16 (end) = r20 (end)				•
42	m16 (end) = r20 (end)				•
	m17 = r21				•
	m18* = r22 (beginning)				•
43	m19* = r11 (end)	•	•	Struck	Struck
	m20 = r12	•	•	•	•

Note: Plus signs (+) mean "pasted over"; asterisks (*) mean "struck through."

2. Substantive changes from version to version

Each successive version introduced substantive changes. They may be described as follows:

- Ampère composed the first version as part of the manuscript for his lecture of September 18. In r8–r11 he was concerned with formulating a general rule for the behavior of the magnetic needle, while in the process developing a terminology that made this formulation possible. In this context he began referring to "right" and "left" relative to the direction of a current.
- Ampère's next step was to paste over r10 with m14 = r10a, which contains both the claim previously made in r10 and a new insight into the possibility of a unified treatment of two previously disparate cases: the current flowing through the conductor, as well as that flowing through the pile itself. This treatment involved the idea of a single circular current encompassing both wire and battery. For the first time, Ampère used the term "circuit." In support of his proposal, he cited experiments in which electrolysis cells were inserted into the circuit. The composition of the gases generated at either end of such a cell confirmed the uniform direction of current. Relative to this direction, the definitions of "right" and "left" remained valid.
- In a third step, the attempt to define "right" and "left" was abandoned without replacement.
- In the fourth and final step, new paragraphs were inserted. They contained, first, an assessment of Ørsted's results in the light of the two faits généraux, interpreting these results as "complicated" (*compliqués*) on grounds that they involve a "combination" of the two "actions" in which the relevant circumstances and laws become difficult to discern. At the same time, the text was reorganized. The descriptions of instruments, originally intended for a separate section (§2), were now attached directly to the descriptions of the associated "actions."

3. Chronology

The terminology provides important points of reference.

The fourth step is the easiest to date. The relocation of pages 41–42 from their original position in §2 must have taken place during the assembly of the second part of the *ACP* mémoire, that is, during the period I have

Table B5. Ampère's terminology in r8–r11 of the lecture manuscript

Page	Paragraphs	Terminology
39	m10 = r8	G → E
	m11 = r9	G → E
40	m12+ = r10	G
	m13*+ = r11 (beginning)	G
	m14 = r10a	G → E
	m15	E
	m16 (beginning)	E
41	m16 (end) = r20(end)	indeterminate
42	m16 (end) = r20 (end)	indeterminate
	m17 = r21	indeterminate
	m18* = r22 (beginning)	G
43	m19* = r11 (end)	indeterminate
	m20 = r12	G → E

Note: Plus signs (+) mean "pasted over"; asterisks (*) mean "struck through."

called phase IV. By that time Ampère had already devised his general theory of electrical currents, having replaced the G terminology of *galvanic* currents with the E terminology of *electrical* currents in phase III, as discussed earlier. Consequently, the new paragraphs composed at this stage (m15 and the beginning of m16) used the E terminology from the outset. To the extent to which a fact of the matter can be established for the remaining paragraphs in r8–r11, all of them initially employed the G terminology.

In those paragraphs surviving from earlier versions into the fourth, and only in those paragraphs, we find the G terminology replaced by the E terminology. It follows that in all likelihood these changes also took place as part of the fourth step. At the same time, we may infer that versions (1) and (2) date to before phase III, when Ampère transitioned from G to E terminology. Version (1) was composed as part of the manuscript of the lecture of September 18 and as such belongs to phase I. Version (2) mentions experiments in which electrolysis cells were used to determine the direction of current within the battery, experiments also conducted during phase I.

Hardest to date is Ampère's third step, which consisted only in striking r11. Since he added nothing new, the terminological shift is of no help in dating this deletion. Of greater utility is the reflection that Ampère must have had some concrete reason for revising his manuscript. Attaining some new insight would not by itself have prompted this move. For the period prior to September 18, such grounds can easily be imagined: the original manuscript was to be the basis for a lecture on that date, and there is ample evidence that he did not compose its text all at once but rather reworked it repeatedly, each time inserting his latest discoveries. But after he had read the text (as he certainly read the portions of the manuscript currently under discussion) he had no discernible reason to continue updating it to reflect the present state of his knowledge, or otherwise revisit it, until it was time for him to prepare the *ACP* manuscript. That leaves phases I and IV as the only plausible candidates for the date of this change.

A further consideration emerges when we reflect on the semantic significance of this step. In taking it, Ampère abandoned the attempt to define "right" and "left" relative to the direction of current. These terms occur in his formulation of the general rule for *action directrice*, as well as in paragraph m28 = r35, in which he addresses the analogy to his hypothesized terrestrial currents. The fact that he never revived his fully worked out explication of the terms may be explained by his subsequent discovery of a new way of formulating the alignment effect without relying on notions of "right" and "left" relative to direction of current. In the *first* part of the *ACP* mémoire—in the very place where he formulated his rule for the alignment effect in print for the very first time—he offered a much stricter way of expressing it, one in which the terms "right" and "left" no longer appear (Ampère 1820d, 67); in this form, the rule remains known as "Ampère's swimming rule." This latter formulation served as sufficient grounds for striking the definition of "right" and "left" from his lecture manuscript in preparing the *second* part of the *ACP* mémoire, even though the published text retained rule 10 = r8, along with the aforementioned m28 = r35, in their original form.

If this assumption holds, the deletion of r11 must have taken place later, and certainly after September 25. Reflecting on the motives that may have driven Ampère to rework his manuscript in the first place, I thus conclude that the third step also took place as he was preparing the manuscript for press. He presumably struck the paragraph first, and then later, perhaps during a further round of revisions, it occurred to him to insert additional paragraphs, with different content, at this point in his text. In sum, we may consign versions (3) and (4) to phase IV and versions (1) and (2), to phase I.

B7. Paragraph r33

1. The currently surviving text of pages 63–64, paragraph m44, of the ACP manuscript

For ease of exposition, I have divided this paragraph into two parts, r33 and r33a.

r33
je fais dans ce moment executer plusieurs autres appareils, deux entre autres, où le {j'ai construit un autre appareil où le fil} conducteur {est} galvanique étant plié en hélice {à pas très fins} autour d'un tube de carton {verre}, xxx de verre doit, d'après la théorie que {je} me suis faite de ces sortes de phénomènes, {ce conducteur doit} presenter quand on y fera passer le courant galvanique {électrique}, une action xxx //(64) semblable à celle d'un aiguille ou d'une barreau aimantée, dans toutes les circonstances, où ceux-ci agissent sur d'autres corps, ou sont mus par le magnétisme terrestre.

r33a
j'ai {déja observé} dxxx deux appareils [:] {une partie des effets que j'attendais de l'emploi d'un conducteur plié en helice} les [courans?] d'helice galvanique fixe et d'helice galvanique mobile. Je vais donner quelques détails sur les expériences que j'ai faites avec ces instruments, effets que j'attendais de cet instrument; {et je ne doute pas que plus on xxx conformement {variera les expériences fondées sur} l'analogie qu'etablit la théorie entre cet instrument et un aimant {barreau aimantée}, plus on obtiendra de preuves que l'existence des courans electriques est la cause {unique} de tous les phénomènes magnetiques.}

2. Prior versions

Reconstruction yields three different versions of the text.

First Version

r33
je fais dans ce moment executer plusieurs autres appareils, deux entre autres, où le conducteur galvanique étant plié en hélice autour d'un

tube de carton ou de verre, xxx doit, d'après la théorie que je me suis faite de ces sortes de phénomènes, presenter quand on y fera passer le courant galvanique, un action xxx semblable à celle d'un aiguille ou d'un barreau aimantée, dans toutes les circonstances, où ceux-ci agissent sur d'autres corps, ou sont mus par le magnétisme terrestre.

r33a
j'ai xxx deux appareils [:] les [courans?] d'helice galvanique fixe et d'helice galvanique mobile. je vais donner quelques détails sur les expériences que j'ai faites avec ces instruments.

In this first version, Ampère refers to two devices, the "fixed" and "mobile" galvanic helices, candidate analogs to bar magnet and magnetic needle, respectively. Ampère's intention was to test the analogy between magnet and appropriately configured conductor in greater depth. These are also the same two devices announced in the "to-do" list in manuscript 205(b). The first version of the paragraph was undoubtedly composed during phase I, before September 18.

Second Version

r33
j'ai construit un autre appareil où le fil conducteur est plié en hélice à pas très fins autour d'un tube de verre; d'après la théorie que je me suis faite de ces sortes de phénomènes, ce conducteur doit presenter quand on y fera passer le courant galvanique, un action semblable à celle d'un aiguille ou d'un barreau aimantée, dans toutes les circonstances, où ceux-ci agissent sur d'autres corps, ou sont mus par le magnétisme terrestre.

r33a
j'ai déja observé un partie des effets que j'attendais de l'emploi d'un conducteur plié en helice, effets que j'attendais de cet instrument.

The fact that in this version Ampère had already struck and replaced the last sentence of the first version of r33a is made clear by the fact that, whereas the first version had referred to at least two instruments, the second began by announcing only one. The final sentence of r33a has been changed accordingly. Other differences between the two versions concern details of the experiments (the tube of cardboard has been replaced by one of glass) and their results. Ampère had clearly already obtained some of the experimental results he had merely anticipated in the first version, though he does not present these results in any detail. With respect to the chronology, it strikes me as likely that the changes took place prior to September 18. The entire paragraph belongs to §2 of the lecture manuscript, which was read on that date, and there is no evident reason for Ampère to have modified the text of a lecture already delivered so as to include further experiments with the helix. Still, I believe the change to have taken place a relatively short time before the lecture, as it did not make it into list of topics prepared for in advance (list B in manuscript AS 205[b]).

Third Version

While the first part (r33) remains as above, r33a has once again been modified and extended.

> r33a
> j'ai déja observé un partie des effets que j'attendais de l'emploi d'un conducteur plié en helice, et je ne doute pas que plus on xxx conformement {variera les expériences fondées sur} l'analogie qu'etablit la théorie entre cet instrument et un aimant {barreau aimantée}, plus on obtiendra de preuves que l'existence des courans electriques est la cause {unique} de tous les phénomènes magnetiques.

The terminology alone is enough to show that this extension took place considerably later, after September 25. I assume it was made when the text was edited for print, during phase IV, when *galvanique* was replaced by *électrique* throughout the text retained for publication. This third version can no longer be considered to belong to the academy lecture. While extending the text, or perhaps even later, Ampère inserted a third footnote, in which he considered ways of compensating for the "longitudinal" effects of the coil he had originally neglected to take into account (see also the published *ACP* mémoire).

B8. Paragraphs r37–r39

As noted above, the text of these paragraphs is scattered throughout the *ACP* manuscript (pp. 55a, 55c, and 55e) and manuscript AS 205bis(j) (pp. 55b,

55d). Ampère appears to have originally written r37 through r39 consecutively, then realized that this topic had exceeded the scope of his lecture. He thus decided to broach the matter only briefly in the lecture, working it out in its own right in a separate text. So as to retain the material drafted for the lecture, he cropped the page (p. 55) at the end of the passage intended for the lecture, preserving the remainder rather than striking it. It is this remnant, AS 205bis(j), that I located in the archive. Ampère pasted the upper half of the page (what I have called p. 55a) onto a fresh sheet, continuing the text of page 55a onto 55f, followed by page 56. On page 55f he concluded the sentence interrupted at the end of page 55a with the words, "l'électricité galvanique."

The topic excised from the lecture manuscript so as to be treated separately is the question of a unified theory of the phenomena of open and closed piles, such as Ampère envisioned as following from his concept of a circuit of electrical current and in which the notion of *action électromotrice* (electro-motive action) would play a central role. His first attempt at a systematic account may be found in AS 206bis(m) (see app. A), but it was not presented in any detail until the first part of the *ACP* mémoire.

B9. Paragraph r46

1. The extant text at m39, pages 60–61 of the ACP manuscript

For ease of exposition, I divide the paragraph into two parts, r46 and r46a:

> r46
> maintenant si c'est là la cause directrice de la part de la terre ce le sera aussi de la part d'un aimant d'où il suit qu'un aimant est un ~~xxx~~ assemblage de courans ~~galvaniques~~ {électriques} qui ont lieu dans des plans perpendiculaires à son axe, dirigés de manière que le pôle austral de l'aimant, qui se porte du coté du nord se trouve à droite de ces courans puisqu'il est toujours à ~~xxx~~ gauche d'un courant placé hors de l'aimant[.]

> r46a
> {et qui lui fait face dans une direction parallele, ou plutôt ces courans s'établissent d'abord dans l'aimant suivant les courbes fermées les plus courtes soit de gauche à droite, soit de droite à gauche et alors la ligne perpendiculaire aux plans de ces courans devient //(61) l'axe de l'aimant,

et ses extremités en font les deux poles.} Ainsi à chacun des poles d'un aimant les courans galvaniques dont il se compose sont dirigés suivant des courbes fermées concentriques, j'ai imité cette disposition autant qu'il était possible avec un courant ~~galvanique~~ {électrique} en en pliant le fil ~~en spirale~~ conducteur en spirale. ~~xxx que~~ cette spirale {était} formée avec un fil de laiton ~~dont les extremites xxx xxx xxx~~ {et terminée par deux portions rectilignes de ce même fil, qui étaient renfermées dans} deux tubes de verre afin qu'elles ne communiquassent pas entre elles et pussent être attachées aux deux extrémités de la pile.

2. *Prior versions*

The text that survives today was revised several times. Some of the revisions may clearly be dated to before September 25, 1820. Careful reconstruction yields three different versions.

First Version

r46
maintenant si c'est là la cause directrice de la part de la terre ce le sera aussi de la part d'un aimant d'où il suit qu'un aimant est un ~~xxx~~ assemblage de courans galvaniques qui ont lieu dans des plans perpendiculaires à son axe, dirigés de manière que le pôle austral de l'aimant, qui se porte du coté du nord se trouve à droite de ces courans puisqu'il est toujours à ~~droite~~ {gauche} d'un courant placé hors de l'aimant

In this passage Ampère sketches a generalization of his hypothesis of circuits of terrestrial currents to circuits in general. This is the generalization that motivated his experiments with coils, experiments described at the end of §2 (see above, r33), which he read on September 18. It may thus be inferred that this first version was composed before that date.

There are indications within the manuscript itself that this passage concluded the first version of the entire manuscript and that its extension, along with all of the paragraphs that followed, were added in a second session rather than drafted with the rest. At this line Ampère's script became noticeably more cramped than before. Toward the end of the page it became still more compressed, until he was finally forced to continue on a new page. This, in turn, called for a new leaf, not part of the bound notebook of four-sided folded sheets in which he had been writing.

Second Version (extension)

> 146a
> et qui lui fait face dans une direction parallele, ou plutôt ces courans s'établissent d'abord dans l'aimant suivant les courbes fermées les plus courtes soit de gauche à droite, soit de droite à gauche et alors la ligne perpendiculaire aux plans de ces courans devient l'axe de l'aimant, et ses extremités en font les deux poles. Ainsi à chacun des poles d'un aimant les courans galvaniques dont il se compose sont dirigés suivant des courbes fermées concentriques, j'ai imité cette disposition autant qu'il était possible avec un courant galvanique en en pliant le fil ~~en spirale~~ conducteur en spirale. ~~xxx que~~ cette spirale {était} formée avec un fil de laiton ~~dont les extremites xxx xxx xxx~~ {et terminée par deux portions rectilignes de ce même fil, qui étaient renfermées dans} deux tubes de verre afin qu'elles ne communiquassent pas entre elles et pussent être attachées aux deux extrémités de la pile.

Considerations of content lend further confirmation and precision to the conjecture that this passage was added later. In it Ampère describes instruments and experiments, contrary to his original intention for this part of the text. Such descriptions were to have taken place in §2. But as that had already been read on September 18, on September 25 he had no choice but to report on instruments completed during the preceding week in §III. I infer that this paragraph and those that follow were added between September 18 and 25.

Third Version

The text itself remained essentially unchanged, except for the replacement of *galvanique* by *électrique*, which presumably took place as the manuscript was being prepared for press, in phase IV.

NOTES

Abbreviations for Collected Materials

AC	Ampère's correspondence. In Louis de Launay, ed., *Correspondance du Grand Ampère*, 3 vols. (Paris: Gauthier-Villars, 1936–43).
AS	Académie des Sciences, Paris, collection of Ampère's files (Dossier Ampère).
FC	Faraday's correspondence. In Frank A. J. L. James, ed., *The Correspondence of Michael Faraday, Volume 1, 1811–31* (London: Institution of Electrical Engineers, 1991).
LR	Lectures reconstructed from Ampère's manuscript for his presentations of September 18 and 25, 1820, transcribed in full in appendix B.
OC	Ørsted's correspondence. In M. C. Harding, ed., *Correspondance de H. C. Örsted avec divers savants*, 2 vols. (Copenhagen: H. Aschehoug & Co., 1920).
RI	Royal Institution archive (London)

Abbreviations for Journals

ACP	*Annales de chimie et de physique* (Paris)
AGSP	*Annales générales des sciences physiques* (Brussels)
Annalen	*Annalen der Physik* (Leipzig)
Ann.Phil.	*Annals of Philosophy* (London)
Bibl.Univ.	*Bibliothèque universelle des sciences, belles-lettres, et arts* (Geneva)
Edin.Phil.J.	*Edinburgh Philosophical Journal*
JCP	*Journal für Chemie und Physik* (Halle)
J.Phys.	*Journal de physique, de chimie, d'histoire naturelle et des arts* (Paris)
Phil.Mag.	*Philosophical Magazine* (London)
Phil.Trans.	*Philosophical Transactions of the Royal Society of London*
Proc.Verb.	*Institut de France: Procès-Verbaux des séances de l'Académie des Sciences*
QJS	*Quarterly Journal of Science* (London)

Introduction

1. See Polanyi's ([1958] 1994) well-known *Personal Knowledge* or his (1962) "Republic of Science."

2. Using one of Einstein's notebooks, for example, Jürgen Renn and Tilman Sauer (1997, 1999) have investigated Einstein's practice during his search for the field equations of general relativity theory. For an example from mathematics, see the work of Pickering and Stephanides (1992).

3. See, for example, *Science as Practice and Culture* (Pickering 1992b) and *Scientific Practice* (Buchwald 1995a).

4. Mirko Grmek (1973, 1991) has explored this distinction in the work of Claude Bernard with particular insight, as has Gerald Geison (1995) in the work of Pasteur.

5. This methodological approach may be found, for example, in the work of F. L. Holmes (1987, 1990). On the growing attention to research practice and appropriate historical resources, see for example, the contributions to *Reworking the Bench* (Holmes, Renn, and Rheinberger 2003).

6. Hacking's (1983) analysis gave rise to numerous responses, as well as to other studies, including, for example, those in various collections (Achinstein and Hannaway 1985; Batens and Bendegem 1988; Gooding, Pinch, and Schaffer 1989; Le Grand 1990; Pickering 1992b; Heidelberger and Steinle 1998; and Radder 2003). I offer more detailed discussion in chapter 7.

7. See also Hentschel's (2000) overview.

8. Heidelberger (1998) offers an instructive sketch of the progressive marginalization of experiment over the course of the history of the philosophy of science.

9. I have placed "empirical" in scare quotes so as to make it clear that, in my view, the

picture of science history as furnishing the facts hitherto processed by the philosophy of science is much too naïve, even when it comes to the supposed furnishing of the "facts." Writing history is anything but an "innocent" compilation of empirical data.

10. Talk of "epistemic things" goes back to Hans-Jörg Rheinberger (1992), who contrasts them with "technical objects," whose status is no longer in question.

11. Such distinctions were already proposed by early advocates (e.g., Hacking 1983; Cartwright 1983) of the "new experimentalism."

12. I am here using the term "language" to refer to the level on which concepts are questioned and subjected to revision. By no means am I speaking of language (or even the specialized language of a science) as a whole.

13. There are numerous studies of particular individuals, developments, and subfields of nineteenth-century electrodynamics, as well as an excellent overview (Darrigol 2000).

14. Using psychosociological categories and with a longer time perspective in mind, Ken Caneva (1981) has sketched a larger frame within which such analysis could take place.

15. In Ampère's case, the works of Blondel (1978, 1992), Williams (1983), and Hofmann (1987a, 1987b, 1988, 1995) stand out, as do the works of Gooding (1985) on Faraday and on experimentation more broadly (1989, 1990a).

16. Heilbron's (1979) comprehensive study presents the now classic overview of this field in the seventeenth and eighteenth centuries but only touches on a few aspects of the early nineteenth century.

17. For debates about micro- and macrohistory, see, for example, the work of Schlumbohm and Gribaudi (1998).

18. Recent analyses of the relationship between history and philosophy of science have been offered by Radder (1997) and by Pinnick and Gale (2000). See also the contributions to the 2002 special issue (vol. 10, no. 4) of *Perspectives on Science*, as well as the work of Schickore (2011). The fact that the international initiative Integrated History and Philosophy of Science (known as IHPS) had by 2014 already held its fifth international conference, with participants growing in number and prominence each year, indicates a rising interest in this cross-disciplinary field.

19. The question of how to deal with historical case studies is taken up by Sauer and Scholl (2015).

20. See the works of Steinle (2002b), Steinle and Burian (2002), and Schickore and Steinle (2006). For illustrative examples of what can be achieved, see works by Chang (2004, 2012), Arabatzis (2006, 2011b), and Schickore (2007), among others.

21. The slogan was also used by Hanson (1962, 580) and by Lakatos (1971, 91).

Chapter 1. Electricity and Galvanism in the Early Nineteenth Century

1. Lichtenberg (1972, 2:344 [no. 1912], 2:471 [no. 382]) makes similar remarks elsewhere.

2. See, for example, Lichtenberg's (1972, 2:344 [no. 1912]) comments on the Geneva naturalist Jean Senebier.

3. Home and Aepinus (1979) is an English translation.

4. Gillmor (1971, 192–94) has studied Coulomb's work, with the history of the force law on the pages noted.

5. Gillmor (1971, 232–33) provides a listing of mémoires.

6. See Coulomb's second mémoire (Joubert 1884, 593).

7. This issue has been explored most recently, and most thoroughly, by Peter Heering (1998), who reconstructed the torsion balance and repeated the experiments, renewing old concerns. Various aspects of this problem are discussed in contributions to the volume edited by Blondel and Dörries (1994). A new analysis (Martínez 2006) makes clear that Coulomb could well have obtained the three results but also that this would have required extremely specific arrangements and materials.

8. See, for example, the work of Frankel (1977, 35–38) for a brief discussion.

9. More on the Paris academy's involvement is in the work by Home and Aepinus (1979). Haüy was a typical transitional figure. While he was perfectly prepared to draw on and credit the work of amateurs in his mineralogical and crystallographic studies, in the field of electricity he was an energetic proponent of the Coulomb program, which excluded amateurs (Blondel 1997b).

10. Later, both Jean-Baptiste Biot and James Clerk Maxwell would emphasize Coulomb's central role in this process (Gillmor 1971, viii, 26n11).

11. Coulomb cites Aepinus as early as 1777, in a mémoire on magnetization. In his eulogy for Coulomb, Jean-Baptiste Delambre (1806) would place him as successor to Aepinus (Home and Aepinus 1979).

12. On Cassini, see the work of Licoppe (1994). On the culture of expertise, see that of Weiss (1995).

13. Several studies address the "quantifying spirit" (Crosland 1972; Frängsmyr, Heilbron, and Rider 1990; Heilbron 1993; and Wise 1995, esp. contributions to the last by Andrea Rusnock, Ken Alder, and Jan Golinski).

14. Barnard (1969) provides a thorough account of the educational reforms; Gillmor (1971, 222–26) offers a summary. Crosland (1992) and Garber (1999) discuss the sciences, and Fischer (1998) addresses the Napoleonic era in particular. Fox and Weisz (1980a) provide a most helpful introductory overview in their collected volume, although those essays are mostly focused on the period after 1830.

15. For a Franco-German comparison, see the work of Lundgreen (1980).

16. This view was expressed by the editor of the influential *Annalen der Physik,* Ludwig Wilhelm Gilbert (1820c).

17. Crosland (1967) provides the most comprehensive study of this institution, though that of Arnold (1983–84) is also exceptionally clear. Fox (1973, repr. in Turner 1976) offers a brief précis, focusing on patronage relations.

18. See the introduction to Laplace's (1814, 2) work, "De la probabilité."

19. The most thorough account of this period remains that of Fox (1974). See also Heilbron's (1993) relevant passage.

20. First published in four volumes (Biot 1816), Biot's physics textbook was soon released in an abbreviated two-volume edition (Biot 1817), which he updated in two

subsequent editions (Biot 1821b, 1824). Fechner translated it into German (Biot 1824–25).

21. The term "potential" would be introduced by George Green in 1826 (Buchwald 1977, 119).

22. For Coulomb's efforts, see his fifth (1787) and sixth (1788) mémoires on electricity (Joubert 1884, 183–272). See also comments by Heilbron (1979, 494–99) and Grattan-Guinness (1990, §7.5.3).

23. For more on Poisson's work and his environment, see material by Arnold (1983–84), Costabel (1978), Métivier, Costabel, and Dugac (1981), and Grattan-Guinness (1990, 1291).

24. For goings on at the Académie des Sciences, see the article by Home (1983).

25. Poisson's (1812a) studies were confined to conductive bodies. My sketch draws on this mémoire and on the work of Grattan-Guinness (1990, §§7.6.2, 7.6.4) and Heilbron (1979, 498–500).

26. For the general result and the critique of Biot, see Poisson's (1812a, 27–28) piece.

27. Poisson (1826, 247) would later claim that his calculations had been "in perfect accord" (*sont parfaitement d'accord*) with Coulomb's measurements. He offered no more analysis than Coulomb, or indeed any of their contemporaries, of margins of error or the like. Grattan-Guinness (1990, §7.6.4) offers a detailed discussion of his mathematical reasoning.

28. Much later it would become clear that Poisson (1826, 251) subscribed to the widespread view that the electrical fluid was held within a charged body by external air pressure. He ignored experimental results that showed that the vacuum was also an insulator.

29. The summary published in the *Annales de chimie et de physique* (Fourier 1816) was presumably written by Fourier himself, according to Fox (1974, 112, with reference to comment by Grattan-Guinness [1972, 460]). The manuscript of his 1807 lecture was published and analyzed by Grattan-Guinness (1972); see also the article by Friedman (1977). On Fourier himself, see the biography by Herivel (1975). Crosland (1994) has written an account of the history of the hugely important *Annales*.

30. On early wave theory, see Buchwald's (1989) monograph.

31. See, for example, Arago's (1985, chap. 53) account of these events in his autobiography or Fox's (1974, 126) comment.

32. Volta's letter, written in French, was read to the Royal Society and rushed into print in the *Philosophical Transactions* (Volta 1800c). An English translation appeared shortly thereafter in the *Philosophical Magazine* (Volta 1800b). Dibner (1964) offers a more recent English translation.

33. Sudduth (1980) offers a good overview of the voltaic pile.

34. On William Nicholson, see the biographical sketch by Thackray (1974). On Anthony Carlisle, see the sketches by Thackray (1971) and Cole (1952).

35. Ritter's objections were taken seriously by chemists, even by Frédéric Cuvier in Paris. The debate surrounding the decomposition theory has been described in brief (Williams 1965, 227–38) and at greater length in other works (Ostwald 1896, chap. 7; Knight 1978, chap. 2).

36. On Davy, see the biography by Knight (1992) or Kendall's (1953) older, more hagiographic effort.

37. On Davy's relationship to the romantic movement, see, for example, the work of Lawrence (1990).

38. See, for example, the architect Thomas Webster's report (Bence Jones 1871, 192–95).

39. There are several general works on the Royal Institution (Berman 1978; Caroe 1985; James 2002). For a broader study of the social context, see the book by Morus (1998).

40. On Grotthus and Berzelius, see, for example, the discussion by Ostwald (1896, 307–22). Davy's theory contains significant elements of that of Grotthus, and it seems unlikely he was unaware of the other's work; see Williams's (1965, 234) discussion.

41. Somewhat more detail may be found in works by Crosland (1967, 23–24) and Fischer (1988, 147, 204).

42. I return to the general ramifications of this question in chapter 7.

43. Golinski (1990) offers a detailed analysis. Davy's (1808) second Bakerian Lecture paper was published in *Philosophical Transactions.*

44. Chilton and Coley (1980) discuss the development of the Royal Institution's laboratory. Sviedrys (1976) and the essays collected by James (1989b) offer views on the overall state of laboratory research at the time.

45. On Gilbert and his *Annalen,* see, for example, the article by Schimank (1963).

46. There are numerous works about Ritter himself (e.g., Schimank 1933; Wetzels 1973; Richter 1988; Strickland 1992). Herrmann (1968, 11–13, 44–45) provides a sketch of Ritter's relationship with Schelling. There is a substantial recent literature on "romantic" natural science, of which Cunningham and Jardine's compilation (1990) is a good exemplar.

47. For one example, see an episode described by Ostwald (1896, 360).

48. Ritter's status as the founder of electrochemistry was acknowledged early on by Ørsted, later reinforced by Ostwald (1896) and others, and in the twentieth century by Schimank (1933) and Herrmann (1968). Opinions about whether Davy or Ritter deserves the ultimate credit for launching electrochemistry continue to be divided even now, surprisingly enough often along lines of nationality and language. The question does not strike me as especially meaningful or significant: both of them brought forth far-reaching innovations, each in his own scientific milieu and with his own unique posterity. Understanding the local exigencies or even contingencies of epistemic processes, processes that themselves aim at generalities, seems to me a much more worthwhile goal.

49. Ritter's article published in 1800 was reprinted in a bound volume of his essays (1806, 1:245ff).

50. For the Paris debacle, see the works of Meyer (1920, vol. 1, p. xxxi) or Christensen (1995, 162–65). Ritter's taint was lasting, to the extent that it was still remembered by many as late as 1820; see, for example, Davy's (1821d, 7) introductory remarks on the prehistory of electromagnetism or Weiss's allusion to it in his August 2, 1820, letter to Ørsted (OC 2:267).

51. For others' means of measuring, see, for example, Volta's (1800c) and Biot's (Biot and Cuvier 1801) efforts.

52. Strickland (1992) offers a pointed analysis but underestimates the extent to which such techniques were common procedure, specific neither to Ritter nor to the romantics.

53. See also Marum's (1801) letter to Volta.

54. For more on Biot's career, see Frankel's (1978) article. For background on Biot's first studies, see Sutton's (1981, 345) comment.

55. For more detailed discussion, see Sutton's commentary (1981, 345–48) or that of Ostwald (1896, 345). Ostwald also mentions Gilbert's remarks on Biot's results.

56. For an overview, see Fischer's (1988, 135–40) description.

57. For a poignant account, see the description by Sutton (1981, 354–57). Napoleon continued to protect Volta in years to come, awarding him a life pension and political appointments in the new republic.

58. On the financial aspect, see, for example, Grattan-Guinness's (1991, §19.2.4) article. The three-thousand-franc annual prize was in fact not actually awarded annually. Davy received it in 1806. The grand prize was never awarded.

59. The report was not published until several years later, in the *Mémoires* (Biot 1804).

60. In the interest of brevity, I compile these conditions more systematically than in Biot's (1804) text. Brown (1969, 69–74) offers a good discussion.

61. Brown (1969, 73–74) offers several examples of difficulties encountered along the way.

62. See also comment by Brown (1969, 76).

63. See, for example, Ørsted's (1820g) report, but see also the first popular overview of electromagnetism to appear in England (Watkins 1828, 2).

64. For example, Étienne-Gaspard Robertson in Paris offered proposals as early as September 1800, followed shortly thereafter by Berlin professor Paul L. Simon. See Sutton's (1981, 334) mention or Ostwald's (1896, 287–91) more detailed description.

65. Ostwald (1896) downplays such applications as sheer quackery, most likely underestimating their true extent and the earnestness with which they were pursued.

66. Extracts from Soemmering's (1859) diary describe his work. Ostwald (1896, 280–87) offers a detailed account of Soemmering's attempt at a telegraph.

67. See, for example, Pfaff's letters of July 27, 1820 (OC 2:466–67) and September 1821 (OC 2:468).

68. On Romagnosi and Mojon, see works by Dibner (1962) and Andrade Martins (2001).

69. See, for example, Ørsted's (1830) later historical précis. To be sure, after Ørsted's discovery some would remember these reports; see, for example, Pictet's (1821) notice in the *Bibliothèque universelle* and its mention in Gilbert's *Annalen der Physik* (68 [June 1821]: 208).

Chapter 2. Electromagnetism

1. Kirstine Meyer (1920) offers a detailed biography of Ørsted in an introduction to his papers. Other works (e.g., Dibner 1962; Williams 1974; Franksen 1981; Jansen, Snorrason, and Lauritz-Jensen 1987, in Danish, with an English summary) provide biographical sketches. The study by Christensen (2013) is very rich in material but has serious historiographical deficits, according to Wilson (2014).

2. Meya and Sibum's (1987, chap. 6) characterization of Ørsted is typical. Caneva's (1997a, 1997b) assessments are more subtle. By considering the influence of both "romantic" physics and "modern," technologically oriented science, Christensen (1995) arrives at a more nuanced reading of Ørsted, though barely touching on his experimental work. For the latter, see Steinle's (2003) work. Christensen (2013) does much to disentangle Ørsted from romantic physics.

3. There is as yet no systematic account of Ørsted's scientific work, though that task has been simplified by the appearance of an English translation of many of his works (Jelved, Jackson, and Knudsen 1998). However, Ørsted's contributions to the philosophy of nature and philosophy of science are not as well represented in that compilation as in his actual corpus.

4. The most thorough account is Meyer's (1920, lxvii–xcvii). Dibner (1962) offers a synthesis and Snelders (1990), a brief sketch.

5. Herschel (1830, 340) would later compare Ørsted's unerring, targeted quest for the electromagnetic effect with Columbus's stubbornness in his own quest for land across the Atlantic. Alexander von Humboldt concurred in this assessment (OC 2:411).

6. Andrade Martins (1999) has conducted a partial survey.

7. The text is reprinted in Meyer's (1920, 2:214–18) compilation and may be found in facsimile in the centenary volume (Ørsted 1920, 11–14) or in Dibner's (1962, 23–26) monograph. Jelved, Jackson, and Knudsen (1998, 413–16) present an English translation.

8. "His ita comparatis, acus magnetica movebitur, et quidem sub ea fili conjungentis parte, quae electricitatem proxime a termino negativo apparatus galvanice accipit, occidentem versus declinabit" (Ørsted (1820o, 2). The page number(s) here refer to Ørsted's Latin circular.

9. Ørsted's word is *gummi-lac,* which Gilbert translated as *Gummilack* (Ørsted 1820n, 302). From the absence of any further remark, we may assume that his fellow experimenters knew what this was. The contemporary English term, "shellac," is inferred from Kauffmann's (1805, 205) dictionary.

10. "Polus super quem intrat electricitas negativa ad occidentem, infra quem ad orientem veritur" (Ørsted 1820o, 3).

11. Ørsted (1821b) offers a very brief sketch at the beginning of his article on electromagnetism.

12. In an essay composed ten years later, Ørsted (1830, quoted in Jelved, Jackson, and Knudsen 1998, 547, emphasis added) claimed to have clearly recognized and expressed the conjecture, as early as July 1820, "that the magnetical effect of the electrical current has a circular *motion* around it." But this was formulated in hindsight; there had been no talk of circular *motion* in the 1820 report.

13. The only surviving cover letters, to Berzelius and Christopher Hansteen, bear the July 22 date. That the letter was sent in printed form may be inferred, for example, from the remark with which Thomson prefaced the print edition of the English translation (Ørsted 1820g, 273). The fact that Pfaff replied on July 27 and Weiss, on August 2 (OC 2:466–67, 2:266–67, respectively) suggests that many recipients had received their copies by the end of July.

14. Letter to Hansteen of July 22, 1820 (OC 2:120): "Det forekommer mig at Følgerne af min Opdagelse kunne vaere (t) meget vidtudstrakte."

15. See Johann L. G. Meinecke's letter to Ørsted, January 18, 1821 (OC 2:431). The editors of Ørsted's correspondence (Jelved, Jackson, and Knudsen 1998) claim the essay was composed in German but do not provide any evidence.

16. One such response came from Christian Samuel Weiss (OC 2:270).

17. In the title of his publication of Ørsted's letter, Blainville gives the date of receipt as September 29 (Ørsted 1820j, 78). Schweigger (1821c, 7) reports that it reached him in Halle "before the end of September."

18. See Mayer's letter to Ørsted, October 17, 1820 (OC 2:430).

19. Grattan-Guinness (1990, 919) offers a different assessment of reactions, without evidence, which in my view totally misses the mark.

20. Meyer (1920, ci–cxii) offers a preliminary sketch, and Caneva (1981) provides a quite general attempt, based on psychosociological categories.

21. Pictet also provides a brief account of goings-on in Geneva. See also Arago's account in his preface to Ørsted's letter in the *ACP* (Ørsted 1820f, 417–18).

22. Van Mons himself was convinced that his journal was the very first to publish the report (OC 2:459).

23. Arago gave those remarks in a detailed introductory footnote on p. 417 of Ørsted's piece (1820f). While the issue bears the date of August 1820, it was available only in October that year (see table 4.1, in chapter 4).

24. The issue in question (July 1820) was mailed on November 11.

25. See Brewster's (1821a) remarks in his January 1821 overview.

26. Writing from London, J. G. Forchhammer told Ørsted, "Dr. Wollaston was thrilled by the discovery, and impressed by its significance" (OC 2:330).

27. See Pfaff's letter to Ørsted, July 27, 1820 (OC 2:466–67).

28. It is worth noting that when Weiss received Ørsted's report, he immediately recalled Ørsted's (actually Ritter's) "failed experiments," conducted during his 1803 trip to Paris. Weiss saw the new findings as bringing felicitous closure to that episode. See Weiss's letter to Ørsted, August 2, 1820 (OC 2:267).

29. There is no evidence that Seebeck might have learned of this result from Ørsted, directly or indirectly (from correspondence, third-party report, etc.). Seebeck (e.g., 1822, 290) did not express himself clearly on this point.

30. For events in Berlin, see Weiss's letter to Ørsted, August 2, 1820 (OC 1:266–70). Seebeck (1821, 38; 1822, 290) detailed the course of his own research, including dates.

31. On Erman, see du Bois-Reymond's (1853) eulogy and Erman's (1821b) monograph, which had already been announced in the February issue of Gilbert's *Annalen der Physik* (67:220–21).

32. For an account of Poggendorff's work, see the essays by Chipman (1964, 132–33) and Hackmann (1984, 213–14).

33. Erman dispatched the diploma, issued on November 21, 1820, to Ørsted on December 14 (OC 2:317–18).

34. The July 1820 issue, in which both of Ørsted's letters (1820d, 1820h) were published, was in press toward the end of September, when Meinecke received the second letter. It presumably did not appear until the end of October, arriving in Frankfurt on November 1 (Buch 1821b, 18). Meinecke claimed the honor of having been the first to report on Ørsted's findings in a German journal (see his letter to Ørsted, January 18, 1821, OC 2:431–32).

35. Like Ørsted, Schweigger began his work with a very powerful pile. But then, he reports, "I soon saw that the electromagnetic effects belonged not to the pile, but to the simple wire, which permitted the construction of livelier versions of the experiments" (Schweigger 1821c). Schweigger's (1821a) lecture did not appear in print until the January 1821 issue of *JCP*, where it was published along with several addenda.

36. This issue of the *Annalen* was presumably not released until the early months of 1821.

37. Gilbert's publication strategy drew criticism, especially by comparison with that of *JCP*. See, for example, Soemmering's letter to Ørsted, February 19, 1821 (OC 2:552).

38. See also Mayer's letter to Ørsted, October 17 (OC 2:430–31) and Stromeyer's letter to Ørsted (OC 2:553–54).

39. Yelin delivered a lecture to the Munich academy's mathematical and physical class on November 11, 1820 (see Yelin's letter to Gilbert, November 17, 1820, excerpted in Gilbert's *Annalen* 66 [1820]: 323–24) and published his first contribution shortly thereafter (Yelin 1820). Muncke's (1820) first report presumably appeared in early 1821. By November, Böckmann (1821a) was already in correspondence with the *JCP*. A detailed article (Böckmann 1821b) appeared in the May issue of Gilbert's *Annalen*, presumably printed toward the middle of 1821. See Prechtel's published (1821) letter to Gilbert, written on December 20, 1820.

40. An abridged German translation appeared in the January issue of *JCP* (Berzelius 1821).

41. The issue had been received by the Académie des Sciences in Paris by the end of January.

42. Marcet reports on Gazzeri and Ridolfi's progress in his letter to Berzelius, January 18, 1821 (Söderbaum 1912–32, 3:214).

43. Configliachi (1821b) also reported on his public lectures and cited with approval the views of the "French physicists" (*physiciens français*). For a timeline, see Gilbert's remarks in his German translation (Configliachi 1821a).

44. Surveys may be found in compilations by Meyer (1920, 2:214) and Jelved, Jackson, and Knudsen (1998, 413).

45. "Arago n'en croyait un mot avant d'essayer, mais nous avons obtenu des résultats si distinctes qu'il n'a pas pu hésiter de s'y rendre." So Marcet informed Berzelius in a letter of September 15, 1820 (Söderbaum 1912–32, 3:210).

46. The aforementioned case of Pfaff is the only one known to me.

47. See, for example, an excerpt from a letter composed on November 12, 1820 (Böckmann 1821a).

48. Gilbert (1820b, 332) drew attention to this point, emphasizing that simple means of verifying the find were now at hand.

49. Lehot was a *répétiteur de physique* at the École Polytechnique. See Dhombres's comment in the appendix to Fourcy's ([1828] 1987, 176) history of the school.

50. Letter from H. C. Schumacher to Ørsted, November 21, 1821 (OC 2:516).

51. In the original, "le champ, que vous aves [*sic*] défriché." Blainville's letter to Ørsted, November 3, 1820 (OC 2:271).

52. Meinecke's preface to *JCP* 31 (1821) (Schweigger 1821b); similar phrasing may also be found in his letter to Ørsted, January 18, 1821 (OC 2:431–32). Other factors, to which Meinecke only alludes, doubtless also played an important role in the decision to launch a new series.

53. With this passage from the preface of his newest volume of the *Annalen* (1821), in which Gilbert first reported on electromagnetism, he already reflected back on the first months of research.

54. See, for example, Trumpler's (1992) detailed discussion.

55. The authoritative source on events at the Académie des Sciences is volume 7 (1916) of the proceedings of the Institut de France: *Procès-Verbaux des séances de l'Académie des Sciences* (cited as *Proc. Verb.*). For Arago's report on the September 4, 1820, session, see *Proc. Verb.*, 90.

56. "La nouvelle a d'abord été reçue très froidement ici. On avait cru que c'etait encore une rêverie allemande" (Söderbaum 1912–32, 4:18). This letter was presumably drafted around October 20. The date of October 2 given in the print edition cannot be right, for Dulong mentions events that took place somewhat later, including Biot's return, which was sometime between October 9 and 16. It seems likely that a digit was omitted either in the manuscript or in transcription, so that "2" represents only the first digit of the date.

57. "Quand M. Arago parla de ces nouveaux phénomènes, on rejeta cela comme on avait rejeté les pierres tombées du ciel. . . . Ils décidaient tous que c'était impossible." Ampère's letter to his friend Jacques Roux-Bordier, February 21, 1821 (AC 2:566).

58. See, for example, Westrum's (1978) article for an account of the factors leading to the eventual acceptance of the existence of meteorites.

59. "Elle est dans l'hypothèse de Coulomb sur la nature de l'action magnétique; on croyait à cette hypothèse comme un fait; elle écartait absolument toute idée d'action entre l'électricité et les prétendus fils magnétiques" (Ampère's letter to Roux-Bordier, February 21, 1821, AC 2:566).

60. See the aforementioned October 1820 letter from Dulong to Berzelius (Söderbaum 1912–32, 4:18). Biot was absent from sessions of the Paris academy for some time. His name does not reappear on the attendance rolls until October 16.

61. Dulong's aforementioned October 1820 letter to Berzelius (Söderbaum 1912–32, 4:18).

62. "M. Ampère read a report on *certain new properties of the principal axes of bodies*" (Minutes of the Académie des Sciences, June 18, 1821, *Proc.Verb.*, 205, my translation).

63. For a more comprehensive biography, see Hofmann's (1995) work. Williams (1970–81) offers a shorter overview.

64. The manuscript has survived. For a discussion and transcription, see the work by Blondel (1982, 61–63, 175–76). See also Hofmann's (1995, 50–53) monograph.

65. A good overview of Ampère's (1814, 1815b, 1816a, 1816b) optical polarization work is in Hofmann's (1995, chaps. 2, 6) monograph. Scheidecker-Chevallier and Locqueneux (1992) discuss these works with particular attention to their connection with Ampère's metaphysical reflections. On Ampère's chemical work, see Mauskopf's (1988) essay.

66. The death of the mathematician Charles Bossut had left a vacancy.

67. On the question of Ampère's familiarity with Ørsted's work, see Caneva's (1980, 129) comment.

68. The correspondence began with Ampère's letter of November 1, 1810, congratulating Davy on the discovery of the alkaline earth metals (AC 1:355).

69. The information on Maine de Biran is taken from the work of Hoefer (1852). The true scope of his correspondence with Ampère is evident only in the most recent edition of his correspondence (Robinet and Bruyère 1993). On Dégerando, see the work by Robert and Cougny (1889–91, name entry Dégerando). Silliman (1970–81, 167) claims that Fresnel also moved in Maine de Biran's circle but provides no evidence.

70. Hofmann's (1995, chap. 5) monograph offers a detailed discussion; see also his earlier (1982, 1987a, 1987b, 1988) works. In my view, as later noted, Hofmann greatly overstates the influence of such reflections on Ampère's scientific work. For a study of Ampère's philosophy and its treatment of Kant, see the article by Segala (1997).

71. "Ampère . . . vécut dans deux sociétés distinctes. Elles avaient pour unique trait de ressemblance la célébrité de leurs membres" (eulogy for Ampère, in Barral 1854–62, 2:34). Arago also noted the academic scientists' disdain for metaphysicians.

72. See, for example, Ampère's letter to his son, September 25, 1820 (AC 2:561–62) or his letter to Roux-Bordier, December 18, 1820 (AC 2:563).

73. "Ampère a été seul pendant un mois, et il a su habilement tirer parti de sa position" (Söderbaum 1912–32, 4:18).

74. The most thorough study is Blondel's (1982, chap. 3) excellent work. Williams (1983) has focused on the early period, while Hofmann (1987a, 1987b) pays special attention to the equilibrium experiments. Hashimoto (1983) gives a sketch of Ampère's theoretical reasoning.

75. Hofmann (1995, chap. 7) relies on Williams (1983), adopting his argument.

76. Ampère does indeed refer to them in later periods. In chapter 4, I consider the background to such allusions.

77. Blondel (1985) has traced some of Williams's missteps; his reply (Williams 1985b) fails to repair them.

78. For example, Williams (1983, 499) operates on the assumption that the refutation of the electrostatic theory of the pile was an important goal for Ampère. I find no evidence for this, even in the considerably expanded suite of source materials on which I have drawn.

79. For example, Mirko Grmek (1991) has pointedly illustrated this problem for the case of Claude Bernard.

Chapter 3. Ampère's First Studies of Electromagnetism

1. Hofmann accepts Williams's reconstruction uncritically.

2. Appendix B contains the complete French text of these reconstructions, with paragraphs labeled r1, r2, etc. In what follows, I refer to this text as "LR" (lecture reconstruction).

3. For materials in the Académie des Sciences archival collection of Ampère's files, I use the initials "AS," followed by the archival folder number, together with a letter in parentheses (or brackets if part of a parenthetical notation) designating a particular document in a given folder. See appendix A, §A1, for a detailed description of how these materials are identified.

4. AS 205(b). Appendix A, §A10, provides a transcription and discussion of the full text.

5. I thank Christine Blondel for drawing my attention to this.

6. In the original, "sans complication avec . . . le magnétisme terrestre" (LR r19–r20).

7. LR r20.

8. AS 180(e), (f). Earlier versions differed in several technical details from the one eventually published. Here, too, Ampère thoroughly explained the procedure for properly adjusting the needle.

9. My use of the term "battery" is anachronistic. Ampère always called it a "pile," even when, as was usually the case, he was actually using a trough device. In his time, a "battery" would mean a number of Leyden jars wired in parallel. I use "battery" to denote any voltaic device, because the slightly archaic English word "pile" typically refers only to a battery with the upright, columnar form of Volta's original device. By 1820, these were no longer in general use.

10. LR r8. The cumbersome language and long sentences are true to Ampère's style.

11. LR r10.

12. Brown (1969) has discussed the origins of the current concept in detail. But he remains unaware of the depths of Ampère's concept or of the consequences that followed from it.

13. In his first lecture at the Académie des Sciences Ampère explicitly emphasized the instrumental character of his terminology; see the anonymous report of one participant (Anonymous 1821, 83).

14. LR r11.

15. LR r9.

16. Ampère never read out the quoted passages on the alignment effect in that form; we have them only from early versions of his manuscript. Before his first academy lecture, he would make some small-seeming but far-reaching changes (discussed later).

17. The setup for these experiments is described in AS 205(b), list A, point 6. A detailed description is also in LR r30, r31.

18. AS 205(b), list A, point 7; LR r10a.

19. AS 205(b), list A, point 4.

20. Experiments with needles made of copper, brass, wood, and steel were also performed in Geneva (Pictet 1821, 282).

21. Wollaston and van Marum, for example, had found indications that conventional (i.e., "static") electricity could bring about chemical decomposition (Brown 1969, 75).

22. See my discussion of the mémoire in the penultimate section of chapter 2. See also the monograph by Blondel (1982, 61–63, 175–76).

23. In the original, "à travers un demi conducteur, afin que le courant galvanique dure." AS 205(b), list A, points 5 and 12.

24. LR r10a.

25. He described these experiments further in LR r39.

26. I have previously discussed this difference in greater detail (Steinle 2000).

27. LR r12.

28. LR r14.

29. In the original, "lorsque je la communiquai à M.r Arago, il me fit remarquer avec raison, que cette attraction et cette répulsion entre un aimant et un conducteur galvanique placés à angles droits dans la direction où ils tendent à se mettre mutuellement dans la direction opposée, pouvait [*sic*] seule rendre raison des résultats //(46) publiés par l'auteur de la découverte, dans le cas où l'aiguille aimantée étant horizontale, on en approche un conducteur galvanique dans une situation verticale, et qu'on pouvait même déduire aisement cette loi, de l'une des expériences de M.r Oersted, celle qu'il enonce ainsi: *posito autem filo* ~~(accipiente~~ (cuius extremitas superior electricitatem à termino négativo apparatûs galvanici accipit) *è regione puncto inter polum et medium acùs sito, occidentem versus agitur.*" LR r14.

30. On Pfaff, see OC 2:466–67; see also the publication by Pictet (1821, 283).

31. LR r15.

32. LR r16.

33. With his hint as to the "general result," Ampère might be referring to the end of LR r8, describing the needle alignment (from right to left relative to the current), or the end of LR r11, in which he introduced the right-left terminology. In neither case can I discern any relevance to the specific findings of Ørsted now at issue. The long quote is from LR r17.

34. LR r19, r14, r17, respectively.

35. LR r15, r16.

36. LR, e.g., r3, r6, and esp. r14 and r16.

37. LR r14. It should be noted that these "general facts" (*faits généraux*) and the mode of explanation that goes along with them differ markedly from what Ampère would later call a "primitive fact" (*fait primitif*). For further discussion, see chapter 4, in the section "Theory of Magnetism and the Force Law."

38. LR r19.

39. LR r13.

40. In the original, "appareil pour les attractions et répulsions magnético-galvaniques à conducteur mobile." LR r29.

41. LR r24, r25 (original emphasis).

42. I return to Ampère's collaboration with Pixii in greater detail in the section after next.

43. LR r26.

44. LR r26.

45. Ampère depicted these devices frequently throughout his numerous publications (e.g., Ampère 1824a).

46. For further discussion, see chapter 5.

47. See Marcet's letter to Berzelius, September 15, 1820 (Söderbaum 1912–32, 3:210).

48. The priority of Ampère's discovery of the magnetic activity of the battery itself would never be disputed.

49. LR r32 (original emphasis).

50. Ørsted (1820e, 275), for example, characterized the piles he used in producing electromagnetic effects by reference to their ability to cause a metal wire to glow red. Schweigger (1821c, 1) described an especially powerful voltaic pile by reference to its capacity "to easily produce metallic potassium, even on the second and third days after its construction."

51. AS 205(b), list A, point 13.

52. The École Polytechnique did have laboratories for students, but only in chemistry. On the laboratory situation in general, see Fourcy's ([1828] 1987, 17, 57) history of the school, Crosland's (1992, 85) study of the Paris academy, and Bradley's (1976, 427–29) article on the polytechnic's early facilities.

53. Conseil d'instruction, issue 5, fol. 824, 82r, archive of the École Polytechnique. See also Dhombres's note in the appendix to Fourcy's ([1828] 1987, 159, 165) history.

54. Fresnel tutored physics and Despretz, chemistry, under Gay-Lussac.

55. LR r33; AS 205bis(h); AS 205(b), list A, points 11 and 17.

56. Though the role of instrument builders, traditionally underestimated, has come increasingly to the fore (e.g., Blondel 1997a or Hentschel 2008), it remains extremely difficult to study in any detail, as the requisite source materials (workshop archives, correspondence, etc.) are almost always missing.

57. By September 18, the August issues of both the *AGSP* and the *Bibl.Univ.* had been delivered to the Paris academy (*Proc.Verb.*, 91). These issues contained independent French translations of Ørsted's (1820e, 1820l) letter.

58. See the "Addition des Rédacteurs" that Pictet (1821) included with his translation of Ørsted's report.

59. LR r35.

60. Blondel (1982), for example, takes the sketch of the "bonhomme," as she names the man in the sketch, as the root of the alignment rule and not its refinement in a specific direction.

61. LR r35 (in-text quote), r36 (block quote).

62. LR r36, r41.

63. LR r36, r37, r40–r45. See also Ampère's letter to his friend Roux-Bordier, February 1821 (AC 2:567).

64. LR r46.

65. In the original, *imiter* (LR r46) or *singer* (AS 205[b], list B).

66. For example, LR r33, r46.

67. LR r33. On the various revisions, see my detailed discussion in appendix B.

68. See Ampère's letter of the evening of September 25, 1820 (AC 2:562).

69. Further discussion is in chapter 4.

70. Maurice Daumas (1953, 379–380) assumes that Dumotier's workshop was also Pixii's workshop. But according to the historian of instrumentation Paulo Brenni (personal communication; I am indebted to him for numerous hints), this overlap cannot be established with any certainty. Until approximately 1818, there were two Dumotier workshops in Paris, one at 31 rue Copeau and the other at 2 rue de Jardinet. Around 1820, the latter began to bear Pixii's name. See also Blondel's (1997a, 165) article.

71. I obtained this figure from a list of instruments and their prices in the Dossier Dulong at the Académie des Sciences in Paris. This ledger mentions a large voltaic pile, to be used in demonstrations, for 600 francs. I thank Matthias Dörries for drawing my attention to the list. Ampère's annual wages were approximately 11,500 francs: 5,000 from his professorship at the École Polytechnique, 6,000 from his position as *inspecteur général de l'université,* and another 500 from a pension at the Institut de France (see his letter of May 22, 1816, to the administration of the École Polytechnique, in his personnel dossier at the archive of the school). This was a respectable income. Further information is found in various general studies of the financial conditions of French academics (e.g., Bradley 1979; Grattan-Guinness 1990, 1272–74; Fox 1974, 135) or in Dhombres's comments, in the appendix to Fourcy's ([1828] 1987, 85, 176, 184) history of the École Polytechnique.

72. LR r48.

73. Ampère mentions this demonstration in a letter written to his son on the same day (AC 2:562) and in his later short report in the "Notes" (Ampère 1820h, 167).

74. Ampère's letter to his son, evening of September 25, 1820 (AC 2:562).

75. LR r48.

76. For the "single electrical fluid," see AS 206(b), list A, point 9. See also appendix A, §A10.

77. Blondel (1989) has investigated this matter thoroughly.

78. LR r37–r38.

79. LR r37–r38.

80. AS 206bis(m). The notation "20.7.[bre]" means September 20. The happy circumstance of this annotation offers the historian one of the very rare opportunities to directly date an item in the Ampère archive. Appendix A contains a transcription of the relevant passage.

81. LR r39.

82. AS 206bis(m).

83. Notions of tension and current remained vague and coincided only partially

with their modern counterparts. The French word *tension* was often used at the time, for example, by Biot (1801a, 1803). For further discussion, see Brown's (1969) article.

84. Though only a very few events may be dated precisely, good arguments for their sequence may be inferred from the source materials. The particular inferences contributing to those arguments range from technical judgments on which pieces of paper fit together to the substantive analysis of cross-references and mentions of particular experiments. A few detailed examples may be found in the appendixes. In addition to Ampère's lecture manuscript in its various versions, manuscript AS 205(b) is of particular importance to the chronology.

85. AS 205(b), list B in the third version; see my discussion of the text in appendix A, §A10.

86. See chapter 7 for a discussion of the exploratory methodology.

87. On my conception of "theory," see chapter 7.

88. Though my reconstruction of the contents of the two lectures goes well beyond any previously existing reports, it is entirely consistent with the few accounts we have of them. See Ampère's (1820h) "Notes" and the report by "Anonymous" (1821) report in the *Bibl.Univ.*

89. LR r6. The first section encompasses passages r6 through r17.

90. LR r18. The second section encompasses LR passages r18 through r33, with passage r33 presumably not yet in its final form at the time of its presentation.

91. I am drawing on Arago's (1820a, 94) own account, which, as he explicitly notes, presents his activities in chronological order.

92. See, for example, Gilbert's (1820c, 293) article. However, Erman's (1821a, 220) reference to the "Ampère-Arago results" made the relationship out to be more equal than it actually was.

93. LR r34. Section 3 includes passages r34–r49.

94. LR r35–r45.

95. LR r46–r47.

96. That Ampère inserted the "Conclusions" only shortly before the academy lecture may be inferred from the fact that, unlike the short account of the positive outcome of that morning's experiment, they were not alluded to elsewhere in the lecture manuscript.

97. This series is paraphrased from the published French text (Ampère 1820b, 1820c, 1820j, 209–12). For details see note 104 below.

98. For example, in the detailed report of the academy session in *Bibl.Univ.*, several of the points in the published text of the "Conclusions" are mentioned explicitly. Though it did not appear until the May 1821 issue, the report was clearly drafted shortly after the session, for its description of Ampère's apparatus contains peculiarities explicable only by recourse to the sheer novelty of his experiments, of the sort likely to be observed only firsthand, by someone actually in attendance. For example, the author compared the device for demonstrating the attraction between two spirals with an electroscope, for both exhibit pendular oscillations (Anonymous 1821, 83).

99. On the concept of an "epistemic thing," see Rheinberger's (1992, 69–72) discussion.

100. Ampère would explicitly acknowledge the programmatic character of his "Conclusions," albeit only retrospectively, in a letter to a friend (August de la Rive, March 23, 1823, AC 2:627) in which he speaks of a "kind of divination" that must have been at work.

101. AC 2:562.

102. AC 2:562.

103. The receipt of Ampère's letter of October 11, along with two copies of his paper, was logged at the Royal Society on November 9, following the end of the summer recess under way since July; see *Phil.Trans.* (1821): 443, and the archives of the Royal Society, AP.9.2. The offprints themselves have not been preserved in London, though copies may be found in Paris, in the library of the Muséum d'Histoire Naturelle and as part of Ampère's manuscript, AS 208bis(f).

104. The first journal issue containing the "Conclusions" was delivered at the beginning of November 1820 (Ampère 1820b). Reprints of this text may be found in the libraries of the Muséum d'Histoire Naturelle and the École Polytechnique. Shortly thereafter it was published in two more venues (Ampère 1820c, 1820j, 209–12).

105. In a later letter to Davy (1825) he stressed that it was only his supplemental income as inspector of schools that allowed him to build the equipment necessary for his research without endangering his family's subsistence (AC 2:679–80).

106. Classes for the lower cohort at the École Polytechnique began on November 10, 1820, and for the upper cohort, on November 13 (Conseil d'instruction, Carton 1: Report of September 14, 1820, archive of the École Polytechnique).

Chapter 4. Competing Pursuits in Paris, 1820–1821

1. Ampère presumably mentioned this experiment in his academy lecture of October 9, that is, AS 205(e).

2. See also my sketch in the last section of chapter 1.

3. Working with the History of Physics Working Group (now sadly defunct) at the Physics Department of the University of Oldenburg, Jörg Lühr has constructed an operational current balance to Ampère's original specifications. I am grateful to him for the information he has provided, and for the opportunity to work extensively with a device so similar to the original.

4. AS 180(c). The drawing is accompanied by a detailed description of the device in a clear hand, AS 180(d).

5. Crosland (1994, 248) also stresses this point.

6. For Berzelius, see Dulong's letter to Berzelius of October 1820 from the compilation by Söderbaum (1912–32, 4:18). We may infer from another letter that Dulong sent to Berthollet on December 19 (Söderbaum 1912–32, 1:75, 100) that Dulong had attached the "Conclusions" to the October letter. For Ørsted, see his letter of November 21, 1820, to Johan Georg Forchhammer (OC 2:335).

7. The resulting report appeared only in January 1821 (Ampère 1820a). That a significant portion of this report rested on very early material may be inferred from

several factors, especially from its terminology. It is not known how this material might have reached the journal.

8. In the title of this essay, Ampère claims to have "presented" it to the academy on October 2. This date is presumably in error. There is no other evidence that he either appeared before the academy in any capacity on that day or delivered any written report. In any case, in all probability the essay could not have been finished, for it contains a detailed discussion of "current balance." If Ampère had had an operational instrument at his disposal by October 2, he would surely have spoken of it to the academy at that day's session; after all, he was strongly inclined to announce his results as quickly as possible. The essay was published in the September issue of the *ACP*, delivered in the second week of November.

9. He repeatedly emphasized this point; see, for example, his letter of February 1821 to Roux-Bordier (AC 2:566).

10. See Brown's (1969, 82–86) article for a more thorough discussion.

11. This is plain from manuscript AS 206bis(f), and it presents a marked contrast with the earlier lecture manuscripts. Elsewhere (Ampère 1820j, 197), he explicitly stressed the transition. This issue is explored more thoroughly in appendix A.

12. His early reflections are known to us only from manuscripts. Blondel (1978) has carefully analyzed manuscript AS 158(a), though not AS 158(b). Here I offer only a brief sketch. Blondel (1982, 83–85) has provided a more thorough discussion.

13. Ampère gave this formula in AS 158(b).

14. Blondel (1982, 82–85) dates manuscript AS 158(a), in which the cosine dependency is explicitly formulated, to as early as October 9. But from the portion of the manuscript not transcribed in her monograph (1982), it appears more probable that he composed it for October 30. The same holds for manuscript AS 158(b), which Blondel does not consider.

15. In the original, "Il nous semble qu'on ne peut donner ce nom qu'aux lois de l'action mutuelle, qu'il faut admettre entre deux points . . . ; dès-lors les faits simples ne peuvent être observés immédiatement, mais seulement conclus des observations à l'aide du calcul" (Ampère 1821b, 371). Remaining quotations in the main text paragraph are from this source.

16. AS 205(o), which also contains the ensuing considerations.

17. Ampère (1820j, 182) alluded to this setup only briefly. Blondel (1978, 56) provides a schematic sketch.

18. Ampère's (1820j, 186–87) published account offers only vague indications on this procedure. But archived notes (AS 205[h]) make clear that he considered such questions very carefully.

19. AS 205(h).

20. This is the only reference to Poisson in accounts of research on electromagnetism. It is possible that the negative outcome of the experimental search for an "action at a distance" that he requested of Boisgiraud contributed to his quick loss of interest.

21. Charles (1746–1823) had become famous for his work on aerostatics, especially for his spectacular balloon flights. See Hoefer's (1854, col. 929–33) biographical sketch.

22. In its essence, the report of November 13 (*Proc. Verb.*, 101) was presumably written by Ampère. The extant manuscript is in Ampère's hand but bears Charles's signature.

23. For the minutes of the academy session of November 6, see *Proc. Verb.*, 100. Fresnel had deposited his text as a sealed note that was immediately opened, however, upon Ampère's request (Blondel 1982, 89–90). The search for chemical effects of magnetism had a long history, largely ignored by historians; see Andrade Martins's (2012) piece.

24. Dulong's letter to Berzelius, end of October 1820 (Söderbaum 1912–32, 4:19).

25. Reprinted in Biot's "précis" (Joubert 1885, 83).

26. The details of the measurement procedure, the number of actual measurements performed and their reproducibility, and the methods of data reduction—in short, Biot's research practice—have not been investigated to date. Even Frankel's (1977, 1978) thorough studies offer no hints in this regard. Biot himself did not publish a detailed account of this work until much later, in the third edition of his textbook (Biot 1824, vol. 2), reprinted in Joubert's (1885, 80–127) collection, and one cannot, of course, assume that the careful reflections found there were already worked out during those two hectic weeks in October 1820.

27. It is not clear whether Ampère received direct hints from Laplace or was simply offering his own reconstruction. See also Blondel's (1982, 57) comment.

28. Blondel (1982, 58–59) has also speculated along these lines, being the only one to have given much attention to this problem. A satisfactory explanation is still lacking and will presumably have to await a detailed study of Biot's research practice. Biot himself (1824, 742) did not acknowledge the mistake until much later.

29. Biot's description also appeared in English (Biot 1821a, 282–83).

30. David Gooding (1990b, 135) takes note of the thoroughness of Biot's description, reading it as an instance of "virtual witnessing," as Shapin and Shaffer (1985) called Boyle's rhetorical technique. This interpretation strikes me as misleading, entirely missing the point at issue. The reason for Biot's thorough description was not, as it had been for Boyle, the absence of direct witnessing among his readers—for these experiments could be easily replicated nearly anywhere. The problem, which was not unique to Biot, consisted rather in the lack of any conceptual scheme in which the relevant geometric configurations might be expressed briefly. Instead, because they had to refer back to elementary spatial concepts, descriptions necessarily became very verbose. This problem did not appear exclusively in the context of communication; it afflicted laboratory research, too.

31. The title of his public academy lecture of April 1821 is indicative: "Sur l'aimantation imprimé aux métaux par l'électricité en mouvement" (Biot 1821c), or, in English translation (Biot 1821a), "On the Magnetization Impressed on Metals by Electricity in Motion."

32. For a brief discussion, see, for example, Frankel's (1977, 53–56) article. Frankel does not connect this study to Biot's later work on electromagnetism.

33. See, for example, Biot's (1820) brief commentary.

34. This judgment remains provisional, pending some future thorough investigation of Biot's laboratory practice. In contrast with Ampère, however, Biot's reports offer not the slightest hint of any phase of insecurity and exploration. In Hasimoto's (1983)

sketch of Biot's reasoning, I find some oversimplifications. Among others, Biot was never "defeated" (51) but delivered a still valid result.

35. On the Société Philomatique, see the works by Mandelbaum (1980, 1988). For a list of its members, see the *Bulletin* masthead for 1820.

36. The issue is recorded in the *Bibliographie de la France* 4356 (November 25, 1820): 664.

37. The report bears no author name, but given the editorial circumstances and the competence required, Biot's authorship seems to me beyond question. In any case, the report passed through his hands before going to press.

38. Receipt of this issue of the *Bulletin* was logged at the academy on January 2, 1821 (*Proc. Verb.*, 125).

39. In the original, "La loi dont il s'agit consiste en que le courant dirigé suivant la résultante exerce . . . une action attractive où répulsive égale à la somme des actions qui résulteraient dans la même direction des deux portions de courants dirigés suivant les composantes" (AS 160[a], AS 160[b]). Another formulation may be found in, for example, Ampère's (1820g, 226) note.

40. Blondel (1982, 91–96) provides an excellent analysis of the reflections and hypotheses Ampère considered, also drawing on archival material (AS 162[b]).

41. Though the drawings were not published until later (Ampère 1822b, plate 6), he presented the corresponding argument in his lecture of December 4 (Ampère 1820g). See also Blondel's (1982, 92–94) discussion.

42. Blondel (1982, 91–92) provides a thorough analysis.

43. See Ampère's lecture of December 11 (Blondel 1982, 94–96).

44. In the original, "Comme c'est de cette loi . . . qu'il a déduit l'expression analytique de l'action mutuelle de deux portion infiniment petites de courants électriques . . . on sent qu'il a dû chercher les moyens de la vérifier par des expériences directes et susceptibles de précision" (Ampère 1821a, 160).

45. He never presented this argument as an integrated whole, instead bringing out various aspects of it at various junctures. Even his contemporaries could really only have made sense of it with the help of a comprehensive summary. Hofmann (1995, 253–69) offers a detailed and persuasive reconstruction.

46. For further discussion, see the comments by Blondel (1982, 92) or Hofmann (1995, 254–56). Hofmann's (1995) work also contains a reproduction of the drawing in Ampère's archives. Overlooking the fact that this first instrument was *exclusively* suited to mathematical measurements, Hofmann also missed a significant point (discussed later). Neither he nor Blondel make clear the extent to which, in devising this instrument, Ampère drew on Biot's work.

47. Following the practice of the time, Ampère never discussed the fact that even equilibrium experiments provoked reflections on sensitivity, margins of error, and the like. Wilhelm Weber (1846, 7–9) would later criticize Ampère for this defect.

48. The significance Hofmann attaches to Ampère's epistemological reflections is also quite clear in his article on "Ampère's invention of equilibrium apparatus" (Hofmann 1987b).

49. Talk of "inductive" reasoning strikes me as ill suited relative to either Ampère or Biot.

50. AS 206bis(j); see also Blondel's (1982, 98) view.

51. In her comment, Blondel (1982, 98–99) quotes the title of Ampère's lecture and draws attention to the fact that it indicates the great significance Ampère attached to these questions: "Calculation of the action exerted on a cylindrical magnet whose dimensions are very small, and that may only rotate in a horizontal plane, by a conductive wire inclined to the horizon and placed in a vertical plane intersecting the magnet's inertial center." He was referring directly to Biot's experimental arrangement, without mentioning him by name.

52. Hofmann (1995, 273) is the first to draw attention to this experiment and to recognize its significance (271–74). My presentation draws heavily on his.

53. The extent of Ampère's continued engagement with the conflict between the two theories is made clear from numerous manuscript sketches. See, for example, AS 189(u), AS 205bis(I), AS 206bis(I).

54. The fact that many of the articles on the list with nominal publication dates in 1820 did not appear until 1821 is due to the significant delay between the month assigned to a given issue of a periodical and the time by which the issue was actually written, edited, and printed. The "official" dates assigned to journal issues are of little use in studies of research practice. Only by carefully researching the actual dates of publication can we avoid gross errors of the kind all too common in Ampère scholarship to date.

55. Letter to Roux-Bordier, February 21, 1821 (AC 2:566).

56. Most of the differences between the *ACP* text and that of the stand-alone publication are primarily matters of format.

57. *Proc.Verb.*, 119; *Phil.Trans.* (1821): 444.

58. In the original, "Le travail de M. Ampère se divise naturellement en deux parties bien distinctes: l'une est purement expérimentale . . . l'autre se compose de considérations sur les phénomènes . . . dont il conclut que ces phénomènes . . . sont uniquement dus à l'électricité agissant dans les corps aimantées" (Ampère 1820f, 138).

59. *Le Moniteur universel,* November 10, 1820, 1491; December 7, 1820, 1599.

60. AS 162(d).

61. Yelin's (1820, 406) piece explicitly refers to articles in *Le Moniteur universel.*

62. Arago's explanation was in his eulogy for Ampère (Barral 1854–62, 2:1–116).

63. Biot's revised text was reprinted in Joubert's (1885, 80–127) compilation.

64. At the last minute, Ampère learned that Biot had scheduled a lecture on electromagnetism, and he composed a short "note" in great haste (Ampère 1821d, 1821e). For details, see works by Blondel (1982) and Hofmann (1995, 279–82).

65. See also Ampère's letter to Roux-Bordier, February 21, 1821 (AC 2:565).

66. AC 2:627.

67. A clear sense of this procedure may be gained from the preserved manuscripts, as described in appendix A.

68. The transcription in appendix A, §A6, provides only a limited sense of the

extent and vehemence of these editorial interventions that are so evident in the original manuscript.

69. This sort of activity also reveals a characteristic property of exploratory periods, to which I return in chapter 7.

70. Other researchers, including, for example, Ørsted (1820i) and Pictet (1821), did discuss this sort of experiment in print.

71. To cite one noteworthy example, in his accounts of Ampère's early works, Williams sometimes neglects even to mention the "general facts." Hofmann (1995, 229), in his biography of Ampère, discusses them, but not in their own right, despite his professed interest in "Ampère's early experimental methods." Blondel (1982) also does not recognize the larger context in which the "general facts" operated.

72. See my discussion at the end of chapter 2.

73. This description was reprinted in Biot's "précis" (Joubert 1885, 83).

Chapter 5. Electromagnetism in London

1. For an overview, see the monograph by Garber (1999).

2. See Miller's (1983) article for a thorough presentation. Snyder (2011) highlights some of these developments.

3. For the evolution of the physical laboratory out of the chemical laboratory in nineteenth-century England, see the articles by Sviedrys (1976, which emphasizes the second half of the century) and Roberts (1976, on chemical laboratories), or the essays collected by James (1989b). Observatories have only recently gained the attention of historians. See, for example, the conference papers delivered at "The Heavens on Earth: Observatory Techniques in Nineteenth Century Science," Max Planck Institute for the History of Science, December 2002, as initiated by Charlotte Bigg, David Aubin, and Otto Sibum.

4. On the London Institution, see the works by Hays (1974) and Morus (1998).

5. By contrast, the decades-long financial struggles of the Royal Institution could be traced in part to its maintenance of a large research laboratory and permanent faculty.

6. To mention only one example, even a relatively minor periodical such as the *Quarterly Journal of Science* had its Paris subscribers. The Bibliothèque Nationale has a complete run of its issues.

7. On Faraday there are numerous biographical accounts, with varying degrees of thoroughness; see those of Williams (1965); Cantor (1991); Cantor, Gooding, and James (1991); or Lemmerich (1991). Frank James's (1991) sketches in the introductions to Faraday's correspondence (hereafter cited as FC) are excellent.

8. There is a significant literature on this voyage; see, for example, the edition of Faraday's journal prepared by Bowers and Symons (1991).

9. Faraday's compensation included a salary of seventy-eight pounds per annum and an apartment in the Royal Institution. This remuneration constituted an improvement over his previous situation, but, given his newfound interests and lifestyle, it was rather tight. See the essay by Forgan (1985, 53).

10. The steel project, which he had been pursuing in collaboration with James Stodart since 1818, brought him broad international recognition (see, e.g., his correspondence [FC, vol. 1] on the subject), not to mention one of his first articles in the *Philosophical Transactions* (Faraday and Stodart 1822). See also Williams's (1965, 109–15) biography. The legal case in which Faraday served as expert witness also gained broad attention; see, for example, FC 1, letter 110, or Fullmer's (1980) article.

11. Williams (1965, 153) rightly draws attention to this record.

12. I discuss these experiments in greater detail below, in the section on Davy.

13. Prior to October 1821, the *Quarterly Journal*'s only pieces on electromagnetism were a short overview by Brande (1821a, January issue) and a translation of Biot's (1821a) essay (July issue).

14. Forgan (1985, 55) offers a colorful description, including the duties required of Faraday in his new position.

15. John Tyndall, Faraday's friend and successor at the Royal Institution, would much later remark on the latter's wellspring of personal stability: "I think that a good deal of Faraday's weekday strength and persistency might be referred to his Sunday Exercises. He drinks from a fount which refreshes his soul for the week" (Tyndall's diary entry for October 24, 1852, Royal Institution [hereafter, RI] MS JT/2/6, p. 164). Jost's (1995, 120) brief characterization strikes me as quite accurate.

16. Table 5.1 is structured differently from table 2.1, which details the events in Paris. Whereas table 2.1 provides only an overview of activities at the Académie des Sciences, table 5.1 encompasses not only such official notice as was taken at the Royal Society but also the laboratory activities of Wollaston, Davy, and Faraday. These efforts probably represent only a part of the laboratory research actually taking place in London but will serve to lend a first impression. Table 5.1 also lists publications, which I had treated separately for the Parisian case (table 4.1). The difference in my approach to the two cases derives mostly from the very different scope of available source materials.

17. Faraday (1836, 521) would later report that an excited Davy brought the issue into the laboratory on October 1.

18. Davy (e.g., 1821d, 18) emphasized this point.

19. The preceding session had convened on July 7.

20. Davy's essay (1812d, quotation on 8) was dated November 12.

21. See Brande's (1821b, 379–80) report. Young had written to officially inform Ørsted of his receipt of the Copley Medal on November 17 (OC 2:516).

22. This report was very Anglo-centric, devoting more space to Wollaston than to Ampère. In his German paraphrase in the *Annalen der Physik*, Gilbert expresses some annoyance on this point: "There is *not a word* on Arago's experiments, nor on anything appearing in these *Annalen*" (Gilbert and Brande 1821, original emphasis).

23. The January issue of the *Philosophical Magazine* was delivered at the beginning of February. It lists the author of the article as "M. Hatchett." This attribution has been overlooked in most accounts, and it may not have been clear to most readers that

it did not refer to the English chemist Charles Hatchett but to the Paris academician J. N. P. Hachette, his surname having been garbled. See, for example, the compilation of Gooding and James (1985). In another (unpublished) essay, Bradley (1989, 5) correctly attributes the article.

24. As mentioned in chapter 3, Cumming (1821a, 1821b) would later, in 1821, publish a proposal for a galvanometer.

25. See Faraday's private chronology in RI FSB, p. 152. The date "April 1821" was entered by Wollaston. On Faraday's participation, see Faraday's (1823, 289) historical statement.

26. See Faraday's (1823, 152) chronology. The lecture would be published toward the end of the year (Davy 1821a).

27. Davy's formulation was extraordinarily vague, especially with regard to the second case. He also appears to have made a mistake in specifying the direction, presumably the result of a simple confusion of upper and lower disk halves. The vagueness has led to erroneous interpretations. David Gooding (1985, 111–13; 1990a, 54) did not realize that Davy was attempting to express experimental results, instead understanding him as describing an experimental procedure involving needles suspended by threads. However, this interpretation does not succeed in making sense of Davy's text. There is also not the slightest indication in the report of any such experiments being done.

28. I have elsewhere given a more systematic analysis of this situation (Steinle 2011).

29. Faraday's *Diary* offers further details on these experiments (Martin 1932, 1:45–46). The London Institution possessed the largest battery available in London at the time (Ockenden 1937).

30. The importance of these studies in clarifying the concepts of current and tension has yet to be fully appreciated.

31. Davy's letter to Ampère, February 10, 1821, AC 2:564.

32. See Phillips's letters to Faraday, August 11, 1821 (FC 1, letter 145), and especially of September 4, 1821 (FC I, letter 147). In the latter, in consideration of Faraday's heavy workload, Phillips apologizes "that I should have been the cause of so much trouble and annoyance to you."

33. He would later speak of the "great labour and fatigue" it had cost him (Faraday 1823, 291; reprinted in Faraday 1839–55, 2:161).

34. His assertion that he had drawn on as many journals "as I could obtain access to" ([Faraday] 1821–22, 195) must thus be taken with this caveat.

35. Ludwig Wilhelm Gilbert, publisher of the *Annalen der Physik,* complained of this lacuna rather colorfully in his letter to Faraday of June 3, 1821 (FC 1, letter 136; see also Faraday's response of September 18, 1821 [FC 1, letter 153]). This phenomenon and its causes demand nuanced historical study, with due attention to the differences among disciplines. Concerning exchanges with Great Britain I find nothing comparable to Kai Kanz's (1997) thorough and thoughtful investigation of the Franco-German exchange.

36. Faraday's copy is in RI F6, vol. 4. This is most likely the copy sent to him by Hachette, with whom he was in communication regarding the project for improving steel (FC 1, letter 471).

37. Subsequent page references refer to the *ACP* (Ampère 1820e) offprint.

38. See Phillips's letter of September 4, 1821 (FC 1, letter 147).

39. In the aforementioned letter of September 4 (FC 1, letter 147), Phillips mentions having read the first two parts. The September issue of the *Annals* was delivered at the beginning of October and the October issue, at the beginning of November. On the chronology, see also Faraday's overview in RI F3B, p. 152.

40. On the plate, the wire was mistakenly labeled NS. As the text makes clear, it should have been NP.

41. Gooding (1990a, 53–55) also ascribes experiments of this sort to Davy but does so without providing any source for his attribution. There are no such indications in Davy's published texts.

42. Williams (1985a, 86–88) argues that Faraday's research interests were decisively influenced by a debate that had been ignited with an article by John Herapath and playing itself out in the pages of the *Annals of Philosophy* in the fall of 1821. In the course of that debate the priority of experimental laws over speculative theories had been repeatedly asserted. It is certain that Faraday shared this view, as indeed his organization of the "Historical Sketch" demonstrates. But it strikes me as implausible that he would have come to this view only in response to the Herapath debate. On the contrary, by the time he learned of the debate the first two parts of the "Historical Sketch" had already been written. The attitude in question can also be discerned, with as much clarity as could be wished, in his earlier annotations to Ampère's article.

43. Faraday (1821–22, 107) explicitly stressed this chronological extension of his overview at the beginning of the third part.

44. This is particularly conspicuous, as Schweigger had conceived of his own galvanometer, independently of Ampère and Poggendorff.

45. De la Rive's letter to Faraday has not been preserved, but Faraday's response of September 12 has (FC 1, letter 148). As discussed below, this device would play an important role in Faraday's own research.

46. See Faraday's letter to Stodart for his estimation of Wollaston's authority and the high opinion he held of him (FC 1, letter 152).

47. See also Gooding's (1985, 1989, 1990a, 1990b, 1998) studies, which I discuss in the next chapter.

48. Faraday himself would later describe his work situation as ideal: "[At the Royal Institution] I have the advantage of being upon the spot; of possessing a perfect laboratory with an assistant in constant occupation; and of having the command of an instrument maker & his men" (letter of July 29, 1829, FC 1, letter 404). Even taking into account that Faraday gave this description after he had become director of the laboratory and that it occurs in a context in which he sought to portray his position to best advantage, certain elements obtained even in 1821. For a brief description of London's other laboratories, see Forgan's (1985, 53–54) essay.

Chapter 6. Faraday's First Studies

1. On September 4, 1821, Richard Phillips had already read the texts of the "Historical Sketch" (FC 1, letter 147).

2. Martin's (1932) edition runs to seven volumes, with the material discussed here coming exclusively from volume 1. Henceforth I refer to this volume simply as *Diary*, and full citations to its text include a date, followed by a parenthetical number preceded by the letter "D" referring to the individual entry number, followed by *Diary* and the page number. In the main text, brief references to the location of material are indicated by the individual *Diary* entry number.

3. For example, "and the following experiments made . . . in the following order," as entered on October 1, 1831 (D36), *Diary* p. 372.

4. James's (1991) compilation is cited throughout as FC 1.

5. One exception is Faraday's essays on electromagnetic induction (see Steinle 1996).

6. In Faraday's collection of offprints (RI F6, vol. 6) we find a brochure (Hare 1819) describing the instrument in detail, emphasizing its exceptional utility in the study of thermal effects, and conceptualizing it as a single pair of plates. Faraday's sketch in the margins of the *Diary* bears great resemblance to the schematic drawing in Hare's (1819, fig. 2) brochure. We may assume that Faraday consulted this text directly. That spring, Hare (1821a, 334) had published an article describing an experiment performed with this device and explicitly emphasizing that its effects could be assessed in the same order of magnitude as the large battery at the London Institution, though its construction and operation were significantly less demanding.

7. This description is §3 in my counting of the paragraphs of Faraday's (1821d) article.

8. Ørsted had distinguished only the different effects at the needle's center and at its end.

9. Faraday does not make this explicit, for he had published the "Historical Sketch" anonymously.

10. In February 1996 and on several other occasions I had the opportunity to experiment extensively with such equipment in the laboratory of the Working Group for Physics Education and History of Science at the University of Oldenburg. I am grateful to Falk Rieß and especially to Dietmar Höttecke for granting me these opportunities and for their multifaceted support and conversation. Gooding (1989, 75–76) also draws attention to the extreme sensitivity of the experiment, as well as the extensive practice required before stable results can be obtained.

11. Ørsted (1820g, 276) had asserted, in the English translation, "that this conflict performs circles."

12. In particular, it becomes clear that Faraday was by no means working toward simply realizing the sort of circular motion of which others had spoken. I stress this point because some previous historical accounts sometimes give the (misleading) impression that Faraday had been looking for circular motion from the outset.

13. Faraday (1821d, 75–76, ¶4; reprint, 1839–55, 2:128) describes this experiment only

in his article. Its absence from the *Diary* (a very unusual circumstance) may perhaps be attributed to Faraday's excitement over his new discovery.

14. In his article Faraday (1821d, 76, ¶8; 1839–55, 2:129) explicitly draws attention to the problem.

15. As I discuss in greater detail below, with respect to the position of the magnet, Gooding (e.g., 1989, 77) always assumes it was oriented vertically. Based on the *Diary*, however, Faraday must without a doubt have held the magnet horizontally in this experiment and those that followed.

16. Gooding (1990a, §7.2) has displayed Faraday's account graphically by means of a formalization grounded in artificial intelligence theory.

17. See Gooding's (1985, 110–23; 1989; 1990a, chaps. 5, 6; 1990b, 136–39; 1998, 312–20) works as mentioned, respectively.

18. This reading may be found as early as Gooding's (1985, 117–18) work, but see also Gooding's (1989, 76–77; 1990a, 125; 1990b, 138) later comments. A different interpretation may be found where Gooding (1998, 314) asserts that the D6 sketch is rotated 90° relative to that presented in D5. This does nothing to clarify matters.

19. One cause for Gooding's misinterpretation may be the fact that in the D6 sketch Faraday initially drew two needles but indicated the supposed circular motion only for one of them. I conjecture that he began the sketch as a complete analog to D5, in which there were also two needles, but realized along the way that he could further compress his representation by indicating the positions for both poles on one and the same needle. The second sketched needle was a holdover, remaining unused.

20. Even leaving aside the misinterpretation, Gooding's (1998, 316) argument for the thesis that "Faraday constructed an image of circular motion by the process I have called dimensional enhancement" strikes me as neither clear nor persuasive. In Gooding's Tables 3 and 4 (1998, 317, 319), proffered as illustration and support, Faraday's D5 sketch is presented as both a two-dimensional and three-dimensional construct, while the D6 sketch is both a three-dimensional structure and a four-dimensional process. I also do not see why Faraday's perspective drawing in entry D3 should count as a two-dimensional representation, while his diagram in D4 counts as three-dimensional. By contrast with Gooding's other example (the investigation and representation of geological formations), the question of dimensionality strikes me as being of little importance here.

21. J. Bradley (1989, 7–8) also objects to treating Faraday's work on September 3, 1821, in isolation but still underestimates the continuity with the "Historical Sketch."

22. For example, in his September 12, 1821, letter to de la Rive, Faraday described *QJS* as "our Institution Journal" (FC 1, letter 148).

23. Faraday was also deeply involved with organizing the typesetting and mailing of the journal and thus intimately familiar with the relevant deadlines. See, for example, Brande's letter to Faraday of September 1820 (FC 1, letter 150).

24. Issues of the *Annals of Philosophy*, publisher of the "Historical Sketch," were usually delivered in the first or second week of the month to which they were assigned.

25. Letter to de la Rive, September 12, 1821 (FC 1, letter 148).

26. In the manuscript of his later "Historical Statement" (RI F3B, p. 154), Faraday had remarked that the slightest hesitation would have led to a delay of three months, when the next issue of *QJS* was scheduled to appear. This passage was deleted in the print version, however—presumably by Wollaston. Regarding the publication date, see Faraday's (1823, 160) comment or his remark in the September 12 letter to de la Rive that the article would be printed in one to two weeks (FC 1, letter 148), or Brande's letter to Faraday discussing galleys of the issue and the woodblocks for Faraday's figures (FC 1, letter 150).

27. Davy (1821d, 14) had mentioned Wollaston's speculations on a "revolution of magnetism."

28. See Faraday's letter to Wollaston of October 30, 1820 (FC 1, letter 154) and his later account (Faraday 1823, 289–90; 1839–55, 2:160). On Davy, see FC 1, letters 146, 149.

29. Gooding (1985, 120–23) briefly discusses Faraday's rotation apparatus, and J. Bradley (1989, 8–10) sketches an overview. Williams (1965, 161–66) concentrates on Faraday's engagement with Ampère, though this is surely only part of what motivated Faraday and certainly not the most important part.

30. Even Faraday's contemporaries were irritated by the absence of any divisions in his text. In an effort to "help the reader[s] and facilitate [their] survey," Gilbert's German translation of the article in his *Annalen der Physik* (Faraday and Gilbert 1822) introduced a division into five "chapters" (*Kapitel*), each of which was further subdivided with various numbered "experiments" (*Versuche*). Gilbert's five "chapters" correspond to my sections I–II, III–V, VI–IX, X–XIII, and XIV–XV, respectively.

31. Faraday mentions Wollaston's name in the *Diary* but not in the article. In his later "Historical Statement" (Faraday 1823, 289; 1839–55, 2:160) he would explain this absence on the grounds that Wollaston had never ventured his prediction either in public or in print.

32. The loss of the article manuscript is particularly painful here.

33. Regarding the polarity of the battery, in both the *Diary* and the article Faraday is silent on currents, just as Ampère had been reticent on this matter at the beginning. Faraday avoids the term entirely, instead referring to the polarity of the battery, as indicated in his sketches. If I sometimes speak of current-bearing wires, this is only for ease of exposition.

34. It should go without saying that the curves in these drawings always indicate the paths of magnet or wire. Because they have sometimes been taken instead as the first representations of lines of force (e.g., Williams 1986, 307), it becomes necessary to make the point. Throughout this whole of Faraday's work there is no intimation of magnetic curves, let alone lines of force.

35. Faraday provides no technical details on his implementation of this experiment. Unlike the following experiment (D18), in which both wires are submerged in the basin, in this case they cannot both come into contact with the same volume of mercury. One of the two wires must presumably be insulated.

36. FC 1, letter 148.

37. Faraday published a brief notice on this device in "his" section of the October issue of *QJS* (Faraday 1821b).

38. He conducted the experiments on September 4 (D17–D18), 6 (D38–D39, D44), 8 (D58), and 10 (D63).

39. September 6, 1821 (D36–D38).

40. September 7, 1821 (D49–D50).

41. Iwan Morus (1992, 8) stresses that Faraday's "strategy" was "to establish rotation as the primitive fact of electromagnetism and to draw attention away from the apparatus through which the fact made itself visible." The first part of this claim touches on an important point, while the second is false. By focusing exclusively on sociological conditions, and especially on Faraday's troubled relationship with William Sturgeon, Morus completely misses the point that Faraday's talk of "simple facts" was bound up with an epistemic problem that had little to do with his humble upbringing, rivalry with Sturgeon, or the like. For Faraday, the importance of reduction to "simple facts" provided an epistemic standard in itself and was independent of issues of public presentation. As essential and indispensable as sociological studies are, they must not be taken as sufficient for understanding a scientific development in its entirety.

42. Letter to Ampère of September 3, 1822 (FC 1, letter 179, emphasis added).

43. The term "power" has a wide range of different meanings in Faraday's article. Sometimes it refers, quite generally, to a capacity or the strength of a capacity, as in the "power" of a pair of plates in a battery (¶2) or a "powerful needle" (¶29). Elsewhere (¶37), "power" appears to be a synonym for "force," describing the varying strength of the mechanical force exerted on a magnet pole rotating about a wire. Finally, there is a very specific meaning, central to the argument at issue.

44. Both text and figure reproduced in Faraday's "Historical Sketch" (1821–22, 110).

45. *Diary* (D48), as printed in Faraday's article (1821d, 79–80, ¶17; 1839–55, 2:132).

46. In the first column of table 6.3 I have noted the paragraph numbers corresponding to Faraday's discussion of the corresponding effects. In every case he either states the law governing the path of motion or refers the reader to its statement elsewhere in the text.

47. In view of the sequence discernible in the *Diary,* the fact that in the article the "powers"-based system is presented right at the transition between the two main themes would appear to be no coincidence.

48. FC 1, letter 148.

49. In a study of Faraday's work on electromagnetic induction (Steinle 1994), I have shown that there, too, he pursued different epistemic goals and different procedures simultaneously, though the weight given to each shifted decisively over the course of his labors.

50. *Diary* (D69).

51. Faraday did not continue his numbering of *Diary* entries on September 8 and 10. For ease of exposition, I assume continuous numbering.

52. The corresponding experiments are related in *Diary* entries D35–D41, D46, D51, D63–D65.

53. *Diary* entries D39–D42, D53–D58, D60–D62, D67–D69.

54. FC 1, letter 148.

55. The successful reduction is presented in ¶¶37–40.

56. Letter to de la Rive, September 12, 1821 (FC 1, letter 148).

57. The motto is reported by Gladstone (1872, 122). Faraday did not always strictly adhere to it. In situations of conflict, he was certainly prepared to publish unfinished work, as in his study of electromagnetic induction (Steinle 1996).

58. It is not clear whether Faraday paid for this device out of his own pocket or was able to commission it on behalf of the Royal Institution. The illustration of the device (shown in fig. 6.11) was not published until three months later.

59. He makes this point explicitly in a later letter to de la Rive, November 16, 1820 (FC 1, letter 158).

60. Letter to Wollaston, October 30, 1821 (FC 1, letter 154).

61. Letter to Stodart, October 8, 1821 (FC 1, letter 152).

62. "As to the opinions which others may have of your conduct, that is your concern, not mine." Wollaston's reply, November 1820 (FC 1, letter 156), to Faraday's letter, October 30, 1821 (FC 1, letter 154).

63. Years later he would reflect on the publication of this first article, writing, "I have regretted ever since I did not delay the publication, that I might have shewn it first to Dr. Wollaston" (Faraday 1823, 290; 1839–55, 2:160).

64. Prior to its publication he gave his "Historical Statement Respecting Electro-Magnetic Rotations" (Faraday 1823) to Wollaston for approval. The authorized manuscript of that article has been preserved (RI F3B, pp. 153–57). He discloses his authorship of the "Historical Sketch" in a public letter to Richard Phillips (FC 1, letter 200), quickly published in *Ann.Phil.* 6 (1823): 67. On January 8, 1824, he was elected a Fellow of the Royal Society.

65. On this episode, see his confrontation with John Davy, brother of the by then deceased Humphry Davy (Faraday 1836; 1839–55, 2:229–34).

66. For an account of the whole affair, see the biography by Williams (1965, 158–60).

67. See Faraday's expression of thanks in his letter to de la Rive, September 12, 1821 (FC 1, letter 148).

68. FC 1, letter 153. Whether the device was included or not cannot be inferred either from the letter or from Gilbert's later report (Faraday and Gilbert 1822, 136). In any case, Gilbert did not publish Faraday's article until the middle of 1822, by which time Ampère's reaction had persuaded him that Faraday was to be trusted. See Gilbert's introduction to the report (124–25).

69. The letter to Hachette has not been preserved, but the date and contents may be inferred from Hachette's reply of October 30 (FC 1, letter 155). For Wollaston and de la Rive, see FC 1, letters 154 and 158. De la Rive quickly published a report on Faraday in the *Bibliothèque universelle,* also reporting his own successful replication (de la Rive and Faraday 1821).

70. See Gilbert's comparative description in the aforementioned report (Faraday and Gilbert 1822, 136).

71. His later description of an illustration of the reduced rotator as "not drawn to any scale" may be taken as an indication of the varying sizes of these devices. For a

much more far-reaching interpretation of this description, which, while plausible, goes well beyond anything in the text, see Höttecke's (1997b) essay.

72. Dietmar Höttecke (1997a, 1997b, 2000) has reconstructed Faraday's rotation devices in various forms and performed experiments with them.

73. As this text enjoyed Brande's editorial approval as a full-fledged article, it could be accompanied by illustrations.

74. Hachette's letter to Faraday, October 30, 1821 (FC 1, letter 155).

75. Hachette's letter to Faraday, November 19, 1821 (FC 1, letter 159).

76. *Proc.Verb.*, Minutes of the Session of November 19, 1821, p. 247.

77. See Hachette's letter to Faraday, December 12, 1821 (FC 1, letter 160). This would also lead to the publication of a report in the February 1822 issue of the proceedings of the Société Philomatique.

78. See Ampère's (1822c, 329–33) article, along with his Figures 19–22 on its plate 3. The November issue of the *ACP*, in which these materials appeared, was circulating in Paris by January 7, 1822. Hashimoto (1983) provides an investigation of Ampère's theoretical analysis.

79. The translation (Faraday 1821e) included the illustrations, on plate 4. On Gay-Lussac's involvement, see Hachette's letter to Faraday, November 19, 1820 (FC 1, letter 159). Receipt of the *ACP* December issue was not logged at the Paris academy until March 18, 1822 (*Proc.Verb.*, p. 292).

80. FC 1, letter 160.

81. Although, unlike many others, this notice lacks the "M.F." byline, Faraday's authorship is indisputable. Besides him, no one else in the tight editorial circle of the *QJS* had yet learned of Ampère's device.

82. On the delivery date, see the "Historical Statement" (Faraday 1823, 289; 1839–55, 2:160).

83. In a private inventory of his important publications in the front of the manuscript of the "Historical Sketch," Faraday lists only these two essays for the period in question (RI F3B, p. 152).

84. With the exception of December 23, which, as a Sunday, was set aside for Faraday's participation in Sandemanian community activity, this work occupied December 21 through 25. The corresponding entries in the *Diary* may be found, unnumbered, on pages 61–63.

85. *Diary*, entry for December 21, 1821, p. 61.

86. *Diary*, entry for December 22, 1821, p. 61.

87. *Diary*, final sentence of the entry for December 22, 1821, p. 62.

88. *Diary*, entry for December 24, 1821, p. 62; see also Faraday's (1822c, 421) note.

89. *Diary*, entry for December 25, 1821, p. 63.

90. Faraday emphasizes that his exclusive attention to the north pole involves a simplification. Strictly speaking, one ought also to take the south pole into account, treating the entire arrangement as involving not the interaction between a wire and *one* magnet pole but between a wire and *two* magnet poles, as he had discussed in greater detail for laboratory-scale effects (Faraday 1822c, 418; 1839–55, 2:155).

91. It is not clear precisely which experiments Faraday had in mind. Of the two

devices demonstrating the terrestrial alignment effect Ampère (1820j) presented, only one involves a circular wire loop, or ring. In the other, the loop is rectangular (see figs. 4.7 and 4.8). For the purposes of Faraday's argument, the rectangular loop is even easier to accommodate.

92. Faraday's religious background also strikes me as playing a significant role in that it suggested that empirical breadth and maintaining theoretical openness were moral obligations; see Faraday's (1854) later lecture, "Observations on Mental Education." Even Cantor's (1991) detailed study does not pay sufficient attention to the consequences of this aspect of Faraday's religiosity for his scientific practice. See my biographical essay (Steinle 2004).

Chapter 7. Experiment and Concept Formation

1. Citron (1998) relates the impressions of one of the participants in the emergence of nuclear physics.

2. Gooding (1989) has also retraced Faraday's rotation experiments. See my discussion in chapter 6.

3. The Working Group for Physics Education and History of Science at the University of Oldenburg has pursued this strategy most consistently; see, for example, works by Rieß (1998) and Heering, Rieß, and Sichau (2000). For a discussion and evaluation of this approach, see also the essay by Pestre (1994).

4. Schaffer (1989) has conducted a similar study of the reception of Newton's optical experiments with the prism, though the far-reaching claims at which he arrived have been strongly contested by Shapiro (1996).

5. See Cantor's (1989) and Morus's (1992) pieces on Faraday, Shapin and Schaffer's book (1985) on the establishment of experimental research in the early modern period, Gundlach's (1992, 1996) works on experimental psychology in the nineteenth century, and Morus's (1998) study for the presentation of electrical experiments to broader society during the same period in English history.

6. Pestre (1995) offers a perspicacious overview of the "social and cultural history of the sciences."

7. Holmes (1992) made this point in his reaction to Shapin and Schaffer (1985).

8. See, for example, the programmatic account in the essay by Hagner and Rheinberger (1998).

9. See also Rheinberger's (1999) incisive commentary.

10. Collins's (1985) book is the locus classicus.

11. The subtitle of their book points sharply to their program: *The Social Construction of Scientific Facts.*

12. For example, works by Bloor (1976), Barnes and Shapin (1979), as well as those by Collins and Pinch (1982), Pickering (1987), and by Barnes, Bloor, and Henry (1996).

13. See, for example, the works of Knorr-Cetina (1981, 1984), Pickering (1982, 1984), and Latour (1990). See Pickering's (1992b) collection for various positions and research projects. Pickering's introductory essay (1992a) to that volume provides an excellent overview.

14. The extent of their differences becomes clear in a debate between Latour (1999) and Bloor (1999a, 1999b).

15. As Collins (1994) correctly points out in his reply, however, Franklin's explicit restriction of his source materials to published texts significantly weakens the persuasive power of his argument.

16. The only exception concerned large batteries like the one initially used by Ørsted; those were to be found only in better-equipped facilities. But when it quickly became known that the effect could also be achieved with single-plate pairs, this constraint vanished.

17. The full significance of this point only becomes apparent when the device is reconstructed (see chap. 6), making it a telling example of the fruitfulness of the replication method in historical research.

18. For a general account of the role of replication and replicability in experimental research, see my essay on the topic (Steinle 2016).

19. Considering only monographs, these studies include those by Cartwright (1983, 1989b), Galison (1987), and A. Franklin (1986, 1990).

20. See works by Rheinberger (1992, 1997), Rheinberger and Hagner (1993), and Burian (1997b).

21. See works by Latour (1990, 1994), Mayo (1996), Heidelberger (1998), and Graßhoff, Casties, and Nickelsen (2000).

22. This has held in fields far removed from those Rheinberger had in mind when proposing these notions. Moritz Epple (1999) has used them with considerable success in his history of the mathematical theory of knots.

23. Elsewhere we find entirely different inferences regarding our understanding of reality, for example, in the works of Gooding (1990a, chap. 7), Hacking (1992, 59–60), or Latour (1994).

24. See also the essay by Cartwright (1989a, 352).

25. Dingler himself carried this through for the concepts of geometry (see also the work of Holzkamp [1968] and contributions to the volume edited by Böhme [1976]). On the concept of mass, see works by Janich (1980, 1985, 1997). Tetens (1987) has considered how a constructivist approach might accommodate relativity theory, which Dingler had strongly rejected as "Jewish."

26. See, for example, my critique of Peter Janich's analysis of the mass concept (Steinle 1992).

27. For a sketch of the problems, see works by Radder (1998) and Buchwald (1998); see also Radder's (1995) essay and his more recent (2003) collection.

28. This caveat also holds for Mayo's (1994) claim that significant statistical reflection always goes into the design and evaluation of experiments. This claim makes sense only given a narrow focus on recent physics. Mayo's (1996) own work is essentially geared toward such statistical methods.

29. This expression is not meant as an appeal to the phenomenological tradition in philosophy, which would imply a host of issues not pertinent to the questions at hand.

30. This claim is at the core of Justus Liebig's (1863, 49) argument against the possibility of empiricism, for example.

31. On this point I differ with Hacking (1983, 153–54), who also distinguishes weak and strong versions of the theory-ladenness thesis. Though I largely agree with Hacking on what constitutes a weak (hence uninteresting) version, my emphasis is different. When it comes to the strong version, the most urgent issue seems to me, contra Hacking, not to be testing or testability but the concept of theory itself.

32. The notion of law has its own history, which is by no means always tied to the specific sense discussed here. Its use in reference to seventeenth-century science is often anachronistic (see, e.g., works by Ruby [1986] and Steinle [1995b, 2008]).

33. John Stuart Mill (1843, Book II, chap. 5) defended this view in his debate with Whewell.

34. I do not claim, of course, that all processes of classification and concept generation have to do with experiments. There are many examples in which the crafting of concepts and categories turned more on strictly theoretical conjectures.

35. See also Mill's (1843, Book II, chap. 5) dispute with Whewell.

36. The notion of a construal bears strong similarities to what Hacking (1992, 44–45) has called a "topical hypothesis." Unlike Hacking, Gooding works through his conception on the basis of a detailed historical case.

37. Significantly, the remark I have cited on the importance of studying early phases in scientific research was drawn from a *footnote*.

38. Faraday to de la Rive, September 12, 1821 (FC 1, letter 148); compare Faraday's (1821d, 78, ¶11) published statement.

39. In purely formal terms, explanation by reduction to simple cases may be understood as a special case of Hempel-Oppenheim-style, deductive-nomological (D-N) explanation (see, e.g., the book by Hempel [1965]), on which understanding a given phenomenon requires drawing on both general laws and particular matters of fact. At the same time, however, this reveals both the indeterminate character of the D-N explanatory scheme and how little it tells us. It does as little justice to the character of laws as phenomenological if-then rules as to the fact that deduction from the simple case is not, formally, a matter of logical subsumption but the demonstration of a chain of phenomena linking the simple case and the case to be explained.

40. Frank James (1989a, 39) has also stressed the importance of non-theory-guided experimentation in Faraday's case.

41. See, for example, Goethe's (1798) "Erfahrung und Wissenschaft"; see also articles by Steinle (2002a) and Steinle and Ribe (2002).

42. Thus did Friedrich Wöhler describe it to Berzelius in 1832 (Wallach and Braun 1901, 2:604).

43. See my characterization in the above section.

44. In substance, the distinction between contexts of discovery and justification was made much earlier, by Duhem in 1908, for example. See Hoyningen-Huene's (1987) detailed discussion of the various versions and aspects of this distinction, and also his more recent account (Hoyningen-Huene 2006).

45. For an account of the debates, see the volume compiled by Schickore and Steinle (2006).

46. The work referred to is *Vom Ursprung der Geometrie* [On the origins of geometry] (Husserl 1956–2002, 6:365–78, esp. 372). As remarked above, the substance of this notion may also be found in the work of Whewell (1840).

47. Reichenbach's attention was focused on the specialized and highly theoretical domain of relativity theory.

48. I have no reservations about speaking of exploration and formative phases in theoretical fields and in mathematics. Studies of Einstein's field equation for general relativity provide a salient example (Renn and Sauer 1997, 1999). To be sure, it has yet to be worked out what, in such cases, takes the place of the empirical criteria invoked in experimentation.

Appendix A. Source Materials for the Early Phases of Ampère's Research in Electrodynamics

1. In English, "galvanic current, a designation that I have felt obliged to replace by that of electrical current" (Ampère 1820j, 197); see also my transcription, paragraph m10 (= r8).

2. AC 2:562.

3. The manuscript for the text appearing in print as §II has been filed separately in another box, AS 159(a).

4. The sheet containing pages 37–38 is a later insertion.

5. This paragraph is written in cramped script on the lower margin of page 37 and the upper margin of page 38. It was presumably inserted after paragraphs r1–r4 had already been written on the following page.

6. The final sentences of the paragraph are written in very cramped script in the lower margin of the page, obviously with a view toward fitting it all on this page so as to ensure a smooth transition to the text of page 41, an insertion.

7. The sheet containing pages 41–42 has been taken from another context (see AS 208bis[g]) and inserted here.

8. This paragraph fragment, struck through in its entirety, is written in the lower margin of page 42. The text of the paragraph continues on the page of AS 208bis(g) that I have numbered page 89.

9. The final sentence of this paragraph, "je reviendrai . . . entre elles" is a later insertion, squeezed onto the bottom of the page.

10. The sheet containing pages 49–50 is a later insertion. On the top of the side numbered page 49, it contains only paragraph m26. The rest of this side and the reverse (p. 50) are blank.

11. Because it was later struck through, only the first digit of this date, a "2," remains recognizable, along with the month, "7.bre." But there can be no question that the date originally read "25. 7.bre"; in September Ampère presented only on the 18th and 25th.

12. The "(1)" appears in the margin to the left of this paragraph; its significance is unclear.

13. An asterisk appears in the margin to the left of this paragraph; its significance is unclear.

14. There is no continuous transition between the preceding text and paragraph r37. The left margin also clearly shifts from one line to the next, marking the beginning of r37. But this paragraph is by no means a later insertion between earlier inscriptions. Its text continues uninterrupted for some time. It was presumably inserted after Ampère composed r43, as a supplemental explication.

15. Paragraph r48 was not written until September 25, when Ampère first obtained the result reported there.

16. Once again, there is a continuity break between the preceding text and this paragraph. The left margin of the inscription clearly shifts from one line to the next. This is presumably another later addition. The paragraph occupies the lower half of page 62, itself the last page of a notebook. Toward the bottom, Ampère's script becomes very cramped, evidently so as not to have to start a new page, and with it a separate sheet.

17. The "(1)" refers to a footnote inserted at the bottom of the page (paragraph m45). From the arrangement it is clear that the note was added only after the remainder of the paragraph had been written down in its present form.

18. The *ACP* typesetter inserted a mark here, along with the marginal note, "tome 15, t. 14, page 209."

19. There is a break in the height of the lines here. This text must have been added later.

20. The typesetter inserted a mark here to indicate a transition to a new page, reflected in the printed text of the article.

21. At this point the manuscript page had been filled. The text continues on page 71.

22. Between the pages I have numbered 88 and 89 there was originally an additional sheet. Ampère removed it in the course of revision, inserting it into the published text; it may be found as pages 41–42 in AS 208bis(f). This explains the non sequitur between pages 88 and 89.

23. The left margin of this paragraph has shifted relative to the preceding text, and the script is larger. The two paragraphs were evidently not written in a single continuous session; r27 appears to have been added later.

24. The character of Ampère's script changes from here on; presumably, he had resumed writing following an interruption.

25. Here the size of the script changes noticeably.

26. In the following transcription, the headings Ampère put at the side of the list, grouping several points together, are given in {brackets} and printed in *italics*.

27. There has as yet been no proper investigation of the kinds of battery used in this research and the names given to them. Ampère rarely discusses the matter, especially in his published work.

28. As noted earlier, this issue is carefully analyzed in the article by Blondel (1989).

Appendix B. Reconstruction of the Manuscript of Ampère's Lectures of September 18 and 25, 1820

1. This paragraph has been written in cramped script in the lower margin of page 37 and the upper margin of page 38. It may have been inserted after paragraphs r1–r4 had already been written and the next page filled.

2. I can find no clear basis for dating the final sentence, clearly a later insertion. Nowhere else does Ampère mention the experiment here alluded to.

3. The left margin of this paragraph has shifted clearly relative to the preceding, and the script is larger. This may indicate that the two paragraphs were not written in continuous succession, and that r27 may have been added later.

4. The following paragraph is not continuous with the preceding text, its left margin having shifted. But it has not been inserted between previously written passages, for it continues uninterrupted for some time. I assume that Ampère, after completing the text through r43, came back to it only later.

5. The following paragraph (r48) was clearly written on September 25, when Ampère first obtained the result he reports. This paragraph contains no breaks, appearing to have been written as a whole, without interruption. We may also infer that r49 was not written before September 25 and presumably no later.

6. Once again the left margin of this paragraph has shifted relative to the preceding, suggesting that it must have been a later insertion. The paragraph occupies the second half of page 62, which is the last page of a notebook. In the lower margin Ampère's script becomes very cramped, evidently in an effort to avoid having to begin a new page, and with it a separate sheet of paper.

7. On the portion reporting the lectures of September 18 and 25, see the first part of Ampère's (1820a, 238–42) analyses.

8. But he did not do so at the very outset. The announcement of the third section was presumably added later; see my remark on r5.

REFERENCES

Achinstein, Peter, and Owen Hannaway, eds. 1985. *Observation, Experiment, and Hypothesis in Modern Physical Science.* Cambridge: MIT Press.

Ackermann, R. 1989. "The New Experimentalism." *British Journal for the Philosophy of Science* 40:185–90.

Ampère, André-Marie. 1802. *Considerations sur la Théorie mathématique de jeu.* Lyon: Frères Perisse.

Ampère, André-Marie. 1814. "Lettre de M. Ampère à M. le comte Berthollet, sur la détermination des proportions dans lesquelles les corps se combinent d'après le nombre et la disposition respective des molécules dont leurs particules intégrantes sont composées." *Annales de chimie* 90:43–86.

Ampère, André-Marie. 1815a. "Considérations générales sur les intégrales des équations aux différentielles partielles." *Journal de l'École Polytechnique* 10 (17): 549–611.

Ampère, André-Marie. 1815b. "Démonstration de la relation découverte par Mariotte, entre les volumes des gaz et les pressions qu'ils supportent à une même température." *Annales de chimie* 94:145–60.

Ampère, André-Marie. 1816a. "Démonstration d'un théorême d'ou l'on peut déduire toutes les lois de la réfraction ordinaire et extraordinaire." *Mémoires de la Classe des Sciences Mathématiques et Physiques de l'Institut de France* 14 (1813–15): 235–48.

Ampère, André-Marie. 1816b. "Essai d'une classification naturelle pour les corps simples." *Annales de chimie et de physique* 1:295–308, 373–94; 2:5–32, 105–25.

Ampère, André-Marie. 1820a. "Analyse des mémoires lus par M. Ampère à l'Académie des Sciences, dans les séances des 18 et 25 septembre, des 9 et 30 octobre 1820." *Annales générales des sciences physiques* 6 (November): 238–57.

Ampère, André-Marie. 1820b. "Conclusions d'un Mémoire sur l'Action mutuelle de deux courans électriques, sur celle qui existe entre un courant électrique et un aimant, et celle de deux aimans l'un sur l'autre; lu a l'Academie royale des Sciences, le 25 septembre 1820; par M Ampère." *Journal de physique* 91 (July): 76–78.

Ampère, André-Marie. 1820c. "'Conclusions.'" *Annales générales des sciences physiques* 6 (October): 129–31.

Ampère, André-Marie. 1820d. "Mémoire présenté à l'Académie royale des Sciences, le 2 octobre 1820, ou se trouve compris le résumé de ce qui avait été lu à la même Académie les 18 et 25 septembre 1820, sur les effets des courans électriques." *Annales de chimie et de physique* 15 (September): 59–76.

Ampère, André-Marie. 1820e. *Mémoires sur l'action mutuelle de deux courans électriques, sur celle qui existe entre un courant électrique et un aimant ou le globe terrestre, et celle de deux aimans l'un sur l'autre, lus à l'Académie royale des sciences, par M. Ampère (Extrait des Annales de chimie et de physique).* Paris: Imp. de Feugueray.

Ampère, André-Marie. 1820f. "Note sur les expériences de M. Ampère, relatives à l'action mutuelle de deux fils conjonctifs, et à celle qui a lieu entre un fil conjonctif et un aimant ou le globe de la terre." *Bulletin des sciences par la Societé Philomatique* (September): 138–40.

Ampère, André-Marie. 1820g. "Note sur un Mémoire lu à l'Académie royale des Sciences dans le séance du 4 decembre 1820." *Journal de physique* 91 (September): 226–30.

Ampère, André-Marie. 1820h. "Notes de M. Ampère sur les lectures qu'il a faites à l'Académie des Sciences." *Journal de physique* 91 (September): 166–69.

Ampère, André-Marie. 1820i. *Notes de M. Ampère sur les lectures qu'il a faites à l'Académie des Sciences (Anhang an den Sonderdruck Hachette [1820a]).* Paris: Imprimerie de Huzard-Courcier.

Ampère, André-Marie. 1820j. "Suite du Mémoire sur l'Action mutuelle entre deux courans électriques, entre un courant électrique et un aimant ou le globe terrestre, et entre deux aimants." *Annales de chimie et de physique* 15 (October): 170–218.

Ampère, André-Marie. 1821a. "Note sur deux mémoires lus par M. Ampère à l'Académie royale des Sciences, le premier dans le séances du 26 décembre 1820, le second dans les séances des 8 et 15 janvier 1821." *Journal de physique* 92 (February): 160–65.

Ampère, André-Marie. 1821b. "Notes relatives au Mèmoire de M. Faraday." *Annales de chimie et de physique* 18 (December): 370–79.

Ampère, André-Marie. 1821c. "Notes, by M. Ampere, of the Communications which he made to the Academy of Sciences." *Philosophical Magazine* 57 (January): 47–49.

Ampère, André-Marie. 1821d. "Notice sur les expériences electro-magnétiques de MM. Ampère et Arago, lue à la séance publique de l'Académie Royale des Sciences de Paris le 2 avril 1821." *Annales générales des sciences physiques* 8 (April): 101–5.

Ampère, André-Marie. 1821e. "Notice sur les expériences electro-magnétiques de MM. Ampère et Arago, lue à la séance publique de l'Académie Royale des Sciences de Paris le 2 avril 1821." *Bibliothèque universelle* 17 (May): 16–20.

Ampère, André-Marie. 1821f. "Suite d'analyse des mémoires de M. Ampère." *Annales générales des sciences physiques* 7 (February): 252–58.

Ampère, André-Marie. 1821g. "Ueber die gegenseitigen Wirkungen, welche aufeinander ausüben zwei elektricische Ströme, ein electrischer Strom und ein Magnet oder die Erdkugel, und zwei Magnete, von Ampère, frei bearbeitet von Gilbert: erste Hälfte." *[Gilbert's] Annalen der Physik* 67 (February): 113–63.

Ampère, André-Marie. 1822a. "Notes relatives au Mémoire de M. Faraday." *Annales de chimie et de physique* 18 (December): 370–79.

Ampère, André-Marie. 1822b. *Recueil d'observations électro-dynamiques*. Paris: Crochard.

Ampère, André-Marie. 1822c. "Suite de la Note sur un Appareil à l'aide duquel on peut vérifier toutes les propriétés des conducteurs de l'électricité voltaïque, découvertes par M. Ampère." *Annales de chimie et de physique* 18 (November): 313–33.

Ampère, André-Marie. 1824a. *Déscription d'un appareil électro-dynamique*. Paris: Crochard et Bachelier.

Ampère, André-Marie. 1824b. *Précis de calcul différentiel et de calcul intégral*. Paris: privately published.

Ampère, André-Marie. 1827. "Mémoire sur la théorie mathématique des phénomènes électro-dynamiques, uniquement déduite de l'expérience, dans lequel se trouvent réunis les Mémoires que M. Ampère a communiqués à l'Académie royale des Sciences, dans les séances des 4 et 26 décembre 1820, 10 juin 1822, 22 décembre 1823, 12 septembre et 21 novembre 1825." *Mémoires de l'Académie Royale des Sciences* 6 (1823): 175–387.

Ampère, André-Marie, and Gillet de Laumont. 1821. "Note sur les Expériences Électro-magnetiques de MM Ørsted, Ampère et Arago, relative à l'identité de l'aimant avec l'électricité." *Annales des mines* 5, no. 4 (1820): 535–58.

Andrade Martins, Roberto de. 1999. "Resistance to the Discovery of Electromagnetism." Paper presented at the conference "Volta and the History of Electricity," Como and Pavia, Italy, September 1999.

Andrade Martins, Roberto de. 2001. "Romagnosi and Volta's Pile: Early Difficulties in the Interpretation of Voltaic Electricity." In *Nuova Voltiana: Studies on Volta and His Times*. Volume 3, edited by Fabio Bevilacqua and Lucio Fregonese, 81–102. Pavia: Università degli Studi di Pavia / Milano: Ulrico Hoepli.

Andrade Martins, Roberto de. 2012. "The Rise of Magnetochemistry from Ritter to Hurmuzescu." *Foundations of Chemistry* 14 (2): 157–82.

Anonymous. 1821. "Notice des séances de l'Académie Royale des Sciences de Paris pendant le mois de septembre." *Bibliothèque universelle* 17 (May): 78–84.

Arabatzis, Theodore. 2006. *Representing Electrons: A Biographical Approach to Theoretical Entities*. Chicago: University of Chicago Press.

Arabatzis, Theodore. 2011a. "Hidden Entities and Experimental Practice: Renewing the Dialogue between History and Philosophy of Science." In *Integrating History and Philosophy of Science: Problems and Prospects*, edited by Seymour Mauskopf and Tad Schmaltz, 125–39. Dordrecht: Springer.

Arabatzis, Theodore. 2011b. "On the Historicity of Scientific Objects." *Erkenntnis* 75 (3): 377–90.

Arago, Dominique F. J. 1820a. "Expériences relatives à l'aimantation du fer et de l'acier par l'action du courant voltaïque." *Annales de chimie et physique* 15 (September): 93–102.

Arago, Dominique F. J. 1820b. "Note à: Experimenta circa effectum, etc. Expériences sur l'effet du conflict électrique sur l'aiguille aimantée. Par Mr. J.Chr. Ørsted, Professeur de physique dans l'Université de Copenhague. Traduction." *Annales de chimie et de physique* 14 (August): 417–18.

Arago, Dominique F. J. (1854) 1985. *Histoire de ma jeunesse*. Paris: Christian Bourgois.

Arago, François, and A.-M.-T Petit. 1816. "Sur les puissances réfractives et dispersives de certains liquides et des vapeurs qu'ils forment." *Annales de chimie et de physique* 1:1–9.

Arnold, David H. 1983–84. "The Mécanique Physique of Siméon Denis Poisson: The Evolution and Isolation in France of His Approach to Physical Theory (1800–1840)." *Archive for History of Exact Sciences* 28:243–367; 29:37–94, 287–307.

Barlocci, Saverio. 1821. "Rapporto del professor Saverio Barlocci sopra alcune sperienze elettro-magnetiche istituite nel gabinetto fisico dell' Università di Roma." *Giornale Arcadico di scienze, lettere e arti* 10 (June): 426–35.

Barnard, Howard C. 1969. *Education and the French Revolution*. London: Cambridge University Press.

Barnes, Barry, David Bloor, and John Henry. 1996. *Scientific Knowledge: A Sociological Analysis*. Chicago: University of Chicago Press.

Barnes, Barry, and Steven Shapin, eds. 1979. *Natural Order: Historical Studies of Scientific Culture*. London: SAGE.

Barral, Jean A., ed. 1854–1862. *Oeuvres complètes de François Arago*. 17 vols. Paris: Gide et J. Baudry.

Batens, D., and J. P. van Bendegem, eds. 1988. *Theory and Experiment: Recent Insights and New Perspectives on Their Relation*. Dordrecht: Kluwer.

Belhoste, Bruno. 1991. *Augustin-Louis Cauchy: A Biography*. Translated by Frank Ragland. New York: Springer.

Bence Jones, Henry. 1871. *The Royal Institution: Its Founder and Its First Professors*. London: Longmans, Green.

Berman, Morris. 1978. *Social Change and Scientific Organization: The Royal Institution, 1799–1844*. Ithaca: Cornell University Press.

Berzelius, Jacob. 1821. "Vom magnetischen Zustande in den Körpern, welche die elektrische Säule entladen." *[Schweigger's] Journal für Chemie und Physik* 31 (1): 94–102.

Biot, Jean-Baptiste. 1801a. "Sur la mouvement du fluide galvanique." *Bulletin des sciences par la Societé Philomatique* 3:45–48.

Biot, Jean-Baptiste. 1801b. "Sur un problème de physique, relatif à l'électricité." *Bulletin des sciences par la Societé Philomatique* 3:21–23.

Biot, Jean-Baptiste. 1803. "Recherches physiques sur cette question: Quelle est l'influence de l'oxidation sur l'électricité dévelopée par la colonne de Volta?" *Annales de chimie* 47:5–42.

Biot, Jean-Baptiste. 1804. "Rapport sur les expériences du citoyen Volta." *Mémoires de l'Institut National des Sciences et Arts: Sciences mathematiques et physiques* 5:195–222.

Biot, Jean-Baptiste. 1816. *Traité de physique expérimentale et mathématique*. 4 vols. Paris.

Biot, Jean-Baptiste. 1817. *Précis élémentaire de physique expérimentale*. 2 vols. Paris: Deterville.

Biot, Jean-Baptiste. 1820. "Sur le magnetisme de la pile de Volta." *Bulletin des sciences par la Societé Philomatique* (August): 128.

Biot, Jean-Baptiste. 1821a. "On the Magnetism impressed on metals by Electricity in Motion, read at the public Sitting of the Academy of Sciences, 2d April 1821." *Quarterly Journal of Science* 11 (July): 281–90.

Biot, Jean-Baptiste. 1821b. *Précis élémentaire de physique experimentale.* Volume 2. 2nd ed. Paris: Déterville.

Biot, Jean-Baptiste. 1821c. "Sur l'aimantation imprimé aux métaux par l'électricité en mouvement; lu à la séance publique de l'Académie des sciences le 2 avril 1821." *Journal de savans* (April): 221–35.

Biot, Jean-Baptiste. 1824. *Précis élémentaire de physique expérimentale.* 2 vols. 3rd ed. Paris: Déterville.

Biot, Jean-Baptiste. 1824–25. *Lehrbuch der Experimental-Physik oder Erfahrungs-Naturlehre, von J B Biot (3. Aufl. 1824).* Translated by Gustav Theodor Fechner. 4 vols. Leipzig: L. Voss.

Biot, Jean-Baptiste, and Frederic Cuvier. 1801. "Sur quelques propriétés de l'appareil galvanique." *Annales de chimie* 39:242–50.

Biot, Jean-Baptiste, and Alexander von Humboldt. 1804. "Sur le variations du magnétisme terrestre à différentes latitudes." *Journal de physique, de chimie, et de l'histoire naturelle* 59:429–50.

Biot, Jean-Baptiste, and Félix Savart. 1820. "Note sur le Magnétisme de la pile de Volta." *Annales de chimie et de physique* 15 (October): 222–23.

Blainville. 1820. "Nouvelles scientifiques. Physique. Expériences électro-magnétiques." *Journal de physique* 91 (August): 151–52.

Blondel, Christine. 1978. "Sur les premières recherches de formule électrodynamique par Ampère (octobre 1820)." *Revue d'histoire des sciences et de leurs applications* 31:53–65.

Blondel, Christine. 1982. *A.-M. Ampère et la création de l'électrodynamique (1820–1827).* Paris: Bibliothèque Nationale.

Blondel, Christine. 1985. "Ampère and the Programming of Research." *Isis* 76:559–61.

Blondel, Christine. 1989. "Vision physique 'etherienne,' mathématisation 'laplacienne': L'électrodynamique d'Ampère." *Revue d'histoire des sciences et de leurs applications* 42:123–37.

Blondel, Christine. 1994. "La 'mécanisation' de l'électricité: Idéal de mesures exactes et savoir-faire qualitatifs." In Blondel and Dörries 1994, 99–119.

Blondel, Christine. 1997a. "Electrical Instruments in 19th Century France: Between Makers and Users." *History and Technology* 13:157–82.

Blondel, Christine. 1997b. "Haüy et l'électricité: De la démonstration-spectacle à la diffusion d'une science newtonienne." *Revue d'histoire des sciences* 50:265–82.

Blondel, Christine, and Matthias Dörries, eds. 1994. *Restaging Coulomb: Usages, controverses et réplications autour de la balance de torsion.* Florence: Olschki.

Bloor, David. 1976. *Knowledge and Social Imagery.* Chicago: University of Chicago Press.

Bloor, David. 1999a. "Anti-Latour." *Studies in History and Philosophy of Science* 30:81–112.

Bloor, David. 1999b. "Reply to Bruno Latour." *Studies in History and Philosophy of Science* 30:131–36.

Böckmann. 1821a. "Bemerkung über denselben Gegenstand." *[Schweigger/Meinecke's] Journal für Chemie und Physik* 31 (January): 34.

Böckmann. 1821b. "Kurzer Bericht von seinen Versuchen über die Wirkungen des geschlossenen Voltaisch-elektrischen Kreises auf die Magnetnadel, und über die Erregung des Magnetismus im Stahl durch gewöhnliche Elektricität." *[Gilbert's] Annalen der Physik* 68 (May): 1–16.

Böhme, Gernot, ed. 1976. *Protophysik*. Frankfurt: M. Suhrkamp.

Boisgiraud, [Jean-Pierre]. 1820. "De l'action de la pile sur aiguille aimantée." *Annales de chimie et physique* 15 (November): 279–89.

Boisgiraud, [Jean-Pierre]. 1821a. "On the Action of the Voltaic pile upon the Magnetic Needle." *Philosophical Magazine* 57 (March): 203–6.

Boisgiraud, [Jean-Pierre]. 1821b. "Sequel of the Experiments on the Action of the Voltaic pile upon the Magnetic Needle." *Philosophical Magazine* 57 (April): 257–58.

Bowers, Brian, and Lenore Symons. 1991. *"Curiosity Perfectly Satisfyed": Faraday's Travels in Europe 1813–1825*. London: Peter Peregrinus.

Bradley, John K. 1989. "The Early Electromagnetic Experiments of Faraday and Their Replication." MPhil diss., Cambridge University.

Bradley, John K. 1991. "Repeating the Electromagnetic Experiments of Michael Faraday." *Physics Education* 26:284–88, 312.

Bradley, Margaret. 1976. "The Facilities for Practical Instruction in Science during the Early Years of the 'École Polytechnique.'" *Annals of Science* 33:425–46.

Bradley, Margaret. 1979. "The Financial Basis of French Scientific Education and Scientific Institutions in Paris 1790–1815." *Annals of Science* 36:451–91.

Brande, William T. 1821a. "On the Connexion of Electric and Magnetic Phaenomena." *Quarterly Journal of Science* 10:361–64.

Brande, William T. 1821b. "Proceedings of the Royal Society: November/December." *Quarterly Journal of Science* 10 (January): 378–87.

Brewster, David. 1821a. "Account of the discoveries of M. Ørsted, respecting the connections between magnetism and galvanism and the subsequent researches of Sir Humphry Davy, Bart, M. Ampère, and M. Biot." *Edinburgh Philosophical Journal* 4 (January): 167–75.

Brewster, David. 1821b. "Account of the New Galvano-Magnetic Condenser invented by M. Poggendorff of Berlin." *Edinburgh Philosophical Journal* 5 (July): 112–13.

Brewster, David. 1821c. "Ampere's Electro-Magnetic Telegraph." *Edinburgh Philosophical Journal* 4 (April): 435.

Brewster, David. 1821d. "Description and Use of the Apparatus employed by M. Ampere in his Electro-Magnetic researches." *Edinburgh Philosophical Journal* 4 (April): 406–16.

Brewster, David. 1821e. "Dr. Wollastons Explanation of the Electro-Magnetic Phenomena." *Edinburgh Philosophical Journal* 4 (April): 435.

Brown, Theodore M. 1969. "The Electric Current in Early Nineteenth-Century French Physics." *Historical Studies in the Physical Sciences* 1:61–103.

Buch. 1821a. "Sur l'électro-magnétisme, par le Dr. Buch, à Francfor: Extrait d'une note du 9 Novembre 1820 (Schweigger u. M. neues Journal)." *Bibliothèque universelle* 16 (February): 119–21.

Buch. 1821b. "Ueber Elektromagnetismus." *[Schweigger/Meinecke's] Journal für Chemie und Physik* 31 (January): 18–31.

Buchwald, Jed Z. 1977. "William Thomson and the Mathematization of Faraday's Electrostatics." *Historical Studies in the Physical Sciences* 8:101–36.

Buchwald, Jed Z. 1989. *The Rise of the Wave Theory of Light: Optical Theory and Experiment in the Early Nineteenth Century.* Chicago: University of Chicago Press.

Buchwald, Jed Z. 1994. *The Creation of Scientific Effects: Heinrich Hertz and Electric Waves.* Chicago: University of Chicago Press.

Buchwald, Jed Z., ed. 1995a. *Scientific Practice: Theories and Stories of Doing Physics.* Chicago: University of Chicago Press.

Buchwald, Jed Z. 1995b. "Why Hertz Was Right about Cathode Rays." In Buchwald 1995a, 151–69.

Buchwald, Jed Z. 1998. "Issues for the History of Experimentation." In Heidelberger and Steinle 1998, 374–91.

Burian, Richard M. 1997a. "Comments on Complexity and Experimentation in Biology." *Philosophy of Science* 64 (Supplement): S279–91.

Burian, Richard M. 1997b. "Exploratory Experimentation and the Role of Histochemical Techniques in the Work of Jean Brachet, 1938–1952." *History and Philosophy of the Life Sciences* 19:27–45.

Burian, Richard M. 2007. "On Micro RNA and the Need for Exploratory Experimentation in Post-Genomic Molecular Biology." *History and Philosophy of the Life Sciences* 29 (3): 285–312.

Caneva, Kenneth L. 1978. "From Galvanism to Electrodynamics: The Transformation of German Physics and Its Social Context." *Historical Studies in the Physical Sciences* 9:63–159.

Caneva, Kenneth L. 1980. "Ampère, the Etherians, and the Ørsted Connexion." *British Journal for the History of Science* 13:121–38.

Caneva, Kenneth L. 1981. "What Should We Do with the Monster? Electromagnetism and the Psychosociology of Knowledge." In *Sciences and Cultures: Anthropological and Historical Studies of the Sciences*, edited by Everett Mendelsohn and Yehuda Elkana, 101–31. Dordrecht: Reidel.

Caneva, Kenneth L. 1997a. "Colding, Ørsted, and the Meaning of Force." *Historical Studies in the Physical and Biological Sciences* 28 (1): 1–138.

Caneva, Kenneth L. 1997b. "Physics and Naturphilosophie: A Reconaissance." *History of Science* 35:35–107.

Cantor, Geoffrey N. 1989. "The Rhetoric of Experiment." In Gooding, Pinch, and Schaffer 1989, 159–80.

Cantor, Geoffrey N. 1991. *Michael Faraday: Sandemanian and Scientist; A Study of Science and Religion in the Nineteenth Century.* Basingstoke: Macmillan.

Cantor, Geoffrey N., David C. Gooding, and Frank A. J. L. James. 1991. *Faraday.* London: Macmillan.

Cardwell, Donald S. L. 1957. *The Organisation of Science in England: A Retrospect*. London: Heinemann.

Caroe, Gwendy. 1985. *The Royal Institution: An Informal History*. London: John Murray.

Carrier, Martin. 1998. "New Experimentalism and the Changing Significance of Experiments: On the Shortcomings of an Equipment-Centered Guide to History." In Heidelberger and Steinle 1998, 175–91.

Cartwright, Nancy. 1983. *How the Laws of Physics Lie*. Oxford: Clarendon Press.

Cartwright, Nancy. 1989a. "Capacities and Abstractions." In *Scientific Explanation*, edited by Phillip Kitcher and Wesley C. Salmon, 349–56. Minneapolis: University of Minnesota Press.

Cartwright, Nancy. 1989b. *Nature's Capacities and Their Measurement*. Oxford: Oxford University Press.

Chang, Hasok. 2004. *Inventing Temperature: Measurement and Scientific Progress*. New York: Oxford University Press.

Chang, Hasok. 2012. *Is Water H_2O? Evidence, Pluralism and Realism*. Dordrecht: Springer.

Chasles, Michel. 1839. "Énoncé de deux théorèmes généraux sur l'attraction des corps et la théorie de la chaleur." *Comptes rendus de l'Académie des Sciences* 8:209–11.

Chilton, Donovan, and Noel G. Coley. 1980. "The Laboratories of the Royal Institution in the Nineteenth Century." *Ambix* 27:173–203.

Chipman, Robert A. 1964. "The Earliest Electromagnetic Instruments." In *Contributions from the Museum of History and Technology*, 121–36. United States National Museum, Bulletin 240. Washington: Smithsonian Institution.

Christensen, Dan C. 1995. "The Ørsted-Ritter Partnership and the Birth of Romantic Natural Philosophy." *Annals of Science* 52:153–85.

Christensen, Dan C. 2013. *Hans Christian Ørsted: Reading Nature's Mind*. Oxford: Oxford University Press. http://site.ebrary.com/lib/alltitles/docDetail.action?docID=10670443.

Citron, Anselm. 1998. "Experiment und Experimentator im Wandel der Technik am Beispiel der Kern- und Teilchenphysik." In Heidelberger and Steinle 1998, 115–22.

Cole, R. J. 1952. "Sir Anthony Carlisle, F.R.S. (1768–1840)." *Annals of Science* 8:255–70.

Collins, Harry M. 1985. *Changing Order: Replication and Induction in Scientific Practice*. Beverly Hills: SAGE.

Collins, Harry M. 1994. "A Strong Confirmation of the Experimenters' Regress." *Studies in History and Philosophy of Science* 25:493–503.

Collins, Harry M., and Trevor Pinch. 1982. *Frames of Meaning: The Social Construction of Extraordinary Science*. London: Routledge & Kegan Paul.

Configliachi. 1821a. "Elektrisch-magnetische Versuche." *[Gilbert's] Annalen der Physik* 69 (10): 203–5.

Configliachi. 1821b. "Expériences d'électricite magnétique." *Annales générales des sciences physiques* 7 (February): 250–51.

Configliachi. 1821c. "Lettera del Prof. Configliachi a C. Ridolfi, etc. Lettre du Prof. Configliachi au Marquis Ridolfi sur les expériences Voltaïco-magnétique; communiqué au Prof. Pictet (Traduction). Paris, 26 Dec 1820." *Bibliothèque universelle* 16 (January): 72–74.

Costabel, Pierre. 1978. "Poisson, Siméon Denis." In *Dictionary of Scientific Biography*, edited by Charles Gillispie, 15:480–90. New York: Scribner.

Coulomb, Charles-Augustin. 1788a. "Deuxieme mémoire sur l'électricité et le magnétisme (1785)." *Histoire et mémoires de l'Académie Royale des Sciences*, 578–611.

Coulomb, Charles-Augustin. 1788b. "Premier mémoire sur l'électricité et le magnétisme (1785)." *Histoire et mémoires de l'Académie Royale des Sciences*, 569–77.

Crawford, Elspeth. 1985. "Learning from Experience." In Gooding and James 1985, 211–27.

Crosland, Maurice. 1967. *The Society of Arcueil: A View of French Science at the Time of Napoleon I.* Cambridge: Harvard University Press.

Crosland, Maurice. 1972. "'Nature' and Measurement in Eighteenth-Century France." *Studies on Voltaire and the Eighteenth Century* 87:277–309.

Crosland, Maurice. 1992. *Science under Control: The French Academy of Sciences 1795–1914.* Cambridge: Cambridge University Press.

Crosland, Maurice. 1994. *In the Shadow of Lavoisier: The "Annales de Chimie" and the Establishment of a New Science.* Oxford: British Society for the History of Science.

Cumming, James. 1821a. "On the Application of Magnetism as a Measure of Electricity." *Transactions of the Cambridge Philosophical Society* 1:281–86.

Cumming, James. 1821b. "On the Connexion of Galvanism and Magnetism." *Transactions of the Cambridge Philosophical Society* 1:269–78.

Cunningham, Andrew, and Nicholas Jardine, eds. 1990. *Romanticism and the Sciences.* Cambridge: Cambridge University Press.

Darrigol, Olivier. 2000. *Electrodynamics from Ampère to Einstein.* Oxford: Oxford University Press.

Daston, Lorraine. 1997. "The Cold Light of Facts and the Facts of Cold Light: Luminescence and the Transformations of the Scientific Fact, 1600–1750." In *Signs of the Early Modern, Part 2: 17th Century and Beyond.* Volume 3, edited by David L. Rubin, 17–44. Charlottesville: Rookwood Press.

Daumas, Maurice. 1943. *Arago 1786–1853: La jeunesse de la science.* 2nd ed. Paris: Belin.

Daumas, Maurice. 1953. *Les instruments scientifiques aux 17e et 18e siècle.* Paris: Presses Universitaires de France.

Davy, Humphry. 1800a. "Account of some Experiments made with the Galvanic Apparatus of Signor Volta." *[Nicholson's] Journal of Natural Philosophy, Chemistry and the Arts* 4 (September): 275–81.

Davy, Humphry. 1800b. "Additional Experiments on Galvanic Electricity." *[Nicholson's] Journal of Natural Philosophy, Chemistry and the Arts* 4 (October): 326–28.

Davy, Humphry. 1800c. "An Account of some Additional Experiments and Observations on the Galvanic Phenomena." *[Nicholson's] Journal of Natural Philosophy, Chemistry and the Arts* 4 (December): 394–402.

Davy, Humphry. 1800d. "Notice of some Observations on the Causes of the Galvanic Phenomena, and on certain Modes of increasing the Powers of the Galvanic Pile of Volta." *[Nicholson's] Journal of Natural Philosophy, Chemistry and the Arts* 4 (November): 337–42.

Davy, Humphry. 1801. "An Account of Some Galvanic Combinations." *Philosophical Transactions* 91:397–402.

Davy, Humphry. 1807. "The Bakerian Lecture, on Some Chemical Agencies of Electricity." *Philosophical Transactions* 97 (part 1): 1–56.

Davy, Humphry. 1808. "The Bakerian Lecture, on Some New Phenomena of Chemical Changes Produced by Electricity." *Philosophical Transactions* 98 (part 1): 1–44.

Davy, Humphry. 1820a. "Nouvelles expériences galvano-magnétiques, par M. H. Davy. [Übersetzung aus dem Phil. Mag., november 1820]." *Journal de physique* 91 (November): 394.

Davy, Humphry. 1820b. "Sur les effets magnétiques produits par l'électricité." *Annales générales des sciences physiques* 6 (November): 258–59.

Davy, Humphry. 1821a. "Farther researches on the magnetic phaenomena produced by electricity; with some new experiments on the properties of electrified bodies in their relations to conducting powers and temperature." *Philosophical Transactions* 111 (part 2): 425–39.

Davy, Humphry. 1821b. "On the magnetic Phaenomeny produced by Electricity, in a letter from Sir H. Davy . . . to W. H. Wollaston. . . . " *Philosophical Magazine* 58 (July): 43–50.

Davy, Humphry. 1821c. "On the magnetic phenomena produced by Electricity." *Annals of Philosophy*, n.s., 2 (August): 81–88.

Davy, Humphry. 1821d. "On the Magnetic Phenomena Produced by Electricity." *Philosophical Transactions* 111 (part 1): 7–19.

de la Rive, Charles-Gaspard, and Michael Faraday. 1821. "Mémoire sur quelques nouvelles expériences électro-magnétiques et en particulier sur celles de Mr. Faraday." *Bibliothèque universelle* 18 (December): 269–86.

Delambre. 1806. "Eloge historique de M. Coulomb." *Mémoires de l'Institut National des Sciences et Arts (sciences mathematiques et physiques)* 7:212 (Histoire).

Di Bucchianico, Marilena. 2014. "A Matter of Phronesis: Experiment and Virtue in Physics; A Case Study." In *Virtue Epistemology Naturalized: Bridges between Virtue Epistemology and Philosophy of Science*, edited by Abrol Fairweather, 291–312. Cham: Springer.

Dibner, Bern. 1962. *Ørsted and the Discovery of Electromagnetism*. New York: Blaisdell.

Dibner, Bern. 1964. *Alessandro Volta and the Electric Battery*. New York: Franklin Watts.

Dingler, Hugo. 1928. *Das Experiment: Sein Wesen und seine Geschichte*. Munich: Reinhardt.

Doebereiner, Johann. 1820. "Extrait d'une lettre de M. Doebereiner aux rédacteurs. I. Sur quelques résultats magnétiques obtenus par une chaine galvanique simple. II. Construction d'une nouvelle chaine galvanique simple." *Annales générales des sciences physiques* 6 (December): 399–400.

Dörries, Matthias. 1998. "Vicious Circles, or, the Pitfalls of Experimental Virtuosity." In Heidelberger and Steinle 1998, 123–40.

du Bois-Reymond, Emil. 1853. *Gedächtnisrede auf Paul Erman, gehalten in der öffentlichen Sitzung der königlichen Akademie der Wissenschaften Berlin am 7. Juli 1853*. Berlin: Reimer.

Duhem, Pierre. (1908) 1978. *Ziel und Struktur der physikalischen Theorien*. Reprint, Hamburg: Meiner / Leipzig: Barth.

Elliott, Kevin C. 2007. "Varieties of Exploratory Experimentation in Nanotoxicology." *History and Philosophy of the Life Sciences* 29 (3): 313–36.

Epple, Moritz. 1999. *Die Entstehung der Knotentheorie: Kontexte und Konstruktionen einer modernen mathematischen Theorie*. Braunschweig: Vieweg.

Erman, Paul. 1821a. "Auszug aus einem Brief an Gilbert, den elektrischen Magnetismus betreffend." *[Gilbert's] Annalen der Physik* 68 (February): 220–21.

Erman, Paul. 1821b. *Umrisse zu den physischen Verhältnissen des von Prof. Ørsted entdeckten elektro-chemischen Magnetismus*. Berlin.

Faraday, Michael. 1821a. "Contact in Voltaic Electricity." *Quarterly Journal of Science* 12 (October): 185.

Faraday, Michael. 1821b. "Electro-Magnetic Instruments invented by M. de la Rive." *Quarterly Journal of Science* 12 (October): 184–85.

Faraday, Michael. 1821c. "New Electro-Magnetic Apparatus." *Quarterly Journal of Science* 12 (October): 186–87.

Faraday, Michael. 1821d. "On Some New Electro-Magnetical Motions, and on the Theory of Magnetism." *Quarterly Journal of Science* 12:74–96.

Faraday, Michael. 1821e. "Sur les Mouvemens électro-magnétiques et la théorie du magnétisme." *Annales de chimie et de physique* 18 (December): 337–70.

[Faraday, Michael]. 1821–22. "Historical Sketch of Electro-Magnetism." *Annals of Philosophy*, n.s., 2 (September): 195–200; 2 (October): 273–90; 3 (February): 107–21.

Faraday, Michael. 1822a. "Description of an Electro-Magnetical Apparatus for the Exhibition of Rotatory Motion." *Quarterly Journal of Science* 12 (January): 283–85.

Faraday, Michael. 1822b. "Electro-magnetical Rotation." *Quarterly Journal of Science* 12 (January): 415–16.

Faraday, Michael. 1822c. "Note on New Electro-Magnetical Motions." *Quarterly Journal of Science* 12 (January): 416–21.

Faraday, Michael. 1823. "Historical Statement respecting Electro-Magnetic Rotation." *Quarterly Journal of Science* 15 (July): 288–92.

Faraday, Michael. 1836. "On the History of the Condensation of the Gases, in reply to Dr. Davy, introduced by some Remarks on that of Electro-magnetic Rotation." *Philosophical Magazine* 8:521–29.

Faraday, Michael. 1839–55. *Experimental Researches in Electricity*. 3 vols., 1839, 1844, and 1855. London: Taylor.

Faraday, Michael. 1854. "Observations on Mental Education." In *Lectures on Education Delivered at the Royal Institution of Great Britain*, 39–88. London: John W. Parker and Son.

Faraday, Michael, and Ludwig W. Gilbert. 1822. "Ueber electrisch—magnetische Bewegungen, und die Theorie des Magnetismus." *[Gilbert's] Annalen der Physik* 71 (June): 124–71.

Faraday, Michael, and James Stodart. 1822. "On the Alloys of Steel." *Philosophical Transactions*, 112:253–70.

Feest, Uljana, and Friedrich Steinle, eds. 2012. *Scientific Concepts and Investigative Practice*. Berlin: De Gruyter.

Feest, Uljana, and Friedrich Steinle. 2016. "Experiment." In *The Oxford Handbook of Philosophy of Science*, edited by Paul Humphreys. Oxford: Oxford University Press.

Fischer, Joachim. 1988. *Napoleon und die Naturwissenschaften*. Stuttgart: Franz Steiner Verlag.

Fleck, Ludwik. (1935) 1979. *Genesis and Development of a Scientific Fact*. Translated by Fred Bradley and Thaddeus J. Trenn. Chicago: University of Chicago Press.

Fleck, Ludwik. (1935) 1980. *Entstehung und Entwicklung einer wissenschaftlichen Tatsache: Einführung in die Lehre vom Denkstil und Denkkollektiv*. Frankfurt: Suhrkamp.

Forgan, Sophie. 1985. "Faraday—From Servant to Savant: The Institutional Context." In Gooding and James 1985, 51–67.

Fourcy, A. (1828) 1987. *Histoire de L'Ecole Polytechnique: Nouveau impression, avec une introduction par Jean Dhombres*. Paris: Belin.

Fourier, Joseph. 1816. "Théorie de la chaleur." *Annales de chimie et de physique* 3:350–75.

Fourier, Joseph. 1824. "Théorie du mouvement de la chaleur dans les corps solides." *Mémoires de l'Académie Royale des Sciences* 4 (1819–20): 185–555.

Fourier, Joseph. 1826. "Suite du mémoire intitulé: Théorie du mouvement de la chaleur dans les corps solides." *Mémoires de l'Académie Royale des Sciences* 5 (1821–22): 153–246.

Fox, Robert. 1973. "Scientific Enterprise and the Patronage of Research in France 1800–1870." *Minerva* 11:442–73.

Fox, Robert. 1974. "The Rise and Fall of Laplacian Physics." *Historical Studies in the Physical Sciences* 4:89–136.

Fox, Robert, and George Weisz. 1980a. "Introduction: The Institutional Basis of French Science in the Nineteenth Century." In Fox and Weisz 1980b, 1–28.

Fox, Robert, and George Weisz, eds. 1980b. *The Organization of Science and Technology in France 1808–1914*. Cambridge: Cambridge University Press.

Frängsmyr, Tore, John L. Heilbron, and Robin E. Rider, eds. 1990. *The Quantifying Spirit in the 18th Century*. Berkeley: University of California Press.

Frankel, Eugene. 1977. "J. B. Biot and the Mathematization of Experimental Physics in Napoleonic France." *Historical Studies in the Physical Sciences* 8:33–72.

Frankel, Eugene. 1978. "Career-Making in the Post-Revolutionary France: The Case of Jean-Baptiste Biot." *British Journal for the History of Science* 11:37–48.

Franklin, Allan. 1986. *The Neglect of Experiment*. Cambridge: Cambridge University Press.

Franklin, Allan. 1990. *Experiment, Right or Wrong*. Cambridge: Cambridge University Press.

Franklin, Allan. 1994. "How to Avoid the Experimenters' Regress." *Studies in History of Philosophy and Science* 25:463–91.

Franklin, Allan, and Harry M. Collins. 2016. "Two Kinds of Case Studies and a New Agreement." In Sauer and Scholl 2016, 95–122.

Franklin, Laura R. 2005. "Exploratory Experiments." *Philosophy of Science* 72:888–99.

Franksen, Ole I. 1981. *H. C. Ørsted: A Man of the Two Cultures*. Birkeroed: Strandberg.

Fresnel, Augustin Jean. 1816. "Mémoire sur la diffraction de la lumière, où l'on examine particulièrement le phénomène des franges colorées que présentent les ombres des corps éclaires par un point lumineux." *Annales de chimie et de physique* 1:239–81.

Fresnel, Augustin Jean. 1820. "Note sur des Essais ayant pour but de décomposer l'eau avec un aimant." *Annales de chimie et physique* 15 (October): 219–22.

Friedman, Robert M. 1977. "The Creation of a New Science: Joseph Fourier's Analytical Theory of Heat." *Historical Studies in the Physical Sciences* 8:73–99.

Fullmer, June Z. 1980. "Technology, Chemistry and the Law in Early 19th-Century England." *Technology and Culture* 21:1–28.

Galison, Peter. 1987. *How Experiments End*. Chicago: University of Chicago Press.

Galison, Peter. 1988. "History, Philosophy, and the Central Metaphor." *Science in Context* 2 (1): 197–212.

Galison, Peter. 1995. "Context and Constraints." In Buchwald 1995a, 13–41.

Garber, Elizabeth. 1999. *The Language of Physics: The Calculus and the Development of Theoretical Physics in Europe*. Boston: Birkhäuser.

Gardiner, K. R., and D. L. Gardiner. 1965. "André-Marie Ampère and His English Acquaintances." *British Journal for the History of Science* 2:235–45.

Gautherot, N. 1801. "Mémoire sur le galvanisme." *Annales de chimie* 39:203–10.

Gazzeri. 1821. "Conferma della recente scoperta, etc. Confirmation de la découverte recente du Prof. Ørsted, par la répétition de son expérience principale, à Florence dans le laboratoire du Marquis Ridolfi. Notice adressée aux Redacteurs de la Biblioth. Univ., par Mr. Gazzeri, Prof. de chimie. (Traduction)." *Bibliothèque universelle* 15 (December [1820]): 260–63.

Gazzeri, Ridolfi, and Antinori. 1821. "Esperimenti elletro-magnetici, etc. Expériences électro-magnétiques faites à Florence par le Prof. Gazzeri, le Marquis Ridolfi, et le Chev. Antinori, du 6 au 18 janv. inclusivement, en presence de quelques amateurs des sciences physiques. (avec fig.) (Traduction)." *Bibliothèque universelle* 16 (February): 101–18.

Geison, Gerald L. 1995. *The Private Science of Louis Pasteur*. Princeton: Princeton University Press.

Gilbert, Ludwig W. 1820a. "Bericht von der Sitzung der Allgemeinen Schweizerischen Gesellschaft für Naturwissenschaften, gehalten zu Genf am 25. bis 28. Juli 1820." *[Gilbert's] Annalen der Physik* 67 (January): 91–108.

Gilbert, Ludwig W. 1820b. "Untersuchungen über die Einwirkung des geschlossenen galvanisch-elektrischen Kreises auf die Magnetnadel." *[Gilbert's] Annalen der Physik* 66 (12): 331–91.

Gilbert, Ludwig W. 1820c. "Vorbericht zu: Entdeckung ausgezeichneter Wirkungen des geschlossenen galvanisch-elektrischen Kreises auf die Magnetnadel und der Kraft der galvanischen Electricität zu magnetisieren." *[Gilbert's] Annalen der Physik* 66 (11): 291–94.

Gilbert, Ludwig W. 1820d. "Vorrede: Zum Band 6 der Annalen der Physik und der physikalischen Chemie." *[Gilbert's] Annalen der Physik* 66:n.p.

Gilbert, Ludwig W., and William T. Brande. 1821. "Ueber die Verbindung zwischen den elektricischen und magnetischen Erscheinungen; etwas aus England." *[Gilbert's] Annalen der Physik* 68 (May): 32.

Gilbert, William. 1600. *De magnete, magneticisque corporibus, et de magno magnete tellure: Physiologia noua, plurimis & argumentis, & experimentis demonstrata*. London: Petrus Short.

Gillmor, C. S. 1971. *Coulomb and the Evolution of Physics and Engineering in Eighteenth-Century France*. Princeton: Princeton University Press.

Gladstone, John H. 1872. *Michael Faraday*. London.

Goethe, Johann W. 1798. "Erfahrung und Wissenschaft." In *Goethes Werke (Hamburger Ausgabe)*, 13:23–25. Hamburg.

Golinski, Jan. 1990. "Humphry Davy and the 'Lever of Experiment.'" In *Experimental Inquiries: Historical, Philosophical and Social Studies on Experimentation in Science*, edited by Homer L. Le Grand, 99–136. Dordrecht: Kluwer.

Gooding, David C. 1985. "'In Nature's School': Faraday as an Experimentalist." In Gooding and James 1985, 105–35.

Gooding, David C. 1986. "How Do Scientists Reach Agreement about Novel Observations?" *Studies in History and Philosophy of Science* 17:205–30.

Gooding, David C. 1989. "History in the Laboratory: Can We Tell What Really Went On?" In James 1989b, 63–82.

Gooding, David C. 1990a. *Experiment and the Making of Meaning: Human Agency in Scientific Observation and Experiment*. Dordrecht: Kluwer.

Gooding, David C. 1990b. "Theory and Observation: The Experimental Nexus." *International Studies in the Philosophy of Science* 4:131–48.

Gooding, David C. 1992. "Putting Human Agency Back into Observation." In *Science as Practice and Culture*, edited by Andrew Pickering, 65–112. Chicago: University of Chicago Press.

Gooding, David C. 1998. "Picturing Experimental Practice." In Heidelberger and Steinle 1998, 298–322.

Gooding, David C., and Frank A. J. L. James, eds. 1985. *Faraday Rediscovered: Essays on the Life and Work of Michael Faraday, 1791–1867*. Basingstoke: Macmillan.

Gooding, David C., Trevor Pinch, and Simon Schaffer, eds. 1989. *The Uses of Experiment: Studies in the Natural Sciences*. Cambridge: Cambridge University Press.

Graßhoff, Gerd. 2011. "Inferences to Causal Relevance from Experiments." In *Explanation, Prediction, and Confirmation*, edited by Dennis Dieks, Wenceslao J. Gonzalez, Stephan Hartmann, Thomas Uebel, and Marcel Weber, 167–82. Dordrecht: Springer Science+Business Media B.V.

Graßhoff, Gerd, Robert Casties, and Kärin Nickelsen. 2000. *Zur Theorie des Experiments: Untersuchungen am Beispiel der Entdeckung des Harnstoffzyklus*. Bern: [Inst. für Philosophie].

Grattan-Guinness, Ivor. 1972. *Joseph Fourier 1768–1830: A Survey of His Life and Work, Based on a Critical Edition of His Monograph on the Propagation of Heat, Presented to the Institut de France in 1807*. Cambridge, MA: MIT Press.

Grattan-Guinness, Ivor. 1990. *Convolutions in French Mathematics, 1800–1840: From the Calculus and Mechanics to Mathematical Analysis and Mathematical Physics*. 3 vols. Basel: Birkhäuser.

Grattan-Guinness, Ivor. 1991. "Lines of Mathematical Thought in the Electrodynamics of Ampère." *Physis* 28:115–29.

Grmek, Mirko D. 1973. *Raisonnement expérimental et recherches toxicologiques chez Claude Bernard*. Geneva: Droz.

Grmek, Mirko D. 1991. *Claude Bernard et la méthode expérimentale*. Paris: Payot.

Gundlach, Horst. 1992. "Psychologie und Experiment um die Mitte des 19. Jahrhunderts." In *Psychologische Forschung und Methode: Das Versprechen des Experiments*, edited by Horst Gundlach. Passau: Passavia Universitätsverlag.

Gundlach, Horst. 1996. "The Hipp Chronoscope as Totem Pole and the Formation of a New Tribe—Applied Psychology, Psychotechnics and Rationality." *Teorie & Modelli*, n.s., 1:65–85.

Hachette, Jean N. P. 1820a. "Sur les expériences électro-magnétiques de MM. Ørsted et Ampère." *Journal de physique* 91 (September): 161–66.

Hachette, Jean N. P. 1820b. *Sur les expériences électro-magnétiques de MM. Ørsted et Ampère. Par M. Hachette: Extrait du* Journal de Physique, *septembre 1820*. Paris: Imprimerie de Huzard-Courcier.

Hachette, Jean N. P. 1821. "On the Electro-Magnetic Experiments of MM. Ørsted and Ampère." *Philosophical Magazine* 57 (January): 40–49.

Hacking, Ian. 1983. *Representing and Intervening: Introductory Topics in the Philosophy of Natural Science*. Cambridge: Cambridge University Press.

Hacking, Ian. 1992. "The Self-Vindication of the Laboratory Science." In Pickering 1992b, 29–64.

Hackmann, William D. 1984. "Leopoldo Nobili and the Beginnings of Galvanometry." In *Leopoldo Nobili e la cultura scientifica del suo tempo*. Volume 11, edited by Gino Tarozzi, 203–33. Bologna: Nuova Alfa Editoriale.

Hagner, Michael, and Hans-Jörg Rheinberger. 1998. "Experimental Systems, Objects of Investigation, and Spaces of Representation." In Heidelberger and Steinle 1998, 355–73.

Hall, Marie B. 1984. *All Scientists Now: The Royal Society in the Nineteenth Century*. Cambridge: Cambridge University Press.

Hanson, Norwood R. 1962. "The Irrelevance of History of Science to Philosophy of Science." *Journal of Philosophy* 59:574–86.

Hare, Robert. 1819. *A new theory of galvanism, supported by some experiments and observations made by means of the calorimotor, a new galvanic instrument*. Philadelphia: Carey and Son.

Hare, Robert. 1821a. "A Memoir on Some new Modifications of Galvanic Apparatus, with Observations in Support of His Theory of Galvanism." *Annals of Philosophy* 1 (May): 329–40.

Hare, Robert. 1821b. "A Memoir on Some new Modifications of Galvanic Apparatus, with Observations in Support of His Theory of Galvanism." *Philosophical Magazine* 57 (April): 284–94.

Hashimoto, Takohiko. 1983. "Ampère vs. Biot: Two Mathematizing Routes to Electromagnetic Theory." *Historia Scientarium: International Journal of the History of Science Society of Japan* 24:29–51.

Hauksbee, Francis. 1754. *Experiences physico-mechaniques sur differens sujets: Et principalement sur la lumiere et l'electricité, produites par le frottement des corps Physico-mechanical experiments on various subjects*. Paris: Cavelier.

Haüy, René-Just. 1787. *Exposition raisonnée de la théorie de l'électricité et du magnétisme s'après les principes de M. Aepinus*. Paris.

Haüy, René-Just. (1803) 1806. *Traité élémentaire de physique.* Rev. 2nd ed. 2 vols. Paris: Delance et Lesueur.

Hays, J. N. 1974. "Science in the City: The London Institution, 1819–40." *British Journal for the History of Science* 7:146–62.

Heering, Peter. 1998. *Das Grundgesetz der Elektrostatik: Experimentelle Replikation und wissenschaftshistorische Analyse.* Florence: Deutscher Universitätsverlag.

Heering, Peter, Falk Rieß, and Christian Sichau, eds. 2000. *Im Labor der Physikgeschichte: Zur Untersuchung historischer Experimentalpraxis.* Oldenburg: BIS-Verlag.

Heidelberger, Michael. 1998. "Die Erweiterung der Wirklichkeit im Experiment." In Heidelberger and Steinle 1998, 71–92.

Heidelberger, Michael, and Friedrich Steinle, eds. 1998. *Experimental Essays—Versuche zum Experiment.* Baden-Baden: Nomos Verlag.

Heilbron, J. L. 1979. *Electricity in the 17th and 18th Centuries: A Study of Early Modern Physics.* Berkeley: University of California Press.

Heilbron, John L. 1976. "Alessandro Volta." In *Dictionary of Scientific Biography*, edited by Charles Gillispie, 14:69–82. New York: Scribner.

Heilbron, John L. 1993. *Weighing Imponderables and Other Quantitative Science around 1800.* Berkeley: University of California Press.

Heilbron, John L., Robert W. Seidel, and Bruce R. Wheaton. 1981. *Lawrence and His Laboratory: Nuclear Science at Berkeley, 1931–1961.* Berkeley: Lawrence Berkeley Laboratory and Office for History of Science and Technology, University of California.

Hempel, Carl G. 1965. *Aspects of Scientific Explanation and Other Essays in the Philosophy of Science.* New York: Free Press.

Hentschel, Klaus. 1997. "The Interplay of Instrumentation, Experiment, and Theory: Patterns Emerging from Case Studies on Solar Redshift, 1890–1960." *Philosophy of Science* 64 (Supplement): S53–S64.

Hentschel, Klaus. 2000. "Historiographische Anmerkungen zum Verhältnis von Experiment, Instrumentation und Theorie." In *Instrument—Experiment: Historische Studien*, edited by Christoph Meinel, 13–51. Berlin: GNT-Verlag.

Hentschel, Klaus, ed. 2008. *Unsichtbare Hände: Zur Rolle von Laborassistenten, Mechanikern, Zeichnern u.a. Amanuenses in der physikalischen Forschungs- und Entwicklungsarbeit.* Diepholz: Verlag für Geschichte der Naturwissenschaften und der Technik.

Herivel, John. 1975. *Joseph Fourier: The Man and the Scientist.* Oxford: Clarendon Press.

Herrmann, Armin, ed. 1968. *Die Begründung der Elektrochemie und Entdeckung der ultravioletten Strahlen von Johann Wilhelm Ritter.* Frankfurt: Akademische Verlagsgesellschaft.

Herschel, John F. W. 1830. *A Preliminary Discourse on the Study of Natural Philosophy.* London: Longman.

Hesse, Mary. 1986. "Changing Concepts and Stable Order [Essay review of *Changing Order*, by Harry Collins]." *Social Studies of Science* 16:714–26.

Hiebert, Erwin N. 1995. "Electric Discharge in Rarefied Gases: The Dominion of Experiment. Faraday. Plücker. Hittorf." In *No Truth Except in the Details: Essays in Honor of Martin Klein*, edited by A. J. Kox and Daniel M. Siegel, 95–134. Dordrecht: Kluwer.

Hoefer, J. C. F. 1854. *Nouvelle biographie générale.* Volume 9. Paris: Firmin Didot.

Hofmann, James R. 1982. "The Great Turning Point in André-Marie Ampère's Research in Electrodynamics: A Truly 'Crucial' Experiment." Doctoral thesis, University of Pittsburgh.

Hofmann, James R. 1987a. "Ampère, Electrodynamics and Experimental Evidence." *Osiris*, 2nd ser., 3:45–76.

Hofmann, James R. 1987b. "Ampère's Invention of Equilibrium Apparatus: A Response to Experimental Anomaly." *British Journal for the History of Science* 20: 309–41.

Hofmann, James R. 1988. "Ampère's Electrodynamics and the Acceptability of Guiding Assumptions." In *Scrutinizing Science: Empirical Studies of Scientific Change*, edited by Arthur Donovan, Larry Laudan, and Rachel Laudan, 201–17. Dordrecht: Kluwer.

Hofmann, James R. 1995. *André-Marie Ampère: Enlightenment and Electrodynamics.* Oxford: Blackwell.

Holmes, Frederick L. 1981. "The Fine Structure of Scientific Creativity." *History of Science* 19:60–70.

Holmes, Frederick L. 1987. "Scientific Writing and Scientific Discovery." *Isis* 78:220–35.

Holmes, Frederick L. 1990. "Laboratory Notebooks: Can the Daily Record Illuminate the Broader Picture?" *Proceedings of the American Philosophical Society* 34:349–66.

Holmes, Frederick L. 1992. "Do We Understand Historically How Experimental Knowledge Is Acquired?" *History of Science* 30:119–36.

Holmes, Frederick L. 1993a. *Hans Krebs: Architect of Intermediary Metabolism 1933–1937.* Oxford: Oxford University Press.

Holmes, Frederick L. 1993b. "The Old Martyr of Science: The Frog in Experimental Physiology." *Journal of the History of Biology* 26:311–28.

Holmes, Frederick L., Jürgen Renn, and Hans-Jörg Rheinberger, eds. 2003. *Reworking the Bench: Research Notebooks in the History of Science.* Dordrecht: Kluwer.

Holton, Gerald, Hasok Chang, and Edward Jurkowitz. 1996. "How a Scientific Discovery Is Made: A Case History." *American Scientist* 84:364–75.

Holzkamp, K. 1968. *Wissenschaft als Handlung.* Berlin: De Gruyter.

Home, Roderick W. 1983. "Poisson's Memoirs on Electricity: Academic Politics and a New Style in Physics." *British Journal for the History of Science* 16:239–59.

Home, Roderick W., and Franz U. Aepinus. 1979. *Aepinus's Essay on the Theory of Electricity and Magnetism.* Translated by P. J. Connor. Princeton: Princeton University Press.

Höttecke, Dietmar. 1997a. "Experimentieren wie ein Faraday oder: Wie man durch die Replikation historischer Experimente Naturwissenschaft verstehen kann. Eine Fallstudie." In *Didaktik der Physik—Physikertagung 1997*, edited by DPG-Fachverband Didaktik, 706–11. Berlin: Technische Universität.

Höttecke, Dietmar. 1997b. "Und es bewegt sich . . . : Was man aus der Arbeit mit Replikationen historischer Experimente lernen kann; Eine Fallstudie." In *Geschichte und Theorie der Naturwissenschaften im Unterricht: Ein Weg zur Naturwissenschaftlich-Technischen Alphabetisierung*, edited by Andreas Dally, Telsche Nielsen, and Falk Rieß, 282–302. Rehburg-Loccum: Evang. Akad. Loccum.

Höttecke, Dietmar. 2000. "How and What Can We Learn from Replicating Historical Experiments? A Case Study." *Science and Education* 9 (4): 343–62.

Hoyningen-Huene, Paul. 1987. "Context of Discovery and Context of Justification." *Studies in History and Philosophy of Science* 18:501–15.

Hoyningen-Huene, Paul. 2006. "Context of Discovery versus Context of Justification and Thomas Kuhn." In Schickore and Steinle 2006, 119–31.

Husserl, Edmund. 1956–2002. *Gesammelte Werke*. Edited by Samuel Ijsseling. Dordrecht: Kluwer.

Institut de France, Académie des Sciences, ed. 1916. *Procès-Verbaux de séances de l'Académie, tome 7, années 1820–1823*. Hendaye. Imprimerie de l'Observatoire d'Abbadia.

Jacob, François. 1998. *Of Flies, Mice, and Men*. Translated by Giselle Weiss. Cambridge: Harvard University Press.

James, Frank A. 1989a. "Michael Faraday's First Law of Electrochemistry: How Context Develops New Knowledge." In *Electrochemistry, Past and Present*, edited by John T. Stock and Mary V. Orna, 32–49. Washington, DC: American Chemical Society.

James, Frank A., ed. 1989b. *The Development of the Laboratory: Essays on the Place of Experiment in Industrial Civilisation*. London: Macmillan.

James, Frank A., ed. 1991. *The Correspondence of Michael Faraday, Volume 1, 1811—December 1831, Letters 1–524*. London: Institution of Electrical Engineers.

James, Frank A. J. L., ed. 2002. *The Common Purposes of Life: Science and Society at the Royal Institution of Great Britain*. Aldershot: Ashgate.

Janich, Peter. 1980. "Newton ab onmi naevo vindicatus." *Philosophia Naturalis* 18:243–55.

Janich, Peter. 1985. "Die Eindeutigkeit der Massenmessung und die Definition der Trägheit." *Philosophia Naturalis* 22:85–103.

Janich, Peter. 1997. *Das Maß der Dinge: Protophysik von Raum, Zeit und Materie*. Frankfurt: M. Suhrkamp.

Jansen, F. J. B., Egill Snorrason, and C. Lauritz-Jensen, eds. 1987. *Hans Christian Ørsted*. Copenhagen: Nyt Nordisk Forlag Arnold Buschk.

"J.C." [James Cumming?]. 1821. "Electro-magnetic Experiments." *Annals of Philosophy*, n.s., 1 (February): 137–38.

Jelved, Karen, Andrew D. Jackson, and Ole Knudsen, eds. 1998. *Selected Scientific Works of Hans Christian Ørsted*. Princeton: Princeton University Press.

Jost, Res, ed. 1995. *Das Märchen vom Elfenbeinernen Turm: Reden und Aufsätze; Herausgegeben von Klaus Hepp und Walter Kohn*. Berlin: Springer.

Joubert, Jacques, ed. 1884. *Collection de mémoires relatifs à la physique*. Volume 1, *Mémoires de Coulomb*. Paris: Gauthier-Villars.

Joubert, Jacques, ed. 1885. *Collection de mémoires relatifs à la physique*. Volume 2, *Mémoires sur l'électrodynamique, première partie*. Paris: Gauthier-Villars.

Kanz, Kai T. 1997. *Nationalismus und internationale Zusammenarbeit in den Naturwissenschaften: Die deutsch-französischen Wissenschaftsbeziehungen zwischen Revolution und Restauration, 1789–1832*. Stuttgart: Franz Steiner Verlag.

Karaca, Koray. 2013. "The Strong and Weak Senses of Theory-Ladenness of Experimentation: Theory-Driven versus Exploratory Experiments in the History of High-Energy Particle Physics." *Science in Context* 26 (1): 93–136.

Kauffmann, C. H. 1805. *The dictionary of merchandise, and nomenclature in all languages: for the use of counting-houses: containing, the history, places of growth, culture, use, and marks of excellency, of such natural productions, as form articles of commerce; with their names in all European languages*. Philadelphia: James Humphreys.

Kendall, James. 1953. *Humphry Davy: "Pilot" of Penzance*. New York: Roy.

Klein, Ursula. 1998. "Paving a Way through the Jungle of Organic Chemistry—Experimenting within Changing Systems of Order." In Heidelberger and Steinle 1998, 251–71.

Knight, David M. 1978. *The Transcendental Part of Chemistry*. Folkestone: Dawson.

Knight, David M. 1992. *Humphry Davy: Science & Power*. Oxford: Blackwell.

Knorr-Cetina, Karin D. 1981. *The Manufacture of Knowledge: An Essay on the Constructivistic and Contextual Nature of Science*. Oxford: Pergamon Press.

Knorr-Cetina, Karin D. 1984. *Die Fabrikation von Erkenntnis: Zur Anthropologie der Naturwissenschaft*. Frankfurt: Suhrkamp.

Lakatos, Imre. 1971. "History of Science and Its Rational Reconstructions." In *PSA 1970: Proceedings of the 1970 Biennial Meeting of the Philosophy of Science Association*. Volume 7, edited by R. C. Buck and Robert S. Cohen, 91–136. Dordrecht: Reidel.

Laplace, Pierre-Siméon de. 1814. *Essai philosophique sur les probabilités*. Paris: Courcier.

Laplace, Pierre-Siméon de, and Antoine Lavoisier. 1783. "Mémoire sur la chaleur." *Mémoires de l'Académie Royale des Sciences* 1780:355–405.

Latour, Bruno. 1990. "The Force and Reason of Experiment." In Le Grand 1990, 49–80.

Latour, Bruno. 1994. *Nous n'avons jamais été moderne: Essai d'anthropologie symétrique*. Paris: Découverte.

Latour, Bruno. 1999. "For David Bloor . . . and Beyond: A Reply to David Bloor's 'Anti-Latour.'" *Studies in History and Philosophy of Science* 30:113–29.

Latour, Bruno, and Steve Woolgar. 1979. *Laboratory Life: The Social Construction of Scientific Facts*. Beverly Hills: SAGE.

Lawrence, Christopher. 1990. "The Power and the Glory: Humphry Davy and Romanticism." In Cunningham and Jardine *1990*, 213–27.

Le Grand, Homer L., ed. 1990. *Experimental Inquiries: Historical, Philosophical and Social Studies on Experimentation in Science*. Dordrecht: Kluwer.

Lehot, C. J. 1820. "[Brief an die Académie des Sciences, verlesen am 13. November 1820]." *Annales de chimie et de physique* 15 (December): 406–8.

Lemmerich, Jost. 1991. *Michael Faraday 1791–1867, Erforscher der Elektrizität*. Munich: Beck.

Lenard, Philip. (1906) 1920. *Über Kathodenstrahlen (Nobel-Vortrag)*. 2nd ed. Berlin: De Gruyter.

Leonelli, Sabina. 2014. "What Difference Does Quantity Make? On the Epistemology of Big Data Biology." *Big Data & Society* 1 (1): 1–11.

Lichtenberg, Georg C. 1972. *Schriften und Briefe, herausgegeben von Wolfgang Promies*. 6 vols. Munich: Hanser.

Licoppe, Christian. 1994. "Coulomb et la physique experimentale: Pratique instrumentale et organisation narrative de la preuve." In Blondel and Dörries 1994, 67–83.

Liebig, Justus. 1863. *Über Francis Bacon von Verulam und die Methode der Naturforschung*. Munich: Cotta.

Lundgreen, Peter. 1980. "The Organization of Science and Technology in France: A German Perspective." In Fox and Weisz 1980b, 311–32.

Mandelbaum, Jonathan. 1980. "La Societé Philomatique de Paris de 1788 à 1835: Essai d'histoire institutionelle et de biographie collective d'une societé scientifique parisienne." Doctoral diss., l'École des Hautes études en Sciences Sociales (EHESS).

Mandelbaum, Jonathan. 1988. "Science and Friendship: The Societé Philomatique de Paris 1788–1835." *History and Technology* 5:179–92.

Martin, Thomas, ed. 1932. *Faraday's Diary: Being the various philosophical notes of experimental investigation made by Michael Faraday, DCL, FRS, during the years 1820–1862 and bequeathed by him to the Royal Institution of Great Britain*. 7 vols. and index. London: G. Bell & Sons.

Martínez, Alberto A. 2006. "Replication of Coulomb's Torsion Balance Experiment." *Archive for History of Exact Sciences* 60 (6): 517–63. doi:10.1007/s00407-006-0113-9.

Marum, Martinus v. 1801. "Lettre de M. van Marum, à M. Volta, professeur à Pavie, contenant des expériences sur la colonne électrique, faites par lui et le professeur Pfaff dans le laboratoire de Teyler, à Harlem, en novembre 1800." *Annales de chimie* 40 (3): 289–334.

Matteucci, Carlo. 1837. "Sur le courant électrique ou propre de la grenouille; second mémoire sur l'électricité animale, faisant suite à celui sur la torpille." *Annales de chimie et de physique* 68:93–106.

Mauskopf, Seymour H. 1988. "Molecular Geometry in 19th-Century France: Shifts in Guiding Assumptions." In *Scrutinizing Science: Empirical Studies of Scientific Change*, edited by Arthur Donovan, Larry Laudan, and Rachel Laudan, 125–44. Dordrecht: Kluwer.

Maxwell, James C. 1873. *A Treatise on Electricity and Magnetism*. 2 vols. Oxford: Clarendon Press.

Mayer, Johann T. 1820. "Kopenhagen [Bericht über Ørsteds Rundbrief]." *Göttingische gelehrte Anzeigen* 171 (23): 1705–11.

Mayo, Deborah G. 1994. "The New Experimentalism, Topical Hypotheses, and Learning from Error." In *PSA 1994: Proceedings of the 1994 Biennial Meeting of the Philosophy of Science Association,* edited by David Hull, Micky Forbes, and Richard M. Burian, 1:270–79. East Lansing: Philosophy of Science Association.

Mayo, Deborah G. 1996. *Error and the Growth of Experimental Knowledge*. Chicago: University of Chicago Press.

Mayr, Ernst. 2004. *What Makes Biology Unique? Considerations on the Autonomy of a Scientific Discipline*. Cambridge: Cambridge University Press..

McLaughlin, Peter. 1993. "Der neue Experimentalismus in der Wissenschaftstheorie." In *Die Experimentalisierung des Lebens: Experimentalsysteme in den biologischen Wissenschaften 1850/1950*, edited by Hans-Jörg Rheinberger and Michael Hagner, 207–18. Berlin: Akademie-Verlag.

Merz, John T. 1904. *A History of European Thought in the Nineteenth Century*. 4 vols. Edinburgh: Blackwood.

Métivier, Michel, Pierre Costabel, and Pierre Dugac, eds. 1981. *Siméon Denis Poisson et la science de son temps*. Paris: École Polytechnique.

Meya, Jörg, and Heinz O. Sibum. 1987. *Das fünfte Element: Wirkungen und Deutungen der Elektrizität*. Hamburg: Rowohlt & Deutsches Museum.

Meyer, Kirstine, ed. 1920. *H. C. Ørsted: Naturvidenskabelige Skrifter—Scientific Papers*. 3 vols. Copenhagen: Hoest & Soen.

Mill, John Stuart. 1843. *A System of Logic, Ratiocinative and Inductive: Being a Connected View of the Principles of Evidence and the Methods of Scientific Investigation*. 2 vols. London: Parker.

Miller, David P. 1983. "Between Hostile Camps: Sir Humphry Davy's Presidency of the Royal Society of London, 1820–1827." *British Journal for the History of Science* 16:1–42.

Morrell, Jack, and Arnold Thackray, eds. 1981. *Gentlemen of Science: Early Years of the British Association for the Advancement of Science*. Oxford: Clarendon Press.

Morus, Iwan R. 1992. "Different Experimental Lives: Michael Faraday and William Sturgeon." *History of Science* 30:1–28.

Morus, Iwan R. 1998. *Frankenstein's Children: Electricity, Exhibition, and Experiment in Early-Nineteenth-Century London*. Princeton: Princeton University Press.

Mottelay, P. F. 1922. *Bibliographical History of Electricity and Magnetism Chronologically Arranged*. London: Charles Griffin.

Muncke. 1820. "Einiges die Polarisierung des Lichtes und die Ørsted'schen Versuche betreffend." *[Gilbert's] Annalen der Physik* 66 (December): 412–15.

Neeff. 1821. "Nachschreiben zu dem vorhergehenden." *[Schweigger/Meinecke's] Journal für Chemie und Physik* 31 (January): 32–34.

Nicholson, William, and Anthony Carlisle. 1800. "Account of the New Electrical or Galvanic Apparatus of Sig. Alex. Volta, and Experiments Performed with the Same." *[Nicholson's] Journal of Natural Philosophy, Chemistry and the Arts* 4 (July): 179–87.

Ockenden, L. C. 1937. "The Great Batteries of the London Institution." *Annals of Science* 2:183–84.

O'Malley, Maureen A. 2007. "Exploratory Experimentation and Scientific Practice: Metagenomics and the Proteorhodopsin Case." *History and Philosophy of the Life Sciences* 29 (3): 337–58.

Ørsted, Hans C. 1812. *Ansicht der chemischen Naturgesetze, durch die neueren Entdeckungen gewonnen*. Berlin: Realschulbuchhandlung.

Ørsted, Hans C. 1813. *Recherches sur l'identité des forces chimiques et électriques*. Paris: J. G. Dentu.

Ørsted, Hans C. 1820a. "Esperienze intorno all'effetto del conflitto elettrico sull' ago calamitato, del Sig. I. Chr. Ørsted, Prof. di Fisica nell' Università di Copenaghen." *Giornale di fisica, chimia e storia naturale*, decade II, tomo III: 335–39.

Ørsted, Hans C. 1820b. "Expériences sur un effet que le courant de la Pile excite dans l'Aiguille aimantée." *Journal de physique* 91 (July): 72–76.

Ørsted, Hans C. 1820c. "Experimenta circa affectum ec. Esperienze sull'effetto dell' elettricità sopra l'ago magnetico: Del Sig. Ørsted. Estratto." *Giornale Arcadico di scienze, lettere e arti* 8 (November): 174–78.

Ørsted, Hans C. 1820d. "Experimenta circa effectum Conflictus electrici in Acum magneticam." *[Schweigger's] Journal für Chemie und Physik* 29 (July): 275–81.

Ørsted, Hans C. 1820e. "Experimenta circa effectum, etc.: Experiences sur l'effet du conflict électrique sur l'aiguille aimantée, par Mr. J.Chr. Ørsted, Prof. de physique dans l'université de Copenhague. (Traduction)." *Bibliothèque universelle* 14 (August): 274–80.

Ørsted, Hans C. 1820f. "Experimenta circa effectum, etc. Expériences sur l'effet du conflict électrique sur l'aiguille aimantée. Par Mr. J.Chr. Ørsted, Professeur de physique dans l'Université de Copenhague. Traduction." *Annales de chimie et de physique* 14 (August): 417–25.

Ørsted, Hans C. 1820g. "Experiments on the Effect of a Current of Electricity on the Magnetic Needle." *Annals of Philosophy* 16 (October): 273–76.

Ørsted, Hans C. 1820h. "Neuere electro-magnetische Versuche von Ørsted in Kopenhagen." *[Schweigger's] Journal für Chemie und Physik* 29 (3): 364–69.

Ørsted, Hans C. 1820i. "New Electromagnetic Experiments." *Annals of Philosophy* 16 (November): 375–77.

Ørsted, Hans C. 1820j. "Nouvelles expériences électro-magnétiques; par J.C. Ørsted, communiquées par l'auteur au Prof. Pictet." *Bibliothèque universelle* 15 (October): 137–41.

Ørsted, Hans C. 1820k. "Nouvelles expériences électro-magnétiques; par M. J.C. Ørsted, Communiquées dans une lettre recue à Paris, le 29 septembre 1820." *Journal de physique* 91 (July): 78–80.

Ørsted, Hans C. 1820l. "Sur les effets de l'explosion électrique sur l'aiguille aimantée." *Annales générales des sciences physiques* 5 (August): 259–64.

Ørsted, Hans C. 1820m. "Ulteriori sperienze elettro-magnetiche." *Giornale Arcadico di scienze, lettere e arti* 8 (December): 343–53.

Ørsted, Hans C. 1820n. "Versuche über die Wirkung des elektrischen Conflicts auf die Magnetnadel." *[Gilbert's] Annalen der Physik* 66 (11): 295–304.

Ørsted, Hans C. 1820o. "Experimenta circa effectum Conflictus electrici in Acum magneticam." Copenhagen, privately printed, 4 pages. [Hafniae]: Typis Schultzianis. Reprint in Ørsted 1920, 11–14.

Ørsted, Hans C. 1821a. "New Electro-Magnetic Experiments." *Philosophical Magazine* 57 (January): 44–47.

Ørsted, Hans C. 1821b. "On Electro-Magnetism." *Annals of Philosophy*, n.s., 2 (November): 321–37.

Ørsted, Hans C. 1830. "Thermo-Electricity." In *Edinburgh Encyclopaedia*. Volume 18, edited by David Brewster, 573–89. Edinburgh.

Ørsted, Hans C. 1852. *The Soul in Nature, with Supplementary Contributions*. Translated from the German by Leonora Horner and Joanna B. Horner. London: Henry G. Bohn.

Ørsted, Hans C. 1920. *La Découverte de l'électromagnétisme faite en 1820 par J.-C. Ørsted / The Discovery of electromagnetism made in the year 1820 by H.C. Ørsted: Publ. for the Ørsted Committee at the expense of the state by Absalon Larsen*. Copenhagen: H. H. Thiele.

Ostwald, Wilhelm. 1896. *Elektrochemie: Ihre Geschichte und Lehre*. Leipzig: Veit & Co.

Peschard, Isabelle. 2012. "Modeling and Experimenting." In *Models, Simulations, and*

Representations, edited by Paul Humphreys and Cyrille Imbert, 42–61. New York: Routledge.

Pestre, Dominique. 1994. "La pratique de reconstitution des expériences historiques, une toute première réflexion." In Blondel and Dörries 1994, 17–30.

Pestre, Dominique. 1995. "Pour une histoire sociale et culturelle des sciences: Nouvelles définitions, nouveaux objets, nouvelles pratiques." *Annales: Histoire, sciences sociales* 3:487–522.

Petit, Alexis M. T., and Pierre-Louis Dulong. 1819. "Recherches sur quelques points importants de la théorie de la chaleur." *Annales de chimie et de physique* 10:395–413.

Pickering, Andrew. 1982. "Elementary Particles: Discovered or Constructed?" In *Physics in Collision: High-Energy ee/ep/pp Interactions*, edited by W. P. Trower and G. Bellini, 439–48. New York: Plenum.

Pickering, Andrew. 1984. *Constructing Quarks: A Sociological History of Particle Physics.* Chicago: University of Chicago Press.

Pickering, Andrew. 1987. "Against Correspondence: A Constructivist View of Experiment and the Real." In *PSA 1986: Proceedings of the Biennial Meeting of the Philosophy of Science Association*, edited by Arthur Fine and Peter Machamer, 2:196–206. East Lansing: Philosophy of Science Association.

Pickering, Andrew. 1992a. "From Science as Knowledge to Science as Practice." In Pickering 1992b, 1–26.

Pickering, Andrew, ed. 1992b. *Science as Practice and Culture.* Chicago: University of Chicago Press.

Pickering, Andrew. 1995a. "Beyond Constraint: The Temporality of Practice and the Historicity of Knowledge." In Buchwald 1995a, 42–55.

Pickering, Andrew. 1995b. *The Mangle of Practice: Time, Agency, and Science.* Chicago: University of Chicago Press.

Pickering, Andrew, and A. Stephanides. 1992. "Constructing Quaternions: On the Analysis of Conceptual Practice." In Pickering 1992b, 139–67.

Pictet, Marc-Auguste. 1821. "Remarque du Traducteur, à: Lettera del Prof. Configliachi a C. Ridolfi, etc. Lettre du Prof. Configliachi au Marquis Ridolfi sur les expériences Voltaïco-magnétique; communiquée au Prof. Pictet (Traduction)." *Bibliothèque universelle* 16:73–74.

Pinnick, Cassandra, and George Gale. 2000. "Philosophy of Science and History of Science: A Troubling Interaction." *Journal for General Philosophy of Science* 31:109–25.

Poggendorff, Johann C. 1821. "Physisch-chemische Untersuchungen zur näheren Kenntnis des Magnetismus der voltaischen Säule." *Isis, oder Encyclopädische Zeitung (Oken)*: 687–710.

Poisson, Siméon-D. 1812a. "Mémoire sur la distribution de l'eléctricité à la surface des corps conducteurs." *Mémoires de la Classe des Sciences Mathématiques et Physiques de l'Institut Imperial de France* 12 (1811): 1–92.

Poisson, Siméon-D. 1812b. "Second mémoire sur la distribution de l'eléctricité à la surface des corps conducteurs." *Mémoires de la Classe des Sciences Mathématiques et Physiques de l'Institut Imperial de France* 12 (1811): 163–274.

Poisson, Siméon-D. 1826. "Mémoire sur la théorie du magnétisme." *Mémoires de l'Académie Royale des Sciences de l'Institut de France* 5 (1821–22): 247–338.

Polanyi, Michael. (1958) 1994. *Personal Knowledge: Towards a Post-Critical Philosophy.* Chicago: University of Chicago Press.

Polanyi, Michael. 1962. "The Republic of Science." *Minerva* 1 (October): 54–73.

Popper, Karl R. (1934) 1969. *Logik der Forschung.* 3rd ed. Tübingen: Mohr.

Prechtel. 1821. "Ansichten über den Magnetismus und dessen Ableitung aus der Electricität." *[Gilbert's] Annalen der Physik* 67 (January): 81–90.

Radder, Hans. 1992. "Experimental Reproducibility and the Experimenters' Regress." In *PSA 1992: Proceedings of the 1992 Biennial Meeting of the Philosophy of Science Association*, edited by David Hull, Micky Forbes, and K. Okruhlik, 1:63–73. East Lansing: Philosophy of Science Association.

Radder, Hans. 1995. "Experimenting in the Natural Sciences: A Philosophical Approach." In Buchwald 1995a, 56–86.

Radder, Hans. 1997. "Philosophy and History of Science: Beyond the Kuhnian Paradigm." *Studies in History and Philosophy of Science* 28:633–55.

Radder, Hans. 1998. "Issues for a Well-Developed Philosophy of Scientific Experimentation." In Heidelberger and Steinle 1998, 392–404.

Radder, Hans, ed. 2003. *The Philosophy of Scientific Experimentation.* Pittsburgh: University of Pittsburgh Press.

Reader, W. J. 1966. *Professional Men: The Rise of the Professional Classes in Nineteenth-Century England.* New York: Basic Books.

Renn, Jürgen, and Tilman Sauer. 1997. "Einsteins Entdeckung des Gravitationslinseneffektes." *Physikalische Blätter* 53:198.

Renn, Jürgen, and Tilman Sauer. 1999. "Heuristics and Mathematical Representation in Einstein's Search for a Gravitational Field Equation." In *The Expanding Worlds of General Relativity.* Volume 7, edited by Hubert Goenner, Jürgen Renn, Jim Ritter, and Tilman Sauer, 87–125. Boston: Birkhäuser.

Rheinberger, Hans-Jörg. 1992. *Experiment, Differenz, Schrift: Zur Geschichte epistemischer Dinge.* Marburg: Basilisken-Presse.

Rheinberger, Hans-Jörg. 1994. "Experimental Systems: Historiality, Narration, and Deconstruction." *Science in Context* 7:65–81.

Rheinberger, Hans-Jörg. 1997. *Toward a History of Epistemic Things: Synthesizing Proteins in the Test Tube.* Stanford: Stanford University Press.

Rheinberger, Hans-Jörg. 1999. "Reenacting History: Review of Pickering, Andrew: The Mangle of Practice, 1995." *Studies in History and Philosophy of Science* 30:163–66.

Rheinberger, Hans-Jörg. 2003. "Carl Correns' Experiments with Pisum, 1869–1899." In Holmes, Renn, and Rheinberger 2003, 221–52.

Rheinberger, Hans-Jörg, and Michael Hagner, eds. 1993. *Die Experimentalisierung des Lebens: Experimentalsysteme in den biologischen Wissenschaften 1850/1950.* Berlin: Akademie-Verlag.

Richter, Klaus. 1988. *Der Physiker des Romantikerkreises Johann Wilhelm Ritter in seinen Briefen an den Verleger Carl Friedrich Ernst Frommann.* Weimar: Hermann Böhlaus Nachfolger.

Ridolfi. 1821. "Lettera del Signor Marchese Ridolfi, etc. Lettre du Marquis Ridolfi au Prof. Pictet, sur une expérience voltaïco-magnétique et les idées qu'elle suggère sur la composition du fluide électrique (Traduction). Florence 16 janvier 1821." *Bibliothèque universelle* 16 (January): 75–78.

Rieß, Falk. 1998. "Erkenntnis durch Wiederholung—eine Methode zur Geschichtsschreibung des Experiments." In Heidelberger and Steinle 1998, 157–72.

Ritter, Johann W. 1798. *Beweis, dass ein beständiger Galvanismus den Lebensprocess in dem Thierreich begleite: Nebst neuen Versuchen und Bemerkungen über den Galvanismus.* Weimar: Industrie-Comptoir.

Ritter, Johann W. 1800a. "Galvanische Versuche über die chemische Natur des Wassers." *Crell's Chemische Annalen* 1:41ff.

Ritter, Johann W. 1800b. "Versuche und Bemerkungen über den Galvanismus der Voltaischen Batterie, in Briefen an L. W. Gilbert: Erster Brief." *[Gilbert's] Annalen der Physik* 7:431–46.

Ritter, Johann W. 1805. "Neue Versuche und Bemerkungen über den Galvanismus, in Briefen an L. W. Gilbert: Zweyter Brief." *[Gilbert's] Annalen der Physik* 19:1–44.

Ritter, Johann W. 1806. *Physisch-chemische Abhandlungen in chronologischer Folge.* 3 vols. Leipzig.

Robert, A., and G. Cougny. 1889–91. *Dictionnaire des parlementaires français.* 5 vols. Paris.

Roberts, Gerrylinn K. 1976. "The Establishment of the Royal College of Chemistry: An Investigation of the Social Context of Early-Victorian Chemistry." *Historical Studies in the Physical Sciences* 7:437–85.

Robinet, André, and Nelly Bruyère, eds. 1993. *Maine de Biran: Oeuvres tome XIII/1; Correspondance philosophique Maine de Biran–Ampère.* Paris: Vrin.

Romo, J., and Manuel García Doncel. 1994. "Faraday's Initial Mistake Concerning the Direction of Induced Currents, and the Manuscript of Series I of His Researches." *Archive for History of Exact Sciences* 47:291–385.

Ruby, Jane E. 1986. "The Origin of Scientific 'Law.'" *Journal of the History of Ideas* 47:341–51.

Sargent, Rose-Mary. 1995. *The Diffident Naturalist: Robert Boyle and the Philosophy of Experiment.* Chicago: University of Chicago Press.

Sauer, Tilman, and Raphael Scholl, eds. 2016. *The Philosophy of Historical Case Studies.* Berlin: Springer.

Schaffer, Simon. 1989. "Glass Works: Newton's Prisms and the Uses of Experiment." In Gooding, Pinch, and Schaffer 1989, 67–104.

Scheidecker-Chevallier, Myriam, and Robert Locqueneux. 1992. "Liens entre philosophie et sciences physiques dans l'oeuvre d'André-Marie Ampère." *Archives Internationales d'Histoire des Sciences* 42:227–68.

Schickore, Jutta. 2007. *The Microscope and the Eye: A History of Reflections, 1740–1870.* Chicago: University of Chicago Press.

Schickore, Jutta. 2011. "More Thoughts on HPS: Another 20 Years Later." *Perspectives on Science* 19:455–81.

Schickore, Jutta, and Friedrich Steinle, eds. 2006. *Revisiting Discovery and Justification: Historical and Philosophical Perspectives on the Context Distinction.* Dordrecht: Springer.

Schimank, Hans. 1933. "Johann Wilhelm Ritter, der Begründer der wissenschaftlichen Elektrochemie: Ein Lebensbild aus dem Zeitalter der Romantik." In *Deutsches Museum: Abhandlungen und Berichte*, 175–206. Berlin: VDI-Verlag.

Schimank, Hans. 1963. "Ludwig Wilhelm Gilbert und die Anfänge der 'Annalen der Physik.'" *Sudhoffs Archiv für Geschichte der Medizin und der Naturwissenschaften* 47:360–72.

Schlumbohm, Jürgen, and Maurizio Gribaudi, eds. 1998. *Mikrogeschichte—Makrogeschichte: Komplementär oder inkommensurabel?* Göttingen: Wallstein-Verl.

Schweigger, Johann S. C. 1821a. "Noch einige Worte über diese neuen elektromagnetischen Phänomene." *[Schweigger's] Journal für Chemie und Physik* 31 (1): 35–41.

Schweigger, Johann S. C. 1821b. "Vorwort." *[Schweigger's] Journal für Chemie und Physik* 31:10–11.

Schweigger, Johann S. C. 1821c. "Zusätze zu Ørsteds elektromagnetischen Versuchen, vorgelesen in der naturforschenden Gesellschaft zu Halle den 16. September 1820." *[Schweigger's] Journal für Chemie und Physik* 31 (1): 1–17.

Seebeck, Thomas J. 1821. "Ueber Elektromagnetismus: Auszug aus einer Abhandlung, welche in der Sitzung der Königlichen Akademie der Wissenschaften zu Berlin vom 14. Dezember 1820 vorgetragen wurde." *[Schweigger's] Journal für Chemie und Physik* 32 (May): 27–38.

Seebeck, Thomas J. 1822. "Über den Magnetismus der galvanischen Kette." *Abhandlungen der Königlichen Akademie der Wissenschaften zu Berlin* 1820–21:289–346.

Segala, Marco. 1997. "Ampère Filosofo." *Nuncius* 12:145–60.

Shapin, Steven, and Simon Schaffer. 1985. *Leviathan and the Air-Pump: Hobbes, Boyle and the Experimental Life.* Princeton: Princeton University Press.

Shapiro, Alan E. 1996. "The Gradual Acceptance of Newton's Theory of Light and Color, 1672–1727." *Perspectives on Science: Historical, Philosophical, Social* 4:59–140.

Sibum, Heinz-Otto. 1995. "Reworking the Mechanical Value of Heat: Instruments of Precision and Gestures of Accuracy in Early Victorian England." *Studies in History and Philosophy of Science* 26:73–106.

Sibum, Heinz-Otto. 1998. "Die Sprache der Instrumente: Eine Studie zur Praxis und Repräsentation des Experimentierens." In Heidelberger and Steinle 1998, 141–56.

Silliman, Robert H. 1970–81. "Fresnel, Augustin-Jean." In *Dictionary of Scientific Biography*, edited by Charles Gillispie, 5:165–71. New York: Scribner.

Silliman, Robert H. 1974. "Fresnel and the Emergence of Physics as a Discipline." *Historical Studies in the Physical Sciences* 4:137–62.

Snelders, H. A. M. 1990. "Ørsted's Discovery of Electromagnetism." In Cunningham and Jardine 1990, 228–40.

Snyder, Laura J. 2011. *The Philosophical Breakfast Club: Four Remarkable Friends Who Transformed Science and Changed the World.* New York: Broadway Books.

Söderbaum, Hendrik G., ed. 1912–32. *Jacob Berzelius: Bref—Lettres.* 14 vols. Uppsala: Almqvist & Wiksells boktryckeri.

Soemmering, Samuel T. 1810. "Über einen elektrischen Telegraphen." *Denkschriften der Königlichen Akademie der Wissenschaften zu München* 1809–10:401–14.

Soemmering, W. 1859. "Auszüge aus dem Tagebuch von Samuel Thomas Soemmering, als Beitrag zur Geschichte der Erfindung des elektrischen Telegraphen." *[Poggendorff's] Annalen der Physik* 107:644–47.

Steinle, Friedrich. 1992. "Was ist Masse? Newtons Begriff der Materiemenge." *Philosophia Naturalis* 29:94–117.

Steinle, Friedrich. 1994. "Experiment, Speculation and Law: Faraday's Analysis of Arago's Wheel." In *PSA 1994: Proceedings of the 1994 Biennial Meeting of the Philosophy of Science Association*, edited by David Hull, Micky Forbes, and Richard M. Burian, 1:293–303. East Lansing: Philosophy of Science Association.

Steinle, Friedrich. 1995a. "Looking for a 'Simple Case': Faraday and Electromagnetic Rotation." *History of Science* 33:179–202.

Steinle, Friedrich. 1995b. "The Amalgamation of a Concept—Laws of Nature in the New Sciences." In *Laws of Nature: Essays on the Philosophical, Scientific and Historical Dimensions*, edited by Friedel Weinert, 316–68. Berlin: De Gruyter.

Steinle, Friedrich. 1996. "Work, Finish, Publish? The Formation of the Second Series of Faraday's 'Experimental Researches in Electricity.'" *Physis* 33:141–220.

Steinle, Friedrich. 1997. "Entering New Fields: Exploratory Uses of Experimentation." *Philosophy of Science* 64 (Supplement): 65–74.

Steinle, Friedrich. 1998. "Exploratives vs. theoriebestimmtes Experimentieren: Ampères erste Arbeiten zum Elektromagnetismus." In Heidelberger and Steinle 1998, 272–97.

Steinle, Friedrich. 2000. "Experiment, Instrument und Begriffsbildung: Ampère, das Galvanometer und der Stromkreis." In *Instrument—Experiment: Historische Studien*, edited by Christoph Meinel, 98–108. Berlin: GNT-Verlag.

Steinle, Friedrich. 2002a. "'Das Nächste ans Nächste reihen': Goethe, Newton und das Experiment." *Philosophia Naturalis* 39:141–72.

Steinle, Friedrich. 2002b. "Experiments in History and Philosophy of Science." *Perspectives on Science* 10 (4): 408–32.

Steinle, Friedrich. 2003. "¿Experimentos románticos? El caso de la electricidad [Romantic experiment? The case of electricity]." In *Ciencia y Romanticismo*, edited by José Montesinos, Javier Ordóñez, and Sergio Toledo, 185–227. La Orotava: Fundación Canaria Orotava de Historia de la Ciencia.

Steinle, Friedrich. 2004. "Michael Faraday." In *Lexikon der bedeutenden Naturwissenschaftler*, edited by Dieter Hoffmann, Staffan Müller-Wille, and Hubert Laitko, 2:3–11. Heidelberg: Spektrum Akademischer Verlag.

Steinle, Friedrich. 2006. "Concept Formation and the Limits of Justification: 'Discovering' the Two Electricities." In Schickore and Steinle 2006, 183–95.

Steinle, Friedrich. 2008. "From Principles to Regularities: Tracing 'Laws of Nature' in Early Modern France and England." In *Natural Law and Laws of Nature in Early Modern Europe: Jurisprudence, Theology, Moral and Natural Philosophy*, edited by Lorraine Daston and Michael Stolleis, 215–31. Aldershot: Ashgate.

Steinle, Friedrich. 2009. "How Experiments Make Concepts Fail: Faraday and Magnetic Curves." In *Going Amiss in Experimental Research*, edited by Friedrich Steinle, Giora Hon, and Jutta Schickore, 119–35. Dordrecht: Springer.

Steinle, Friedrich. 2010. "Concepts, Facts, and Sedimentation in Experimental Science." In

Science and the Life-World: Essays on Husserl's "Crisis of European Sciences," edited by David Hyder and Hans-Jörg Rheinberger, 199–214. Stanford: Stanford University Press.

Steinle, Friedrich. 2011. "Die Entstehung der Feldtheorie: Ein ungewöhnlicher Fall der Wechselwirkung von Physik und Mathematik?" In *Mathematics Meets Physics: A Contribution to Their Interaction in the 19th and the First Half of the 20th Century,* edited by Karl-Heinz Schlote and Martina Schneider, 441–85. Frankfurt: Verlag Harri Deutsch GmbH.

Steinle, Friedrich. 2012. "Goals and Fates of Concepts: The Case of Magnetic Poles." In Feest and Steinle 2012, 105–25.

Steinle, Friedrich. 2013. "Wie Begriffe entstehen: Der Fall der Elektrodynamik." In *Quanten,* edited by Konrad Kleinknecht, 45–96. Stuttgart: S. Hirzel.

Steinle, Friedrich. 2016. "Stability and Replication of Experimental Results: A Historical Perspective." In *Reproducibility: Principles, Problems, Practices, and Prospects,* edited by Harald Atmanspacher and Sabine Maasen. New York: Wiley.

Steinle, Friedrich, and Richard Burian. 2002. "Introduction: History of Science and Philosophy of Science." *Perspectives on Science* 10 (4): 391–97.

Steinle, Friedrich, and Neil Ribe. 2002. "Exploratory Experimentation: Goethe, Land, and Color Theory." *Physics Today* 55 (July): 43–49.

Strickland, Stuart W. 1992. "Circumscribing Science: Johann Wilhelm Ritter and the Physics of Sidereal Man." PhD diss., Harvard University.

Sturgeon, William. 1826. "Account of an Improved Electro-magnetic Apparatus." *Annals of Philosophy,* n.s., 12 (November): 357–61.

Sudduth, William M. 1980. "The Voltaic Pile and Electro-Chemical Theory in 1800." *Ambix* 27:26–35.

Sutton, Geoffrey V. 1981. "The Politics of Science in Early Napoleonic France: The Case of the Voltaic Pile." *Historical Studies in the Physical Sciences* 11 (2): 329–66.

Sviedrys, Romualdas. 1976. "The Rise of Physics Laboratories in Britain." *Historical Studies in the Physical Sciences* 7:405–36.

Tatum, John. 1821. "Electro-Magnetic Experiments." *Philosophical Magazine* 57 (June): 446–47.

Tetens, Holm. 1987. *Experimentelle Erfahrung: Eine wissenschaftstheoretische Studie über die Rolle des Experiments in der Begriffs- und Theoriebildung der Physik.* Hamburg: Meiner.

Thackray, Arnold. 1971. "Anthony Carlisle." In *Dictionary of Scientific Biography,* edited by Charles Gillispie, 3:67–68. New York: Scribner.

Thackray, Arnold. 1974. "William Nicholson." In *Dictionary of Scientific Biography,* edited by Charles Gillispie, 10:107–9. New York: Scribner.

Trumpler, Maria. 1992. "Questioning Nature: Experimental Investigations of Animal Electricity in Germany, 1791–1810." PhD diss., Yale University.

Trumpler, Maria. 1997. "Verification and Variation: Patterns of Experimentation in Investigations of Galvanism in Germany, 1790–1800." *Philosophy of Science* 64 (Supplement): S75–S84.

Trumpler, Maria. 1999. "From Tabletops to Triangles: Increasing Abstraction in the Depiction of Experiments in Animal Electricity from Galvani to Ritter." In *Luigi*

Galvani: International Workshop—Proceedings, edited by Marco Bresadola and Guiliano Pancaldi, 115–45. Bologna: CIS, Dipartimento di Filosofia, Università de Bologna.

Turner, Gerald L., ed. 1976. *The Patronage of Science in the Nineteenth Century*. Leyden: Nordhoff.

van Fraassen, Bas C. 1981. "Theory Construction and Experiment: An Empiricist View." In *PSA 1980: Proceedings of the Biennial Meeting of the Philosophy of Science Association*, edited by P. D. Asquith and Ronald N. Giere, 2:663–77. East Lansing: Philosophy of Science Association.

Volta, Alessandro. 1800a. "De l'électricité dite galvanique." *Annales de chimie* 40 (3): 225–72.

Volta, Alessandro. 1800b. "On the Electricity excited by the mere Contact of Conducting Substances of Different Kinds." *Philosophical Magazine* 7:298–311.

Volta, Alessandro. 1800c. "On the Electricity excited by the mere Contact of Conducting Substances of Different Kinds." *Philosophical Transactions* 90 (part 2): 403–31.

Wallach, O., and J. von Braun, eds. 1901. *Briefwechsel zwischen J. Berzelius und F. Wöhler*. 2 vols. Leipzig: Engelmann.

Waters, C. K. 2007. "The Nature and Context of Exploratory Experimentation: An Introduction to Three Case Studies of Exploratory Research." *History and Philosophy of the Life Sciences* 29 (3): 275–84.

Watkins, Francis. 1828. *A Popular Sketch of Electro-Magnetism or Electro-Dynamics*. London: John Taylor.

Watts, Isaac. 1809. *The Improvement of the Mind*. London.

Weber, Wilhelm. 1846. *Elektrodynamische Maassbestimmungen*. Leipzig: Weidmann'sche Buchhandlung.

Weingarten, Michael. 1997. "Was ist der Gegenstand biologischer Forschung: Das Leben oder der Organismus?" In *Repraesentatio mundi: Festschrift zum 70. Geburtstag von Hans Heinz Holz*, edited by Hermann Klenner, 503–26. Cologne: Dinter.

Weiss, Burghard. 1995. "Experiment und Kontext: Die elektrische Waage des C. A. Coulomb als Gegenstand historischer Kontroversen." In *Brückenschläge: 25 Jahre Lehrstuhl für Geschichte der exakten Wissenschaften und der Technik an der Technischen Universität Berlin*, edited by Hans-Werner Schütt and Burghard Weiss, 347–77. Berlin: Engel.

Westrum, Ron. 1978. "Science and Social Intelligence about Anomalies: The Case of Meteorites." *Social Studies of Science* 8:461–93.

Wetzels, Walter D. 1973. *Johann Wilhelm Ritter: Physik im Wirkungsfeld der deutschen Romantik*. Berlin: Walter de Gruyter.

Whewell, William. 1840. *The Philosophy of the Inductive Sciences, Founded upon Their History*. 2 vols. London: Parker.

Williams, Leslie P. 1965. *Michael Faraday: A Biography*. London: Chapman & Hall.

Williams, Leslie P. 1970–81. "Ampère, André-Marie." In *Dictionary of Scientific Biography*, edited by Charles Gillispie, 1:139–46. New York: Scribner.

Williams, Leslie P. 1974. "Ørsted, Hans Christian." In *Dictionary of Scientific Biography*, edited by Charles Gillispie, 10:182–86. New York: Scribner.

Williams, Leslie P. 1983. "What Were Ampère's Earliest Discoveries in Electrodynamics?" *Isis* 74:492–508.

Williams, Leslie P. 1985a. "Faraday and Ampère: A Critical Dialogue." In Gooding and James 1985, 83–104.

Williams, Leslie P. 1985b. "Reply [to Blondel 1985]." *Isis* 76:361.

Williams, Leslie P. 1986. "Why Ampère Did Not Discover Electromagnetic Induction." *American Journal of Physics* 54:306–11.

Wilson, Andrew D. 2014. "Ørsted Triumphator? [Review of *Hans Christian Ørsted: Reading Nature's Mind*, by Dan Charly Christensen]." *Isis* 105 (1): 188–92. doi:10.1086/675562.

Wise, Norton M., ed. 1995. *The Values of Precision*. Princeton: Princeton University Press.

Yelin. 1820. "Über den Zusammenhang der Electricität und des Magnetismus." *[Gilbert's] Annalen der Physik* 66 (12): 395–411.

INDEX OF NAMES

Aepinus, F. U. T., 19, 22, 425n11
Aldini, G., 51
Althaus, baron, 64
Ampère, A. M., 7–13, 19, 50, 61–62, 65–66, 71, 72–201 passim, 203, 211, 213–19, 221–30 passim, 236, 238, 240, 244, 246, 250, 252–53, 263–65, 267–71, 274–82, 285–88, 292–99, 309, 312–15, 320, 322–24, 326, 328, 333, 335, 337, 339–49, 366–67, 373–91, 405–22
Ampère, Albine, 76
Ampère, J. J., 76–77
Ampère, Josephine, 76
Antinori, 221–22
Arago, F., 26, 31–32, 58, 61–62, 64–68, 71–72, 77, 79, 97, 102–4, 110, 112–14, 132–33, 137, 141, 145–46, 159, 162–63, 170, 172, 183, 188–89, 212, 216, 221–22, 347, 355, 366, 378–79, 394, 399

Babbage, C., 203
Banks, J., 32, 34, 202–3
Barlocci, S., 65
Barlow, P., 203
Barnard, S., 206
Beddoes, T., 36
Bernoulli, J., 75
Berthollet, C. L., 24, 26, 43, 221
Berzelius, J. J., 37, 39, 48, 64, 66, 71, 147, 221–22
Biot, J. B., 12, 26–29, 31, 42–48, 64, 68–77 passim, 80–81, 93, 112, 114, 124, 140–41, 143, 148–50, 163–72, 177–78, 181–84, 187–91, 200–201, 214, 221, 236, 250, 298–99, 313, 333, 337
Blainville, 62, 69, 188
Blondel, C., 10, 80–82, 139, 144, 190
Böckmann, 40, 64
Boisgiraud, 72–73, 162, 221
Boyle, R., 76, 303, 316–17

Brande, W. T., 68, 204–6, 211, 223, 236, 248–49, 269–71, 280
Brewster, D., 62, 211, 328
Buchwald, J. Z., 305
Burian, R., 309, 316, 329

Caneva, K., 80–81
Cantor, G., 283
Carlisle, A., 34–35, 37, 41–42, 48
Carrier, M., 311–12
Cartwright, N., 309–11
Casties, R., 5, 321, 329
Cauchy, A. L., 77–78, 80
Collins, H., 306–8
Coulomb, C. A., 18, 20–24, 27–30, 42–43, 46, 71, 82, 158, 164, 173, 303
Cumming, J., 112, 211
Cuvier, F., 63, 78–79
Cuvier, G., 63, 78–79

Davy, H., 13, 24, 34–40, 48, 62, 66, 68–69, 77, 110, 140, 171, 202–7, 211–16, 219–27, 236, 248–49
De la Rive, C. G., 60–61, 64, 110–12, 140, 222, 248, 261, 263–66, 276–83, 304
Delambre, J.-B.-J., 31, 77, 141
Despretz, C. M., 112, 132, 183
Dufay, C., 326, 333
Duhem, P., 318
Dulong, P.-L., 26, 31–32, 71–72, 81, 112, 121, 130, 147, 162, 197
Dumotier, 121, 134, 137

Erman, P., 40, 63, 68
Euler, L., 75

Faraday, M., 7–12, 19, 54, 76, 82, 155–56, 189–90, 203, 205–15, 216, 219–27, 228–99 passim, 304, 309, 312–14, 322–28
Fleck, L., 1, 334, 337
Fox, R., 25
Frankel, E., 25
Franklin, A., 307
Fresnel, A. J., 31–32, 77, 81, 112, 132, 162–63, 183, 188–89, 221, 380

Galison, P., 305, 309–10
Galvani, L., 34, 42, 65, 68, 70, 82
Gay-Lussac, J. L., 26, 31, 62, 73, 148–49, 163, 285
Gilbert, L. W., 40, 44, 50, 59–60, 63, 65, 69, 186, 203, 263, 284, 318
Goethe, J. W. v., 328
Gooding, D. C., 10, 11, 245–47, 266, 309, 311, 313, 316, 318, 320–22, 331
Graßhoff, G., 5, 309, 321

Hachette, J. N. P., 51, 73, 186–87, 211, 283–86
Hacking, I., 4, 302, 305, 309, 311–13, 318, 328
Heidelberger, M., 4, 309, 310, 330
Hentschel, K., 305, 328
Herschel, J. W. F., 40, 203
Hofmann, J., 10–11, 77, 123, 139, 144, 181, 183, 285
Hooke, R., 317
Humboldt, A. v., 40, 130, 136, 138, 144, 165, 168, 197, 327
Husserl, E., 319, 333
Hutton, C., 203

Jacob, F., 1–2
Joule, J. P., 302

Kant, I., 17, 53, 76, 79
Kekulé, F. A., 335
Kepler, J., 317, 335
Klein, U., 329
Koselleck, R., 7
Krebs, H., 329
Krüger, L., 315

La Borne, 74, 221
Laplace, P. S., 19, 22, 24–32, 43–44, 62, 66, 68, 70, 72, 75, 77, 80–81, 114, 132, 136, 138, 140–41, 144, 150, 153, 154, 158, 163–64, 166, 168, 170, 176, 181–82, 187, 200, 313, 333, 337
Latour, B., 302, 306–7, 309

Lavoisier, L. A., 22, 25, 35
Lehot, C. J., 68, 73, 221
Lenard, P., 328
Lichtenberg, G. C., 18
Liebig, J., 329

Maine de Biran, F. P. G., 76
Malus, E. L., 26, 28, 328
Marcet, A., 61, 110
Marum, M. v., 43
Maxwell, J. C., 8, 82, 201
Mayer, T., 61, 64, 70
Mayo, D., 309, 312
McLaughlin, P., 310
Meinecke, J. L. G., 63, 68–69
Merz, J. T., 26
Mill, J. S., 320
Mojon, B., 51
Moll, 222
Morichini, 65, 81
Morveau, G. de, 43
Muncke, 64, 70

Napoleon I, 23–24, 31, 39, 44, 50, 78
Neeff, 64
Newman, J., 280
Newton, I., 19, 21–22, 200, 201, 317, 333–34
Nicholson, W., 34–37, 41, 43, 48, 204
Nollet, J. A., 22
Novalis (alias v. Hardenberg, F.), 40

Ørsted, H.-C., 222–23, 349, 350–52, 354–56, 383, 391, 394–95, 399
Offerhaus, 222

Paschen, F., 318
Petit, A. T., 31–32, 112
Pfaff, C. H., 40, 43, 50, 63, 103, 327
Phillips, R., 216, 218
Pickering, A., 300, 305
Pictet, M.-A., 61, 62, 64–66, 70, 103, 140, 222
Pixii, A. H., 326
Pixii, N.-C., 90, 108–9, 113–14, 123, 145, 151, 156, 157, 161, 326
Planck, M., 316
Plücker, J., 328
Poggendorff, J. C., 63, 112, 210
Poisson, S.-D., 19, 26–31, 69, 129, 154, 162, 172, 197
Polanyi, M., 1
Popper, K., 316
Prechtel, J. J., 64, 74

Radder, H., 307
Reichenbach, H., 332
Rheinberger, H.-J., 124, 136, 191, 304–5, 309, 310, 318
Ridolfi, 64–65, 171, 185, 221, 223–24
Ritter, J. W., 35, 40–43, 51, 53–54, 71, 91, 327
Romagnosi, G. D., 51
Rousseau, J. J., 75
Rumford, 36–37

Sargent, R.-M., 316
Savart, F., 64, 75, 143, 163, 165–67, 170–72, 178, 221
Schaffer, S., 303
Schelling, F. W. J., 40, 53
Schweigger, J. S. C., 61, 63, 64, 68, 109, 112, 140, 141, 171, 221, 222, 223
Seebeck, T. J., 61, 63, 68, 171
Shapin, S., 303
Simon, P. L., 40
Soemmering, S. T., 50, 145
Stapfer, P. A., 78
Stodart, J., 282–83
Sturgeon, W., 50
Sutton, G., 47

Tatum, J., 204–5, 211
Thenard, L. J., 26
Thompson, B. (alias Rumford, C.), 36

Thomson, T., 62
Tilloch, A., 204
Trumpler, M., 327

van Fraassen, B., 313
van Mons, 61
Volta, A., 32–36, 40, 42–46, 48, 52, 65, 70, 93, 124, 135, 148

Watts, I., 205
Weber, W., 7, 145, 181
Weiss, C. S., 63
Whewell, W., 203, 319
Williams, L. P., 10, 11, 82–84, 86, 123, 139, 144, 206, 216, 225
Wollaston, W. H., 35, 62, 203, 207–11, 213, 223, 236, 238, 240, 249, 257, 259, 269–71, 276, 282–83, 287, 340, 345, 347, 381, 382
Woolgar, S., 302, 306

Yelin, J. K., 64, 70, 221
Young, T., 62, 68